CONVERSION FACTORS AND USEFUL RELATIONS

$1\ \text{u} = 1.66 \times 10^{-27}\ \text{kg} = 931.5\ \text{MeV}/c^2$

$1\ \text{eV} = 1.60 \times 10^{-19}\ \text{J}$

$1\ \text{J} = 6.24 \times 10^{18}\ \text{eV}$

$1\ \text{eV/particle} = 23.1\ \text{kcal/mol} = 96.5\ \text{kJ/mol}$

$1\ \text{T} = 10^4\ \text{G}$

$1\ \text{Å} = 0.1\ \text{nm} = 10^{-10}\ \text{m}$

$1\ \text{cal} = 4.18\ \text{J}$

$1\ \text{barn} = 10^{-28}\ \text{m}^2$

$1\ \text{year} = 3.16 \times 10^7\ \text{s}$

$1\ \text{light-year} = 9.46 \times 10^{15}\ \text{m}$

$hc = 1240\ \text{eV·nm}$

$e^2/4\pi\epsilon_0 = 1.44\ \text{eV·nm}$

$hcR_\infty = 13.6\ \text{eV}$

kT at $T = 300\ \text{K} \simeq 0.025\ \text{eV}$

$m_p/m_e = 1836$

Modern Physics

MODERN PHYSICS

Frank J. Blatt

Professor of Physics, University of Vermont

McGRAW-HILL, INC.

New York St. Louis San Francisco Auckland Bogotá Caracas Lisbon London Madrid Mexico Milan Montreal New Delhi Paris San Juan Singapore Sydney Tokyo Toronto

MODERN PHYSICS

1 2 3 4 5 6 7 8 9 0 VNH VNH 9 0 9 8 7 6 5 4 3 2

ISBN 0-07-005877-6

This book was set in Times Roman by York Graphic Services, Inc.
The editors were Susan J. Tubb, Denise T. Schanck, and Holly Gordon;
the designer was Leon Bolognese;
the production supervisor was Janelle S. Travers.
The photo editor was Safra Nimrod;
the photo researcher was Debra Hershkowitz.
Von Hoffmann Press, Inc., was printer and binder.

Library of Congress Cataloging-in-Publication Data

Blatt, Frank J.
Modern physics / Frank J. Blatt.
p. cm.
Includes bibliographical references and index.
ISBN 0-07-005877-6
1. Physics. I. Title.
QC21.2.B56 1992
530—dc20 91-29223

ABOUT THE AUTHOR

FRANK J. BLATT earned B.S. and M.S. degrees in electrical engineering from M.I.T. and the Ph.D. in physics from the University of Washington. After three years as Research Associate with Professors Seitz and Bardeen at the University of Illinois, he joined the faculty of the physics department of Michigan State University. Since 1987 Dr. Blatt has been a member of the faculty of the University of Vermont. Professor Blatt spent a year at Oxford University on an N.S.F. post-doctoral fellowship and has served as visiting professor at the Swiss Federal Institute of Technology (E.T.H., Zurich), Simon Fraser University (Vancouver, B.C.), Université de Louvain (Belgium), Leeds University (England), University of North Sumatra (Indonesia), Institute of Technology (Shah Alam, Malaysia), and the United States Military Academy (West Point). Professor Blatt has written a number of books—*Physics of Electronic Conduction in Solids* (McGraw-Hill, 1968), *Thermoelectric Power of Metals* (with C. L. Foiles, D. Greig, and P. A. Schroeder, Plenum Press, 1972), and *Principles of Physics* (Allyn & Bacon, 1983, 1986, 1989)—as well as over 80 research articles in solid state physics and biophysics.

For JMB

Contents

(Sections marked * are optional.)

Preface

This is a textbook for a one-semester introductory course in modern physics. The topics covered here include relativity, elementary quantum mechanics, and atomic, molecular, solid state, nuclear, and elementary particle physics, as well as a brief account of instrumentation for nuclear physics research. The text assumes that the reader has completed the standard introductory calculus-based physics course.

The primary purpose of a first course in modern physics, often the only course on the subject, is to introduce the student to the concepts and ideas of twentieth-century physics. Although the new, indeed revolutionary, approach to our understanding of nature is now widely accepted, these novel ideas met at first much skepticism and occasionally overt resistance. Understandably, students also find the concepts and axioms of relativity and quantum physics difficult to embrace. In presenting the subject, I have adopted an historical approach because the reader is better able to follow the reasoning that led physicists at the start of the twentieth century to suggest what then seemed preposterous propositions. Almost every chapter begins with a brief but succinct account of events that led to the development of the topic treated in the following pages, and throughout the text the reader will find numerous references to the original literature.

There are several reasons for including historical material and ample references to primary sources in a physics text. First, it helps students appreciate that science is an ongoing human enterprise, that whatever we may today accept as "scientific truth" is subject to question and may well have to be abandoned in light of new data. Second, historical accounts place modern science, a pervasive component of our culture, in proper perspective. And last, the occasional historical or anecdotal paragraphs provide a welcome interlude between more difficult sections of the text.

The first three chapters are devoted to the theory of relativity. Chapter 3, an innovation in an introductory modern physics text, is a largely descriptive account of the general relativity; it is included in light of recent technological advances that have allowed careful and precise experiments and have stimulated new interest in the field.

Quantum theory is the central theme of the next five chapters. Chapter 4 summarizes the experimental findings that ultimately led to broad acceptance of energy quantization. Chapter 5 is an account of the Bohr model of the hydrogen atom. The concept of cross section is introduced and illustrated in connection with Rutherford scattering. The de Broglie hypothesis and experiments that validated it are described in Chapter 6.

Elementary quantum mechanics is the subject matter of Chapters 7 and 8. The Schrödinger equation is introduced in Chapter 7 and the standard one-dimensional examples—infinite and finite square wells, barrier penetration, and the harmonic oscillator—are presented. Chapter 8 addresses primarily the quantum mechanics of the hydrogen atom. It is here that the phrase "it can be shown," studiously avoided elsewhere in the text, will occasionally be found. The formal solution of the Schrödinger equation in spherical coordinates is well beyond the mathematical sophistication of the student whose background is a one-year course in differential and integral calculus. Yet, to present the results, such as angular momentum quantization, in an ad hoc manner without at least a "hand-waving" justification does the student a disservice by suggesting some mystical aspect to quantization and selection rules.

Chapters 9 and 10 concern the quantum theory of atomic and molecular structure. Here intrinsic spin and the Pauli exclusion principle make their first appearance. Spin-orbit splitting, selection rules for dipole transitions, the ground state of helium, and the forces responsible for the bound state of the hydrogen molecule are presented in these chapters. Here, too, is the first of many sections on technological applications of modern physics—a discussion of the principles of maser and laser operation.

Chapter 11, Statistical Physics, reviews the concepts of classical statistical mechanics and then shows how the different counting procedures that must be employed when dealing with indistinguishable particles lead to Bose-Einstein and Fermi-Dirac distributions. The chapter concludes with a section on liquid helium, a convincing experimental demonstration of Bose-Einstein condensation.

Chapters 12 and 13 are devoted to solid state physics, the first to crystal structure, cohesive forces, and a brief survey of crystal defects and their crucial roles in diffusion, plastic flow, and crystal growth. The chapter concludes with a description of the energy states of the quasi-free electron gas in a periodic potential and introduces the concepts of energy bands, energy gaps, and effective masses.

Chapter 13 focuses on electronic properties. Electronic specific heat and conductivity of metals is followed by a discussion of electronic properties of homogeneous semiconductors. The next section is devoted to devices, including *p-n* junction diodes, tunnel diodes, *n-p-n* and field-effect transistors, photovoltaic cells, and junction lasers. In the final section of the chapter the dominant features of superconductivity are described; Josephson junctions and SQUIDs are discussed, and some applications of these sensors are presented.

Nuclear physics is the subject of Chapters 14 and 15. Chapter 14 contains a general description of the properties of nuclei, of nuclear forces, binding energy, nuclear spin and magnetic moment, and of the liquid drop model. The final section of this chapter gives a fairly detailed explanation of nuclear magnetic resonance and one of the most significant practical applications of this technique, NMR imaging.

Chapter 15 is concerned with radioactivity and nuclear reactions. Here again, several sections are devoted to practical matters—fission and fusion as energy sources, and the use of radioisotopes in archeology and medicine.

Chapter 16 contains brief descriptions of accelerators and particle detectors, instruments used in nuclear physics research. The final chapter is a brief survey of elementary particle physics and touches on the more recent theoretical advances in our understanding of the fundamental forces of nature.

In a one-semester course it is probably not possible to cover all the topics presented in the text. Some sections and chapters have been marked as optional and may be omitted without loss of continuity. For example, Chapter 3 (General Relativity) and Chapter 16 could well be omitted, depending to some extent on the interest and background of the students. Most of the "application" sections have also been marked as optional. Generally, in these sections I have opted for detailed and concise presentations, which has meant that to keep the text to reasonable size, some interesting technological spin-offs of modern physics had to be omitted. I have done this because in my experience students are dissatisfied with a cursory gee-whiz summary that fails to help them really understand how a device operates or a technique is employed in practice. Those applications that have been included span a broad spectrum and should appeal to most science and engineering students.

In preparing this text I have benefitted from the insight of many students whose names I no longer recall but who helped unknowingly by interrupting a lecture whenever I failed to explain a topic clearly. I am also privileged to acknowledge the assistance of the following reviewers whose comments and criticism have been most valuable: Gordon Aubrecht, Ohio State University; John Barach, Vanderbilt University; J. Birchall, University of Manitoba; Roger Clapp, University of South Florida; Maj. William E. Eichinger, U.S. Military Academy; Harry E. Grove, University of Rochester; Joseph Kiskis, University of California—Davis; Richard Kouzes, Princeton University; Patrick Labelle, Cornell University; Peter Lindenfeld, Rutgers University; A. E. Livingston, University of Notre Dame; Alan Nathan, University of Illinois; Julian V. Noble, University of Virginia; Lawrence Pinsky, University of Houston; James Trefil, George Mason University; Walter Wales, University of Pennsylvania; James K. Walker, University of Florida; James Whitmore, Pennsylvania State University; E. J. Winhold, Rensselaer Polytechnic Institute; and Tung-Mow Yan, Cornell University.

I am also indebted to John Bullock, Princeton University; David Cowan, University of Missouri; Thomas A. Kaplan, Michigan State University; David Sandison, Cornell University; and Yannis Tsamouranis, University of Houston, who solved end-of-chapter problems, called my attention to ambiguities, and suggested additional problems.

The work was initiated at the suggestion of Malvina Wassermann without whose confidence in my ability and fortitude to bring the project to successful conclusion I would not have begun to write this book. I am also grateful for the help and encouragement of McGraw-Hill Publishing, especially Anne Duffy, Holly Gordon, Jack Maisel, Safra Nimrod, Denise Schanck, and Susan Tubb.

Last, though by no means least, I wish to thank Jane Dahl-Blatt for her expert professional assistance and constant support. Without her help it is doubtful that this book could have been completed.

Frank J. Blatt

Modern Physics

The Theory of Special Relativity I: The Lorentz Transformation

Chapter 1

The relativity theory arose from necessity, from serious and deep contradictions in the old theory from which there seemed no escape. The strength of the new theory lies in the consistency and simplicity with which it solves all these difficulties, using only a few very convincing assumptions.

Albert Einstein and Leopold Infeld, *The Evolution of Physics* (1938)

1.1 INTRODUCTION

The year 1905 has rightly been called the *annus mirabilis* of modern physics. That year, Albert Einstein published five papers of which four have left an indelible imprint on twentieth-century science and culture.

Figure 1.1 Albert Einstein (*a*) in 1905 at the age of 26 (*The Bettmann Archive/Bettmann Newsphotos.*); (*b*) off Long Island, about 1935, enjoying his favorite relaxation (*AP/Wide World Photos.*); (*c*) in his later years, at his home in Princeton, New Jersey (*The Bettmann Archive/Bettmann Newsphotos.*)

(*a*)

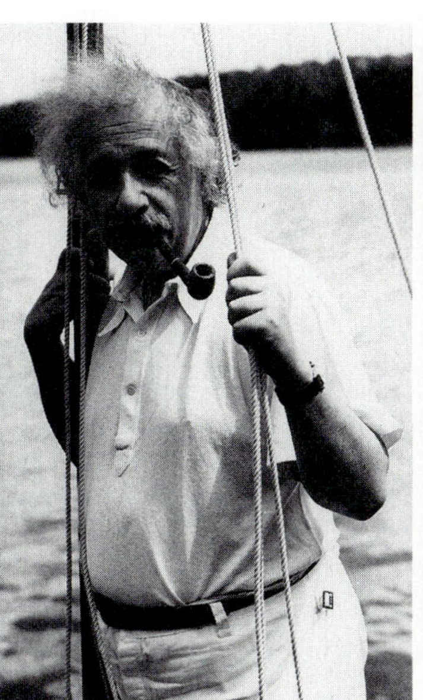

(*b*)

(*c*)

One seldom finds the bound volume of a scientific journal locked carefully in the rare book collection of a library, but that is where one often finds Volume 17 of *Annalen der Physik,* the "Einstein volume," as it is widely known. In it appear the seminal article on the theory of special relativity;[1] the lucid explanation of Brownian motion which provided incontrovertible support for the atomic hypothesis;[2] and the paper on the generation and absorption of light which includes, almost as an aside, the theory of the photoelectric effect that the Nobel committee cited in awarding the 1921 Nobel Prize to Einstein.[3]

Einstein has often been called the Newton of the twentieth century, and the impact of both scientists on society is indeed comparable. Newton, however, was a vain, petty, cantankerous, and egocentric person, whereas all who had the good fortune to know Einstein were as much impressed by his personal warmth, unlimited generosity, integrity, and humanity as by his awe-inspiring intellect.[4]

1.2 CLASSICAL RELATIVITY: THE GALILEAN TRANSFORMATION EQUATIONS

Objects are located in space by means of a coordinate system, a *frame of reference* or *reference frame*. One distinguishes between two types of reference frames: inertial frames, and accelerated frames. An *inertial reference frame* is, by definition, one in which Newton's first law is valid: **Every body continues in its state of rest, or of uniform motion in a straight line, unless it is compelled to change that state by forces impressed on it.** This law of inertia is not valid in an accelerated reference frame. A ball placed on the floor of a car will move to the rear if the car accelerates forward even though no forces act on it; a coin placed on a rotating turntable will slide to the periphery though no visible force pushes it away from the center. The special theory of relativity confines itself to inertial reference frames. Accelerating reference frames are considered in the theory of general relativity (see Chapter 3).

To illustrate, consider the motion of a boat crossing a river flowing at constant velocity $\mathbf{u}$ (Figure 1.2). The boat travels at a velocity $\mathbf{v}'$ over the water. To an observer floating on a raft in the river, the boat appears to travel at right angles to the river bank. An observer on the bank, however, sees that the boat's velocity makes an angle θ with respect to the bank, where $\theta = \tan^{-1}(v'/u)$. If we designate by S the reference frame of O, the stationary observer on the bank, and by S' the reference frame of O', the moving observer on a raft, we see that the velocities are related by

$$\mathbf{v} = \mathbf{v}' + \mathbf{u} \tag{1.1}$$

If we choose our reference frame S so that the velocity of the moving reference frame S' is in the x direction, the coordinates in the two reference frames are related by

1. A. Einstein, *Ann d. Phys.* [4] **17,** 891 (1905).
2. A. Einstein, *Ann d. Phys.* [4] **17,** 549 (1905).
3. A. Einstein, *Ann d. Phys.* [4] **17,** 132 (1905).
4. Some of the better biographies of Einstein are listed at the end of this chapter.

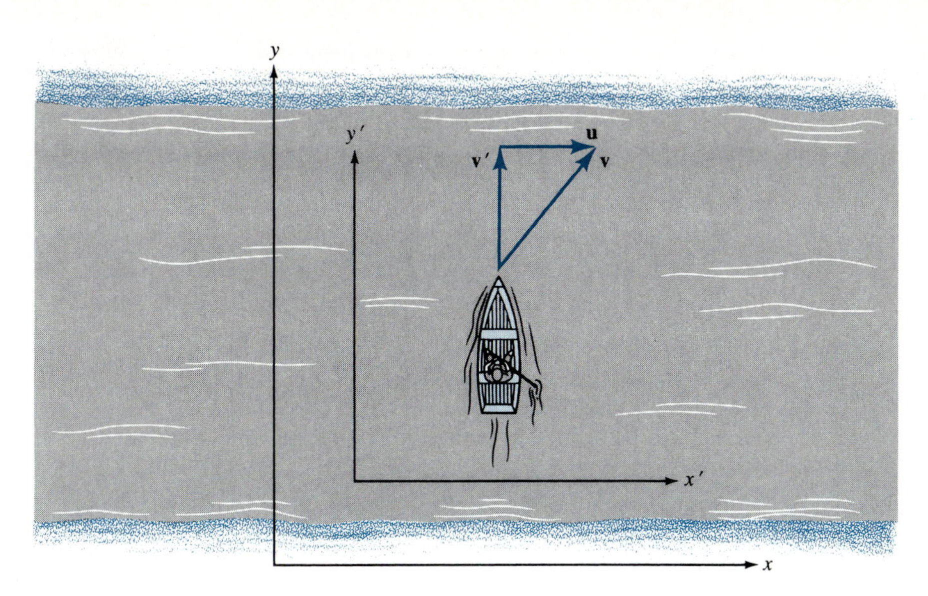

Figure 1.2 In the reference frame of the stream (x', y'), the velocity of the boat is $\mathbf{v}'$. In the reference frame of the shore (x, y), the velocity of the boat is $\mathbf{v} = \mathbf{v}' + \mathbf{u}$, where $\mathbf{u}$ is the velocity of the stream relative to the shore.

$$x' = x - ut \tag{1.2a}$$

$$y' = y \tag{1.2b}$$

$$z' = z \tag{1.2c}$$

and the velocity components are related by

$$v'_x = v_x - u \tag{1.3a}$$

$$v'_y = v_y \tag{1.3b}$$

$$v'_z = v_z \tag{1.3c}$$

Since both S and S' are inertial reference frames, the velocity u must be constant. On differentiating Equations (1.3a) through (1.3c) with respect to time, we see that

$$\mathbf{a}' = \mathbf{a} \tag{1.4}$$

Equations (1.2) to (1.4) are the Galilean transformation equations. Not only are they simple, but they also make eminently good sense. We know that if we move away from a stationary object at some speed, that object appears to recede at the same speed. If we are traveling in a car at constant speed and another approaches from behind, traveling at a greater but also constant speed, the distance between the two cars will diminish at a constant rate until the second car overtakes us, and then will increase at the same constant rate.

Although we have chosen to identity S as the "rest" frame in the preceding discussion, it is surely not a rest frame in an absolute sense. The earth rotates about its axis, it is in orbit about the sun, and the entire solar system revolves about the center of our galaxy. In fact, one cannot identify any reference frame as the "true" rest frame. Newton understood this:

. . . the parts of space cannot be seen, or distinguished from one another by our senses. . . . From the positions and distances of things from any body considered as immovable, we define all places; and then with respect to such places, we estimate all motion. . . . And so, instead of absolute places and motions, we use relative ones.

And yet, Newton also wrote,

The center of the system of the world is immovable. This is acknowledged by all, while some contend that the earth, others that the sun, is fixed in that center.

The reason the true "immovable" center of the world cannot be identified is inherent in Newton's laws of motion. It follows from Equation (1.4) that the second law is not changed by a Galilean transformation; i.e., $\mathbf{F} = m\mathbf{a} = m\mathbf{a}'$. Similarly, the conservation laws of energy and linear and angular momenta retain their validity in the S and S' frames. To use the jargon of physics, the laws of classical mechanics are *invariant* under a Galilean transformation.

It follows that no experiment of classical mechanics can ever tell us whether we, in frame S, are moving and frame S' is at rest, or vice versa, or frames S and S' are both moving at different but constant velocities. To ask "Am I moving?" is, therefore, strictly speaking quite meaningless, because nature precludes ever finding an answer. All one can ask is, "Am I moving relative to the earth's surface, the stern of a ship, or the cockpit of a plane?"

1.3 ELECTROMAGNETIC WAVES AND THE LUMINIFEROUS ETHER

In the latter half of the nineteenth century Maxwell's synthesis of electricity and magnetism—in particular, his prediction of electromagnetic waves that propagate through space at the speed of light—appeared to provide a means for determining the "true" rest frame of the universe.

The wave nature of light had been recognized since the work of Young and Fresnel at the beginning of the nineteenth century. Since all known waves were transmitted by some medium—sound waves by air, surface waves on water by the water—physicists tried to identify the medium that transmits light waves. Failing to do so, scientists and philosophers simply invented the medium, calling it the luminiferous ether, or simply the *ether,* for short.

The trouble was that this ether had to be endowed with truly bizarre properties. Since the speed of propagation of a wave in a medium is greater the greater the stiffness of the medium, the ether, transmitting an electromagnetic disturbance at the speed of light, would have to be incredibly resilient. Yet, the ether did not impede the movement of objects, and, moreover, defied all attempts to remove it from any region.

Maxwell, who had formulated his theory of electromagnetism in terms of electric and magnetic "stress tensors," expressed the prevalent view of the time in these words:

Whatever difficulty we may have in forming a consistent idea of the constitution of the ether, there can be no doubt that the interplanetary and interstellar spaces are not empty, but are occupied by a material substance or body which is certainly the largest, and probably the most uniform body of which we have any knowledge.

Though the ether was tauntingly elusive, sooner or later an ingenious experimenter would discover it. Indeed, Maxwell's work presented that opportunity. Maxwell's equations predicted electromagnetic waves that propagate *through the ether* at the speed of light, $c = 3.00 \times 10^8$ m/s. With all space permeated by the ether, and the earth itself embedded in it, there could be only two possible options: either the earth, in its travels, drags the ether along, just as the earth's atmosphere travels and rotates with the sphere; or the earth moves through the ether as it orbits the sun, though no drag or other effects of that relative motion had ever been observed.

The first possibility is inconsistent with the observation of stellar aberration, the small annual motion of nearby stars relative to the distant background. Hence, the earth must be moving with some finite velocity relative to the ether, at least during most of the year if not at all times.

Now, at last, one could identify the true, Newtonian rest frame, the inertial frame of the ether. In that frame, and only in it, would the measured speed of light agree with Maxwell's result. All that needed to be done was to perform precise measurements of the speed of light on earth along different directions—for example, parallel and perpendicular to the earth's velocity as it orbited the sun—and compare the results.

That is easier said than done. The orbiting speed of the earth is about 30 km/s, or about 1/10,000 of the speed of light. To reveal the difference in the speed of light parallel and perpendicular to the earth's velocity through the ether, the experiment would have to be of great accuracy, because, as we shall see presently, that difference would be no more than 1 part in 100 million. In 1881, Albert A. Michelson devised such an experiment, and in the next few years perfected the design so that this small difference could be detected.

1.4 THE MICHELSON-MORLEY EXPERIMENT[5]

A schematic diagram of the justly celebrated Michelson-Morley experiment is shown in Figure 1.3. A collimated beam of light from the source S is split by the half-silvered mirror M_1. Beam A proceeds to mirror M_2 and is reflected back to M_1; here a portion is reflected down. Beam B is reflected to mirror M_3, and part of this reflected beam is transmitted through M_1 and combines with the reflected portion of beam A. Since beams A and B originate from the same source, they are coherent, and, upon superposing, interfere according to their relative phase. That phase relation is determined by the difference in optical path length for the two beams which, as we now show, depends on the motion of the apparatus relative to the ether.

Let us assume that, at the time of the experiment, the earth moves with velocity $\mathbf{u}$ relative to the ether. During the first part of the light's travel along A (see Figure 1.4), the apparatus is moving in the direction of propagation. According to Equation (1.3a), the velocity of light in the inertial frame of the apparatus is then $c - u$, and the time required for light to travel from M_1 to M_2 is $L/(c - u)$. Similarly, the time

5. A. A. Michelson and E. W. Morley, *Am. J. Sci.* **34,** 333 (1887).

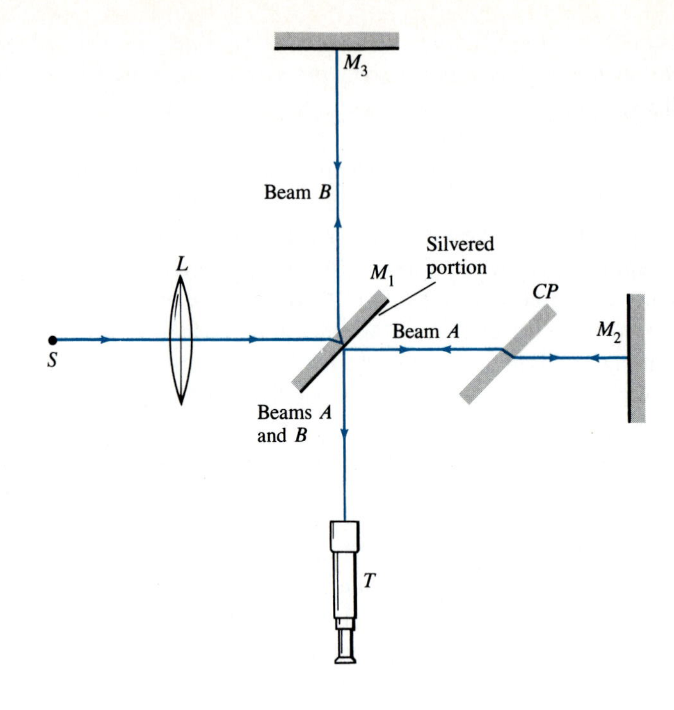

Figure 1.3 The Michelson-Morley apparatus (schematic diagram). Light from the source S is collimated by the lens L and split into two coherent beams by the half-silvered mirror M_1. Beams A and B are reflected back to M_1 and then superposed to create an interference pattern observed with the telescope T. CP is a compensation plate made of glass so that beams A and B both traverse the same optical paths.

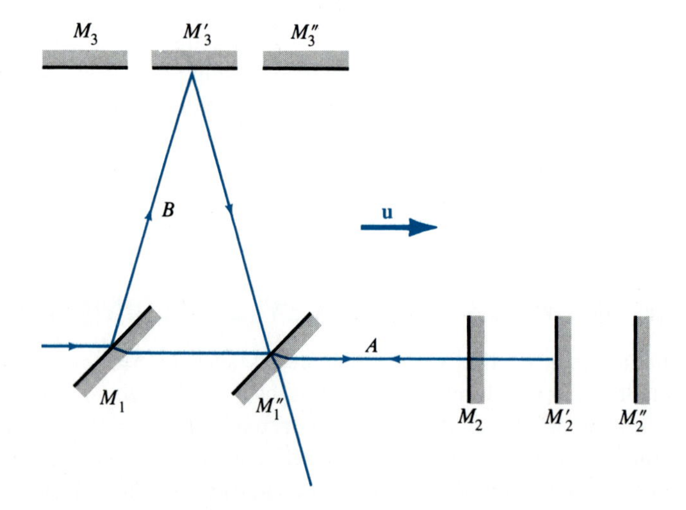

Figure 1.4 The paths of beams A and B of Figure 1.3 as seen from a reference frame at rest relative to the ether. The velocity of the apparatus relative to the ether is **u**.

required for the return to M_1 is $L/(c + u)$. Therefore the total time elapsed in this round trip is

$$T_A = \frac{L}{c-u} + \frac{L}{c+u} = \frac{2Lc}{c^2 - u^2} = \frac{2L/c}{1-\beta^2} \qquad \textbf{(1.5)}$$

where we have introduced the symbol

$$\beta = \frac{u}{c} \qquad \textbf{(1.6)}$$

We next calculate the time T_B for path B, shown in Figure 1.4. During the time interval $T_B/2$, mirror M_3 has moved a distance $uT_B/2$ to the right. Consequently, the total length of path B is

$$L_B = 2\sqrt{L^2 + \left(\frac{uT_B}{2}\right)^2}$$

Since light travels in the ether at the speed c, the time that has elapsed is

$$T_B = \frac{L_B}{c}$$

Solving for T_B one obtains

$$T_B = \frac{2L/c}{\sqrt{1-\beta^2}} \tag{1.7}$$

Evidently, there is a time difference between the two paths due to the motion of the apparatus relative to the ether given by

$$\Delta T = T_A - T_B = \frac{2L}{c}\left(\frac{1}{1-\beta^2} - \frac{1}{\sqrt{1-\beta^2}}\right) \tag{1.8}$$

Since, for the case considered here, $\beta^2 = (u/c)^2 \simeq 10^{-8}$, we can simplify Equation (1.8) using the binomial expansion; i.e., we replace $(1 - \beta^2)^{-1}$ by $1 + \beta^2$ and $(1 - \beta^2)^{-1/2}$ by $1 + \beta^2/2$. Equation (1.8) then reduces to

$$\Delta T = \frac{L}{c}\beta^2 \tag{1.9}$$

If the apparatus is turned through 90°, the roles of paths A and B are interchanged, and consequently the time difference is then the negative of Equation (1.9). The total time shift as a result of such rotation is therefore twice the value of Equation (1.9), or

$$\Delta T_{\text{tot}} = \frac{2L}{c}\beta^2 \tag{1.10}$$

and the corresponding phase shift between the two waves is

$$2\pi\left(\frac{\Delta T_{\text{tot}}}{T}\right) = \frac{2\pi c\, \Delta T_{\text{tot}}}{\lambda} = \left(\frac{4\pi L}{\lambda}\right)\beta^2 \tag{1.11}$$

where T is the period and λ the wavelength of the light wave.

To minimize vibrations and distortions due to stresses during rotation, Michelson and Morley mounted the entire apparatus on a large stone slab which floated on mercury (see Figure 1.5). They also improved the accuracy of the measurement by means of additional mirrors that lengthened the light path L to 11 m. Even though the factor β^2 is only about 10^{-8}, $4\pi L/\lambda = 44\pi/5.5 \times 10^{-7} = 2.5 \times 10^8$ for the yellow light used in the experiment. The phase difference due to a 90° rotation of the

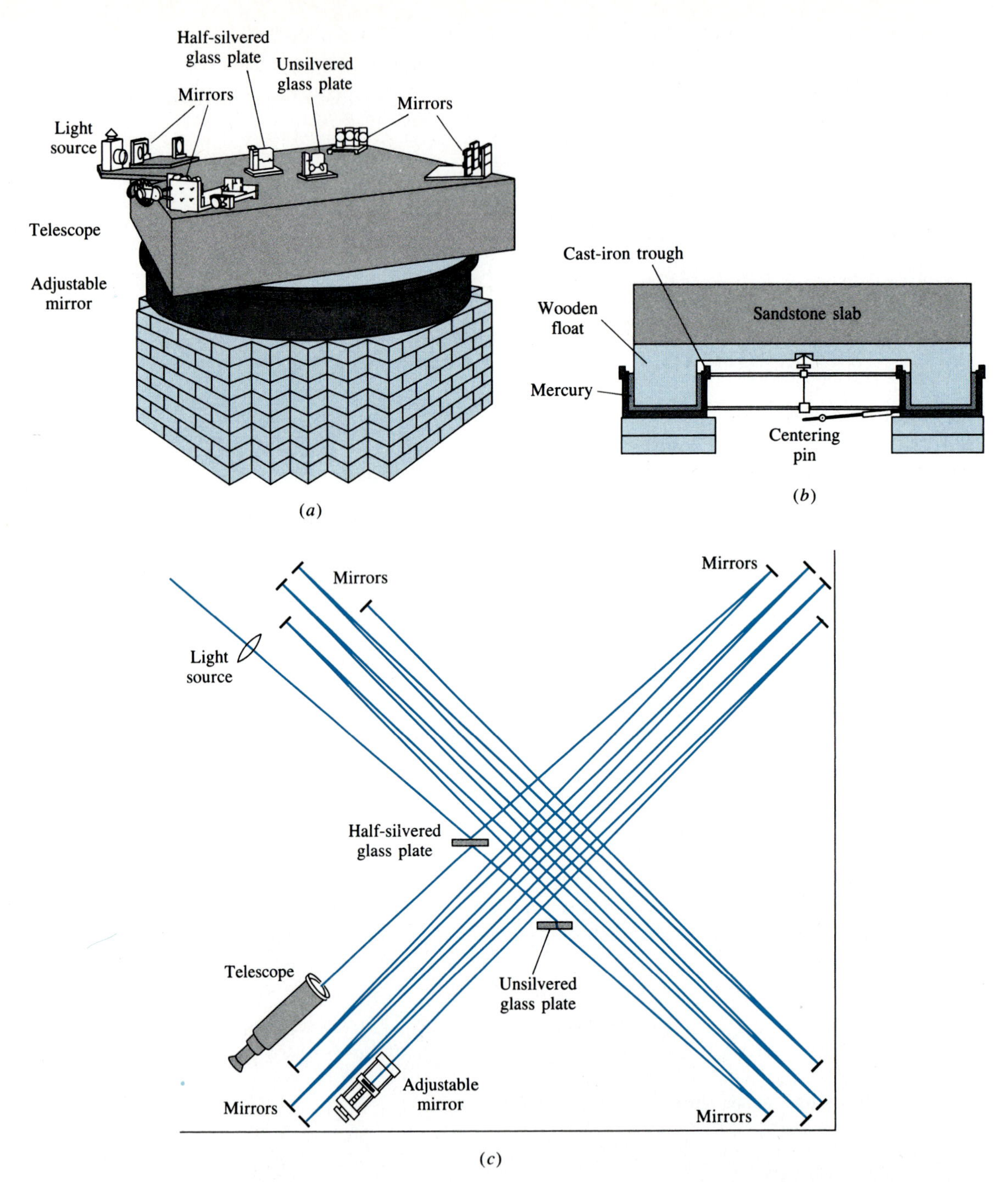

Figure 1.5 Drawing of the apparatus used by Michelson and Morley in 1887. (*a*) The light source, mirrors, and telescope are all mounted on a 5-ft square stone slab, which floats on a pool of mercury (*b*). This permits slow rotation of the entire apparatus without introducing strains. (*c*) A top view of the apparatus shows the additional mirrors used by Michelson and Morley to increase the optical path length.

apparatus is then 2.5 radians, resulting in a readily measurable shift of the interference pattern.

The results of the experiment were a deep disappointment:

> *The displacement to be expected was 0.4 fringe. The actual displacement was certainly less than a twentieth part of this, and probably less than a fortieth part.*

There still remained the remote possibility that the experiment had been performed just when the earth was nearly at rest relative to the ether.

> *The experiment will therefore be repeated at intervals of three months, and thus all uncertainty will be avoided.*

Michelson took no part in these repetitions, which were performed by Morley and D. C. Miller. His disillusionment with this beautifully conceived and superbly executed experiment is evident from the following passage, written in 1902, fifteen years after the event:

> *I think it will be admitted that the problem, by leading to the invention of the interferometer, more than compensated for the fact that this particular experiment gave a negative result.*

Three years later, Einstein published his paper ''On the Electrodynamics of Moving Bodies,'' which set forth the theory of special relativity. That his ''negative'' result confirmed one of the most profound theories of nature did not alter Michelson's attitude. When he and Einstein met briefly in 1931, Michelson remarked that he regretted that his experiment might have been responsible for giving

Figure 1.6 Albert A. Michelson, Albert Einstein, and Robert A. Millikan at the California Institute of Technology in 1931. (*AP/Wide World Photos.*)

birth to such a "monster"—referring to the theory of special relativity. In fact, that theory was not motivated by Michelson's experiment. As Einstein later recalled, "In my development, Michelson's result has not had a considerable influence. I even do not remember if I knew of it at all when I wrote my first paper on the subject."

1.5 THE THEORY OF SPECIAL RELATIVITY

Following publication of the Michelson-Morley experiment in 1887, physicists tried to reconcile its negative result with the fact, deduced from stellar aberration, that the earth does move relative to the ether. George F. FitzGerald[6] and H. A. Lorentz[7] independently suggested that as a result of motion through the ether, the linear dimension of an object contracts along the line of relative motion by a factor $1/\gamma$, where γ is defined by

$$\gamma = \frac{1}{\sqrt{1 - \beta^2}} \tag{1.12}$$

Since a meterstick placed alongside a moving object would suffer the same contraction, it is impossible to confirm or disprove this suggestion.

As we shall see presently, Einstein's theory results in the same length contraction, now known as the *Lorentz-FitzGerald contraction*. However, Einstein was motivated not by the paradox of the Michelson-Morley result, but by the difficulties inherent in Maxwell's equations, and his relativistic length contraction is a consequence of the theory, not an ad hoc device designed to reconcile an inexplicable experimental fact with classical mechanics.[8]

If the equations of electromagnetism are subjected to a Galilean transformation, one finds that, unlike Newton's equation of motion, they do change their form. In the terminology of physics, Maxwell's equations are not *invariant* under a Galilean transformation.

We can exhibit this lack of invariance as follows. Suppose we measure the force between two equal stationary charges separated by a distance d as shown in Figure 1.7. We would find that the two charges repel each other according to Coulomb's law. Now suppose *these same charges* are examined by another observer who is moving with velocity $\mathbf{u}$ along a line perpendicular to that separating the two charges. That observer will see the two charges moving with velocity $-\mathbf{u}$ in his or her inertial reference frame. The two moving charges constitute two currents in the same direction, and two parallel currents exert a mutual attraction. Therefore, this observer will find that the charges not only repel according to Coulomb's law

---

6. G. F. FitzGerald, *Science* **13,** 390 (1889).

7. H. A. Lorentz, *Verls. K. Ak. Amsterdam* **1,** 74 (1892).

8. Lorentz was well aware of the fundamental, essential difference between his and Einstein's work. "The chief cause of my failure," he wrote in 1915, "was my clinging to the idea that the variable t [time in the absolute, ether reference frame] can only be considered as the true time and that my local time t' must be regarded as no more than an auxiliary mathematical quantity. In Einstein's theory, on the contrary, t' plays the same part as t; if we want to describe phenomena in terms of x', y', z', t' we must work with these variables exactly as we would with x, y, z, t."

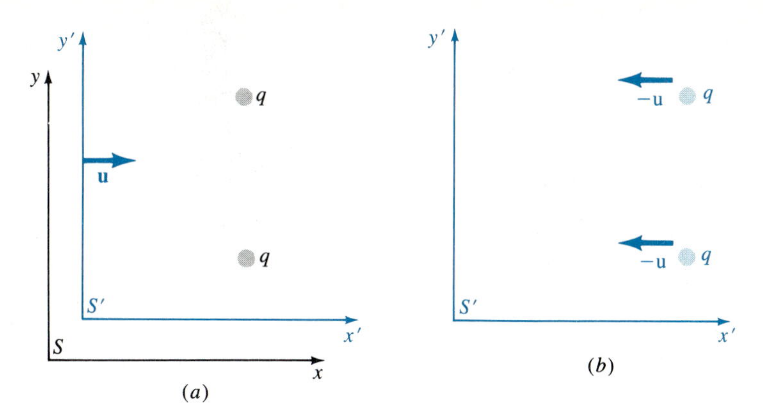

Figure 1.7 (*a*) The two charges q are at rest in reference frame S. Their interaction is expressed by Coulomb's law. (*b*) The same two charges as seen by an observer in reference frame S' that moves with velocity **u** relative to S. In S', the two charges constitute two parallel currents and, therefore, exert an attractive as well as a repulsive force on each other.

but also exert an attractive force due to their velocities *relative to the observer*. We are forced to conclude that observers in two different inertial reference frames will measure *different forces* between the same two charges.

The central problem, as Einstein perceived it, had little to do with the existence of an ether, but hinged on the *principle of relativity:*

The laws of nature are the same in all inertial reference frames.

Either the principle of relativity had to be abandoned; or Maxwell's equations, so eminently successful in explaining all that was known about electromagnetism, are incorrect; or the rules of transformation from one inertial frame to another, Equations (1.2) through (1.4), are not correct. To Einstein, the validity of the principle of relativity was axiomatic, and he was convinced that Maxwell's equations are true and must yield the same result in all inertial reference frames. Einstein therefore proceeded from the following two basic postulates.

1. **The principle of relativity is valid for all natural events.**
2. **The speed of light in vacuum, as measured in any inertial reference frame, is *c* regardless of the motion of the light source relative to that reference frame.**

The second postulate is a drastic departure from previous scientific philosophy. It also appears to violate common sense.

Suppose that in Figure 1.8 the light source E, at rest in the inertial frame S,

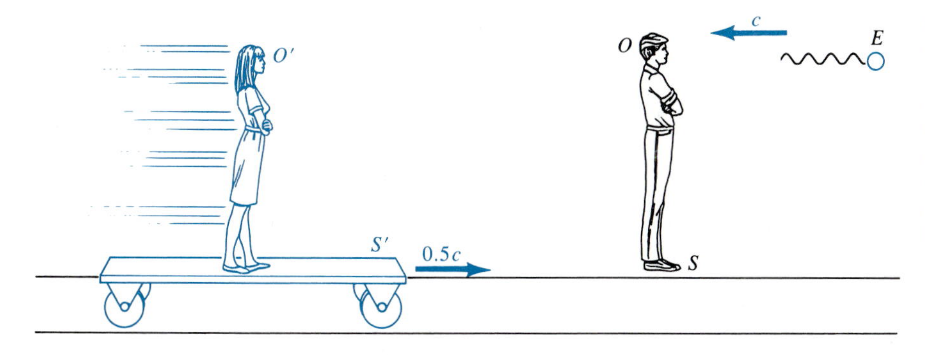

Figure 1.8 Source E emits a pulse of light. Observer O is at rest relative to E; observer O' approaches E with a velocity of $0.5c$. According to the second postulate of the special theory of relativity, both observers will measure the same speed of the light pulse, namely, c.

emits a light pulse. An observer O in S measures the speed of that pulse and finds that it is c. A second observer, O', at rest in the reference frame S' which is approaching E at a speed of $0.5c$, also measures the speed of the same light pulse. We expect that O' will find that the pulse travels past her at the speed $1.5c$. According to Einstein's second postulate, however, she, too, will conclude that the pulse propagates at the speed c.

As we shall see, Einstein's postulate and "common sense" are not in conflict. The point is that relative velocities in everyday experience are orders of magnitude smaller than the velocity of light. When $v \ll c$, the relativistic transformation equations that we shall shortly derive reduce to the familiar Galilean transformation.

Starting from the two fundamental postulates, the results of the theory of special relativity follow from a sequence of simple, logical arguments. Those results are, however, very strange. For example, moving objects contract along the line of motion; moving clocks run slow; if the velocity of an object is $\mathbf{v}$ in frame S, in frame S', moving with velocity $\mathbf{u}$, its velocity is not $\mathbf{v}' = \mathbf{v} - \mathbf{u}$; the inertial mass of an object depends on its speed.

1.6 THE LORENTZ TRANSFORMATIONS

We shall now derive the relativistic transformation equations. We make only two assumptions. First, both reference frames, S and S', are inertial frames. That is, they move at constant velocity. Second, the postulate of Einstein concerning the constancy of the speed of light in vacuum is correct. We make no further assumptions with respect to the spatial or temporal coordinates. We designate the coordinates in S by x, y, z, and t, and the coordinates in S' by x', y', z', and t'.

To arrive at transformation equations that are consistent with these assumptions we consider the propagation of a light pulse in vacuum. We set the coordinate origin of S' equal to zero at the instant the light source, located at the origin of S, emits a brief pulse. To simplify the derivation we also set $t = t' = 0$ at that instant.

Because the light pulse propagates at the speed c in both reference frames and space is isotropic, the wavefront of the pulse is spherical in both inertial reference frames. The location of the wavefront in the two reference frames is given by

$$r = ct \qquad r^2 = x^2 + y^2 + z^2 \tag{1.13a}$$

and

$$r' = ct' \qquad r'^2 = x'^2 + y'^2 + z'^2 \tag{1.13b}$$

We seek transformation equations that relate the primed to the unprimed coordinates, and vice versa. Since $\mathbf{u}$, the velocity of S' relative to S, is along the x direction,[9] it follows that

$$y' = y \qquad \text{and} \qquad z' = z \tag{1.14}$$

The remaining coordinates must be related by equations of the form

9. This condition does not restrict the generality of the results. We can always rotate our coordinate system so that the x axis coincides with the relative motion of S'.

$$x' = ax + bt \tag{1.15a}$$

$$t' = fx + gt \tag{1.15b}$$

The transformation equations must be linear in the variables to ensure the basic equivalence of all inertial reference frames. If terms involving x^2 or t^2 or higher powers of these variables appeared in the transformation equations, motion at constant velocity in S would transform into accelerated motion in S'.

We can rewrite Equation (1.15a) as

$$x' = a\left[x + \left(\frac{b}{a}\right)t\right] \tag{1.16}$$

Dimensionally, b/a must be a velocity; that $b/a = -u$ is apparent if we ask for the coordinates in S of $x' = 0$, the origin of S'. At time t, the origin of S' is at $x = ut$. From Equation (1.16) we now have

$$0 = a\left[ut + \left(\frac{b}{a}\right)t\right] \qquad \frac{b}{a} = -u$$

and

$$x' = a(x - ut) \tag{1.17a}$$

The inverse transformation must be identical to Equation (1.17a) except for a change in sign of u:

$$x = a(x' + ut') \tag{1.17b}$$

To determine the coefficient a, we turn to the formal expression of Einstein's postulate, $r = ct$, $r' = ct'$. We may, without loss of generality, set $y = y' = z = z' = 0$. Then,

$$x' = ct' = a(ct - ut) = act(1 - \beta) \tag{1.18a}$$

$$x = ct = a(ct' + ut') = act'(1 + \beta) \tag{1.18b}$$

From Equation (1.18a) we have $t' = at(1 - \beta)$ and substituting this in Equation (1.18b) we find

$$a = \frac{1}{\sqrt{1 - \beta^2}} = \gamma \tag{1.19}$$

The transformation equations for x and x' and t and t' are then

$$x' = \gamma(x - ut) \qquad x = \gamma(x' + ut') \tag{1.20a}$$

$$y' = y \qquad z' = z \tag{1.20b}$$

$$t' = \gamma\left(t - \frac{u}{c^2}x\right) \qquad t = \gamma\left(t' + \frac{u}{c^2}x'\right) \tag{1.20c}$$

where

$$\gamma = \frac{1}{\sqrt{1 - \beta^2}} \qquad \beta = \frac{u}{c} \tag{1.21}$$

These are the *Lorentz transformations,* so named by Poincaré because H. A. Lorentz had proposed them in 1904 as the transformations that would be consistent with the null result of Michelson and Morley and leave Maxwell's equations invariant.

Before considering some of the consequences that derive from these transformation equations, let us compare them with the Galilean transformation and also examine their structure.

Perhaps the most notable difference between Equations (1.20) and (1.2) is that in the former the length *and the time scales* in S and S' differ. That difference is characterized by the constant γ, which depends on the relative velocity between S and S'. Evidently, Newton's assertion that

> *absolute, true and mathematical time of itself and from its own nature flows equally without relation to anything external*

cannot be maintained. However, if $u \ll c$, $\gamma \simeq 1$, and the relativistic transformations reduce to the Galilean.

The second important feature of Equations (1.20) is their symmetry with respect to interchange of the S and S' frames. The equations are identical except for a change in the sign of u, which is as it should be. We may regard S as the "stationary" frame with S' moving in the positive x direction at a speed u, or consider S' the "stationary" frame, in which case S moves in the negative x direction at the speed u. Frames S and S' are completely equivalent; the only quantity that appears in the transformation equations is the *relative* velocity between S and S'. It follows that we can devise no experiment that will distinguish a particular reference frame as the "true" stationary frame, the ether frame of Maxwell.

1.6(a) Simultaneity, Length Contraction, and Time Dilation

According to Newtonian mechanics, two simultaneous events that occur at two different locations in S will also be simultaneous in a reference frame S' moving with velocity u relative to S. That result follows from the presumption that the time scales in the two reference frames are identical. Since t and t' differ in relativistic mechanics we must reexamine that conclusion.

Suppose that in S two events, A and B, occur at $x_A = -X$ and $x_B = +X$ at time $T_A = T_B = T$. Will an observer in S' also find that the two events are simultaneous?

In S', according to Equation (1.20c), event A will occur at

$$T'_A = \gamma\left(T - \frac{-u}{c^2}X\right) = \gamma T + \gamma\left(\frac{u}{c^2}\right)X$$

and event B at

$$T'_B = \gamma T - \gamma\left(\frac{u}{c^2}\right)X$$

Consequently, to the observer in S' the two events are *not* simultaneous; event B occurs before event A.

Two events that are simultaneous in one inertial reference frame are not simultaneous in another reference frame that is moving with respect to the first.

The reason for this dependence of simultaneity on the reference frames is that the information that an event has taken place is transmitted to the two observers at a finite, albeit very high speed. The situation is illustrated in Figure 1.9. In reference frame S, the two light flashes appear simultaneous to the observer at rest midway between A and B. The observer in S' approaches B and recedes from A at speed u. The signal from B therefore reaches this observer before the signal from A.

Next we consider a simple measurement process, the measurement of length. The length of an object is defined as the difference between the coordinates of its two ends. If the object is moving, it is, of course, essential that the coordinates of the endpoints be measured at the same instant.

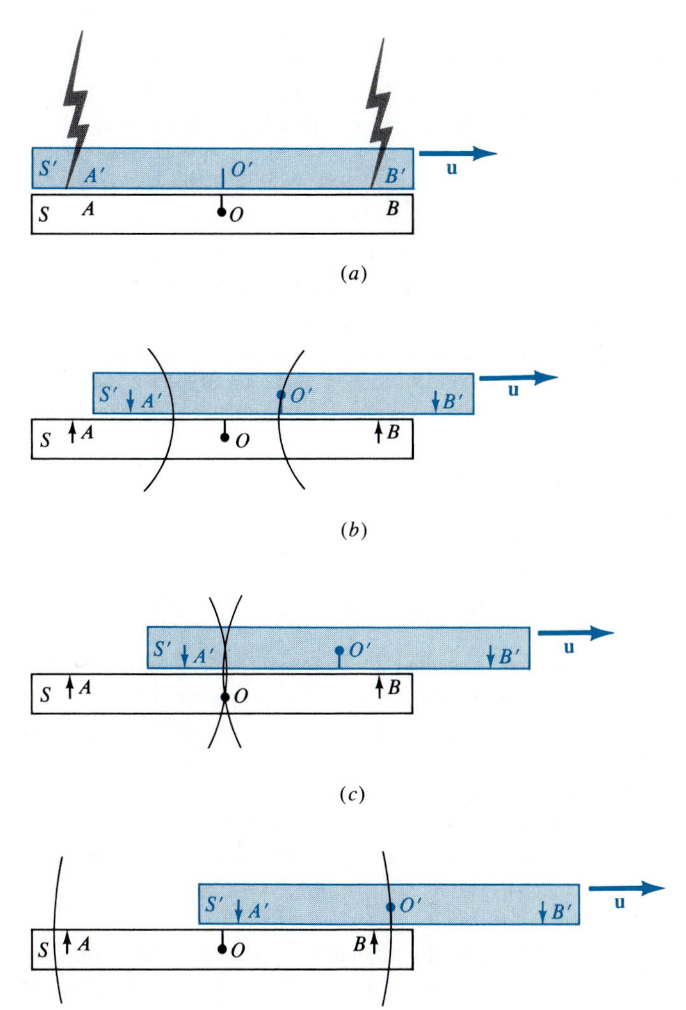

Figure 1.9 Illustration of the relativity of simultaneity. Light flashes strike at A and B. To observer O, the two flashes appear simultaneous. Observer O', in a reference frame moving to the right with velocity **u**, records the flash from B before recording the flash from A; in this observer's reference frame, the two events are not simultaneous.

We assume that a rod rests along the x axis in reference frame S. Its length is L, by which we mean that $x_2 - x_1 = L$, where x_2 and x_1 are the coordinates of its endpoints in S.

From Equation (1.20a) we have

$$x_2 = \gamma(x_2' + ut_2') \qquad x_1 = \gamma(x_1' + ut_1')$$

Hence,

$$L = x_2 - x_1 = \gamma[(x_2' - x_1') + u(t_2' - t_1')] \tag{1.22}$$

The length of the rod in the S' frame is the difference between the coordinates of its ends *measured at the same instant according to clocks synchronized in* S', i.e., when $t_2' = t_1'$. Thus, Equation (1.22) gives

$$L = \gamma L' \qquad \text{and} \qquad L' = \frac{L}{\gamma} \tag{1.23}$$

In other words, the observer in S', who sees the rod moving past at speed u and measures its length, finds the length L', shorter by the factor $1/\gamma$ than the length L measured by the observer in the rest frame of the rod.

Moving objects are contracted along the direction of motion.

This length contraction is exactly the same as that proposed by Lorentz and FitzGerald. It is common practice to refer to the length measured in the rest frame of the object as its *proper length,* although there is, of course, nothing "improper" about the length measured by a moving observer.

Finally, suppose that two events occur in reference frame S' at $x' = 0$ and at times t_1' and t_2'. According to Equation (1.20c) these events will occur in S at times $t_1 = \gamma t_1'$ and $t_2 = \gamma t_2'$. The time interval between the two events in frame S' is

$$\Delta T' = t_2' - t_1'$$

In reference frame S the interval between the same two events is

$$\Delta T = t_2 - t_1 = \gamma \, \Delta T'$$

The "stationary" observer in reference frame S finds that the time that has elapsed between the two events is longer by the factor γ than the time interval $\Delta T'$. The observer in S therefore concludes that the "clock" in the moving reference frame S' runs slow. Conversely, an observer in S' would conclude that the clock in S is the one that runs slow. In other words, if two reference frames are moving with relative velocity u, an observer in one reference frame will always conclude that clocks in the other run slow. Here we must emphasize that by "clock" we mean any system that measures time intervals—a digital watch, a normal heartbeat, or the half-life of a radioactive isotope.

We can illustrate the change in clock rates by examining the behavior of the clocks directly. For this purpose the most convenient clock is the "mirror clock" shown in Figure 1.10. A brief light pulse is emitted from the lower mirror and

Figure 1.10 The "light clock." One tick of the clock corresponds to the time required for light to make a round trip between the two mirrors.

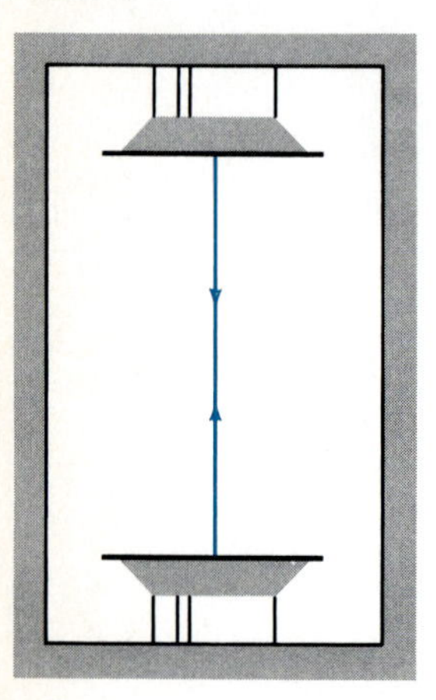

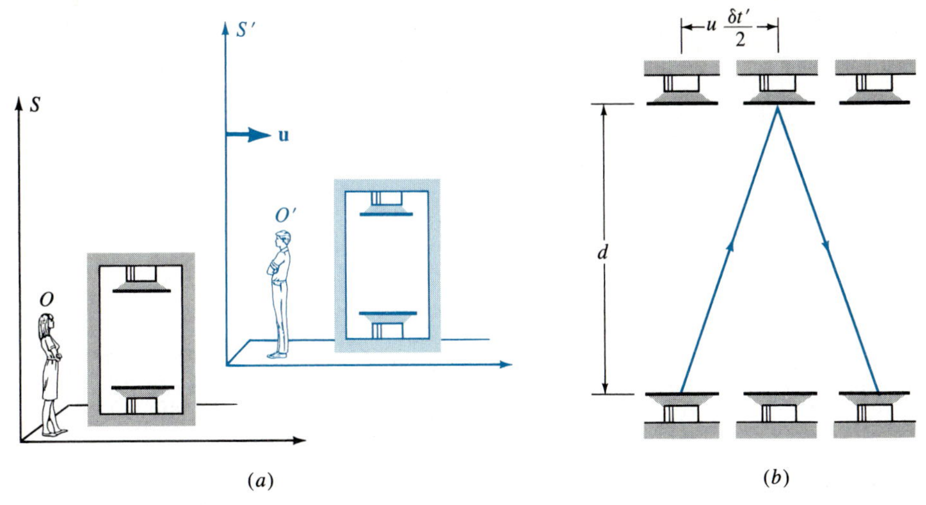

Figure 1.11 According to observer O in reference frame S, the light path in the moving clock in reference frame S' is longer than the light path in the light clock in S. Consequently, the time for one tick of the light clock in S' is longer than that for one tick of the clock in S **as measured by observer O.** According to O, the clock in S' therefore runs slow. Notice, however, that, *mutatis mutandis,* the clock in S runs slow **as measured by observer O' using his light clock.**

reflected back and forth between the two mirrors. One "tick" of this clock is the time interval for one round trip of the light pulse.

We construct two identical mirror clocks and place one in reference frame S, the other in S'. Observer O measures time intervals in units of $\delta t = 2d/c$ and notes that the clock in S' ticks at regular intervals $\delta t'$. The light path in the clock in S', as seen by observer O, is obviously longer than the light path in the clock carried by O. Comparison of Figures 1.11 and 1.4 shows that the light path of the clock in S' is just like that of the light in path B of the Michelson-Morley experiment. That is, the path length, as seen by O, is

$$2\sqrt{d^2 + \left(\frac{u\ \delta t'}{2}\right)^2}$$

Since light travels at the speed c in every reference frame, that light path must equal $c\ \delta t'$. Solving for $\delta t'$, one finds

$$\delta t' = \frac{2d/c}{\sqrt{1 - \beta^2}} = \gamma\ \delta t$$

Accordingly, the time interval $\delta t'$ between ticks of the moving clock, as seen by the observer in frame S, is greater by the factor γ than the time interval δt between ticks of the identical clock at rest in reference frame S. In other words, we again conclude that the moving clock runs slow. This relative change in clock rates is referred to as *time dilation*. Time dilation and length contraction play a crucial rule in high-energy physics.

EXAMPLE 1.1 Charged pions, π^+ and π^-, can be produced in high-energy collisions (see Chapter 15). Charged pions are unstable and decay with a half-life of 1.8×10^{-8} s; that is, if at time $t = 0$ there are N pions present, 1.8×10^{-8} s later only $N/2$ pions will be left.

Pions are produced in an accelerator and emerge from the machine at a speed of $0.996c$. How far do these particles travel in the laboratory before half of them decay?

SOLUTION If we were to forget about time dilation, we would conclude that, in traveling a distance of $0.996(3 \times 10^8 \text{ m/s})(1.8 \times 10^{-8} \text{ s}) = 5.38$ m from the exit slit of the accelerator, half of the pions will have decayed. However, 1.8×10^{-8} s is the half-life of the pions in their rest frame. According to the observer in the laboratory, the "clock" of the pions runs slow, and as measured by the laboratory clock, the half-life of the moving pions is

$$t'_{1/2} = \gamma t_{1/2} = \frac{1.8 \times 10^{-8} \text{ s}}{\sqrt{1 - 0.996^2}} = 20.1 \times 10^{-8} \text{ s}$$

and the pions will travel $0.996(3 \times 10^8 \text{ m/s})(20.1 \times 10^{-8} \text{ s}) = 60.2$ m before half of them decay.

EXAMPLE 1.2 A very rapidly moving spaceship travels between points A and B. According to an observer in the rest frame of A and B, the separation between A and B is 20 light-seconds, the distance light travels in 20 seconds (6×10^9 m). This observer notes that the spaceship passes point A at $T = $ 12:00:00 and point B at $T = $ 12:00:25. What is the speed of the spaceship? As recorded by a passenger in the spaceship, what is the distance between A and B, and how much time is required for the trip?

SOLUTION Since the spaceship travels a distance of $20c$ meters in 25 s, its speed is

$$v = \frac{20c}{25} = 0.8c$$

Let S' be the inertial frame of the spaceship. Thus, $u = v$, and using Equations (1.6) and (1.12), we find

$$\gamma = \frac{1}{\sqrt{1 - (0.8c/c)^2}} = \frac{1}{0.6} = \frac{5}{3}$$

With respect to the clock in S, the clock of a passenger in the spaceship runs slow. Hence, the time interval measured by a passenger is not 25 s but $25/\gamma = 15$ s, and the passenger will see point B passing the window 15 s after seeing point A flash by. According to this passenger, the distance between the two points is not

$20c = 6 \times 10^9$ m but $v\,\Delta T' = (15 \text{ s}) \times (0.8c) = 3.6 \times 10^9$ m. That is just the Lorentz contracted distance

$$L' = \frac{L}{\gamma} = (0.6)(6 \times 10^9 \text{ m}) = 3.6 \times 10^9 \text{ m}$$

EXAMPLE 1.3 Reference frame S' moves with velocity $0.6c$ in the positive x direction with respect to reference frame S. The clocks S and S' are synchronized to $t = t' = 0$ the instant the coordinate origins of the two frames coincide. In frame S, two events, A and B, occur at $x_A = 1.0$ km, $t_A = 3.0$ μs and $x_B = 1.5$ km, $t_B = 4.6$ μs. Determine the time and position coordinates in the S' frame of events A and B. Is there a reference frame in which the two events are simultaneous? If so, what is the velocity of that frame with respect to S?

SOLUTION We transform to the S' reference frame, using Equations (1.20a) and (1.20c):

$$\beta = 0.6 \qquad \gamma = \frac{1}{\sqrt{1-\beta^2}} = \frac{1}{0.8} = 1.25$$

$$x'_A = 1.25(1 \times 10^3 - 0.6 \times 3 \times 10^8 \times 3.0 \times 10^{-6}) = 575 \text{ m}$$

$$t'_A = 1.25\left(3 \times 10^{-6} - \frac{0.6 \times 1 \times 10^3}{3 \times 10^8}\right) = 1.25 \times 10^{-6} \text{ s} = 1.25\ \mu\text{s}$$

$$x'_B = 1.25(1.5 \times 10^3 - 0.6 \times 3 \times 10^8 \times 4.6 \times 10^{-6}) = 840 \text{ m}$$

$$t'_B = 1.25\left(4.6 \times 10^{-6} - \frac{0.6 \times 1.5 \times 10^3}{3 \times 10^8}\right) = 2.0 \times 10^{-6} \text{ s} = 2.0\ \mu\text{s}$$

The two events will be simultaneous in another inertial frame S' if, in that frame, $t'_A = t'_B$; i.e., if

$$t_A - \frac{\beta x_A}{c} = t_B - \frac{\beta x_B}{c}$$

or

$$\beta = \frac{t_B - t_A}{x_B - x_A} c$$

In this case, $t'_A = t'_B$ if $\beta = (1.6 \times 10^{-6}/500)(3 \times 10^8) = 0.96$, or $u = 0.96c$.

1.6(b) The Twin Paradox

The famous (or infamous) twin paradox goes as follows. Jim and Tim are identical twins, aged 20 years. Tim, the more adventurous, sets out on a journey to one of the nearer stars, 35 light-years from Earth. His spaceship accelerates to its cruising speed of $0.99c$ so rapidly that we may neglect the time required to achieve that speed compared with the time required to reach the star. On arriving at his destina-

tion, Tim immediately turns around and heads for home, again at the phenomenal speed of $0.99c$.

As he disembarks on Earth he is greeted by his brother Jim, who, now 90 years old, had almost given up hope of ever seeing Tim alive. Jim, on welcoming his brother, is not surprised that in the more than 70 years since Tim left, he has barely aged ten years. Jim knows about time dilation and realized that Tim's "clock" (heartbeat, metabolism) was slow as measured by the clock on Earth. However, Tim is greatly distressed that his brother looks so close to death when Tim knows that he has been gone barely ten years. Moreover, as far as Tim could tell, the earth was at first receding at $0.99c$ and then approaching at $0.99c$. According to Tim, it was the earth that was "moving" and Jim should be by far the younger of the two brothers. Something is not quite right.

The point is that Tim's reference frame was **not** an inertial reference frame at all times, because to return to Earth, Tim had to decelerate, turn around, and then accelerate once again. The situation is, therefore, not symmetrical with respect to the two brothers, and special relativity is not applicable over the entire time.

Although time dilation was one of the conclusions of Einstein's 1905 paper, *fifty years later* there were still occasionally heated arguments on the twin paradox in the scientific literature. It was argued that, after all, no one had ever done such an experiment. In 1972 an equivalent experiment was finally performed which resolved this apparent paradox once and for all [see Section 1.7(b)].

EXAMPLE 1.4 If we neglect the time required for acceleration at Earth and the distant star, how much has each of the twins aged during Tim's round trip?

SOLUTION Since one light-year is the distance traveled by light in the course of one year, Tim, traveling at $0.99c$ a distance of 70 light-years in Jim's reference frame, will require

$$\frac{70}{0.99} = 70.707 \text{ yr} = 70 \text{ years, 8 months, 15 days}$$

Tim's clock, however, is slow as seen by Jim. The time dilation factor $1/\gamma$ is

$$\sqrt{1 - 0.99^2} = 0.141$$

and, consequently, Tim will have aged only by

$$70.707 \times 0.141 \simeq 10 \text{ years}$$

EXAMPLE 1.5 As seen by Tim during his trip, what is the distance between Earth and the star that was his destination?

SOLUTION In Tim's inertial frame, the distance of 35 light-years as measured by an observer on Earth is diminished by the Lorentz-FitzGerald contraction to

$$\frac{35}{\gamma} = 35 \times 0.141 = 4.935 \text{ light-years}$$

or just under five light-years.

1.6(c) The Velocity Transformations

In the theory of special relativity, the Galilean coordinate transformation, Equations (1.2), is replaced by the Lorentz transformation, Equations (1.20). We expect that the transformation equations for the velocity components will also take on a different form.

The velocities v_x, v_y, v_z, and v'_x, v'_y, v'_z are defined in the usual way as the time derivatives of the corresponding space coordinates; that is,

$$v_x = \frac{dx}{dt} \qquad v_y = \frac{dy}{dt} \qquad v_z = \frac{dz}{dt} \tag{1.24}$$

$$v'_x = \frac{dx'}{dt'} \qquad v'_y = \frac{dy'}{dt'} \qquad v'_z = \frac{dz'}{dt'} \tag{1.25}$$

Taking differentials of both sides of Equations (1.20a) through (1.20c) yields

$$dx = \gamma(v'_x + u)\, dt' \qquad dy = v'_y\, dt' \qquad dz = v'_z\, dt' \tag{1.26}$$

$$dt = \gamma\left(1 + \frac{uv'_x}{c^2}\right) dt' \tag{1.27}$$

Therefore,

$$\frac{dx}{dt} = v_x = \frac{v'_x + u}{1 + v'_x u/c^2} \tag{1.28a}$$

$$\frac{dy}{dt} = v_y = \frac{v'_y}{\gamma(1 + v'_x u/c^2)} \tag{1.28b}$$

$$\frac{dz}{dt} = v_z = \frac{v'_z}{\gamma(1 + v'_x u/c^2)} \tag{1.28c}$$

The inverse transformations are given by interchanging primed and unprimed velocities and replacing u by $-u$ in Equations (1.28).

We shall return to these transformation equations presently. For the moment, we only draw attention to the fact that, although the coordinates perpendicular to **u** do not change under the relativistic transformations, the corresponding velocity components do; the reason is that the time differentials dt and dt' are not identical but transform according to Equation (1.27).

EXAMPLE 1.6 The captain of spaceship *Alpha* notes from his radar that he is on a straight collision course with spaceship *Beta*, closing toward the crash at a relative speed of $0.8c$. In a reference frame in which the two spaceships are traveling at equal speeds, what is the speed of each?

SOLUTION If we applied the Galilean transformation, we would conclude that in a reference frame in which each spaceship is approaching at the same speed, that speed would be $0.4c$. Since the speeds involved are of the order of c, we cannot use the Galilean transformation but must resort to Equation (1.28a).

Let the S' frame be the reference frame of spaceship *Alpha*, and the S frame be that of the observer who sees both ships closing in at equal speeds. In frame S, $v_\alpha = +u$, $v_\beta = -u$. If we write $u = pc$, where p is some number less than 1, the transformation equation gives, for *Alpha* and *Beta*, respectively,

$$p = \frac{0 + p}{1 + 0} = p \qquad -p = \frac{-0.8 + p}{1 - 0.8p}$$

The second equation leads to the quadratic equation

$$p^2 - 2.5p + 1 = 0$$

which has two solutions, $p = 0.5$ and $p = 2.0$. Since p cannot be greater than 1, only the first solution has physical significance.

In a reference frame moving at $0.5c$ with respect to spaceship *Alpha*, an observer will see *Alpha* and *Beta* approaching at $0.5c$.

1.7 CONSEQUENCES OF THE LORENTZ TRANSFORMATIONS

1.7(a) The Relativistic Doppler Effect

We have all heard the change in pitch of a police siren when a speeding patrol car passes. As the car approaches, the pitch is high; as it recedes, the pitch is low. If the receiver rather than the source is moving, there is a similar effect. The classical expressions for this Doppler shift are

$$f_r' = \frac{1}{1 - (u/v)} f_0 \tag{1.29}$$

$$f_s' = \left(1 + \frac{u}{v}\right) f_0 \tag{1.30}$$

where v is the speed of sound, u is the relative speed of approach of source or receiver, f_0 is the frequency of the source, and f' is the frequency at the receiver. Equation (1.29) applies when the receiver is stationary and the source approaches the receiver, while Equation (1.30) applies if the source is stationary and the receiver is approaching the source.

The apparent lack of symmetry between these two equations tells us that they cannot be valid for electromagnetic waves. Consider, for example, the case when $u/v = 0.5$. If the receiver is at rest and the source moving, $f' = 2f_0$, but if the receiver is approaching a stationary source, $f' = 1.5f_0$. The observed frequency therefore depends not only on the relative motion of source and receiver but also on whether it is the source or the receiver that is in motion relative to the transmitting medium. If these relations applied to electromagnetic waves, it would then be possible to determine the "absolute rest frame of the ether."

It is not difficult to see where we must make the relativistic correction. In the nonrelativistic Doppler formula, the time scale is the Newtonian absolute time scale, equal at source and receiver, regardless of the reference frame used to describe the phenomenon.

Suppose the receiver is at rest; that is, we shall work in the reference frame of the receiver. The source is approaching at the speed u. If the frequency at the receiver is f'_r, that means that in a time interval Δt_r, $f'_r\,\Delta t_r$ complete cycles have been recorded. **In the time frame of the receiver,** the number of complete cycles emitted by the source, approaching at the speed u, is then

$$N = f'_r\left(1 - \frac{u}{c}\right)\Delta t_r = f_0\,\Delta t_s \tag{1.31}$$

where f_0 is the frequency of the source as measured in the rest frame of the source and Δt_s is the corresponding proper time interval.

But, as seen by the stationary receiver, the moving clock of the source runs slow, and the time interval Δt_s is less than the interval Δt_r by the factor $1/\gamma$; that is, $\Delta t_s = \Delta t_r/\gamma$. Hence,

$$\begin{aligned} f'_r &= \frac{f_0}{1 - u/c}\left(\frac{\Delta t_s}{\Delta t_r}\right) = \frac{1}{\gamma(1 - u/c)}f_0 \\ &= \frac{\sqrt{1-\beta^2}}{1-\beta}f_0 = \sqrt{\frac{1+\beta}{1-\beta}}f_0 \end{aligned} \tag{1.32}$$

It is left for Problem 1.3 to show that if the receiver is approaching the source, that is, if we view the phenomenon from the reference frame of the source, the Doppler-shifted frequency at the receiver is again given by Equation (1.32).

There is a further consequence of relativity on the Doppler effect. Whereas in the classical case there is no frequency shift if the relative velocity of source and receiver is perpendicular to the line joining them, that is no longer true in the relativistic case (see Problem 1.13). This *transverse Doppler effect,* a consequence of time dilation, is a strictly relativistic phenomenon which has been confirmed by experiment.

The Doppler effect plays a central role in astrophysics. The speed of recession of distant galaxies, and, by inference, their distance from our own, is determined by comparing emission spectra from elements in their atmospheres as recorded on earth with spectra of these elements in the laboratory. Generally, Doppler shifts are expressed as wavelength rather than frequency changes. In the relativistic as well as the classical limit $\lambda'/\lambda_0 = f_0/f'$.

EXAMPLE 1.7 Astronomers usually give the fractional Doppler shift, $\Delta\lambda/\lambda_0$. In 1963, J. Greenstein reported that light from the galactic object 3C 48 is red-shifted by 37 percent. In other words, $\Delta\lambda/\lambda_0 = 0.37$ and the change in wavelength is toward the red, i.e., in the long-wavelength direction. What can one say about the relative motion of 3C 48 with respect to the earth?

SOLUTION We use the relation $\lambda'/\lambda_0 = f_0/f'$. Also, since $\Delta\lambda = \lambda' - \lambda_0$, $\lambda'/\lambda_0 = (\Delta\lambda/\lambda_0) + 1$. Equation (1.32) then gives

$$1.37 = \frac{f_0}{f_r'} = \frac{1-\beta}{\sqrt{1-\beta^2}} = \sqrt{\frac{1-\beta}{1+\beta}}$$

and

$$1.877 = \frac{1-\beta}{1+\beta}$$

with the result $\beta = -0.305$. The galaxy has a speed of $0.305c$ relative to the earth; the negative sign shows that the galaxy is receding from, not approaching, the earth. [Equation (1.32) was based on the assumption that source and receiver were moving toward each other.]

Object 3C 48 is one of numerous *quasars*. The large red shift implies that it is extremely far from Earth. That conclusion appears to be contradicted by its unusually high luminosity. The correct interpretation of quasar data is still a matter of controversy among astrophysicists.

1.7(b) Experimental Evidence of Relativistic Kinematics

One of the earliest direct experimental tests of time dilation was performed by Rossi and Hall in 1941.[10] They measured the flux of muons (μ-mesons) atop Mount Washington in New Hampshire, at an altitude of about 2000 m above sea level, and again at sea level. Muons are unstable elementary particles that decay with a half-life of 1.5 μs into an electron and a neutrino.[11] Even if muons were to travel toward the earth's surface at the speed of light, it would take $2000/c = 6.67\ \mu\text{s}$ to complete a trip of 2000 m. Consequently, if the flux of muons on Mount Washington were I_0, the flux at sea level should, according to nonrelativistic mechanics, be reduced to

$$I = \frac{I_0}{2^{6.67/1.5}} = \frac{I_0}{22}$$

Rossi and Hall found that the ratio of the flux on Mount Washington to that at sea level was 1.4.

This apparent discrepancy is readily understood in light of relativistic correc-

---

10. B. Rossi and D. B. Hall, *Phys. Rev.* **59,** 223 (1941).

11. The meaning of half-life, $t_{1/2}$, is as follows. If, at $t = 0$, there are N_0 radioactive particles, then at time $t = t_{1/2}$, only $N_0/2$ particles remain, $N_0/2$ having decayed in that time. It follows that at any time $t = bt_{1/2}$, the number of particles that remain is given by

$$N(t) = \frac{N_0}{2^b}$$

Half-life is discussed more fully in Chapter 15.

tions. As seen by an observer on earth, the muons' "clock" runs slow. Since $1.4 \simeq \sqrt{2} = 2^{1/2}$, in the inertial frame of the muon only half of one half-life, about 0.75 μs, has elapsed during the trip. It follows that the time dilation factor is $6.67/0.75 = 8.9 = \gamma$. If we now solve for the speed of the muons, we find that

$$u = 0.994c$$

The reduction in the decay of muon flux can also be viewed from the muon's inertial frame. In that reference frame, the earth's surface approaches the muon at the speed of $0.994c$. The distance of 2000 m in the earth reference frame is contracted to a mere $2000/8.9 = 225$ m, a distance that the approaching earth covers in 0.75 μs.

A more dramatic demonstration of time dilation was reported about thirty years later by J. C. Hafele and R. E. Keating.[12] We quote directly from their paper:

> *During October, 1971, four cesium atomic beam clocks were flown on regularly scheduled commercial jet flights around the world twice, once eastward and once westward, to test Einstein's theory of relativity with macroscopic clocks. From the actual flight paths of each trip, the theory predicts that the flying clocks, compared with reference clocks at the U.S. Naval Observatory, should have lost 40 ± 23 nanoseconds during the eastward trip, and should have gained 275 ± 21 nanoseconds during the westward trip. . . . Relative to the atomic time scale of the U.S. Naval Observatory, the flying clocks lost 59 ± 10 nanoseconds during the eastward trip and gained 273 ± 7 nanoseconds during the westward trip, where the errors are the corresponding standard deviations. These results provide an unambiguous empirical resolution of the famous clock "paradox" with macroscopic clocks.*

1.8 THE RELATIVISTIC EXPRESSIONS IN THE CLASSICAL LIMIT

The reason none of the startling, almost bizarre effects we have described in the preceding sections had been observed until recent times is that they do not assume significance until relative speeds comparable to the speed of light are approached. In our normal lives we may travel at speeds of one or a few kilometers per second, and astronauts in spaceships may reach speeds ten times as great relative to the earth, but compared with the speed of light they are almost negligibly small. I have included here the word *almost* because, in recent years, measurement techniques have become so refined that relativistic effects can be observed directly in the laboratory. Putting aside these high-precision experiments, relativistic effects do not play a role in terrestrial mechanics of macroscopic objects.

That conclusion follows from the fact that all of the relativistic transformations and derived expressions, such as the Doppler shift equations, reduce to their classical equivalents when $u \ll c$. The proofs of this statement are the subject of a few problems at the end of this chapter.

---

12. J. C. Hafele and R. E. Keating, *Science* **177,** 166 (1972).

For small values of β the order of magnitude of the relativistic correction is easily estimated. The critical parameter is γ, which is unity when $\beta = 0$, increases slowly at first, but then diverges as β approaches 1, i.e., as u approaches c. For $\beta \ll 1$, $(1 - \beta^2)^{-1/2} \simeq 1 + \beta^2/2$. For example, if $u = 0.01c$, $\beta^2 = 10^{-4}$, and γ differs from 1 by only 0.005 percent. Relativistic corrections can be expected to be of this order of magnitude at that speed. The few exceptions to this rule of thumb are those phenomena that are strictly relativistic, such as the transverse Doppler effect.

SUMMARY

The special theory of relativity was designed to reconcile Maxwell's equations of electromagnetism with the principle of relativity:

The laws of nature are the same in all inertial reference frames.

Since the speed of light in vacuum is an immediate consequence of Maxwell's equations, special relativity is based on the following postulate:

The speed of light in vacuum, as measured in any inertial reference frame, is *c* regardless of the motion of the light source relative to that reference frame.

It follows from this postulate that if reference frame S' is moving at constant speed u in the positive x direction relative to reference frame S, then the coordinates (x, y, z, t) of an event in S are related to the coordinates (x', y', z', t') of the same event in S' by

$$x' = \gamma(x - ut) \qquad x = \gamma(x' + ut') \tag{1.20a}$$

$$y' = y \qquad z' = z \tag{1.20b}$$

$$t' = \gamma\left(t - \frac{u}{c^2}x\right) \qquad t = \gamma\left(t' + \frac{u}{c^2}x'\right) \tag{1.20c}$$

where

$$\gamma = \frac{1}{\sqrt{1 - \beta^2}} \qquad \beta = \frac{u}{c} \tag{1.21}$$

Equations (1.20) are the *Lorentz transformations*. Time dilation (according to which moving clocks run slow) and the Lorentz-FitzGerald length contraction are direct consequences of Equations (1.20). Using these equations, one can derive the velocity transformation equations

$$v'_x = \frac{v_x - u}{1 - v_x u/c^2} \tag{1.28a}$$

$$v'_y = \frac{v_y}{\gamma(1 - v_x u/c^2)} \tag{1.28b}$$

$$v'_z = \frac{v_z}{\gamma(1 - v_x u/c^2)} \tag{1.28c}$$

In the limit $u \ll c$, the Lorentz transformation reduces to the classical (Galilean) transformation.

The predictions of time dilation and length contraction of relativistic kinematics have been confirmed experimentally.

Bibliography

Biographies of Einstein

Jeremy Bernstein, *Einstein,* Viking, New York, 1973.

Ronald W. Clark, *Einstein,* World Publishing, Cleveland, 1971.

Abraham Pais, *"Subtle is the Lord . . . ,"* Oxford University Press, New York, 1982.

K. Sugimoto, *Albert Einstein,* Schocken Books, New York, 1989.

Special Relativity

H. Bondi, *Relativity and Common Sense,* Doubleday, New York, 1964.

J. Bronowski, "The Clock Paradox," *Scientific American,* February 1963, p. 134.

Albert Einstein, *Relativity,* Crown, New York, 1961.

A. P. French, *Special Relativity,* Norton, New York, 1968.

N. David Mermin, *Space and Time in Special Relativity,* McGraw-Hill, New York, 1968.

A. I. Miller, *Albert Einstein's Special Theory of Relativity,* Addison-Wesley, Reading, Mass., 1981.

Robert Resnick, *Introduction to Special Relativity,* Wiley, New York, 1968.

R. S. Shankland, "The Michelson-Morley Experiment," *Scientific American,* November 1964, p. 107.

E. F. Taylor and J. A. Wheeler, *Spacetime Physics,* Freeman, San Francisco, 1966.

PROBLEMS

1.1 The time interval between the ticks of two identical clocks is 2.0 s. One of the two clocks is set in motion so that its speed relative to the observer, who holds the other clock, is $0.8c$. What is the time interval between ticks of the moving clock as measured by the observer with the stationary clock?

1.2 The distance between New York and Los Angeles is about 5000 km. A jet, traveling at a speed of 900 km/h, leaves New York at 12:00:00. At what time will the jet fly over the Los Angeles airport? How much time has elapsed as measured by a passenger on the plane?

1.3 Derive the relativistic Doppler formula for the case in which the source is "stationary" and the receiver is approaching the source at the speed u. Show that the result is identical to Equation (1.32).

1.4 Show that when $\beta \ll 1$, the relativistic transformations for the coordinates and velocities reduce to the Galilean transformation equations.

1.5 Two guns, A and B, located at $x_A = 0$ and $x_B = 1.5$ km, fire at an approaching rocket. Gun A fires at $t = 0$; gun B fires at $t = 1.0\ \mu$s. The detectors on the rocket record the flashes from both guns as simultaneous events. Determine the x component of the rocket's velocity.

1.6 Two spaceships travel in exactly opposite directions as seen from the earth. The speed of each, as measured by an observer on the earth, is $0.84c$. What is the speed of one of the spaceships as measured by a passenger in the other?

1.7 Two spaceships pass each other, traveling in opposite directions. The speed of ship B as measured by a passenger in ship A is $0.96c$. This passenger has measured the length of ship A as 100 m and determines that ship B is 30 m long. What are the lengths of the two ships as measured by a passenger in ship B?

•1.8 Two receivers, A and B, are positioned along the x axis. A source, moving at velocity βc along the x axis between A and B, radiates light of frequency f_0 in the rest frame of the source.

(a) Show that to lowest order in β, the average frequency recorded by the two receivers is f_0.

(b) Show that there is a second-order Doppler effect, i.e., a correction to f' that depends on β^2, and that this correction is the same when source and receiver approach and when they recede from each other.

(c) Derive an expression for the average frequency recorded by the two receivers, including the second-order Doppler effect.

(d) Calculate $(f'_A + f'_B)/2f_0$ when $u = 10^3$ m/s.

1.9 A beam of radioactive particles emerges from the exit slit of an accelerator with a speed of $0.92c$. Detectors placed 2.0 m and 4.0 m from the slit show that the intensity of the beam is 1.0×10^8 particles/cm^2·s at the first detector and 2.5×10^7 particles/cm^2·s at the second detector.

(a) Determine the half-life of the particles in their rest frame.

(b) What is the beam intensity at the exit slit of the accelerator?
(c) Suppose that the accelerator is adjusted so that the particles emerge with a speed of $0.985c$ in the laboratory. If the beam intensity at the exit slit is the same as that calculated in part (b), what will the readings of the detectors be at 2.0 m and 4.0 m from the exit slit?

1.10 In recent years, quasars with red shifts $\Delta\lambda/\lambda_0$ greater than 4 have been discovered. What is the velocity relative to Earth of quasars with red shifts of 4.0?

•1.11 A particle has a velocity $v_r = 0.8c$ in the laboratory. The laboratory is itself moving with velocity $v_l = 0.6c$ relative to an inertial frame S.

(a) What is the velocity of the particle in S if v_r and v_l are parallel? If they are antiparallel?
(b) What is the velocity of the particle in S if v_r and v_l are at right angles?

1.12 A meterstick is positioned so that it makes an angle of 30° with the x axis. Determine the length of the meterstick and its orientation as seen by an observer who is moving along the x axis with a speed of $0.8c$.

•1.13 Derive an expression for the transverse Doppler effect. Consider a source of light moving with velocity βc perpendicular to the line between source and receiver, and determine the frequency received. Compare your result with that of Problem 1.8(b).

1.14 Figure 1.12 shows a sound interferometer consisting of a transducer, a "beam splitter," two reflectors, and a microphone. The distance between the beam splitter and the reflectors is 0.50 m. The speed of sound in air is 300 m/s. The transducer emits sound at a frequency of 150 kHz. There is a steady wind blowing along the x direction. As the apparatus is turned through 90°, the sound intensity as recorded by the microphone, which initially registers a maximum, fluctuates, passing through 17 minima, and registers a maximum once again at the end of the 90° rotation. Determine the wind velocity.

1.15 Show that the null result of the Michelson-Morley experiment is consistent with the Lorentz-FitzGerald contraction.

1.16 Two identical light clocks of the sort described in the text are placed at 90° to each other, at rest in reference frame S. An observer, who moves with speed βc in a direction parallel to the light path of one of the clocks, measures the difference between the rates of the two clocks. Derive an expression for that difference in the clock rates.

•1.17 A modern alternative to the Michelson-Morley experiment is shown in Figure 1.13. Two lasers are tuned to the same frequency (the process of tuning involves a fine adjustment of the end mirrors of the laser tube; in effect, each laser acts like the light clock described in the text). The difference frequency can be determined by a technique known as light-beating spectroscopy. Show that according to the special theory of relativity there should be no frequency difference between the two lasers as the apparatus is turned about its axis. This experiment was performed by Jaseja, Javan, Murray, and Townes [*Phys. Rev.* **133,** A1221 (1964)], and within experimental uncertainty of better than 1 part in 10^{11}, no frequency change was observed.

1.18 At $t = t' = 0$, $x = x' = 0$. Events A and B occur in S at $t_A = 0.3\ \mu s$, $x_A = 150$ m and $t_B = 0.4\ \mu s$, $x_B = 210$ m. Where and when do these events occur in S' if

(a) S' moves at a velocity of $+0.6c$?
(b) S moves at a velocity of $-0.6c$?

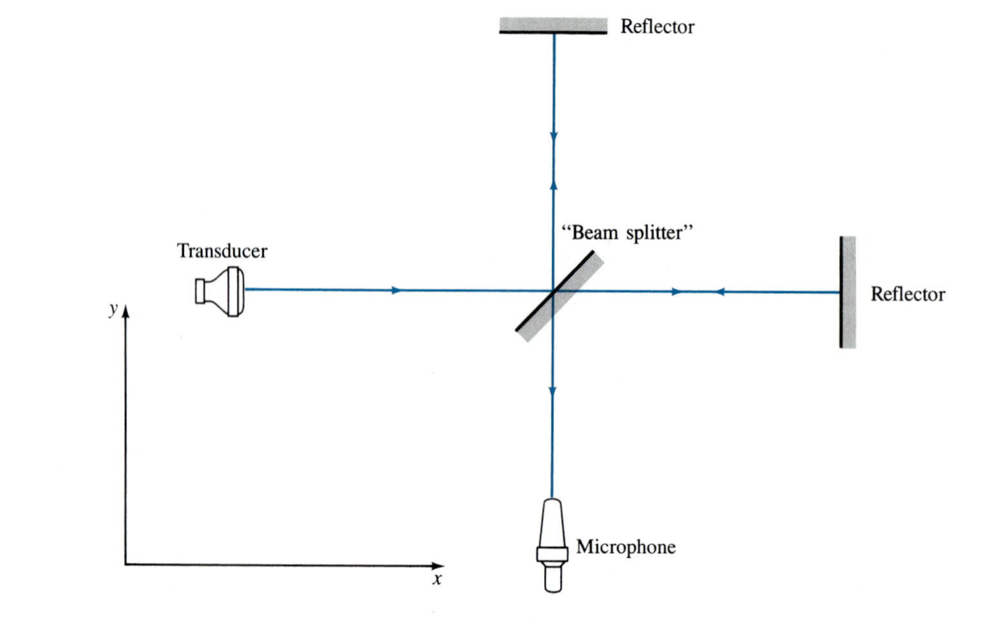

Figure 1.12 Problem 1.14.

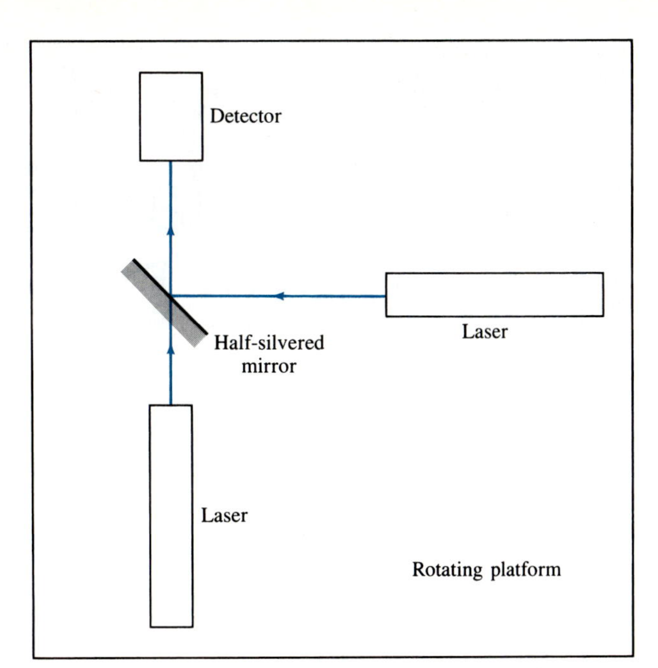

Figure 1.13 Problem 1.17.

(c) Could these two events appear to be simultaneous to an observer in another reference frame? If so, what must be the velocity of that reference frame relative to S?

•**1.19** A spaceship of proper length $L = 200$ m travels at a speed of $0.68c$ past a radio station. At the instant that the tail end of the ship passes the radio station as noted by an observer in the station, a signal is sent by the station's transmitter and subsequently detected by the receiver in the nose of the spaceship. Assume that the instant the nose of the spaceship passes the radio station, the clocks aboard the ship and at the station are synchronized to $t = t' = 0$.

(a) At what time, according to the clock aboard the spaceship, is the signal sent?
(b) At what time, according to the clock at the radio station, is the signal received by the spaceship?
(c) At what time, according to the clock aboard the spaceship, is the signal received?
(d) Where, according to an observer at the radio station, is the nose of the spaceship when the signal is received?

1.20 In frame S' events A and B are simultaneous, and the separation in space between A and B is 5.0 km. What is the time difference between these two events in a reference frame S in which the spatial separation between A and B is 8.0 km?

1.21 Two identical spaceships, each 200 m long in its rest frame, pass one another traveling in opposite directions. According to a passenger in one of the spaceships, the relative velocity of the two ships is $0.58c$.

(a) As measured by this passenger, how long does it take for the other ship to pass by?
(b) What is the length of each spaceship in a reference frame in which both are of equal length?

•**1.22** In reference frame S, a particle has a velocity $\mathbf{v} = (v_x, v_y, v_z)$ and an acceleration $\mathbf{a} = (a_x, a_y, a_z)$. Derive expressions for the acceleration components a'_x, a'_y, and a'_z in a reference frame that has a velocity $\mathbf{u} = (u_x, 0, 0)$ relative to S.

1.23 Three sources, A, B, and C, radiate monochromatic light at a frequency f_0 in their respective rest frames. In the rest frame of A, source B has a velocity $+\beta c$, source C a velocity $-\beta c$. Determine the frequencies of the light received by C from sources A and B if (a) B approaches C, and (b) B recedes from C.

1.24 Frame S' has a velocity $u_x = \beta c$ relative to S. A light pulse is emitted at $t = t' = 0$, $x = x' = 0$ whose direction of propagation in S makes an angle of 30° with the x axis. Determine the angle at which this light pulse propagates in S'.

1.25 Suppose that at the time light emitted by a quasar is received on earth, earth and quasar are receding at $0.94c$. If the light had a wavelength of 640 nm in the rest frame of the quasar, what is the wavelength of this radiation as recorded in the observatory on earth?

•**1.26** A binary star system in our own galaxy rotates in the plane parallel to the line between the star system and the earth. Both stars emit light of the same frequency in the rest frame of each star. The Doppler shift from one star is $\Delta\lambda/\lambda_0 = \pm 0.010$; that from the other star is $\Delta\lambda/\lambda_0 = \pm 0.016$. Determine the ratio of the masses of the two stars. Suppose the lighter of the two stars has a mass equal to that of the sun. What is the separation between the two stars?

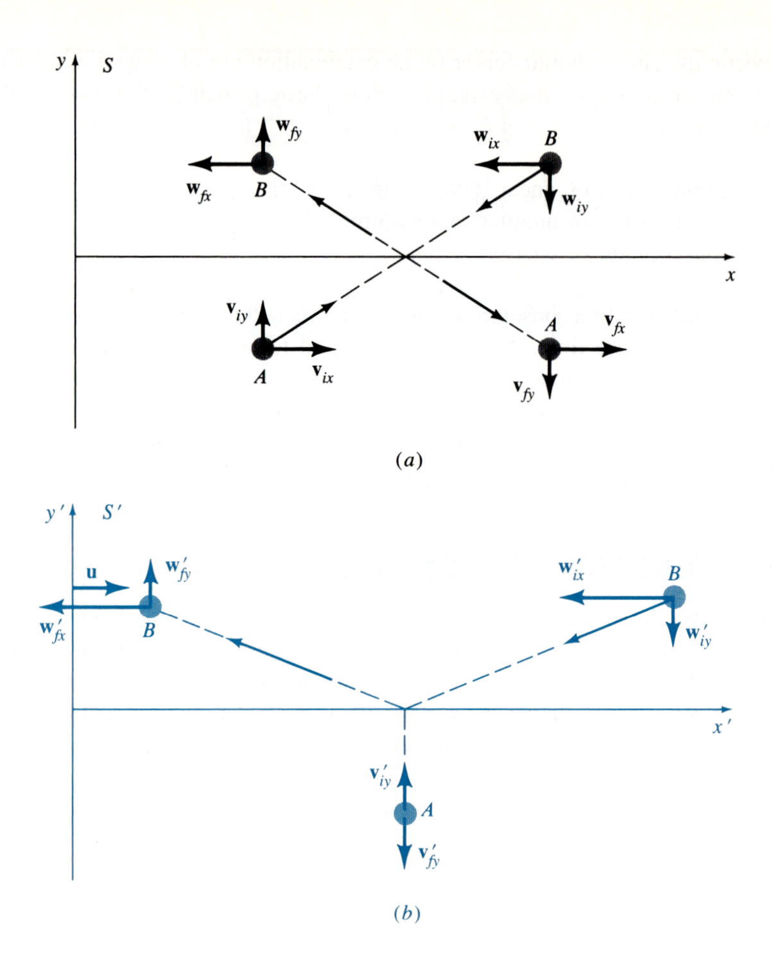

Figure 2.1 An elastic collision between two identical particles. (*a*) In reference frame *S*, the collision is symmetric. (*b*) In reference frame *S'* the *x'* component of the velocity of particle *A* is zero.

employ the Lorentz and not the Galilean transformation. The Lorentz transformation equations affect the y as well as the x components of $\mathbf{v}$ and $\mathbf{w}$. In S', v'_{iy} and w'_{iy} are given by

$$v'_{iy} = v_{iy}\frac{1}{\gamma(1 - \beta^2)} \qquad w'_{iy} = w_{iy}\frac{1}{\gamma(1 + \beta^2)} \tag{2.8}$$

where we have set $u = v_{ix} = -w_{ix}$, so that in the S' frame $v'_{ix} = 0$.

If we now impose the condition $v_{iy} = -w_{iy}$, the y components of velocity in the S' frame are related by

$$v'_{iy} = -w'_{iy}\left(\frac{1 + \beta^2}{1 - \beta^2}\right) \tag{2.9}$$

Similarly,

$$v'_{fy} = -w'_{fy}\left(\frac{1 + \beta^2}{1 - \beta^2}\right) \tag{2.10}$$

Therefore, if we define momentum in the usual way,

$$\mathbf{p} = m\mathbf{v} \tag{2.11}$$

the change in momentum of particle B in the S' frame is

$$\Delta \mathbf{p}'_B = 2mw'_{fy}$$

while that of particle A is

$$\Delta \mathbf{p}'_A = 2mv'_{fy} = -2mw'_{fy}\left(\frac{1+\beta^2}{1-\beta^2}\right)$$

and the sum $\Delta \mathbf{p}'_A + \Delta p'_B$ is evidently not zero. In other words, if we use the Newtonian definition of momentum, Equation (2.11), momentum is conserved in S but not in S'. Thus a fundamental law of physics appears to be violated by the Lorentz transformation, in direct contradiction to the principle of relativity, which maintains that the fundamental laws of physics must be independent of the choice of inertial reference frame.

It is, however, possible to preserve momentum conservation by redefining inertial mass. Momentum is conserved in all inertial reference frames if we define inertial mass by

$$m = \gamma m_0 \tag{2.12}$$

where m_0 is the *rest mass,* that is, the mass of the body in a reference frame in which it is at rest. To demonstrate that momentum is conserved with this redefinition of mass, we now show that the expression for the y component of momentum of particle A is the same in S and S'. (The corresponding proof for particle B is left as Problem 2.9.)

In the S frame, the relativistic mass of particle A is

$$m_A = \gamma m_0 = \frac{m_0}{\sqrt{1 - v^2/c^2}} \qquad v^2 = v_x^2 + v_y^2 \tag{2.13}$$

The y component of its momentum is

$$p_{Ay} = m_A v_y = \frac{m_0 v_y}{\sqrt{1 - v^2/c^2}} \tag{2.14}$$

In the S' frame, the y component of particle A's momentum is

$$p'_{Ay} = m'_A v'_y = \frac{m'_A v_y}{\sqrt{1 - u^2/c^2}} \tag{2.15}$$

where $u = v_x$ and

$$m'_A = \frac{m_0}{\sqrt{1 - v'^2/c^2}} = \frac{m_0}{\sqrt{1 - v_y'^2/c^2}} \tag{2.16}$$

since in the S' frame the x component of the velocity of particle A is zero. We now apply the velocity transformation, Equation (1.28), and obtain

$$\sqrt{1 - \frac{v_y'^2}{c^2}} = \sqrt{1 - \left(1 - \frac{u^2}{c^2}\right)^{-1}\left(\frac{v_y^2}{c^2}\right)}$$

$$= \sqrt{\frac{1 - v^2/c^2}{1 - u^2/c^2}} \tag{2.17}$$

where we have used the fact that $u = v_x$. If we now substitute Equation (2.17) in Equations (2.15) and (2.16), we obtain

$$p'_{Ay} = \frac{m_0 v_y}{\sqrt{1 - v^2/c^2}} \tag{2.18}$$

which is identical to Equation (2.14). We see that if we define momentum relativistically, i.e.,

$$\mathbf{p} = m\mathbf{v} = \gamma m_0 \mathbf{v} \tag{2.19}$$

it is invariant under a Lorentz transformation. Therefore, if momentum conservation holds in S, it will also be obeyed in S' and, by inference, in any other inertial reference frame.

The defining equation for force is still Newton's second law, Equation (2.2), with the relativistic momentum of Equation (2.19) replacing the Newtonian formula $\mathbf{p} = m_0\mathbf{v}$. That is,

$$\mathbf{F} = \frac{d\mathbf{p}}{dt} = \frac{d}{dt}(m\mathbf{v}) = \frac{d}{dt}\left(\frac{m_0\mathbf{v}}{\sqrt{1 - \beta^2}}\right) \tag{2.20}$$

Two features are noteworthy here. First, in the limit $v \ll c$, the relativistic expressions for momentum and mass reduce to the Newtonian values, $\mathbf{p} = m_0\mathbf{v}$ and $m = m_0$. Second, in Newtonian dynamics, application of a constant force results in a constant acceleration and a constantly increasing velocity that would ultimately exceed the velocity of light. In relativistic dynamics, the momentum of the body increases at a constant rate, but, as v approaches c, it is the relativistic mass that increases without limit, not the velocity. The speed of light is the limiting speed in any reference frame.

2.3 ENERGY

In Newtonian dynamics, the kinetic energy (KE) of a body is the work done by the net force that acts on the body in accelerating it from rest to the speed v. That is,

$$\text{KE} = \int \mathbf{F} \cdot d\mathbf{s} = \int \frac{d}{dt}(m_0\mathbf{v}) \cdot d\mathbf{s} = m_0 \int \frac{d\mathbf{v}}{dt} \cdot d\mathbf{s} = m_0 \int \frac{d\mathbf{s}}{dt} \cdot d\mathbf{v}$$

$$= m_0 \int \mathbf{v} \cdot d\mathbf{v} = \frac{m_0 v^2}{2}$$

where we have written m_0 to emphasize that in this well-known expression we use the velocity-independent rest mass.

In relativistic dynamics, we employ the same definition for the kinetic energy but the relativistic expression for the momentum, Equation (2.19). Thus,

$$\mathrm{KE} = \int \mathbf{F} \cdot d\mathbf{s} = \int \frac{d\mathbf{p}}{dt} \cdot d\mathbf{s} = \int \frac{d\mathbf{s}}{dt} \cdot d\mathbf{p} = \int \mathbf{v} \cdot (\mathbf{v}\, dm + m\, d\mathbf{v}) \qquad \textbf{(2.21)}$$

where $m = \gamma m_0$. If we divide Equation (2.12) by γ and square both sides, we obtain

$$m^2c^2 - m^2v^2 = m_0^2c^2 \qquad \textbf{(2.22)}$$

and, taking differentials,

$$2mc^2\, dm - 2mv^2\, dm - 2vm^2\, dv = 0$$

or

$$c^2\, dm = v^2\, dm + mv\, dv = v(v\, dm + m\, dv) \qquad \textbf{(2.23)}$$

The right-hand side of Equation (2.23) is just the integrand of Equation (2.21). Therefore,

$$\mathrm{KE} = \int_{m_0}^{m} c^2\, dm = mc^2 - m_0c^2 \qquad \textbf{(2.24)}$$

The kinetic energy is the difference between two terms, mc^2 and m_0c^2. The first depends on the object's speed through the dependence of m on v. The second term, which is independent of the object's speed, is its *rest energy*, m_0c^2. The *total relativistic energy* is the sum of the kinetic energy and the rest energy.

$$E = m_0c^2 + \mathrm{KE} = mc^2 \qquad \textbf{(2.25)}$$

The total energy of a body is its mass multiplied by the square of the speed of light in vacuum.

Equation (2.25), which since 1945 has become a household phrase, tells us that mass and energy are two facets of the same coin, energy. The corollary of that result is that the two apparently separate conservation laws of classical physics, conservation of energy and conservation of mass, are really one, conservation of mass-energy.

One can arrive at the mass-energy equivalence by another route, as in the following example.

EXAMPLE 2.1 At $t = 0$, a particle, initially at rest, emits two pulses of electromagnetic radiation of equal frequencies in opposite directions. By transforming to a reference frame moving with a velocity perpendicular to the direction of propagation of the light pulses in the rest frame of the particle, show that momentum conservation requires that the mass of the particle diminish by $\Delta M = 2\epsilon/c^2 = \Delta E/c^2$, where ϵ is the energy of each light pulse.

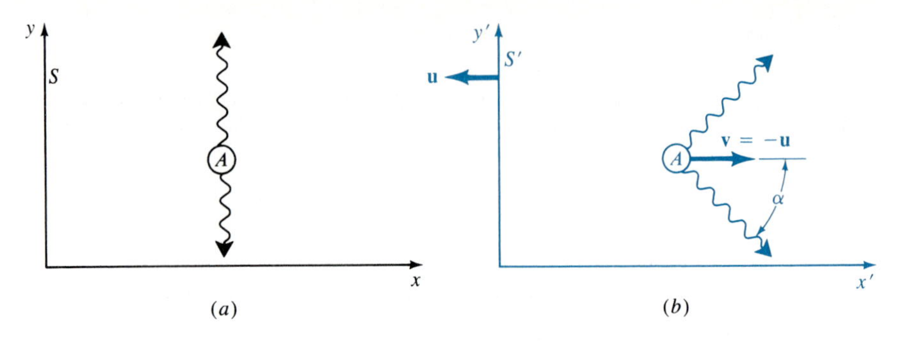

Figure 2.2 A radioactive particle, *A*, emits two photons of equal energy. (*a*) In reference frame *S*, particle *A* is initially at rest; the two photons propagate in opposite directions, and particle *A* is at rest also in the final state. (*b*) In reference frame *S′*, particle *A* moves with velocity **v** before *and* after emitting the two photons. In this frame, the two photons do not propagate in opposite directions.

SOLUTION Figure 2.2 illustrates the event in the two inertial reference frames. In S, the particle of mass M is at rest before the decay. Since the momenta ϵ/c of the two light pulses are oppositely directed, the particle is at rest also for $t > 0$.

In S', the momentum of the particle before the emission of the light pulses is $M_i'\mathbf{v} = -M_i'\mathbf{u}$, where $\mathbf{u}$ is the velocity of S' relative to S. For $t > 0$, the velocity of the particle is also $\mathbf{v}$. (In frame S the particle is at rest before and after the event; because S' moves at velocity $\mathbf{u}$ relative to S, the final velocity of the particle in the S' frame is $-\mathbf{u} = \mathbf{v}$.) Let the mass of the particle after the event be M_f'. Its momentum following the event is $M_f'\mathbf{v}$.

In the S' frame, the momenta of the two electromagnetic pulses do not cancel. In this inertial frame, each pulse has an x component of momentum equal to

$$\frac{\epsilon}{c}\cos\alpha = \left(\frac{\epsilon}{c}\right)\left(\frac{v}{c}\right) = \frac{\epsilon v}{c^2} \tag{2.26}$$

where we have made use of the basic postulate of relativity that the speed of light is c in every inertial reference frame.

The y components of the momenta cancel as they did in the S frame. The total momentum of the two pulses in S' is therefore

$$\frac{2\epsilon v}{c^2} = \frac{\Delta E v}{c^2} \tag{2.27}$$

Momentum conservation for the system requires that

$$M_i'v = M_f'v + \frac{\Delta E v}{c^2}$$

or

$$M_i' - M_f' = \Delta M = \frac{\Delta E}{c^2} \qquad \Delta E = \Delta M c^2 \tag{2.28}$$

QED

The example demonstrates that, to quote Einstein,

If a body gives off energy E in the form of radiation, its mass diminishes by E/c^2. The fact that the energy withdrawn from the body becomes energy of radiation evidently makes no difference, so that we are led to the more general conclusion that the mass of a body is a measure of its energy content.

In ordinary mechanical, electrical, and chemical actions in which energy is exchanged, the energies involved are so small compared with the rest energy of the system that the concomitant change in rest mass is unobservable. Still, with cautious foresight, Einstein concluded his paper with these words:

It is not impossible that with bodies whose energy content is variable to a high degree (e.g., with radium salts) the theory may be successfully put to the test.

EXAMPLE 2.2 Determine the change in the rest mass of a system consisting of a proton and an electron as the two particles combine to form a hydrogen atom. The ionization energy of hydrogen is 13.6 eV.

SOLUTION The rest mass of the proton is 1.672649×10^{-27} kg; that of the electron is 9.109534×10^{-31} kg. When the electron and proton combine to form a hydrogen atom, the ionization energy of 13.6 eV is released as ultraviolet radiation. The rest mass of the hydrogen atom is therefore smaller than the sum of the electron and proton rest masses by the amount

$$\Delta M = \frac{(13.6 \text{ eV})(1.6 \times 10^{-19} \text{ J/eV})}{(3 \times 10^{8} \text{ m/s})^2} = 2.42 \times 10^{-35} \text{ kg}$$

This is a change of less than 2 parts in 10^8 of the rest mass of the hydrogen atom, about a thousandth of the uncertainty in the mass of the proton. It would be futile to attempt to measure such a mass difference directly.

Not only does the mass of a system change as it absorbs or releases energy, but the entire rest mass of a body may be converted to energy. Conversely, rest mass can be created by electromagnetic energy provided that fundamental conservation laws—conservation of charge, energy, and linear and angular momentum—are not violated. The most common event of this kind is the creation of an electron-positron pair,[1] generally referred to as pair production. When a photon of sufficient energy interacts with a charged particle, for example, a proton, the entire energy of the

1. The *positron* is the "antiparticle" of the electron; it has the same mass as an electron but carries a positive charge *e*. See also Chapters 14 and 17.

$$E^2 - c^2p^2 = E'^2 - c^2p'^2 = E_0^2 \tag{2.34}$$

Although in arriving at Equation (2.34) we considered only a single body, the expression is equally valid for a system of two or more bodies. In that case, the invariant E_0 is the total energy of the system in that inertial reference frame in which the total momentum vanishes. That is, E_0 is the total energy of the system in the reference frame in which the center of mass (CM) of the system is at rest. The quantity E_0 is the sum of the rest energies of the separate constituents *only* if in the CM reference frame these constituents are all at rest. In general, E_0 is *not* the sum of the rest energies.

EXAMPLE 2.4 Determine the threshold energy for the creation of a proton-antiproton pair in an interaction in which a proton from an accelerator interacts with a stationary proton in the target. An antiproton is a particle whose mass is the same as that of the proton.

SOLUTION Let us write the reaction as follows:

$$\mathrm{p}_1 + \mathrm{p}_2 \rightarrow \mathrm{p}_1 + \mathrm{p}_2 + \mathrm{p}_3 + \bar{\mathrm{p}}$$

where $\bar{\mathrm{p}}$ represents the antiproton.[2] In a reference frame S' in which the CM of the system is at rest, the threshold energy is simply the sum of the rest energies of the four particles that constitute the end products of the reaction, since in this reference frame the particles produced are at rest. That is,

$$E' = 4m_0c^2 = E_i' = 2mc^2$$

where E_i' is the initial energy of the colliding particles in the CM frame. Here m_0 is the rest mass of a proton (and an antiproton), and m is the relativistic mass of each of the colliding protons, p_1 and p_2. In S' the velocities of the protons p_1 and p_2 are therefore given by

$$m = 2m_0 = \gamma m_0$$

or

$$\gamma = (1 - \beta^2)^{-1/2} = \frac{m}{m_0} = 2 \qquad \beta^2 = \frac{3}{4}$$

Let us now transform to the laboratory reference frame, S. Since in S, p_2 is initially at rest while in S', p_2 is moving with a speed βc, it follows that S' must have a velocity βc relative to S. To determine the speed of p_1 in S, we use the velocity transformation Equation (1.28a), and obtain

$$v_1 = \frac{\beta c + \beta c}{1 + (\beta c)^2/c^2} = \frac{2\beta c}{1 + \beta^2} \qquad \beta_1 = \frac{v_1}{c} = \frac{2\beta}{1 + \beta^2}$$

---

2. Antiparticles are represented by a bar over the symbol for the particle.

Hence,

$$\gamma_1 = (1 - \beta_1^2)^{-1/2} = \sqrt{\frac{1}{1 - 4\beta^2/(1 + \beta^2)^2}} = \frac{1 + \beta^2}{1 - \beta^2} = 7$$

In the laboratory, the total energy of the incident proton must be at least $7m_0c^2$ in order that when it collides head on with a stationary proton in the target a proton-antiproton pair can be produced in this inelastic collision. The kinetic energy of the incident proton must, therefore, be greater than $6m_0c^2 = 5.62$ GeV.

Proton-antiproton pairs were first produced artificially in 1955 at the University of California, using the proton beam from the Bevatron, a machine designed to produce protons of 6 GeV (BeV) kinetic energy.

As is apparent from the preceding example, only a fraction of the energy of the incident particle is used to create the new particles. In this particular case, at threshold, the incident proton has a kinetic energy of $6m_0c^2$, of which only $2m_0c^2$ is used to create the proton-antiproton pair. The process is evidently not energy efficient. The energy efficiency diminishes drastically as the rest energy of the particles that are produced increases. This can be seen from the following general relation (Problem 2.6):

$$\text{KE}_{\text{th}} = \frac{(\Sigma m_{\text{in}} + \Sigma m_{\text{fin}})(\Sigma m_{\text{fin}} - \Sigma m_{\text{in}})}{2m_{\text{tar}}}c^2 \qquad \textbf{(2.35)}$$

where KE_{th} is the threshold kinetic energy of the bombarding particle, Σm_{in} is the sum of the rest masses of the initial particles, Σm_{fin} is the sum of the rest masses of the final particles of the reaction, and m_{tar} is the rest mass of the target particle.

For example, suppose one wanted to produce a particle whose rest mass is 100 times that of a proton in the reaction

$$\text{p} + \text{p} \rightarrow M + \text{p} + \text{p}$$

where M is the rest mass of the massive new particle. Using a stationary target of protons, the kinetic energy of the incident proton would have to be at least

$$\frac{(104m_{\text{p}})(100m_{\text{p}})c^2}{2m_{\text{p}}} \simeq 5000 \text{ GeV} = 5 \text{ TeV}$$

Not only is the energy efficiency of the reaction less than 2 percent, but, worse yet, there are no machines capable of producing a proton beam of that energy.

One way around that obstacle is to construct the accelerator so that the laboratory frame is the CM reference frame of the experiment. In that case all of the kinetic energy of the two colliding protons can be converted into rest mass; the energy efficiency is now 100 percent. During the past decade, several accelerators have been constructed in which two beams of particles are made to collide, and more are currently under construction or in the planning stage (see Chapters 16 and 17). The Z^0 particle, whose rest mass is about $100m_{\text{p}}$, was discovered in 1983 in

(*a*)

(*b*)

Figure 2.3 The colliding-beam accelerator at CERN. (*a*) Aerial view. The large circle in the foreground shows the location of the 28-GeV proton synchrotron, and the circular area in the background shows the 300-m rings in which beams of protons circulate in opposite directions. (*b*) Schematic diagram; the protons collide at eight points around the rings. *(Courtesy CERN.)*

collisions of protons and antiprotons in the intersecting storage rings at the European Organization for Nuclear Research (CERN).

2.5 FORCE AND ACCELERATION

In classical mechanics, $\mathbf{F} = d\mathbf{p}/dt$ reduces to $\mathbf{F} = m\mathbf{a}$, where m is the rest mass. Since m is a scalar, $\mathbf{F}$ and $\mathbf{a}$ are parallel. In relativity, force is again defined as the rate of change of momentum, and the acceleration is the rate of change of velocity. Since the mass now depends on the velocity, we have

$$\mathbf{F} = \frac{d}{dt}(m\mathbf{v}) = m\,\frac{d\mathbf{v}}{dt} + \mathbf{v}\,\frac{dm}{dt} \tag{2.36}$$

The time derivative of the mass is given by

$$\frac{dm}{dt} = \frac{dE/dt}{c^2} = \left(\frac{1}{c^2}\right)\frac{d}{dt}(\text{KE} + m_0c^2) = \left(\frac{1}{c^2}\right)\frac{d(\text{KE})}{dt} = \frac{\mathbf{F}\cdot\mathbf{v}}{c^2} \qquad \textbf{(2.37)}$$

The acceleration is then

$$\mathbf{a} = \frac{d\mathbf{v}}{dt} = \left(\frac{1}{m}\right)\left(\mathbf{F} - \mathbf{v}\frac{dm}{dt}\right) = \frac{\mathbf{F}}{m} - \left(\frac{\mathbf{v}}{m_0c^2}\right)(\mathbf{F}\cdot\mathbf{v}) \qquad \textbf{(2.38)}$$

Equation (2.38) departs significantly from the classical expression. The radical new feature is that the acceleration is, in general, *not* in the direction of the applied force. While the first term of Equation (2.38) is a vector along **F,** the second is a vector directed along **v,** not **F**. The acceleration vector is the sum of those two vectors. Only in two special cases are **a** and **F** parallel:

1. **F** is collinear with **v**. In this case Equation (2.38) reduces to

$$\mathbf{a} = \frac{(1-\beta^2)^{3/2}}{m_0}\mathbf{F} = \frac{\mathbf{F}}{m_{\parallel}} \qquad \textbf{(2.39)}$$

   where $m_{\parallel}$ is the so-called *longitudinal mass.*
2. **F** is perpendicular to **v**. In this case, $\mathbf{F}\cdot\mathbf{v} = 0$ and the second term of Equation (2.38) vanishes. The relation between force and acceleration is then the familiar one, $\mathbf{F} = m\mathbf{a}$, where, of course, m is now the relativistic mass given by Equation (2.12), sometimes referred to as the *transverse mass*.

Both special cases are of practical interest. The first applies to the dynamics of particles in linear accelerators, where the dominant force is in the direction of the particle's velocity. The second case is represented by a charged particle in a region of uniform magnetic field, where the force acting on the particle is given by $q(\mathbf{v}\times\mathbf{B})$ and is perpendicular to **v**. The great majority of particle accelerators employ magnets to confine the particle beam to a circular or near circular trajectory; in these regions, $\mathbf{a} = q(\mathbf{v}\times\mathbf{B})/m$.

EXAMPLE 2.5 What is the radius of the circular orbit described by an electron in a uniform magnetic field of 1.0 T perpendicular to the electron's velocity if the kinetic energy of the electron is (a) 0.51 MeV? (b) 50 MeV?

SOLUTION The force acting on the electron is

$$\mathbf{F} = q\mathbf{v}\times\mathbf{B}$$

Since that force is normal to **v**, the acceleration is in the direction of the force and is given by

$$\mathbf{a} = \frac{q\mathbf{v}\times\mathbf{B}}{m}$$

where m is the transverse mass, given by Equation (2.12). The electron's trajectory is a circle whose radius is obtained by setting the centripetal acceleration equal to qvB/m:

$$\frac{v^2}{r} = \frac{qvB}{m} \qquad r = \frac{mv}{qB} \qquad rB = \frac{p}{q} \tag{2.40}$$

Equation (2.40) tells us that the product of the radius of curvature and magnetic field equals the momentum per unit charge. Since in this, and nearly all cases, the charge is either e or $-e$, where $e = 1.6 \times 10^{-19}$ C, the product rB is a direct measure of the particle's momentum.

(a) Since the rest energy of the electron is 0.51 MeV, the total relativistic energy is $2m_0c^2$. Its relativistic momentum is given by Equation (2.34):

$$(pc)^2 = E^2 - (m_0c^2)^2 = 3(m_0c^2)^2$$
$$p = \sqrt{3}(m_0c)$$

and the radius of its circular trajectory is

$$r = \frac{p}{qB} = \frac{\sqrt{3}(9.11 \times 10^{-31}\text{ kg})(3.0 \times 10^8\text{ m/s})}{(1.6 \times 10^{-19}\text{ C})(1.0\text{ T})}$$
$$= 2.96 \times 10^{-3}\text{ m} = 2.96\text{ mm}$$

(b) In this case the rest energy is so small compared with the total energy that we may safely neglect it and write

$$p \simeq \frac{E}{c}$$

and

$$r = \frac{p}{qB} = \frac{E}{qBc} = \frac{50 \times 10^6\text{ MeV}}{(1.0\text{ T})(3 \times 10^8\text{ m/s})} = 0.167\text{ m} = 16.7\text{ cm}$$

Of the many predictions of special relativity, the velocity dependence of the relativistic mass was the first to be confirmed experimentally. In beta decay (see Chapter 15), discovered just before the start of the twentieth century, the energies of electrons emitted by the radioactive source range from near zero to a maximum energy that is characteristic of the radioactive source. In the earliest experiments, the electrons passed through a velocity selector and then entered a region of uniform magnetic field. The radius of curvature of their circular paths in this field is, as we have seen, a direct measure of $p = mv$. Once v was determined, the mass of the electrons could then be calculated. The results of such measurements are shown in Figure 2.4.

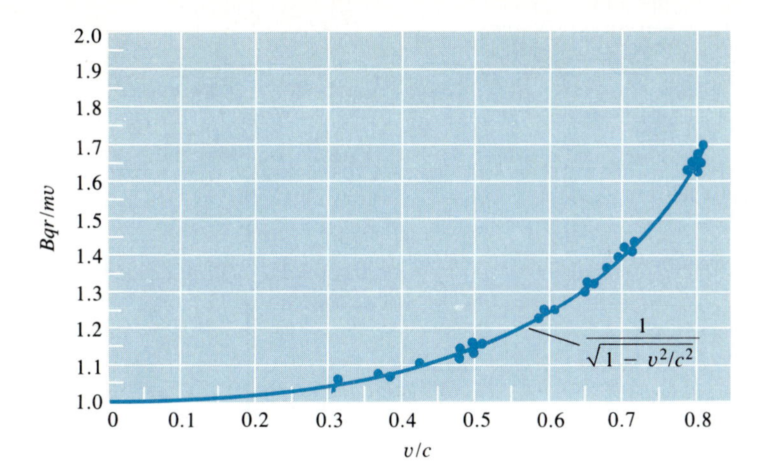

Figure 2.4 Results of an experimental verification of the dependence of electron mass on velocity.

SUMMARY

Conservation of momentum and the Lorentz transformation require a redefinition of linear momentum:

$$\mathbf{p} = m\mathbf{v} \tag{2.19}$$

where m is the *relativistic* mass

$$m = \gamma m_0 \tag{2.12}$$

and m_0 is the *rest mass,* i.e., the inertial mass of the object in the reference frame in which it is at rest.

Similarly, conservation of energy requires the redefinition

$$E = mc^2 = m_0c^2 + \mathrm{KE} = \sqrt{E_0^2 + c^2p^2}$$

It is convenient to define a four-dimensional coordinate system, x, y, z, ict. In this coordinate system the length of the vector $\mathbf{s} = (x, y, z, ict)$ and $E_0 = \sqrt{E^2 - c^2p^2}$ are invariant under a Lorentz transformation.

Bibliography

A. P. French, *Special Relativity,* Norton, New York, 1968.

M. P. Hougan and C. M. Will, "Modern Tests of Special Relativity," *Physics Today,* May 1987, p. 69.

N. David Mermin, *Space and Time in Special Relativity,* McGraw-Hill, New York, 1968.

Julian Schwinger, *Einstein's Legacy,* Scientific American Library/Freeman, San Francisco, 1985.

Victor F. Weisskopf, "The Visual Appearance of Rapidly Moving Objects," *Physics Today,* November 1960, p. 24.

PROBLEMS

2.1 What is the velocity of an electron that has dropped from rest through a potential difference of 10^6 V?

2.2 Calculate the relativistic energy of a proton whose momentum is $2m_0c$.

2.3 What is the speed of a particle whose relativistic mass is twice its rest mass?

2.4 What is the relativistic energy of a proton whose speed is $0.8c$?

2.5 A proton is to be accelerated to a speed of $0.6c$. What is the potential difference through which it must drop?

•2.6 Derive Equation (2.35).

2.7 Calculate the kinetic energy and the relativistic mass of an electron whose speed is (a) $0.01c$; (b) $0.4c$; (c) $0.9999c$.

2.8 Show that for $v \ll c$ the relativistic expression for kinetic energy reduces to the classical result.

2.9 Show that for the elastic collision of Figure 2.1, $p'_{By} = p_{By'}$, where $\mathbf{p}_B$ is the relativistic momentum of particle B.

•2.10 A particle has a total relativistic energy of 5 MeV and momentum of 4 MeV/c. (a) Determine its rest mass. (b) What is the particle's energy in a reference frame in which its momentum is 3 MeV/c? (c) What are the relative velocities of the two reference frames?

2.11 A particle of rest mass m_0 has an energy $E = 4m_0c^2$. What is the momentum of this particle in units of m_0c? What is the energy of this particle in a reference frame in which $p = 2m_0c$?

2.12 A particle of rest mass M_0 decays into two identical particles of rest mass m_0. Obtain an expression for the velocities of the two particles if M_0 is at rest.

•2.13 Repeat Problem 2.12 for the case in which the original particle of rest mass M_0 has an energy of $3M_0c^2$ when it decays and the velocities of the decay products are along the direction of motion of the parent particle.

2.14 An electron is accelerated from rest in a uniform electric field $E = 10^7$ V/m. What is the acceleration of the electron after it has traveled a distance of (a) 3 cm? (b) 10 cm? (c) 1 m?

•2.15 A particle of rest mass m_0 and charge q is accelerated from rest by a uniform electric field E. Derive expressions for the velocity and position of the particle as a function of time.

2.16 When high-energy protons strike protons at rest, neutral pions (π^0) are produced. Determine the threshold energy of the protons of the accelerator beam for this process. (See Table 2.1.)

•2.17 Figure 2.5a shows an elastic proton-proton collision in the CM reference frame. In this CM frame, the kinetic energy of each proton is $2m_0c^2$. (a) What is the velocity of the CM frame relative to the laboratory frame in which one of the protons is at rest prior to the collision? (b) What is the kinetic energy of the incident proton in the laboratory reference frame? (c) What is the angle 2θ between the velocities of the two protons in the laboratory following the collision?

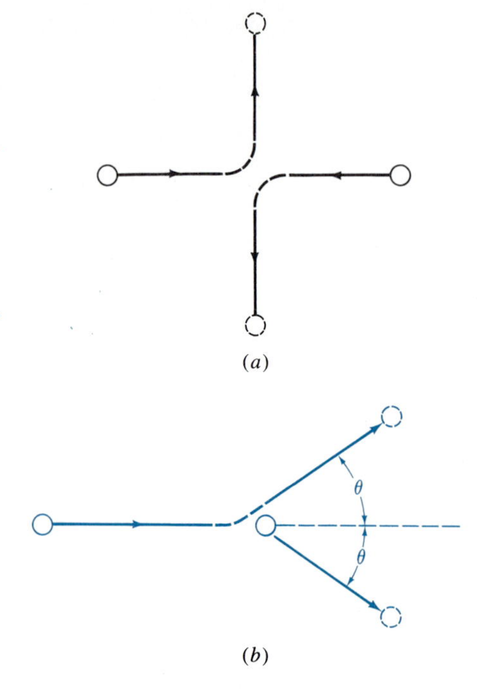

Figure 2.5 Problem 2.17.

•2.18 Repeat Problem 2.17 when the kinetic energy of each proton in the CM frame is $10m_0c^2$.

2.19 A particle of rest mass m_0 with kinetic energy $3m_0c^2$ makes a perfectly inelastic collision with an identical particle that is at rest, forming a particle of rest mass M_0. Determine the energy and momentum of the product particle and its rest mass M_0.

2.20 In *Compton scattering*, a photon of wavelength λ_i makes a collision with a stationary electron. Show that λ_f, the wavelength of the scattered photon, is given by

$$\lambda_f = \lambda_i + \frac{h}{m_e c}(1 - \cos\theta)$$

where θ is the angle between the directions of the incident and the scattered photon. The quantity h/m_ec is called the *Compton wavelength of the electron* and has a value of 2.426×10^{-3} nm.

2.21 Using the result of Problem 2.20, show that when a photon of energy $E \gg m_ec^2$ is backscattered, the energy of the scattered photon is about 0.25 MeV, independent of the energy of the incident photon.

2.22 A particle of rest mass M_0 decays in flight into two particles of equal rest mass m_0. In the laboratory, one of the resulting particles is at rest, while the other has a kinetic energy of $3m_0c^2$. Obtain an expression for the rest mass M_0 in terms of m_0.

2.23 The K^0 particle, whose rest mass is 497.7 MeV/c^2, decays into a π^+ and a π^-. What are the kinetic energies of the pions ($\pi^\pm$) if the K^0 is at rest prior to the decay?

•**2.24** Suppose that following the decay of a K^0 one of the pions is at rest in the laboratory reference frame. Determine the kinetic energies of the other pion and of the K^0 prior to the decay.

2.25 A particle carrying charge e and traveling with a velocity of $0.8c$ follows a circular trajectory of radius 0.518 m in a uniform magnetic field of 1.2 T. Determine the rest mass of the particle.

2.26 The radius of the circular trajectory of a π^+ in a magnetic field of 1.0 T is 0.96 m. Calculate the kinetic energy of the π^+.

2.27 A free neutron is unstable, decaying into a proton, an electron, and an antineutrino. The process is written

$$\mathrm{n} \rightarrow \mathrm{p} + \mathrm{e} + \bar{\nu}$$

The neutrino has zero rest mass but can carry off some energy. Suppose that in a particular decay of the neutron, the neutrino carries off none of the energy. Calculate the energy and velocity of the electron.

2.28 A π^+ at rest decays into a μ^+ and a neutrino ($\pi^+ \rightarrow \mu^+ + \nu$). The rest mass of the μ^+ is the same as that of the μ^-, namely, 105.7 MeV/c^2. Calculate the energy of the neutrino and the kinetic energy of the μ^+.

2.29 Calculate the binding energy of the α particle; i.e., determine the energy that would be required to separate an α particle into two protons and two neutrons.

Chapter 3 The General Theory of Relativity

Hoch geehrter Herr Kollege:

Eine einfache theoretische Ueberlegung macht die Annahme plausibel, dass Lichtstrahlen in einem Gravitationsfelde eine Deviation erfahren. Am Sonnenrande müsste diese Ablenkung $0.84''$ *betragen und wie* $1/R$ *abnehmen (*R = *Entfernung vom Sonnenzentralpunkt).*

[*My dear Colleague:*

On the basis of a simple theoretical argument it appears plausible that light rays experience a deviation in a gravitational field. At the edge of the sun this deflection should be $0.84''$ *and diminish as* $1/R$ (*R = distance from the sun's center).*]

Albert Einstein to Professor George E. Hale, in a letter dated October 14, 1913

3.1 INTRODUCTION

The preceding chapters focused on the central arguments and most important conclusions of the special theory of relativity. We now venture beyond the restrictions of that theory.

The general theory of relativity concerns itself with natural phenomena as they appear to observers in different *noninertial* reference frames. Though not immediately apparent, the general theory thereby develops into a theory of gravitation.

That Newton's theory of gravitation was incomplete and unsatisfactory was recognized already by its author. Its basic flaw is the presumption of action at a distance. The gravitational force is assumed to communicate between bodies with infinite speed, in violation of the basic postulate of special relativity. By contrast, the electrostatic interaction, though similar in form to the gravitational, is transmitted via the electric field at the very high but finite speed c. The general theory of relativity does for gravity what Maxwell's theory achieved for electricity. The gravitational force is mediated by the gravitational field, as the electromagnetic force is mediated by the electromagnetic field. However, because of an inherent nonlinearity, the formal mathematical description of the gravitational field proves to be far more complex than the electromagnetic.

We cannot hope to present the general theory with anything approaching the logical order and completeness of the preceding two chapters. The special theory could be developed using simple algebra and elementary calculus. The general theory calls for an altogether different level of mathematical sophistication, one rarely achieved even by practicing physicists.

In the following pages we first consider the relationship between gravitation and the transformation to noninertial reference frames, which finds expression, for local events, in the principle of equivalence first enunciated by Einstein. We then examine some of the expected consequences of this principle and describe experimental results that confirm them.

The generalization to inhomogeneous gravitational fields is accomplished in terms of the nonlinear field theory alluded to already. In a formal way, this can be rephrased in terms of a geometry of space-time that departs from the familiar Euclidean description in the region of an inhomogeneous gravitational field. These ideas are very briefly sketched. The final section of this chapter summarizes the principal predictions of the general theory and the current status of their experimental verification.

3.2 THE PRINCIPLE OF EQUIVALENCE

The adjective *special,* as applied to relativity, refers to the restriction of that theory to *inertial* reference frames, that is, to reference frames in uniform rectilinear motion. Such reference frames must not accelerate along any direction or rotate about some point. Since the latter restriction eliminates the most widely used reference frame, the earth, as an inertial frame, it is appropriate to ask ourselves how we can determine that the earth is indeed rotating.

There are several indications that reveal the earth's rotation about its axis, quite apart from the diurnal motion of the sun and stars. One is the oblateness of the globe. A drop of liquid in free space, under the influence of its own surface tension and gravitational forces, is a sphere. However, if the drop rotates about its center, it assumes an *oblate* spheroidal shape.

Another evidence of rotation is the motion of the Foucault pendulum. That motion is most easily envisaged for a pendulum freely suspended above a pole. An observer in the rotating frame of the earth will see the pendulum change its plane of swing in the *absence* of any external force.

Yet another consequence of rotation is the Coriolis effect. Imagine a disk in uniform rotation about its axis and an observer standing near the center of that disk. If this observer throws a projectile radially toward the periphery of the disk, it will, in the rotating reference frame, follow a *curved* and not a straight path.

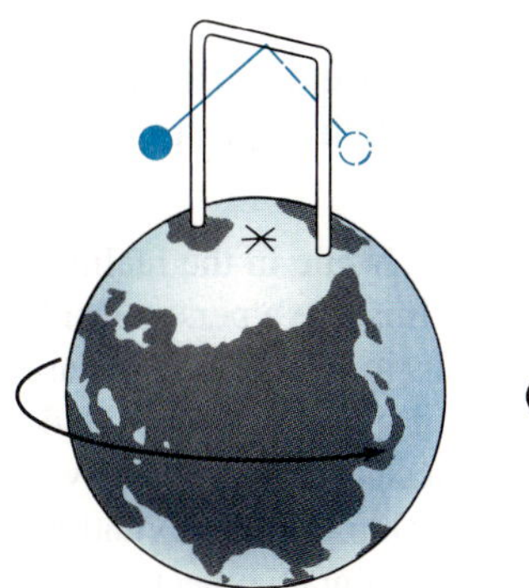
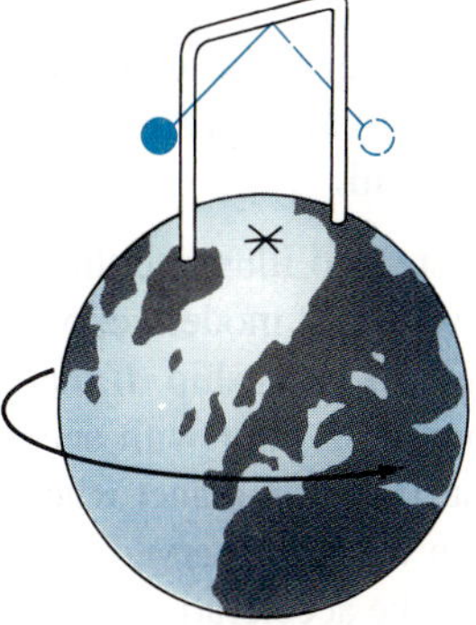
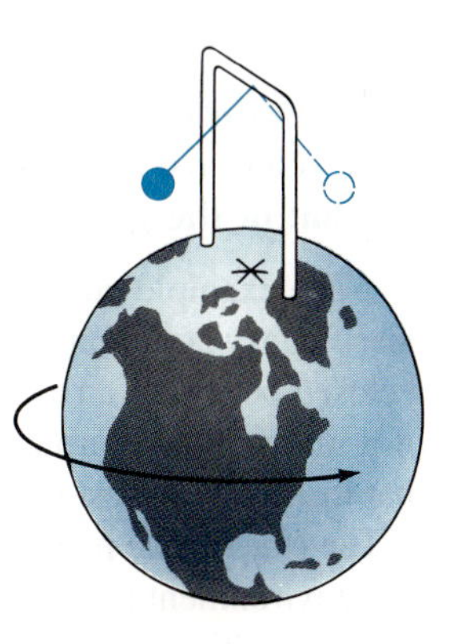

Figure 3.1 The Foucault pendulum at the north pole. The pendulum swings in a fixed plane. To an observer on the earth, the plane of the pendulum appears to change as the earth rotates.

inertial frame (neglecting the earth's rotation), will not be parallel, but will converge toward the center of the planet. It is difficult to conceive of a noninertial reference frame that could replicate this behavior. Similarly, the inertial force that acts on a body when viewed from a rotating reference frame defies replication by a gravitational field. (The centrifugal force increases with distance from the center of rotation; i.e., the equivalent gravitational field should be zero at, and increase linearly with distance from, the center. Moreover, that gravitational field would have to include a velocity-dependent component, equivalent to the Coriolis force.)

Despite this restriction, the principle of equivalence has led to a number of important predictions, some of them quite startling on first encounter. The great power of the principle lies in the fact that it permits us to make deductions about the behavior of objects in a homogeneous gravitational field *without recourse to a theory of gravitation*. We simply study the event as it might be seen by an observer in a gravitation-free environment but in a uniformly accelerating reference frame.

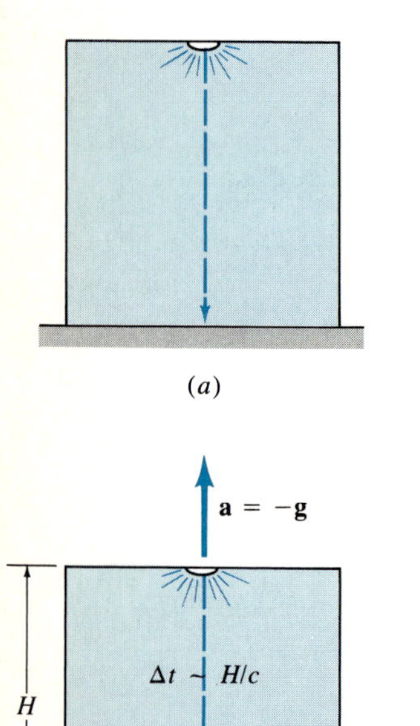

Figure 3.5 The gravitational frequency shift. (*a*) A source of light fixed to the ceiling radiates at the frequency ν. (*b*) The effect of the gravitational field is equivalent to an acceleration $\mathbf{a} = -\mathbf{g}$ of the laboratory. When the light emitted by the source reaches the receiver on the floor, the receiver is moving upward at a speed $v = a\Delta t = gH/c$. The frequency measured by the receiver is, therefore, Doppler-shifted toward higher frequency. From the principle of equivalence it follows that the same Doppler shift will occur if source and receiver are in a gravitational field.

3.3 GRAVITATIONAL RED SHIFT: GRAVITATIONAL TIME DILATION AND LENGTH CONTRACTION

Probably the most direct and transparent prediction of the principle of equivalence is the *gravitational red shift*. Suppose light pulses from a monochromatic source on the ceiling are detected at the floor, as in Figure 3.5. To determine the influence, if any, of the gravitational field on the light, we make use of the equivalence principle and consider what would be observed if the room were in a gravity-free environment but accelerated upward with acceleration g. We assume that at the instant the pulse of light is emitted, the source is at rest.

If the height of the room is H, the time required for light to traverse the distance to the detector is H/c (we are here neglecting the very small distance $\frac{1}{2}gt^2$ by which the floor has risen in that time), and the velocity of the detector at the instant the light pulse strikes it is, therefore, $v = gt = gH/c$. Since the detector approaches the source with this velocity, the wave is Doppler-shifted to higher frequency or shorter wavelength. Provided $v = gH/c \ll c$,

$$\frac{\delta\nu}{\nu} = -\frac{\delta\lambda}{\lambda} = \frac{v}{c} = \frac{gH}{c^2} \tag{3.3}$$

The shift is toward higher frequency or shorter wavelength, i.e., toward the "blue." However, if the source is located on the floor and the detector attached to the ceiling, the frequency measured at the detector is shifted toward lower values, i.e., toward the "red."

EXAMPLE 3.1 The Eiffel Tower in Paris is 300 m tall. What is the fractional gravitational red shift due to this elevation difference?

SOLUTION Equation (3.3) gives

$$\frac{\delta\nu}{\nu} = \frac{gH}{c^2} = \frac{(9.8\ \text{m/s}^2)(300\ \text{m})}{(3 \times 10^8\ \text{m/s})^2} = 3.27 \times 10^{-14}$$

The product gH in Equation (3.3) is just the difference in gravitational potential between source and detector. The principle of equivalence thus leads to the prediction that the frequency of light suffers a fractional change whose magnitude is given by $\delta\phi/c^2$, where $\delta\phi$ is the change in gravitational potential between source and detector.

The magnitude of this gravitational shift is readily calculated. Even for an elevation difference of 8000 m (roughly the altitude of Mount Everest), the fractional shift is less than 10^{-12}. Any attempt to measure this shift in the laboratory was clearly hopeless, or, at least, so it seemed. Early attempts to observe this gravitational shift therefore relied on careful measurements of wavelengths of light from stellar sources, especially very massive, dense stars, so-called *white dwarfs*. Since in each instance, the source is at a much lower gravitational potential than the detector, light should be shifted to a longer wavelength, i.e., toward the red; hence, the origin of the phrase gravitational red shift.

The calculation of stellar gravitational red shifts is not as straightforward as the one just outlined. In its transit to the detector on earth, the light first rises, so to speak, out of the gravitational potential well of the star, and then drops through the gravitational field of the earth to the spectrometer in the laboratory. A simple calculation shows that, compared to the gravitational potential at the star's surface, that at the earth's is negligible. The real difficulty arises because the equivalence principle is valid only in a uniform gravitational field. While that may be a satisfactory approximation over distances small compared to the star's radius, it certainly is not justified in the present case. One way out of the dilemma is to assume that Equation (3.3)—where the right-hand side is interpreted as $\delta\phi/c^2$, the change in gravitational potential divided by c^2—holds even when the gravitational field is nonuniform. With that assumption, one finds for light emitted at the surface of a star

$$\frac{\delta\lambda}{\lambda} = \frac{GM}{Rc^2}$$

where M is the mass and R the radius of the star. For light emitted from the surface of the sun, $\delta\lambda/\lambda \simeq 2 \times 10^{-6}$, a small but measurable wavelength shift. There are, however, other effects that lead to a red shift—notably, the second-order Doppler shift due to the thermal motion of the atoms in the sun's chromosphere—that overshadow the gravitational contribution.

The first white dwarf discovered was Sirius B, the small but extremely massive companion of Sirius A. The displacement of the spectral lines from Sirius B was measured by Adams in 1924.[1] The data agreed with the predicted gravitational red shift, but the reliability of the result was in doubt. The estimate of the radius of Sirius B proved unreliable, and difficulties in separating the spectrum originating from Sirius A from that of its white dwarf companion frustrated precise measurement of spectral lines.

In the end, the definitive test of the gravitational red shift was a laboratory experiment. The stage for the experiment was set in 1958 with the discovery of the Mössbauer effect (see Chapter 15), which permitted the measurement of incredibly small frequency shifts between source and detector. The first successful measure-

1. W. S. Adams, *Proc. Nat. Acad. Sci.* **11,** 382 (1925).

ment[2] by Pound and Rebka[3] at Harvard used an elevation difference between their ^{57}Fe source and detector of only 22.5 m, corresponding to a predicted fractional shift of only 4.92×10^{-15}! The measured value was $(5.13 \pm 0.51) \times 10^{-15}$, and subsequent improvements have reduced the uncertainty from 10 percent to 1 percent.

An alternative interpretation of the frequency shift which focuses directly on the role of the gravitational field is as follows.

If an object of finite rest mass falls through a gravitational potential difference $\delta\phi$, its kinetic energy increases. In the case of light, the particles are photons[4] of zero rest mass, moving at the speed c. Their energy is given by

$$E = h\nu = \frac{hc}{\lambda} \tag{3.4}$$

the Planck-Einstein formula. Here h is Planck's constant.[5]

Now the principle of equivalence is based on the equality of inertial and gravitational mass. Let us assume that we may ascribe to the photon a *gravitational* mass equal to its inertial mass E/c^2. In moving through a gravitational potential difference $\delta\phi = gH$, then, to conserve energy, the photon's frequency must change so that

$$h\nu' = h\nu + \left(\frac{E}{c^2}\right)\delta\phi = h\nu\left(1 + \frac{\delta\phi}{c^2}\right) \tag{3.5}$$

Consequently,

$$\frac{\delta\nu}{\nu} = \frac{\delta\phi}{c^2} = \frac{gH}{c^2} \tag{3.6}$$

which is the result obtained earlier.

Previously we computed the effect of a uniform gravitational field by imagining the field turned off and replaced by an accelerating reference frame. We then calcu-

---

2. A research group at the Atomic Energy Research Establishment, Harwell, England, was also engaged in the same effort and, in fact, was the first to publish. Their result was, however, soon cast in doubt by a young undergraduate student at Cambridge University, Brian D. Josephson, who called attention to the well-known fact that, contrary to the classical Doppler effect, the relativistic expression includes a second-order (quadratic) term as well as the first-order term that is linear in v/c [*Phys. Rev. Lett.* **4,** 341 (1960)]. If source and detector are at rest in an *inertial* frame, the first-order term does not result in a net shift; as many atoms of the emitting source move toward as away from the detector, the former giving rise to a blue and the latter to a red shift, which cancel on averaging.

The quadratic term, however, leads to a *red* shift that is just proportional to temperature. As Josephson showed, this quadratic Doppler effect results in a shift greater than the predicted gravitational shift if source and detector temperatures differ by more than 1 °C. This difficulty had been recognized by Pound and Rebka, who carefully controlled the temperature of source and detector.

Josephson shared the 1973 Nobel Prize in physics for his theoretical work on tunnel currents in superconductor junctions (see Chapter 13).

3. R. V. Pound and G. A. Rebka, Jr., *Phys. Rev. Lett.* **4,** 337 (1960).

4. A *photon* is a quantum of light.

5. These matters are discussed in detail in Chapter 4.

lated what an observer in this accelerating reference frame would conclude about frequencies and wavelengths of light.

We can, however, take another viewpoint, that of two observers in the same reference frame, but located at different gravitational potentials. Initially, both observers compare their transmitting and detecting instruments at the same location and find them to agree exactly with regard to frequency calibration. If now the second observer, O', moves with his or her detector to a location of lower gravitational potential and the first, O, releases a light pulse of frequency ν, as measured by O, observer O' will measure a frequency given by Equation (3.5), i.e., a value different from that reported by O. In other words, in a given time interval, the number of complete cycles counted by the two observers differs. Since in the process of transmission no "cycles" are lost, the clocks by which O and O' measure time must run at different rates. We are led to the conclusion that **clock rates are affected by gravitational fields; clocks run more slowly the lower the gravitational potential.**

EXAMPLE 3.2 An atomic clock is flown in a commercial plane at an altitude of 12,000 m. The plane's cruising speed is 850 km/h. Compare the time dilation contributions due to effects of special and general relativity with respect to a clock in a laboratory at sea level.

SOLUTION From Equation (1.23c), the fractional change in the clock rate due to the motion of the clock is

$$\frac{\Delta t}{t} = \gamma - 1 = \left[1 - \left(\frac{v}{c}\right)^2\right]^{-1/2} - 1 \simeq \frac{1}{2}\left(\frac{v}{c}\right)^2$$

where we have used the binomial expansion valid when $v/c \ll 1$. Substituting the appropriate numerical values, we have

$$\frac{\Delta t}{t} = \frac{[(850 \text{ km/h})/(3.6 \text{ m}\cdot\text{h/km}\cdot\text{s})]^2}{2(3 \times 10^8)^2} = 3.1 \times 10^{-13}$$

From Equation (3.3), reinterpreted as a fractional change in clock rate due to a gravitational time dilation, we have

$$\frac{\Delta t}{t} = \frac{-gH}{c^2} = \frac{-(9.8 \text{ m/s}^2)(1.2 \times 10^4 \text{ m})}{(3 \times 10^8)^2} = -1.3 \times 10^{-12}$$

We see that in this instance the time dilation due to special relativity is less than a third of the effect due to gravitation, which, in this case, is actually a contraction rather than a dilation. In the analysis of the Hafele-Keating experiment [Section 1.7(c)], careful attention had to be paid to effects due to gravitation and special relativity.

Not only do clock rates depend on gravitational potential, but, arguing in terms of wavelength rather than frequency, it follows that rulers must contract on being transported to a location of lower gravitational potential. Einstein saw in these effects the most serious obstacle to a generalization of the special theory of relativity. As he recalled,

> *it took me a long time to see what coordinates in general really meant in physics. I did not find the way out of this dilemma till 1912.*

3.4 THE GENERAL THEORY OF RELATIVITY: GRAVITATION

The ''way out'' was to discard the Euclidean geometry of space and to introduce a new space-time geometry which is dependent on the gravitational field. In a very weak gravitational field, that is, far from a massive object, the geometry approaches the Euclidean space-time geometry in which light propagates along straight lines. In a nonuniform gravitational field the geometry is altered, or space-time is curved or ''warped.'' What is meant thereby is that geometric figures constructed of light rays will not satisfy the criteria of Euclidean geometry.

The usual example that is cited to illustrate the concept of curved space is that of a triangle drawn on the surface of a very large sphere. If the sides of the triangle are very short compared to the sphere's radius, the spherical surface in the ''local'' region so closely approximates a plane that the sum of the vertex angles will appear to add to 180° as required by Euclid. However, if the ''same'' triangle is drawn on a sphere of much smaller radius, so that the sides of the triangle are comparable to the radius, the sum of the vertex angles will exceed 180°.

This ''example'' is, however, a very imperfect representation. It is important to bear in mind that in general relativity it is not ordinary space but four-dimensional space-time that is distorted by the presence of a massive object. It is this distortion, characterized by the set of equations derived by Einstein, that we perceive as a gravitational field.

3.5 PREDICTIONS OF THE GENERAL THEORY OF RELATIVITY

Perhaps the most famous prediction of the general theory is the bending of light by a gravitational field. This effect has an intriguing history which begins with Newton, who envisioned a light ray as a narrow beam of extremely tiny particles that travel in straight trajectories and at enormous speed through space.[6] Now, according to his law of gravitational attraction, the acceleration of gravity experienced by a body at some point in space should be the same regardless of its mass. Therefore, any particle, however light, should experience this acceleration. Hence, his first ''Query'' in *Opticks:*

---

6. Before Newton was thirty-five years old, Roemer had made the first moderately accurate measurement of the speed of light.

Do not Bodies act upon Light at a distance, and by their action bend its rays, and is not this action strongest at the least distance?

In 1801, just a few years before the experiments of Young and Fresnel put the corpuscular theory of light to rest for the century, a German mathematician, Johann Georg von Soldner, computed the trajectory of a "particle" of light that passes close to the periphery of the sun.[7] To perform this calculation one must know only the sun's mass, the universal gravitational constant, and the speed of light. The last was known to be close to 3×10^8 m/s from measurements by Bradley. The missing data were the universal gravitational constant and the sun's mass. When, in 1798, Cavendish's paper on the measurement of G appeared in the *Philosophical Transactions of the Royal Society,* Soldner was able to answer Newton's query. His result is the same as that deduced by application of the principle of equivalence. This was also the first prediction published in 1911 by Einstein,[8] who was quite unaware of Soldner's work. Within a year, however, Einstein had recognized the seriousness of the dilemma posed by gravitational length contraction and time dilation, and, four years later, published his paper on the general theory of relativity, which included the correct prediction for the trajectory of a light ray in the sun's gravitational field. The predicted deflection is just twice as great as that based on the principle of equivalence. That principle is inapplicable here because the light ray traverses a region of nonuniform gravitational field.

The verification of that result involves precise measurements of the position of stars whose light just grazes the sun's disk in reaching an observer on earth. The predicted bending should result in an outward displacement of 1.74 seconds of arc, a very small but measurable amount. Since scattering of sunlight by the atmosphere makes stellar observation during the day impossible, a test of Einstein's theory had to await a total solar eclipse.

In 1914 a German expedition did leave for Russia, where a total eclipse would occur later that year. However, before that natural event an unnatural one, the outbreak of World War I, took place. After some weeks of internment, the scientists were returned to Germany in exchange for some Russian officers who had been taken prisoner.

After the war, at the urging of Sir Arthur Eddington, two British astronomical expeditions were dispatched—to Sobral, Brazil, and to the island of Principe off the coast of west Africa—to make observations during the total solar eclipse of May 29, 1919. Eddington led the Principe expedition.

When totality began, the dark disc of the moon . . . was visible through cloud, much as . . . on a night when no stars can be seen. There was nothing for it but to carry out the arranged programme and hope for the best.

Luckily, the cloud cover was partly broken, and of sixteen photographs taken during the five minutes of totality, one good plate was obtained.

The results from this plate gave a definite displacement, in good accordance with Einstein's theory.

Figure 3.6 A beam of light from a distant star is bent as it passes close to the sun. As a result, the apparent position of the star shifts. The change is greatly exaggerated in this drawing.

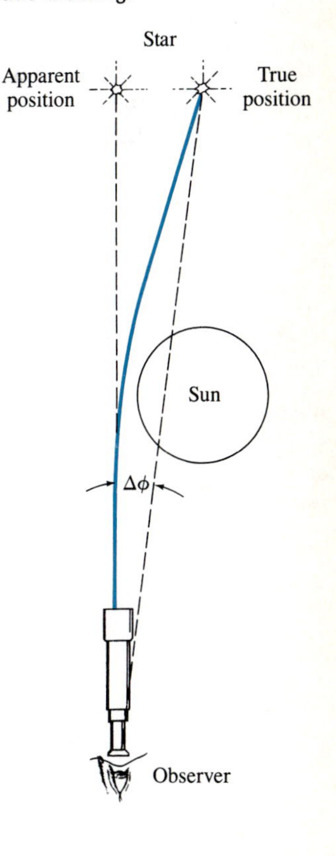

---

7. J. von Soldner, *Berl. Astr. Jahrb.* (1804), p. 161.

8. A. Einstein, *Ann. d. Phys.* [4] **35,** 898 (1911).

The Sobral expedition was more fortunate and returned with several excellent photographs. The observed deflections, 1.61 ± 0.3 seconds (Principe) and 1.98 ± 0.16 seconds (Sobral), were in satisfactory agreement with Einstein's prediction.[9]

The results of the expedition, announced at a joint session of the Royal Society and the Royal Astronomical Society in November 1919, created a worldwide sensation, partly because of their scientific importance but also because it seemed remarkable that so soon after the war's end a British team should have set out to confirm the predictions of a German physicist. Einstein, who became a sensation himself almost overnight, was not unaware of these implications. The *London Times* asked him to write an article on relativity that he concluded, with tragic prescience, as follows:

> *Here is yet another application of the principle of relativity for the delectation of the reader: Today I am described in Germany as a "German savant," and in England as a "Swiss Jew." Should it ever be my fate to be represented as a* bête noire, *I should, on the contrary, become a "Swiss Jew" for the Germans and a "German savant" for the English.*

The second prediction contained in Einstein's 1916 paper on general relativity concerns the precession of the perihelion of Mercury.

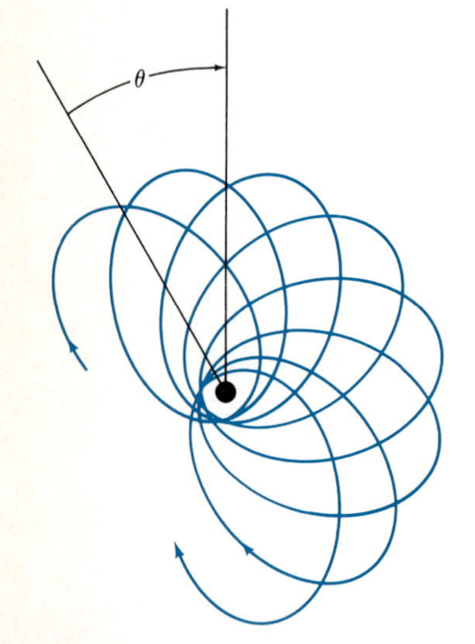

Figure 3.7 Precession of the perihelion (point of closest approach to the sun) of the orbit of Mercury. Here the eccentricity of the elliptic orbit is greatly exaggerated, as is the rate of precession.

In the absence of other planets, each planet would retrace the elliptical path of its previous orbit. About 1840, at the suggestion of Arago, the French astronomer Leverrier analyzed the motion of Mercury, the innermost planet, whose orbit is the most eccentric of the planets. The results were startling. The orbit itself precesses at a rate of about 1.6° per century. After the perturbing influences of all the other planets had been carefully calculated, there still remained 43 seconds of arc per century unaccounted for.[10] Though small, it was nonetheless a discrepancy that troubled astronomers.

Leverrier proposed that there may be an as yet undiscovered planet circling close to the sun whose gravitational pull would cause the small additional precession. He even gave it a name, Vulcan, and for many years astronomers searched for Vulcan without success.

In 1907, in a letter to his friend Conrad Habicht, Einstein wrote that he was busy on "a relativistic theory of gravitation with which I hope to account for the still unexplained secular changes of the perihelion movement of Mercury." Nine years later he obtained the answer: The precession due to the curvature of space-time near the sun is just 43 seconds of arc per century.

In recent years, with the discovery of stellar radio sources, various advanced radar and laser ranging techniques, and artificial satellites and space probes, new precise measurements have added further experimental support to Einstein's theory. One class of experiments is of particular interest because it clearly demonstrates that it is space-time, not just three-dimensional space, that is curved by a massive object. The experiment involves the measurement of the time required for an electromagnetic wave to travel from the Earth to Mars and back when Mars and Earth are just entering a superior conjunction (i.e., when Earth and Mars are on opposite sides

9. F. W. Dyson, A. S. Eddington, and C. Davidson, *Phil. Trans. Roy. Soc. London* **A220,** 291 (1920).

10. U. J. Leverrier, *Ann. Obs. Paris* **5,** 1 (1859).

of the sun). The incremental time delay due to the curved path alone would amount to about 30 nanoseconds. There is, however, an additional contribution, the so-called Shapiro time delay, that has its origin in the curvature of the time coordinate of space-time, and in this instance amounts to about 250,000 nanoseconds.

A radar transponder was put aboard the *Viking* space probe, which was landed on Mars in 1976. When Mars started to pass behind the sun's disk, a radar signal was sent to *Viking* and the time interval between its transmission and the reception of the returned signal carefully measured. The returned signal was delayed by the 250,000 nanoseconds predicted by general relativity.[11]

The bending of light due to the curvature of space-time about a massive star or an entire galaxy is similar to that produced by a converging lens. The first indication of the existence of such gravitational lenses was the "double" quasar Q 0957 + 561 discovered in 1979. The manner in which a gravitational lens can produce a double image is indicated schematically in Figure 3.8. Since then, other astronomical image distortions have been observed and interpreted in terms of gravitational lenses.[12]

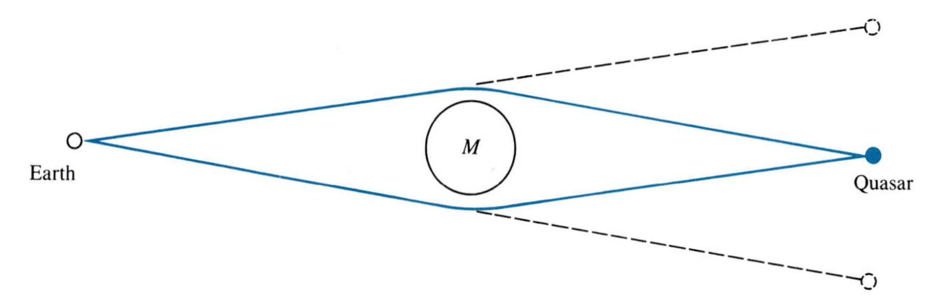

Figure 3.8 Light from a distant source is bent by a massive object. Under suitable conditions, the light appears to originate from two separate but otherwise identical sources, accounting for the observation of the "double" quasar Q 0957 + 561.

We have already mentioned the gravitational red shift, another of the predictions of Einstein's 1916 paper. Closely related to this effect are "black holes," first considered by J. R. Oppenheimer and S. Snyder in 1939.[13] A black hole is an object so dense and massive that the gravitational field at its surface allows nothing, not even light, to escape its clutches.

As we saw, the gravitational red shift may be interpreted in terms of the inertial energy change of light in propagating from a point of low to a point of higher gravitational potential. The gravitational potential at the surface of a star is

$$\phi(R) = \frac{-GM}{R} \tag{3.7}$$

Suppose

$$\frac{GM(E/c^2)}{R} \geq E = h\nu \tag{3.8}$$

11. I. I. Shapiro et al., *J. Geophys. Res.* **82,** 4329 (1977). See also I. I. Shapiro et al., *Phys. Rev. Lett.* **26,** 1132 (1971).

12. A. Stockton, *Ap. J. Lett.*, **242,** L141 (1980).

13. J. R. Oppenheimer and H. Snyder, *Phys. Rev.* **56,** 455 (1939).

In the above, the left-hand side is the negative of the gravitational potential energy of an object of mass E/c^2; the right-hand side is the inertial energy of a photon of that inertial mass at the surface of the star. According to the arguments of the preceding section, a photon cannot escape the gravitational potential well if the inequality (3.8) is satisfied; it simply has insufficient energy. Since the mass of the photon is proportional to its energy, that condition will hold whatever the photon's frequency.

This simple, "hand waving" argument is only qualitatively correct. Though the famous Schwarzschild solution of the Einstein equations leads to a similar conclusion, the critical radius, also known as the Schwarzschild limit, is given by

$$R_c = \frac{2GM}{c^2} \tag{3.9}$$

rather than (3.8).

Direct observation of a black hole is evidently impossible. Such an object will absorb or swallow all matter or energy that impinges on its surface, and radiation emitted from its surface falls back "down" before it can escape. Nevertheless, the presence of a black hole might be inferred by careful examination of the motion of stars. If a binary pair consisted of one visible star and a black hole, the rotation of the visible companion about the common center of mass, as determined from Doppler shift data, should reveal the presence of the black hole and allow an estimate of its mass. It appears that the X-ray source Cygnus X-1 may be the companion of a black hole. There is also substantial empirical evidence that a very massive black hole sits at the center of our own Milky Way galaxy.[14]

EXAMPLE 3.3 Calculate the density of an object of one solar mass whose radius is the critical Schwarzschild radius. Compare this density with the nuclear density of approximately 2.3×10^{17} kg/m^3.

SOLUTION The mass of the sun is 2×10^{30} kg. The universal gravitational constant is 6.67×10^{-11} N·m^2/kg^2.

From Equation (3.9) the Schwarzschild radius of one solar mass is

$$\begin{aligned} R_c &= \frac{2(2 \times 10^{30}\ \text{kg})(6.67 \times 10^{-11}\ \text{N·m}^2/\text{kg}^2)}{(3 \times 10^8\ \text{m/s})^2} \\ &= 2.96 \times 10^3\ \text{m} \simeq 3\ \text{km} \end{aligned}$$

If the sun collapsed to a sphere of 3-km radius without loss of mass, it would then be a black hole. The mass density would be

$$\rho = \frac{M}{V} = \frac{2 \times 10^{30}\ \text{kg}}{(4\pi/3)(2.96 \times 10^3\ \text{m})^3} = 1.84 \times 10^{19}\ \text{kg/m}^3$$

---

14. C. H. Townes and R. Genzel, *Scientific American*, April 1990, p. 46.

a density nearly 100 times greater than even the enormous nuclear density. It follows that our sun could not collapse to a black hole.

If the initial stellar mass is larger, the density of the black hole need not be as great. Astrophysicists have estimated that only stars whose masses are greater than about five solar masses can collapse to black holes.

Yet another prediction of the general theory concerns the manner whereby the gravitational force is transmitted. Already in 1916, Einstein showed that his equations had wavelike solutions, similar to the wavelike solutions of Maxwell's equations. Just as a time-varying charge distribution excites electromagnetic waves, so a time-varying mass distribution will excite gravitational waves. Both types of wave propagate through vacuum at the speed of light.

About 1960, J. Weber set out to detect gravitational waves using two very large aluminum cylinders as "detectors." Eight years later he galvanized the scientific community with the first report of the detection of gravitational waves. However, subsequent measurements at other laboratories failed to confirm his results, and at present it is generally agreed that Weber's announcement was premature. Nevertheless, there is convincing indirect evidence of gravitational waves.

That evidence comes from precise measurements of the period of the binary pulsar PSR 1913 + 16, a system of two dense objects, one a rapidly rotating neutron star.[15] The two objects are so massive and so close that their orbital period is only eight hours. Such a rotating dumbbell should radiate gravitational waves, and, as a result, slowly lose energy. That energy loss manifests itself as a small increase in the orbital angular velocity. Since the pulsar is a very stable clock, even minute changes in orbital period can be detected. Figure 3.9 shows the orbital phase shift as a function of time over a period of ten years. The solid line is the phase shift calculated using general relativity. There can be little doubt that the binary pulsar loses energy as a result of emission of gravitational waves.

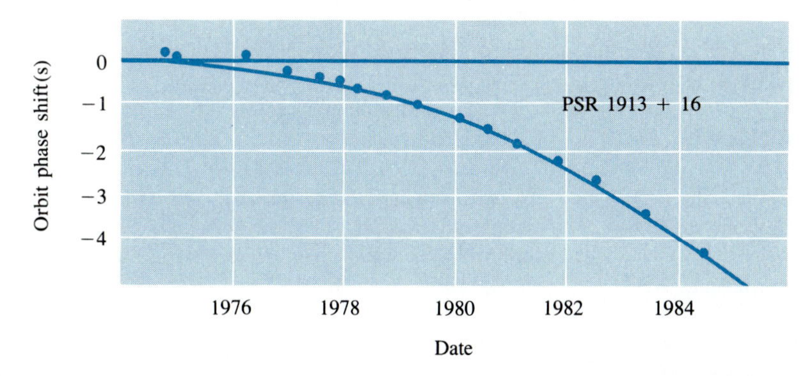

Figure 3.9 The orbital phase shift of the binary pulsar PSR 1913 + 16 as a function of time. The points represent experimental data obtained over one decade. The solid line is the calculated phase shift based on the theory of general relativity. *(After* Modern Physics in America, *American Institute of Physics, 1988, p. 80.)*

---

15. J. H. Taylor and J. M. Weisberg, *Astrophys. J.* **253,** 908 (1982).

SUMMARY

The general theory of relativity is in essence a theory of gravitation.

In a homogeneous gravitational field the results of the general theory can be deduced from the *principle of equivalence:*

> **The effect of a homogeneous gravitational field is equivalent to that of a reference frame in uniform acceleration in a direction opposite to that of the gravitational field.**

The principle of equivalence leads to *gravitational* time dilation and length contraction, effects that are independent of the time dilation and length contraction of special relativity.

In the general theory an inhomogeneous gravitational field is represented by a "curvature" of space-time.

Among the predictions of the general theory are

1. The deviation of light in a gravitational field
2. The gravitational red shift
3. The precession of the perihelion of Mercury
4. The time delay of electromagnetic waves that pass through a gravitational field (Shapiro time delay)
5. The existence of "black holes"
6. Radiation of gravitational waves

Of these predictions, 1, 2, 3, and 4 have been confirmed by direct experimental observation, and 5 and 6 by astronomical evidence.

Bibliography

Peter Bergmann, *Introduction to the Theory of Relativity,* Dover, New York, 1976.

Peter Bergmann, *The Riddle of Gravitation,* New York, 1968.

A. Burrows, "The Birth of Neutron Stars and Black Holes," *Physics Today,* September 1987, p. 28.

F. A. Chaffee, "The Discovery of a Gravitational Lens," *Scientific American,* November 1980, p. 70.

Albert Einstein, *Relativity,* Crown, New York, 1967.

R. Geroch, *General Relativity from A to B,* University of Chicago Press, Chicago, 1978.

Stephen W. Hawking, *A Brief History of Time,* Bantam, New York, 1988.

Stephen W. Hawking and Werner Israel (eds.), *300 Years of Gravitation,* Cambridge University Press, New York, 1987.

R. K. Pathria, *The Theory of Relativity,* Pergamon, New York, 1974.

D. Sciama, *The Unity of the Universe,* Doubleday, New York, 1961.

E. L. Turner, "Gravitational Lenses," *Scientific American,* July 1988, p. 54.

J. M. Weisberg, J. H. Taylor, and L. A. Fowler, "Gravitational Waves from an Orbiting Pulsar," *Scientific American,* October 1981, p. 74.

Clifford M. Will, *Was Einstein Right?,* Basic Books, Inc., New York, 1986.

PROBLEMS

•**3.1** The temperature of the sun's chromosphere is about 5500 K. A hydrogen atom, moving at rms speed, radiates light of 656.2604-nm wavelength.

(a) Calculate the second-order Doppler shift.
(b) Calculate the gravitational red shift. That is, calculate the difference between the wavelength of the radiation detected by a receiver in a laboratory on earth and the wavelength observed in the laboratory using a hydrogen discharge tube.
(c) Compare the results of parts (a) and (b). Could one measure the gravitational red shift reliably using atomic spectral lines emitted in the sun's chromosphere?

3.2 The sun rotates about its axis at a rate of about one revolution every 27 days. Calculate the maximum value of $\delta\lambda/\lambda$ due to the transverse Doppler effect. Compare this Doppler shift with the second-order Doppler shift and the gravitational red shift calculated in Problem 3.1

3.3 A synchronous satellite in orbit above the equator is used to relay microwave transmission from an earth station. To what frequency should the satellite receiver be tuned if the frequency of transmission is exactly 3.0 GHz? (Neglect all Doppler shifts.)

3.4 What is the minimum mass of a star that collapses to a black hole and whose density is the nuclear density? Express your result in solar mass units.

3.5 Calculate the bending of starlight that grazes the sun, using the Newton–von Soldner approach. (See Section 3.5.)

3.6 A quasar is in exact superior conjunction with a dense, spherical globular galaxy. The light from the globular galaxy is red-shifted by 37 percent, and radiation from the quasar is red-shifted by 85 percent. There is an empirical relation between the velocity of recession of an astronomical object and its distance from the solar system, *Hubble's law,*

$$v = Hd$$

where H is Hubble's constant. H has the value 23 km/s·MLy, where MLy stands for "million light-years." Determine the distance from the sun of the globular galaxy and of the quasar.

The mass of the galaxy is 10^{11} solar masses, and its radius is 1×10^4 Ly. Viewed from an observatory on earth, the quasar does not appear as a point source of light but as a ring surrounding the globular galaxy. Explain this observation and calculate the angle subtended by this ring. Assume that the relativistic bending of a photon's trajectory is, in this case, also twice that calculated using the Newton–von Soldner approach.

•**3.7** A light source is placed at the center of a turntable of radius R that rotates at an angular velocity ω. A detector is placed at the rim of the turntable. Obtain an expression for $\Delta\nu/\nu$, where ν is the frequency radiated by the source and $\nu + \Delta\nu$ is the frequency at the detector. Consider first the situation as seen in the reference frame of the source. Next, repeat the derivation by considering the situation in the reference frame of the detector.

•**3.8** The Pound-Rebka experiment was successful because the experimenters maintained careful temperature control at source and receiver. For what temperature difference between source and receiver, assuming both are near room temperature, is the second-order Doppler shift equal to the gravitational red shift associated with an elevation difference of 20 m?

Chapter 4

Roots of the Quantum Theory

Kurz zusammengefasst, kann ich die ganze Tat nur als einen Akt der Verzweiflung bezeichnen.

[*In summary, I can only characterize the entire work as an act of desperation.*]

Max Planck, in a letter to R. W. Wood, July 1931

4.1 INTRODUCTION

Unlike the theory of relativity, the intellectual achievement of a single genius, the quantum theory developed as an intricate mosaic to which, over a period of several decades, a host of scientists made important contributions. Like the theory of relativity, quantum theory not only clarified several unresolved problems but also predicted numerous hitherto unexpected phenomena. Experimental confirmation of these new effects—for example, the Compton effect (see Section 4.6) and tunneling of particles through a classically impenetrable barrier (see Sections 7.8 and 13.4)—converted skeptics to accept the quantum theory even though the interpretation of some of its basic postulates is, even today, a source of controversy.

Experimental results of specific heat measurements on gases and solids were among the earliest unresolved problems of classical physics. Maxwell, a principal architect of the kinetic theory of gases, had pointed out that the experimental heat capacities of diatomic gases ($\frac{5}{2}R$) did not agree with the theory's prediction ($\frac{7}{2}R$), but could find no flaw in the theory. Within a very few years, new anomalies surfaced in the heat capacities of solids.

Early in the nineteenth century, available data suggested that the heat capacities of all solids are equal to $3R = 6$ cal/mol·K, now known as the law of Dulong and Petit, the two French scientists who first recognized that pattern. The simple explanation of this remarkably universal result provided by the kinetic theory was initially hailed as testimony of its validity. However, as the techniques for gas liquefaction improved and heat capacities were measured to ever lower temperatures, each year brought to light more and more departures from the law of Dulong and Petit. Yet, no one could find a flaw in the arguments that led to that law.

There were other unsolved problems, among them the one posed by the spectral lines of the elements. These were first recorded by Wollaston in 1802, and then measured more precisely by Fraunhofer in 1814. Fraunhofer observed not only emission but also absorption lines, that is, sharp, dark lines in the otherwise continuous rainbow spectrum of sunlight, which he labeled A, B, C, D, ... Their origin was suggested by Kirchhoff and Bunsen about fifty years later. They noted the emission of the same Fraunhofer D line (actually a doublet) when ordinary table salt was sprinkled into the flame of a candle. Kirchhoff concluded that these lines

> *always arise from the presence of sodium; the dark D line in the solar spectrum permits us to conclude that sodium is present in the sun's atmosphere.*

Thereafter, atomic spectroscopy became one of the most active fields of experimental physics. Since each element had its own unique spectral signature, and, moreover, since minute quantities could yield resolvable lines, spectroscopy emerged as a valuable analytic tool of chemistry; the identification of spectra became important to the rapidly expanding chemical industry. But the physical process that gave rise to the sharp spectral lines remained an enigma. It seemed natural to assume that somehow each atom might vibrate, like a string, tuning fork, or drumhead, and that the frequencies of the emitted spectral lines corresponded to those of the vibrating systems. However, none of the many models studied ever yielded resonance frequencies in accord with experiment.

In 1885, a Swiss schoolteacher, Johann Balmer, hit upon the following simple expression for the wavelengths of the lines of the hydrogen spectrum:

$$\lambda = 364.6 \frac{m^2}{m^2 - 4} \text{ nm} \qquad (m \text{ an integer} \geq 3) \tag{4.1}$$

Although Equation (4.1) fit the experimental data precisely, neither Balmer nor anyone else could offer an acceptable physical basis for it. Equation (4.1) was just so much numerology; yet, its success was so remarkable that henceforth spectral analysts invariably turned to this or an equivalent equation. The fundamental basis for Balmer's formula was recognized only after Bohr's seminal work on atomic structure.

Curiously, the solution to these and other puzzles came from apparently quite unrelated research, namely, Max Planck's effort, using recent developments in thermodynamics and statistical mechanics, to deduce a theoretical formula for the spectral distribution of blackbody radiation.

4.2 BLACKBODY RADIATION[1]

At any temperature $T > 0$ K, every object radiates electromagnetic waves. The *radiance* R, the power radiated per unit area, depends on the temperature as well as the character of the radiating surface. When radiation strikes an opaque object, some is absorbed and some reflected. Light-colored, shiny surfaces reflect more than do dark, rough surfaces. An object whose surface absorbs all incident radiation is called a *blackbody*. An object in thermal equilibrium with its surroundings radiates as much energy as it absorbs. It follows that a blackbody is a perfect emitter as well as absorber.

The spectral distribution of the emitted radiation, the *spectral radiancy* $R(\lambda)$, is a sensitive function of temperature. Since, by definition, the power radiated per unit

---

1. For a fascinating account of Planck's work, see T. S. Kuhn, *Black-Body Theory and the Quantum Discontinuity, 1894–1912,* Oxford University Press, London, 1978.

area in the wavelength interval $d\lambda$ is $R(\lambda)\ d\lambda$, the radiance and spectral radiancy are related by

$$R = \int R(\lambda)\ d\lambda \tag{4.2}$$

In 1860, Gustav Kirchhoff proved that if a cavity is in equilibrium at temperature T, the intensity distribution of the radiation within that cavity is proportional to the spectral radiancy of a blackbody at the same temperature. Hence, a blackbody radiator is readily constructed by making a small opening in the wall of a hollow enclosure. For such a cavity radiator,

$$R(\lambda) = \left(\frac{c}{4}\right)U(\lambda) \quad \text{or} \quad R(\nu) = \left(\frac{c}{4}\right)U(\nu) \tag{4.3}$$

where $U(\lambda)$ is the intensity distribution (with respect to wavelength) of the radiation within the cavity and $R(\nu)$ and $U(\nu)$ are the corresponding frequency distributions.

In 1884, Ludwig Boltzmann employed thermodynamic arguments to show that

$$R = \sigma T^4 \tag{4.4}$$

This relation is now known as the Stefan-Boltzmann law (Josef Stefan had deduced it empirically). A few years later, Wilhelm Wien extended Boltzmann's analysis and demonstrated that the spectral radiancy must be of the form

$$R(\lambda) = \frac{\phi(\lambda T)}{\lambda^5} \tag{4.5}$$

where $\phi(\lambda T)$ is some arbitrary function of the product of wavelength and absolute temperature. The formula

$$R(\lambda) = B\lambda^{-5}e^{-b/\lambda T} \tag{4.6}$$

and its equivalent in terms of frequency,

$$R(\nu) = A\nu^3 e^{-g\nu/T} \tag{4.7}$$

known as the *Wien distribution law,* seemed to fit the data quite well. However, in a few years, more precise experiments at infrared frequencies revealed significant departures from Wien's law.

In October 1900, Max Planck, who had always been keenly interested in thermodynamics, suggested the relation

$$R(\nu) = \frac{C\nu^3}{e^{g\nu/T} - 1} \tag{4.8}$$

with the modest statement that

> *so far as I can see by quick inspection, [it] represents the hitherto published observational data just as satisfactorily as the best previously proposed distri-*

Figure 4.1 Max Planck (1858–1947). *(Maison Albert Schweitzer/Courtesy AIP, Neils Bohr Library.)*

bution function. . . . I, therefore, feel justified in directing attention to this new formula which . . . I take to be the simplest excepting Wien's.[2]

In fact, Planck's formula was not only "as satisfactory" but clearly superior to others. Like Balmer's formula, it was at that point an empirical rule, unsupported by theory. As Planck recalled, the task of justifying his formula called for "a few weeks of the most strenuous work of my life."

Planck regarded the blackbody cavity as an ensemble of oscillators whose frequencies ranged over the entire spectrum. What he discovered in those weeks of strenuous work was that, in performing the requisite summations of the statistical treatment, the energy of an oscillator could not be taken as a continuous function:

We, however, consider, and this is the essential point, E to be composed of a determinate number of finite parts,

proportional to the oscillator frequency, ν. In short, Planck was forced to assume that the energy of an oscillator must take on the value

$$\epsilon_n = nh\nu \tag{4.9}$$

where n is an integer and h is *Planck's constant*

$$h = 6.626 \times 10^{-34}\ \text{J·s} = 4.136 \times 10^{-15}\ \text{eV·s} \tag{4.10}$$

Planck himself was very uncomfortable with this radical departure from the conventional continuum of classical physics and years later referred to his assumption as an "act of desperation [ein Akt der Verzweiflung]." For a few years he tried, in vain, to fit this mathematical stricture on the energy intervals into a classical mold. Then, in 1905, Einstein and Ehrenfest demonstrated that quantization of harmonic oscillator energy levels, expressed by Equation (4.9), was a necessary consequence of Planck's radiation law, and Planck, almost ruefully, reconciled himself with his own work.

In the following section we demonstrate how Equation (4.9) leads to the Planck distribution, Equation (4.8).

4.2(a) Derivation of the Planck Distribution Law

Our aim is to obtain an expression for $R(\nu)$. Since, except for the factor $c/4$, $R(\nu)$ is identical to $U(\nu)$, the intensity distribution inside a cavity, we focus our attention on $U(\nu)$.

Imagine a cavity in the shape of a rectangular parallelepiped with reflecting metallic walls. At the walls, the electric field $\mathbf{E}$ must vanish, and therefore the radiation within the cavity will be an ensemble of standing waves that satisfy the boundary condition $\mathbf{E} = 0$ at the walls. A standing wave is a superposition of two traveling waves, propagating in opposite directions.

2. M. Planck, *Verh. d. D. Phys. Ges.* **2,** 237 (1900).

We represent a traveling wave in the form

$$\mathbf{E} = \mathbf{E}_0 e^{i(\omega t - \mathbf{k}\cdot\mathbf{r})} \tag{4.11}$$

where $\mathbf{k}$ is the *wave vector* whose components are

$$k_x = \frac{2\pi}{\lambda_x} \qquad k_y = \frac{2\pi}{\lambda_y} \qquad k_z = \frac{2\pi}{\lambda_z} \tag{4.12}$$

The expression for a standing wave is then

$$\mathbf{E} = \mathbf{E}_0 e^{i\omega t}(e^{i\mathbf{k}\cdot\mathbf{r}} - e^{-i\mathbf{k}\cdot\mathbf{r}}) \tag{4.13}$$

To satisfy the boundary conditions, the wavelengths of the traveling waves must be

$$\lambda_x = \frac{2L_x}{n_x} \qquad \lambda_y = \frac{2L_y}{n_y} \qquad \lambda_z = \frac{2L_z}{n_z} \tag{4.14}$$

where L_x, L_y, L_z are the lengths of the cavity and n_x, n_y, n_z are positive integers. Consequently, the wave vectors must have components

$$k_x = \left(\frac{\pi}{L_x}\right)n_x \qquad k_y = \left(\frac{\pi}{L_y}\right)n_y \qquad k_z = \left(\frac{\pi}{L_z}\right)n_z \tag{4.15}$$

To derive the Planck distribution we need to know the number of possible modes in the frequency interval between ν and $\nu + d\nu$. For electromagnetic waves, $\nu = c/\lambda = 2\pi ck$, and so we seek the number of standing waves whose wave vectors fall in the interval between k and $k + dk$. When counting the number of standing waves, subject to condition (4.15), we must remember that n_x, n_y, and n_z have to be *positive* integers to avoid counting the same standing wave twice.

We proceed by first calculating the number of standing waves whose wave vector is less than some value k. The vector of magnitude k defines a sphere in this $\mathbf{k}$-space, a vector space whose Cartesian coordinates are k_x, k_y, k_z. The volume of that sphere is $\frac{4}{3}\pi k^3$. Now the number of wave vectors that terminate in the volume element $dk_x\, dk_y\, dk_z$ is, from Equation (4.15),

$$dn_x\, dn_y\, dn_z = \frac{L_x L_y L_z}{\pi^3}\, dk_x\, dk_y\, dk_z = \left(\frac{V}{\pi^3}\right) dk_x\, dk_y\, dk_z$$

In other words, the density of wave vectors in $\mathbf{k}$-space is V/π^3.

For each value of $\mathbf{k}$ there are two independent modes, corresponding to the two independent directions of polarization of an electromagnetic wave. Consequently, the number of modes in the positive octant (positive n_x, n_y, n_z) of the sphere of radius k is

$$N(k) = 2\left(\frac{1}{8}\right)\left(\frac{4\pi k^3}{3}\right)\left(\frac{V}{\pi^3}\right) = \left(\frac{V}{3\pi^2}\right)k^3 \tag{4.16}$$

The number of modes between k and $k + dk$ is obtained by differentiating $N(k)$. We call this the *density of states* in **k**-space and use the symbol $\mathcal{N}(k)$. Thus

$$\mathcal{N}(k)\, dk = \left(\frac{V}{\pi^2}\right) k^2\, dk \qquad \textbf{(4.17)}$$

is the number of modes in the interval between k and $k + dk$.

Since $k = 2\pi/\lambda = 2\pi\nu/c$, the number of modes per unit volume in the frequency interval between ν and $\nu + d\nu$ is

$$g(\nu)\, d\nu = \frac{8\pi}{c^3}\, \nu^2\, d\nu \qquad \textbf{(4.18)}$$

The energy density in the cavity as a function of frequency is now given by the product of $g(\nu)$ and the average energy of an electromagnetic mode of frequency ν. The probability that at a temperature T a mode of energy ϵ is excited is

$$f(\epsilon) = Ce^{-\epsilon/kT} \qquad \textbf{(4.19)}$$

where $C = 1/kT$ is a normalization constant (see Problem 4.3) and $e^{-\epsilon/kT}$ is the Boltzmann factor. If the energy ϵ were a continuous function, as assumed in classical physics, the average energy of a mode would be

$$\langle\epsilon\rangle_{\text{cl}} = \int_0^\infty \epsilon f(\epsilon)\, d\epsilon = kT \qquad \textbf{(4.20)}$$

If we now multiply $\langle\epsilon\rangle_{\text{cl}}$ by $g(\nu)$, we arrive at the expression first derived by Rayleigh and Jeans,

$$U(\nu) = \left(\frac{8\pi}{c^3}\right)\nu^2 kT \qquad \textbf{(4.21)}$$

Equation (4.21), proposed actually a year or two after Planck's formula, cannot possibly be correct. The difficulty, recognized by Rayleigh and Jeans, is that according to Equation (4.21) the energy density increases without limit as the frequency increases. That is, $U(\nu)$ diverges as $\nu \rightarrow \infty$. This "unphysical" consequence of Equation (4.21) came to be known as the *ultraviolet catastrophe*.

Two points are noteworthy. First, the reason no such disaster befalls the Wien distribution is that it simply asserts, a priori, that the distribution is a universal function of the form $\phi(\lambda T) = \phi'(T/\nu)$, specifically, the exponential $e^{-g\nu/T}$, which approaches zero faster than any power of ν. The corresponding function of the Rayleigh-Jeans expression is T/ν, and this does not fall off rapidly enough with increasing ν to prevent the divergence as $\nu \rightarrow \infty$.

Second, although the Rayleigh-Jeans law is clearly incorrect, it does approach the experimental data more closely in the long-wavelength, low-frequency region than does Wien's expression. Therefore, a satisfactory distribution law should approach the Wien distribution at high frequencies and the Rayleigh-Jeans law at low frequencies. Equation (4.8) does, indeed, meet those conditions (see Problem 4.7).

If, as proposed by Planck, the allowed energy is not a continuous variable, the

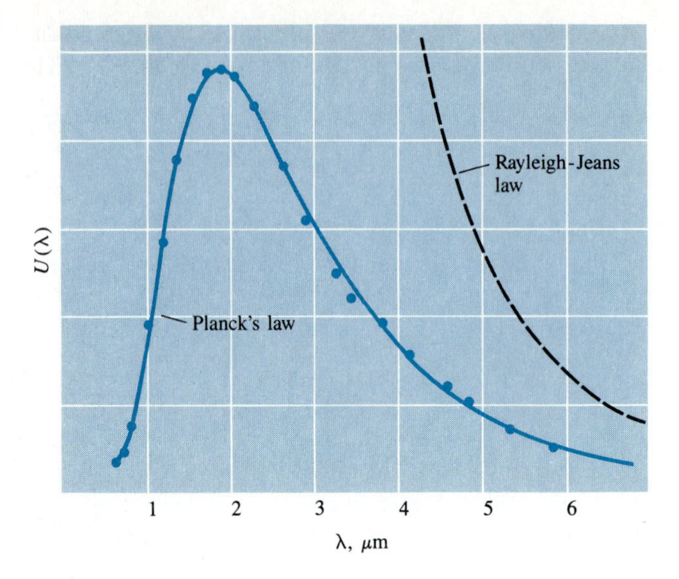

Figure 4.2 The spectral distribution of blackbody radiation. The Planck law is shown in color, the classical Rayleigh-Jeans law in black.

integral form for the average energy, Equation (4.20), is inapplicable. Instead, we should perform a summation over the discrete energy states given by Equation (4.9). The average energy is then

$$\langle\epsilon\rangle = \sum \epsilon_n f_n \tag{4.22}$$

where $\epsilon_n = nh\nu$

and
$$f_n = C'e^{-\epsilon/kT} = C'e^{-nh\nu/kT} \tag{4.23}$$

The normalization constant C' is now obtained from the condition

$$\sum f_n = 1 \tag{4.24}$$

To evaluate that sum we note that we can write

$$\sum f_n = C' \sum e^{-nx} = C'[1 + e^{-x} + (e^{-x})^2 + (e^{-x})^3 + \cdots]$$
$$= \frac{C'}{1 - e^{-x}}$$

where we have set $x = h\nu/kT$ and have made use of the binomial expansion of $(1 + a)^{-1}$. Thus $C' = (1 - e^{-h\nu/kT})$.

To evaluate $\langle\epsilon\rangle$ we again write out the summation terms:

$$\langle\epsilon\rangle = \sum \epsilon_n f_n = (1 - e^{-h\nu/kT})h\nu\sum ne^{-nx}$$
$$= (1 - e^{-h\nu/kT})h\nu[e^{-x} + 2(e^{-x})^2 + 3(e^{-x})^3 + \cdots]$$
$$= (1 - e^{-h\nu/kT})h\nu\left\{-\frac{d}{dx}[e^{-x} + (e^{-x})^2 + (e^{-x})^3 + \cdots]\right\}$$
$$= (1 - e^{-h\nu/kT})h\nu\left[\frac{d}{dx}(1 - e^{-x})^{-1}\right] = \frac{h\nu}{e^{h\nu/kT} - 1} \tag{4.25}$$

On multiplying Equation (4.25) by the density of states, $g(\nu)$, we obtain the Planck distribution

$$U(\nu) = \frac{8\pi\nu^2}{c^3}\left(\frac{h\nu}{e^{h\nu/kT} - 1}\right) \tag{4.26}$$

The excellent agreement of Equation (4.26) with experiment is, of course, the best argument in favor of Planck's formula and of the fundamental assumption that is the basis of that expression. We should, however, probe a little deeper and understand, on physical grounds, why the assumption of discrete energy levels removes the ultraviolet catastrophe and yet gives a result that also agrees with the Rayleigh-Jeans formula at low frequencies.

If ν is small so that $h\nu \ll kT$, quantization of the energy levels is of no consequence, because the number of energy levels contained within an energy range of the order kT is very large; the summation is then closely approximated by an integration over an energy continuum. The result of that integration is the classical Rayleigh-Jeans formula.

On the other hand, if $h\nu \gg kT$, the finite spacing between energy levels assumes crucial significance. For instance, if $h\nu = 5kT$, the probability that the system is excited to the state $n = 1$ compared with the probability that the system is in the ground state, $n = 0$, is given by the Boltzmann factor $e^{-5} \simeq \frac{1}{150}$. In other words, the minimum excitation energy of the radiation field at higher frequencies is so great that the probability for excitation of the system by thermal agitation is minimal. At constant temperature, that excitation probability diminishes exponentially with increasing frequency, and, consequently, the likelihood of excitation of high-frequency modes falls to zero as $\nu \rightarrow \infty$: the ultraviolet catastrophe is obviated.

Similar arguments can be employed in numerous instances to assess the probable importance of quantization. In general, quantum effects are negligible whenever $h\nu \ll kT$, where ν is a characteristic frequency or $\epsilon = h\nu$ is a characteristic energy of the system. Conversely, when $h\nu \gg kT$ one should expect characteristic quantum phenomena.

Alternatively, classical physics may be viewed as quantum physics in the limit $h \rightarrow 0$. For example, if we allow h to approach zero, the summations of Equations (4.22) and (4.23) may then be replaced by integrals, and we again obtain the Rayleigh-Jeans law. It is interesting in this connection to recall Jeans's appraisal of his own and Planck's work:

> *The methods of both are in effect the methods of statistical mechanics and of the equipartition of energy, but I carry the method further than Planck, since Planck stops short of putting* $h = 0$. *I venture to express the opinion that it is not legitimate to stop short at this point, as the hypotheses upon which Planck has worked lead to the relation* $h = 0$ *as a necessary consequence. Of course, I am aware that Planck's law is in good agreement with experiment if* h *is given a value different from zero, while my own law, obtained by putting* $h = 0$, *cannot possibly agree with experiment. This does not alter my belief that the value* $h = 0$ *is the only value which it is possible to take, my view being that the supposition that the energy of the ether is in equilibrium with that of matter is utterly erroneous in the case of ether vibrations of short wavelength under experimental conditions.**

*J. H. Jeans, *Nature,* **72,** 294 (1905).

Like many of his contemporaries, Jeans, rather than abandon his conviction in classical continuity, chose to deny that a system can achieve true thermal equilibrium with its surroundings. It is difficult to comprehend how, under such circumstances, one could ever apply thermodynamic arguments, as Maxwell, Boltzmann, Gibbs, and others had done with consummate skill and remarkable success.

EXAMPLE 4.1 Derive Stefan's law, Equation (4.4), from Planck's law. Using that result, obtain an expression for the Stefan-Boltzmann constant σ in terms of known physical constants and determine its value.

SOLUTION The radiance R is the integral of the spectral radiancy $R(\nu)$ over all frequencies. From Equation (4.26), and recalling that $R(\nu) = (c/4)U(\nu)$, we obtain

$$R = \frac{c}{4}\int_0^\infty \left(\frac{8\pi\nu^2}{c^3}\right)\left(\frac{h\nu}{e^{h\nu/kT} - 1}\right) d\nu$$

We now follow the familiar procedure of changing the variable from ν to $x = h\nu/kT$. Replacing ν by kTx/h and $d\nu$ by $(kT/h)\ dx$, the radiance is then

$$R = \frac{2\pi k^4 T^4}{c^2 h^3}\int_0^\infty \left(\frac{x^3}{e^x - 1}\right) dx \qquad \textbf{(4.27)}$$

The definite integral in Equation (4.27) has the value $\pi^4/15$. Hence,

$$R = \frac{2\pi^5 k^4}{15c^2h^3}T^4 = \sigma T^4 \quad \text{where} \quad \sigma = \frac{2\pi^5 k^4}{15c^2h^3} = 5.67 \times 10^{-8}\ \mathrm{W/m^2\cdot K^4} \qquad \textbf{(4.28)}$$

At the beginning of the twentieth century, the Stefan-Boltzmann constant, as well as c and k, was known to fair precision. Planck's initial calculation of the constant h was based on Equation (4.28).

4.3 SPECIFIC HEAT

4.3(a) The Specific Heat of Crystals

The low-temperature departure of the specific heat of crystalline solids from the Dulong-Petit value of $3R$ also finds a ready explanation within the framework of Planck's hypothesis. The seminal work on this, as on many other problems in quantum physics, was done by Einstein.[3] The importance of his contribution is not so much in the calculation, which, as we shall see presently, poses no real difficulties, but in the pivotal proposition that quantization applies with equal force to any oscillating system, not just electromagnetic radiation. If correct, the application of classical dynamics to any system whose characteristic vibrational energy is less than

---

3. A. Einstein, *Ann. d. Phys.* **22,** 180 (1907).

kT must be abandoned. For instance, kinetic theory, which only a decade earlier had furnished some of the most cogent arguments in support of the atomic hypothesis, would have to be critically reexamined.

It is, therefore, not altogether surprising that it was this work on specific heats that first brought Einstein to the attention of scientists outside the very small group of theoretical physicists who recognized the significance and brilliance of his 1905 papers on relativity. In 1906, Walther Nernst, a respected physical chemist, was engaged in measurements of specific heats, especially at low temperatures, in an effort to confirm his own thermodynamic theories. Nernst soon became aware of Einstein's paper on heat capacities of solids, and in 1909, shortly after Einstein had assumed a junior faculty position at the University of Zurich, Nernst paid the young theorist a visit so that they could discuss problems of common interest. The chemist George Hevesy, who was then an assistant in Zurich, recalls that among his colleagues it was this visit by Nernst that raised Einstein's reputation. He had come "as an unknown man to Zurich. Then Nernst came, and the people in Zurich said, 'This Einstein must be a clever fellow, if the great Nernst comes so far from Berlin to Zurich to talk to him.'"

Figure 4.3 Walther Nernst (1864–1941). *(Francis Simon Collection/ Courtesy AIP, Niels Bohr Library.)*

Einstein's model of a crystal was the same as that used earlier to explain the law of Dulong and Petit. Each atom is assumed to vibrate with the same frequency ν about its equilibrium position in the lattice. Since each atom can vibrate independently along three orthogonal directions, its average vibrational energy is just three times the result given by Equation (4.25). Consequently, the total internal energy per mole is

$$U = N_A\langle\epsilon\rangle = N_A\frac{3h\nu}{e^{h\nu/kT} - 1} \tag{4.29}$$

The molar heat capacity at constant volume is the derivative of Equation (4.29) with respect to temperature:

$$c_v = 3R\frac{x^2e^x}{(e^x - 1)^2} \tag{4.30}$$

Here $x = h\nu/kT = \Theta_E/T$, where Θ_E is the *Einstein temperature*.

A comparison of Equation (4.30) with then available data on diamond is shown in Figure 4.4, page 74. Although the fit is not perfect, quantization of the vibrational energy levels evidently accounts for the approach of c_v to zero as T approaches zero. It is noteworthy that, as in the case of blackbody radiation, here again quantum effects become prominent when $h\nu > kT$, that is, when $T < \Theta_E$; when $T \gg \Theta_E$, Equation (4.30) reduces to the classical result, $c_v = 3R$.

As is apparent from Figure 4.4, the heat capacity does not fall to zero exponentially as predicted by Equation (4.30) but more slowly. The reason for this discrepancy between theory and experiment, recognized by Einstein, lies in the assumption that the atoms of the crystal behave as independent oscillators, each vibrating with the same frequency $\nu_E = k\Theta_E/h$. A few years later, Debye refined this calculation, allowing for a range of vibrational frequencies from $\nu = 0$ to a maximum frequency, the *Debye frequency* ν_D. The reason for the finite cutoff frequency in this case, in contrast to the infinite number of vibrational modes of the blackbody cavity, is that the total number of vibrational modes cannot exceed the number of

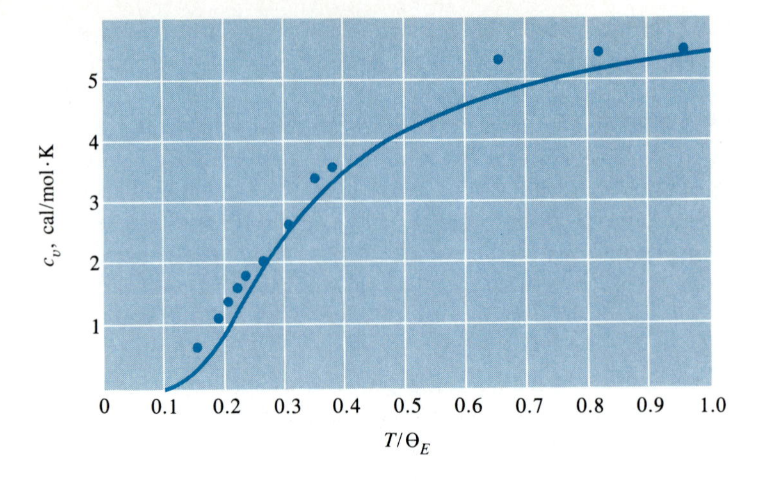

Figure 4.4 Comparison of the molar heat capacity of diamond with the Einstein formula, Equation (4.30). The value of Θ_E was chosen by fitting one point of the experimental curve.

degrees of freedom of the constituent atoms of the crystal; this condition restricts the range of possible frequencies. Debye's expression for the crystal's internal energy is identical to Equation (4.27)—without the factor ($c/4$)—except that the integration is not between zero and infinity but extends from zero to ν_D. Again substituting $x = h\nu/kT$, one obtains

$$U = 9N_A kT\left(\frac{T}{\Theta_D}\right)^3 \int_0^{\Theta_D/T} \frac{x^3}{e^x - 1}\, dx \tag{4.31}$$

where

$$\Theta_D = \frac{h\nu_D}{k}$$

is the *Debye temperature* of the crystal.

As before, we obtain the heat capacity by differentiating Equation (4.31) with respect to T, with the result

$$c_v = 9R\left(\frac{T}{\Theta_D}\right)^3 \int_0^{\Theta_D/T} \frac{x^4 e^x}{(e^x - 1)^2}\, dx \tag{4.32}$$

The integral in Equation (4.32) cannot be evaluated in closed form. However, as $T \rightarrow 0$, Θ_D/T approaches infinity, and the integral may then be approximated by extending the upper limit to ∞. The resulting definite integral has the value $4\pi^4/15$, and, therefore, the specific heat of the crystal at low temperatures is

$$c_v = \frac{12\pi^4}{5} R\left(\frac{T}{\Theta_D}\right)^3 \qquad T \ll \Theta_D \tag{4.33}$$

This T^3 dependence of the low-temperature heat capacity is in excellent agreement with experiment, as illustrated in Figure 4.5.

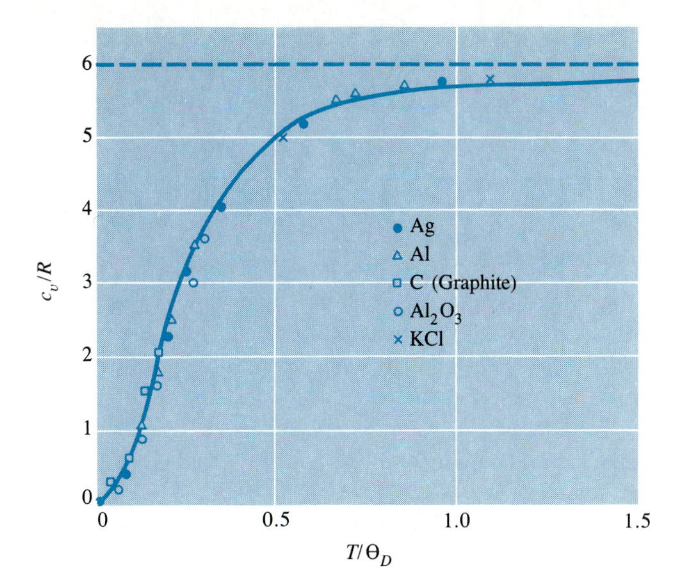

Figure 4.5 The molar heat capacities of several solids as functions of T/Θ_D. The Debye temperature was adjusted for optimum fit to the experimental data. Note that a single universal curve gives an excellent fit to the experimental data.

EXAMPLE 4.2 The Debye temperatures of copper and silver are 309 K and 215 K. What are the Debye frequencies of copper and silver, and what might one expect for the Debye temperature of gold? Note that copper, silver, and gold have identical crystal structures and are very similar in many of their physical properties.

SOLUTION From Equation (4.31),

$$\nu_D(\text{Cu}) = \frac{k\Theta_D(\text{Cu})}{h} = \frac{(1.38 \times 10^{-23}\ \text{J/K})(309\ \text{K})}{6.626 \times 10^{-34}\ \text{J}\cdot\text{s}} = 6.44 \times 10^{12}\ \text{Hz}$$

and

$$\nu_D(\text{Ag}) = \nu_D(\text{Cu}) \left[\frac{\Theta_D(\text{Ag})}{\Theta_D(\text{Cu})}\right] = (6.44 \times 10^{12}\ \text{Hz})\left(\frac{215}{309}\right) = 4.48 \times 10^{12}\ \text{Hz}$$

Since the chemical and physical properties of these two metals are so similar, it is reasonable to expect that the interatomic forces between atoms of copper and of silver might be nearly the same. If that is true, it then follows that the vibrational frequencies of these quasi mass-spring systems should be inversely proportional to the square root of the atomic masses. That is, we anticipate that

$$\frac{\nu_D(\text{Cu})}{\nu_D(\text{Ag})} \simeq \sqrt{\frac{M(\text{Ag})}{M(\text{Cu})}} = \sqrt{\frac{108}{63.5}} = 1.3$$

The ratio of the Debye frequencies is 1.44, which is moderately close to our estimate, based on rather broad assumptions. Proceeding in the same manner, we would estimate that the Debye temperature of gold should be

$$\Theta_D(\text{Au}) \simeq \Theta_D(\text{Ag})\sqrt{\tfrac{108}{197}} = (215\text{ K})(0.741) = 159\text{ K}$$

The measured Debye temperature of gold is 164 K, in fair agreement with this estimate.

4.3(b) Specific Heat of Gases

At the beginning of the chapter we remarked that kinetic theory predicts a heat capacity of diatomic gases of $\frac{7}{2}R$ whereas experimental values are $\frac{5}{2}R$. The reason for this discrepancy is that the vibrational frequency of the equivalent mass-and-spring system is generally so high that $h\nu \gg kT$ near room temperature. The vibrational modes of the molecule are, therefore, only very weakly excited at all but very elevated temperatures and consequently do not contribute significantly to the average energy and heat capacity of the gas.

In fact, not only the vibrational but also the rotational energy of a diatomic molecule is quantized. We shall examine rotational quantization in detail in Chapter 10; here we only point out that, on reducing the temperature sufficiently, even the rotational modes may be "frozen" and may fail to contribute to the heat capacity. For most diatomic gases liquefaction precedes this quenching of rotational energy, but in H_2 one can see the contributions of rotational and vibrational modes quite clearly (Figure 4.6).

We see that energy quantization is a pervasive phenomenon. It applies to electromagnetic radiation, to the vibrational and rotational motion of atoms and molecules, and, as we shall see later, to atomic and nuclear energies, to the energy of a magnet in a magnetic field, and even to the magnetic field itself. Although in nearly all instances this quantization plays an essential role only at the level of atomic events, the consequences of quantization reveal themselves in drastic departures of macroscopic phenomena from the predictions of classical physics. Blackbody radiation and the heat capacities of gases and solids are but a few of many examples. In some instances, quantum effects result in bizarre, new phenomena, such as super-

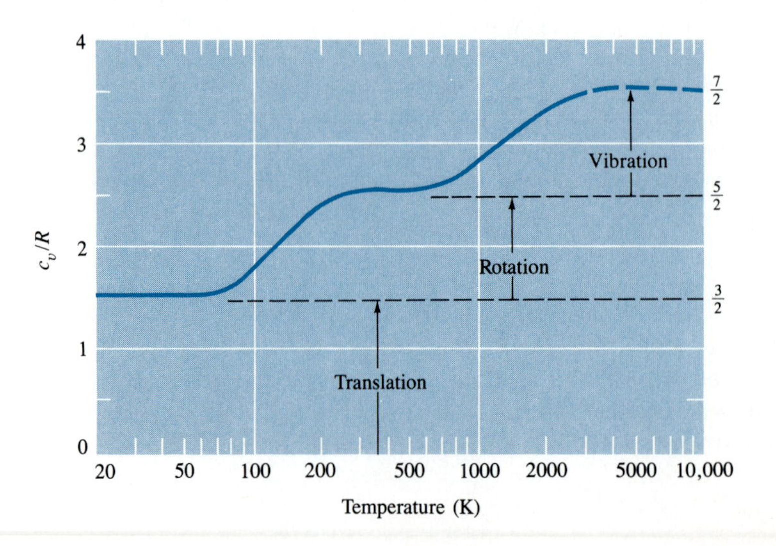

Figure 4.6 The molar heat capacity of H_2 as a function of temperature. For $T < 100$ K, the rotational degrees of freedom do not contribute to the heat capacity; the vibrational modes contribute to the heat capacity only for $T > 1000$ K.

conductivity and the superfluidity of liquid helium, that are quite outside the realm of classical physics.

4.4 THE PHOTOELECTRIC EFFECT

In 1887, Heinrich Hertz succeeded in generating and detecting the electromagnetic waves predicted more than a decade earlier by James Clerk Maxwell. Hertz's paper remains a model for experimenters: he not only showed how to produce and observe the presence of these waves, but also reported on measurements of numerous wave-like properties such as polarization, reflection, refraction, constructive and destructive interference, and the formation of standing waves. Hertz was such a careful and meticulous experimenter that he even took note of an effect that, ironically, was the seed of a modern version of Newton's corpuscular theory of light, a theory which these experiments had presumably put to rest forever. Hertz noted that if light from the initiating spark discharge was allowed to fall on the detecting spark gap, a spark jumped across that gap more readily than if the light was shielded from it.

One of Hertz's students, Phillipp Lenard, pursued that phenomenon further. His results and those of others demonstrated that

1. Electrons are emitted from an illuminated surface.
2. Emission of these *photoelectrons* is practically instantaneous, no matter how weak the intensity of the incident light.
3. There is a threshold frequency for photoelectric emission that depends on the material that is illuminated. If the frequency of the incident radiation is less than the critical value ν_c, no photoelectrons are released no matter how intense the illumination.
4. If the frequency of the incident light is greater than ν_c, the kinetic energy of the emitted electrons spans a finite range, up to a maximum value that is a linear function of the frequency of the incident light.
5. Provided that $\nu > \nu_c$, the number of electrons emitted per second is proportional to the intensity of the incident radiation.

Of these, only the last, the proportionality between light intensity and the number of emitted electrons, was amenable to explanation. The others defied comprehension. For example, order-of-magnitude estimates predicted that under weak illumination of about one milliwatt per square meter, an interval of more than an hour would be required before electrons near the surface of a metal can gain enough energy from the radiation field to escape. Also, the well-defined frequency threshold for photoemission remained mysterious.

Once again, the crucial insight in resolving these puzzles came from Einstein. His paper on the generation and absorption of light makes the fundamental assumption that

> *radiant energy from a point source is not distributed continuously throughout an increasingly larger region, but, instead, this energy consists of a finite number of spatially localized energy quanta which, moving without subdividing, can only be absorbed and created as whole units.*[4]

---

4. A. Einstein, *Ann. d. Phys.* **20,** 199 (1906).

These whole units we now call *photons*. Each photon carries an energy given by the Planck relation

$$\epsilon = h\nu \tag{4.34}$$

The reason one does not normally perceive this quantized, quasi-corpuscular nature of light is that the number of photons striking even a small detecting region, such as the pupil of the eye, is enormous. Hence the "finite number of spatially localized energy quanta" is so great that these appear as a continuous stream, as do the individual molecules in a jet of gas. At extremely low intensities, however, this quantization is readily observed using modern, sensitive photon counters.

EXAMPLE 4.3 A monochromatic point source radiates light of 500-nm wavelength at a rate of 20 mW uniformly in all directions. How many photons will enter the 2-mm-diameter pupil of an observer stationed 2 km from the source?

SOLUTION The energy of a photon of wavelength of 500 nm is

$$\epsilon = h\nu = \frac{hc}{\lambda} = \frac{(6.63 \times 10^{-34}\ \text{J}\cdot\text{s})(3 \times 10^{8}\ \text{m/s})}{5.00 \times 10^{-7}\ \text{m}} = 4.0 \times 10^{-19}\ \text{J}$$

Since the source radiates 2×10^{-2} J/s, the number of photons emitted per second is

$$N = \frac{2 \times 10^{-2}\ \text{J/s}}{4.0 \times 10^{-19}\ \text{J}} = 5 \times 10^{16}\ \text{photons/s}$$

The number of photons that pass through a hole of 2 mm diameter at a distance of 2 km is equal to the total number emitted times the ratio of the area of a hole of 2 mm diameter to the area of a sphere of 2 km radius. Consequently, the number of photons that enter the observer's eye per second is

$$n = \frac{(5 \times 10^{16}\ \text{photons/s})(\pi \times 10^{-6}\ \text{m}^2)}{16\pi \times 10^{6}\ \text{m}^2} = 3.1 \times 10^{3}\ \text{photons/s}$$

Thus, even at this great distance from a very weak source, the number of photons that enter the eye is about 3000 per second. The eye perceives the individual photons as a continuous stream.

Photoelectric emission is readily understood in terms of absorption of photons by the emitting material. As postulated, photons can be absorbed only as "whole units." All of a photon's energy must by transferred to the electron. If some minimum energy is required to release an electron, the threshold frequency for photoemission is then a direct measure of this minimum energy, the so-called *work function* ϕ. If $h\nu < \phi$, no photoemission can occur no matter how intense the radiation, that is, no matter how many photons strike the emitting surface.

If $h\nu > \phi$, only a portion of the photon's energy is used to liberate the electron. The remaining energy now appears as kinetic energy of the photoelectron. Of course, not all electrons that absorb a photon are necessarily in the highest occupied energy state within the emitter, and the kinetic energy of photoelectrons therefore spans a finite range. However, the maximum kinetic energy of the emitted electrons is the difference between the energy of the absorbed photon and that needed to surmount the energy barrier at the surface. Therefore,

$$\mathrm{KE}_{\max} = h\nu - \phi \tag{4.35}$$

This kinetic energy maximum is experimentally determined by measuring the retarding voltage V_0 that must be applied across the phototube to prevent any current flow when the cathode is illuminated. One then arrives at the Einstein equation

$$eV_0 = h\nu - \phi \tag{4.36}$$

According to Equation (4.36) a plot of V_0 versus ν should be a straight line of slope e/h, intersecting the ordinate at $-\phi$ eV. Experiments designed to verify the Einstein equation were first carried out by Millikan about a decade later, and these provided one of the best early determinations of Planck's constant.[5]

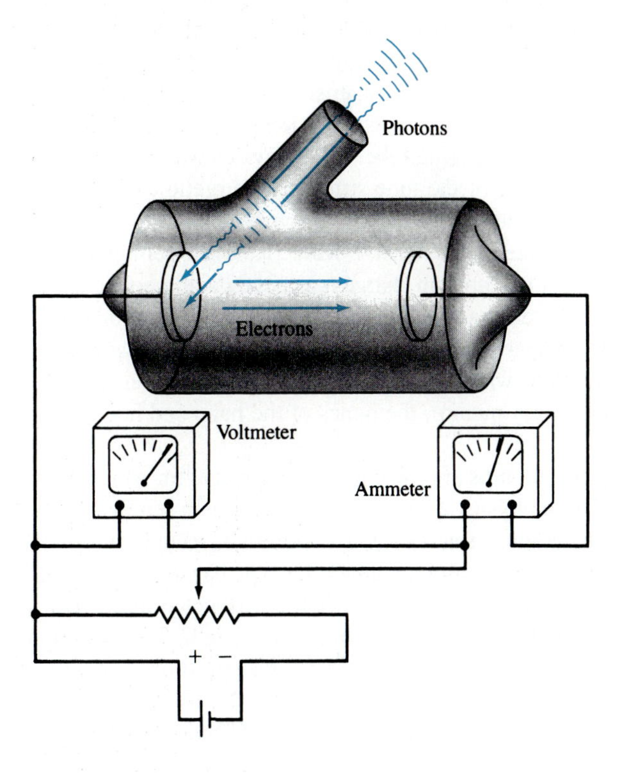

Figure 4.7 Schematic diagram for measuring the critical stopping potential V_0 in the photoelectric effect.

5. R. A. Millikan, *Phys. Rev.* **7,** 355 (1916). The 1923 Nobel Prize was awarded to Millikan in part for his work on the photoelectric effect.

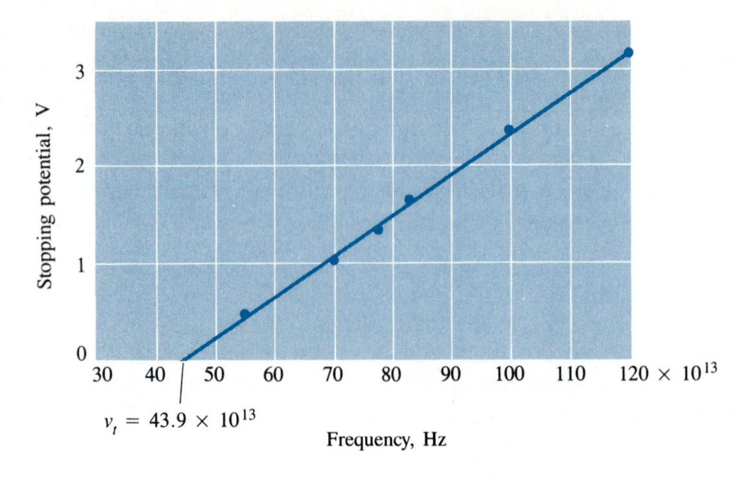

Figure 4.8 Critical stopping potential versus frequency. The data shown here were obtained by Millikan, *Phys. Rev.* **7**, 362 (1916).

4.5 X RAYS

Figure 4.9 Wilhelm Konrad Röntgen (1845–1923). *(Courtesy AIP, Niels Bohr Library.)*

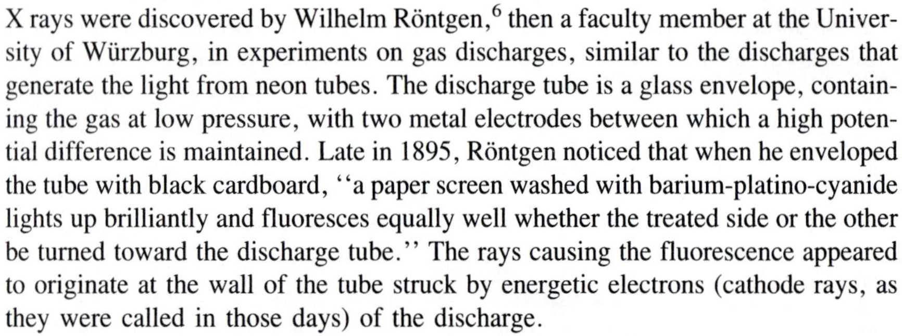

X rays were discovered by Wilhelm Röntgen,[6] then a faculty member at the University of Würzburg, in experiments on gas discharges, similar to the discharges that generate the light from neon tubes. The discharge tube is a glass envelope, containing the gas at low pressure, with two metal electrodes between which a high potential difference is maintained. Late in 1895, Röntgen noticed that when he enveloped the tube with black cardboard, ''a paper screen washed with barium-platino-cyanide lights up brilliantly and fluoresces equally well whether the treated side or the other be turned toward the discharge tube.'' The rays causing the fluorescence appeared to originate at the wall of the tube struck by energetic electrons (cathode rays, as they were called in those days) of the discharge.

Within a few weeks of intensive work, Röntgen demonstrated that these rays fog photographic film; carry no charge, since they could not be deflected by electric or magnetic fields; discharge electrified objects; and are produced whenever energetic electrons strike a solid object. Last, Röntgen had his wife place her hand over a photographic plate while he energized the discharge, and thereby produced the first anatomical X-ray picture, the bones of the hand and her wedding band clearly visible.

Less than two months after Röntgen's discovery, for which he was awarded the first Nobel Prize in physics, physicians used X rays in setting broken limbs. Unfortunately, the dangerous effects of X rays were not recognized for some years, and people often exposed themselves unwittingly to hazardous levels of radiation.

Since these rays were uncharged, it was suspected quite early on that X rays were electromagnetic waves of very short wavelength. This was demonstrated conclusively in 1912 when, following an inspired suggestion by Max von Laue, Friedrich and Knipping demonstrated diffraction of X rays by crystals. The regularly spaced planes of atoms in a crystal serve as a diffraction grating (see Section 12.2), and from the known spacing between crystal planes, the wavelength of the incident radiation could be calculated. The experiments yielded wavelengths between about 0.01 and 0.05 nm. Modern machines produce X rays between 1.0 and 0.001 nm in wavelength.

---

6. Röntgen was the first Nobel laureate in physics (1901). In Germany, X rays are known as *Röntgen Strahlen*.

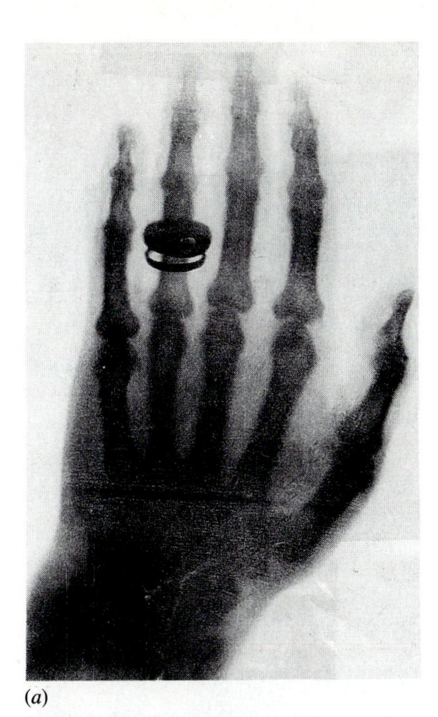

(*a*)

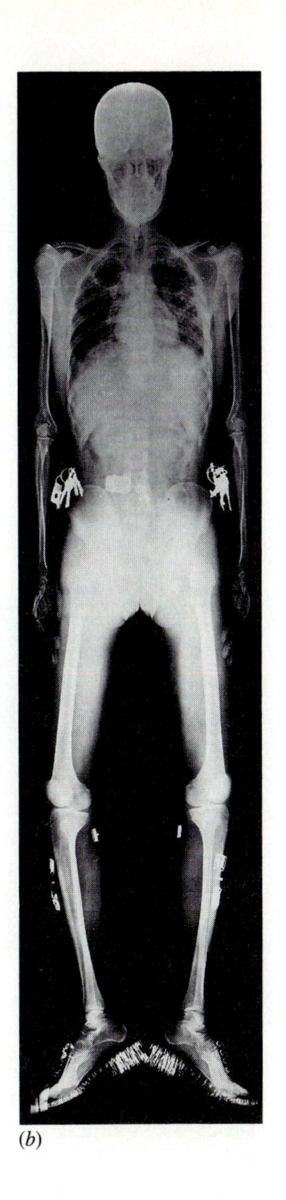

(*b*)

Figure 4.10 Two radiographs taken by Röntgen. (*a*) The hand of Mrs. Röntgen. *(The Bettmann Archive/Bettmann Newsphotos.)* (*b*) Radiograph of a fully clothed man, showing not only the man's skeleton but also the keys in his pockets, the nails of his shoes, and the metal clasps of his garters. *(Deutsches Museum, Munchen.)*

Analysis of the spectrum of radiation emitted by an X-ray tube shows the presence of two quite distinct contributions. There is a continuous spectrum with a well-defined short-wavelength cutoff, and, superimposed on this, there are several intense sharp lines. The continuous spectrum depends on the voltage between cathode and anode, as shown in Figure 4.13, page 83, but does not depend significantly on the target material.[7] The wavelengths of the sharp lines, however, are characteristic of the target material and independent of the applied voltage.

The continuous radiation is known as *Bremsstrahlung,* a German word meaning, literally, braking radiation. The term is appropriate because this radiation is

---

7. The efficiency for production of X rays by electron bombardment depends on the atomic number of the target material. A change from copper to tungsten, for example, increases the intensity of the continuous radiation substantially, but has no pronounced effect on the shape of the curves of Figure 4.13 or on the cutoff wavelength.

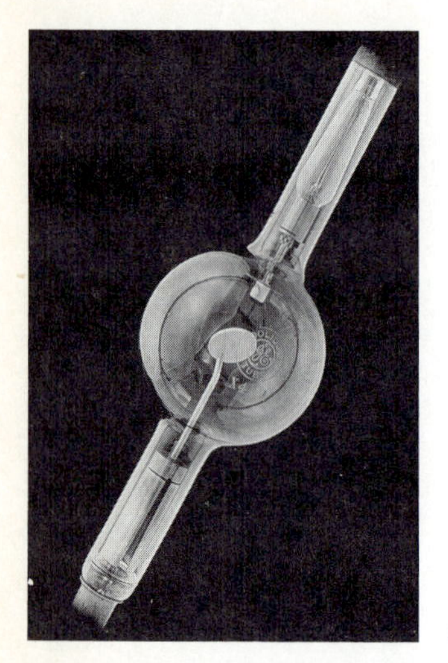

(a)

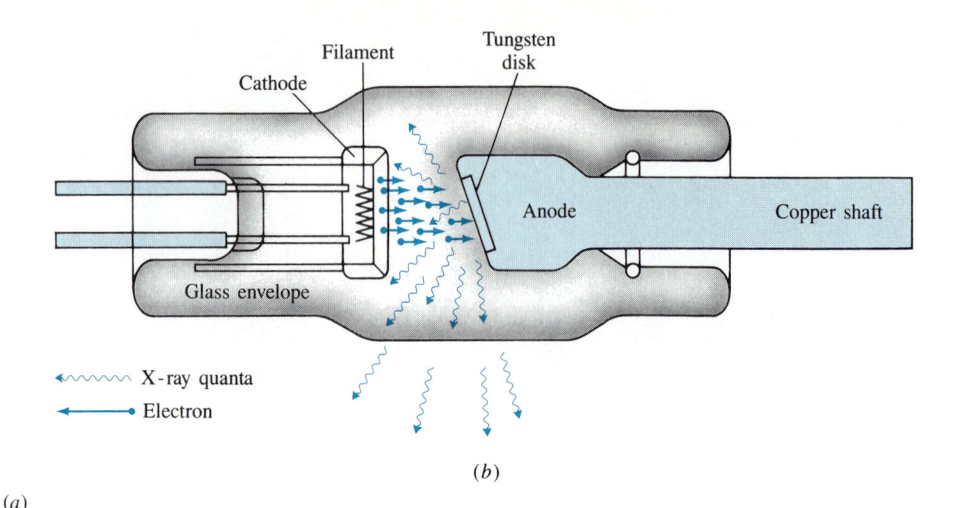

(b)

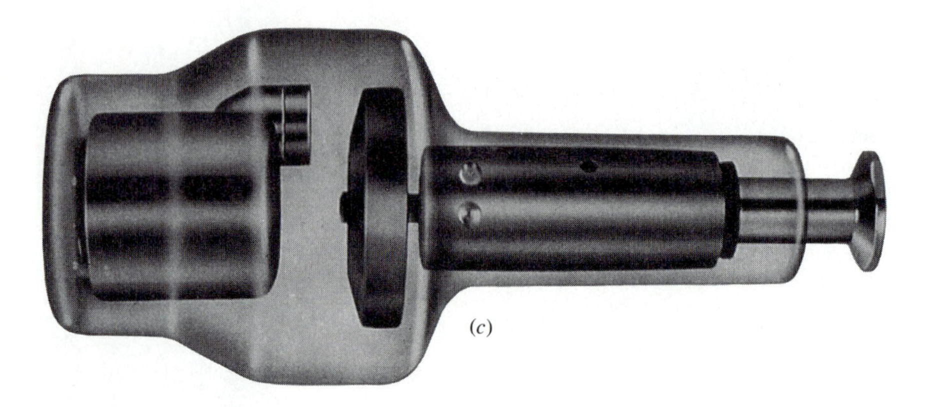

(c)

Figure 4.11 (a) An early X-ray tube. *(Courtesy G. E. Medical.)* (b) Schematic drawing of a modern X-ray tube. Electrons emitted by the heated cathode are accelerated toward the anode and strike the target, a tungsten disk. Only a small fraction of the electrons' energy is released in the form of X rays. Most of the energy of the incident electrons heats the anode, which must be cooled to prevent melting or evaporation of the target. Sometimes a heavy copper shaft is a sufficient heat sink; water cooling is often employed for high-intensity X-ray tubes. (c) Very intense X-ray beams require the use of a rotating anode, so that the electron beam does not constantly strike the same small region of the target surface. The cathode is off center. The anode disk rotates slowly about its axis. *(Courtesy G. E. Medical.)*

caused by the rapid deceleration of the electrons as they strike and enter the anode. Although the explanation of the continuous spectrum in terms of classical electrodynamics is essentially correct, the classical theory does not reproduce the observed spectral distribution. In particular, the short-wavelength cutoff and its dependence on applied voltage are strictly quantum effects.

The Planck-Einstein quantization hypothesis provides a simple explanation of the short-wavelength limit as well as the wavelength distribution of the continuous spectrum. When an electron falls from rest through a potential difference of V_A volts, it acquires a kinetic energy of V_A eV. As it comes to rest in the target, that electron can release no more than V_A eV of energy. If all of that energy is converted to electromagnetic radiation, the wavelength is

$$\lambda_c = \frac{hc}{eV_A} \tag{4.37}$$

We have here the inverse of the photoelectric effect; the maximum energy of the emitted photon equals the energy of the incident electron.

The continuous spectrum arises because the production of photons is not the only mechanism whereby an electron can dissipate energy. As we shall see later,

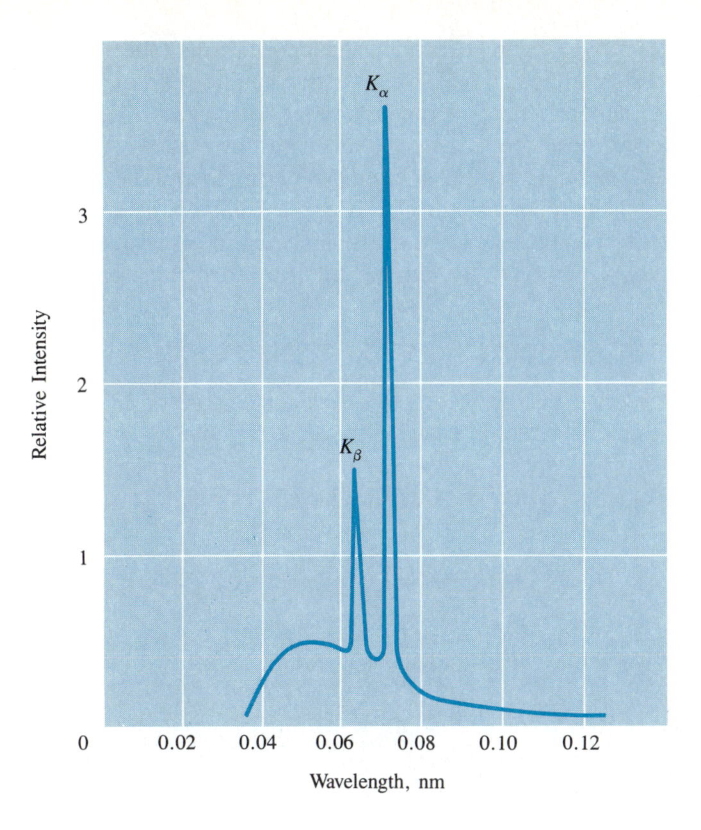

Figure 4.12 A typical X-ray spectrum, showing the continuous portion and two characteristic lines, designated K_α and K_β. The target is molybdenum.

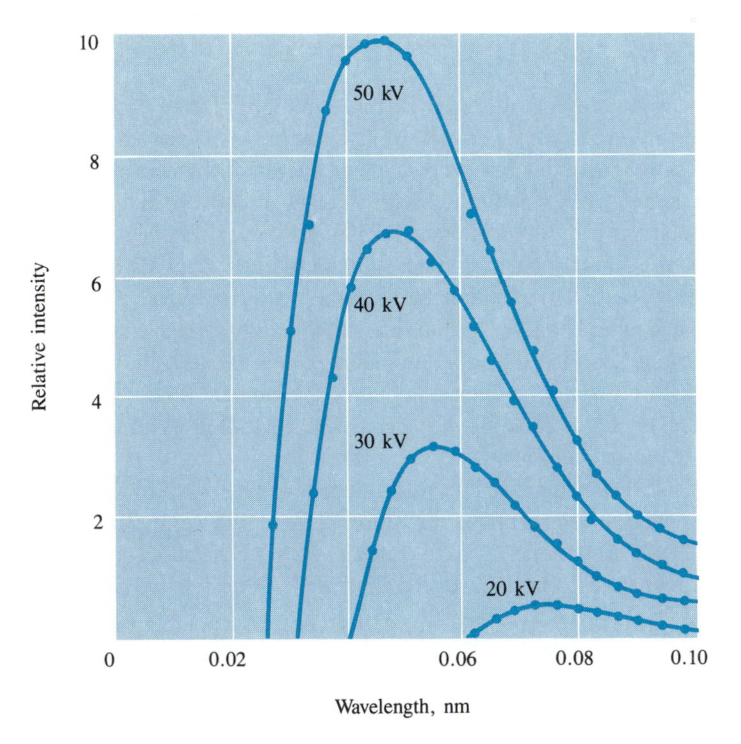

Figure 4.13 The spectral distribution of continuous X rays. Note that the intensity increases dramatically with increasing potential and that the short-wavelength cutoff shifts to lower values.

energetic electrons can cause excitation and ionization of atoms, can displace atoms from their normal lattice sites, can cause large-amplitude vibrations of atoms—in other words, can lose energy through a variety of nonradiative processes. Moreover, an electron need not release all its energy by creating a single photon; in coming to rest, it could emit two or more photons in rapid succession. For instance, the emission of two photons will give rise to a broad distribution of wavelengths because the two photons can share the total energy in an infinite number of ways.

As regards the origin of the line spectrum, we shall consider that in Chapter 5.

EXAMPLE 4.4 What is the short-wavelength cutoff of X rays produced by 10,000-eV electrons striking the target of an X-ray tube?

SOLUTION From Equation (4.37) we have

$$\lambda_c = \frac{hc}{eV_A} = \frac{1240 \text{ eV}\cdot\text{nm}}{10^4 \text{ eV}} = 0.124 \text{ nm}$$

where we have made use of a convenient result, namely,

$$\frac{hc}{e} = \frac{(6.626 \times 10^{-34} \text{ J}\cdot\text{s})(3 \times 10^8 \text{ m/s})}{1.6 \times 10^{-19} \text{ J/eV}} = 1.24 \times 10^{-6} \text{ eV}\cdot\text{m}$$

$$= 1240 \text{ eV}\cdot\text{nm} \qquad \textbf{(4.38)}$$

4.6 COMPTON SCATTERING

Figure 4.14 Arthur Holly Compton (1892–1962). *(Courtesy AIP, Niels Bohr Library.)*

Perhaps the most convincing early evidence that light displays particlelike behavior came from the work of the American physicist A. H. Compton.[8]

If electromagnetic energy propagates as a stream of particlelike photons, then the interaction of these "particles" with an electron must obey the same conservation laws that govern classical scattering events. In particular, when a photon is scattered by a stationary electron, conservation of momentum demands that some of the incident photon's momentum be imparted to the electron, which thereby acquires some kinetic energy. To conserve total energy, the energy of the scattered photon must be less than that of the incident. Consequently, the wavelength of the scattered photon must be greater than that of the incident, and the change in wavelength will depend on the angle through which the photon is scattered, being greatest for a "head-on" collision.

A typical scattering event is illustrated in Figure 4.15. We assume that the electron is initially at rest; this is not a restrictive assumption, since we can always transform to the rest frame of the electron, apply the results we shall derive presently, and then transform back to the laboratory reference frame.

8. A. H. Compton, *Phys. Rev.* **21,** 483; **22,** 409 (1923). Compton was awarded the 1927 Nobel Prize for the "discovery of the effect named after him."

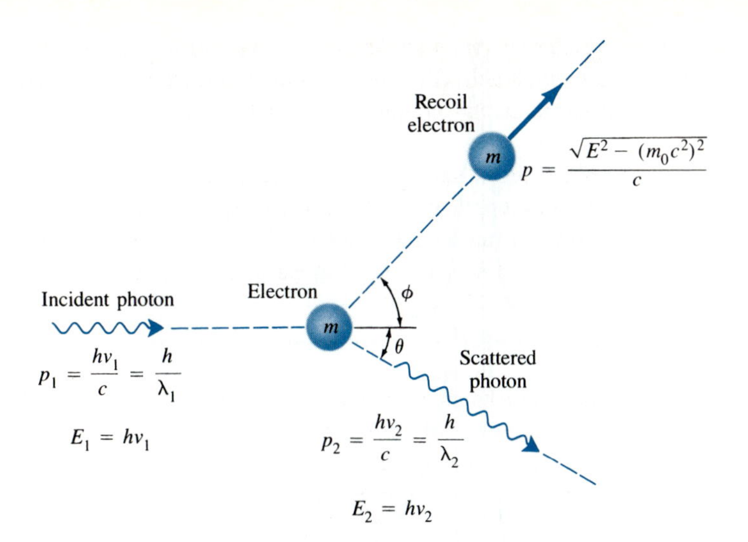

Figure 4.15 The geometry of Compton scattering. The incident photon collides with an electron at rest. In the process, energy and momentum are transferred to the electron.

Energy conservation is expressed by

$$h\nu_i + m_0c^2 - h\nu_f = \sqrt{(m_0c^2)^2 + (p_ec)^2} \qquad \textbf{(4.39)}$$

where ν_i and ν_f are the frequencies of the incident and scattered photon, m_0 is the rest mass of the electron, and p_e is the electron's momentum following the collision.

Momentum conservation requires that

$$\mathbf{p}_i = \mathbf{p}_f + \mathbf{p}_e \qquad \textbf{(4.40)}$$

where $\mathbf{p}_i$ and $\mathbf{p}_f$ are the initial and final momenta of the photon. We now write

$$p_e^2 = (\mathbf{p}_i - \mathbf{p}_f) \cdot (\mathbf{p}_i - \mathbf{p}_f) = p_i^2 + p_f^2 - 2\mathbf{p}_i \cdot \mathbf{p}_f = p_i^2 + p_f^2 - 2p_ip_f \cos\theta \qquad \textbf{(4.41)}$$

Next, we multiply both sides of Equation (4.41) by c^2 and replace p_ic and p_fc by $h\nu_i$ and $h\nu_f$. Thus

$$(p_ec)^2 = (h\nu_i)^2 + (h\nu_f)^2 - 2h^2\nu_i\nu_f \cos\theta \qquad \textbf{(4.42)}$$

If we square Equation (4.39) we obtain another expression for $(p_ec)^2$,

$$(p_ec)^2 = (h\nu_i)^2 + (h\nu_f)^2 - 2h^2\nu_i\nu_f + 2m_0c^2(h\nu_i - h\nu_f) \qquad \textbf{(4.43)}$$

and equating expressions (4.42) and (4.43) we arrive at the result

$$\frac{1}{h\nu_f} - \frac{1}{h\nu_i} = \frac{1}{m_0c^2}(1 - \cos\theta) \qquad \textbf{(4.44)}$$

or

$$\lambda_f - \lambda_i = \Delta\lambda = \frac{h}{m_0c}(1 - \cos\theta) \qquad \textbf{(4.45)}$$

Equation (4.45) is the famous *Compton equation* which relates the change in photon wavelength to the scattering angle and the rest mass of the scattering parti-

In Thomson's arrangement, **v** and **B** were perpendicular to **E**. If **B** is directed so that **v** × **B** is antiparallel to **E**, the net force on the charge vanishes if

$$v = \frac{E}{B} \tag{5.2}$$

Thus, by measuring the potential V and the magnetic field B for which the beam remains undeviated, Thomson could determine the speed of the cathode ray particles in the deflecting region. These speeds were typically about one-tenth of the speed of light.

Thomson then turned off the magnetic field and measured the deflection of the beam due to the field **E** only. The deflection of the spot on the glass envelope is simply related to v, E, the length of the deflecting plates, the distance between the deflecting plates and the end of the glass tube, and the charge-to-mass ratio, q/m (Problem 5.1). Thomson obtained the value $q/m = -0.7 \times 10^{11}$ C/kg. The currently accepted value of the charge-to-mass ratio of the *electron*, the name given to the cathode ray particles, is

$$\frac{e}{m} = -1.76 \times 10^{11} \text{ C/kg} \tag{5.3}$$

Figure 5.4 Robert A. Millikan (1868–1953). *(Courtesy AIP, Niels Bohr Library.)*

By 1897, the charge-to-mass ratio of a number of gaseous ions had already been determined. Thomson concluded from his data that

> *the value of m/e is independent of the nature of the gas, and that its value* 10^{-7} *is very small compared with the value* 10^{-4}, *which is the smallest value of the quantity previously known, and which is the value for the hydrogen ion in electrolysis.*[2]

Attempts to measure the electronic charge directly were immediately initiated in Thomson's laboratory, but he was able to obtain only very approximate values. The first reliable measurement of the charge e was performed by R. A. Millikan in 1909.[3] The justly famous "Millikan oil drop experiment" is today duplicated in many undergraduate laboratories. The result obtained by Millikan, $e = 1.64 \times 10^{-19}$ C, compares well with the currently accepted value

$$e = 1.602 \times 10^{-19} \text{ C} \tag{5.4}$$

Thus, in the early years of the twentieth century, the electron had gained acceptance as a fundamental and ubiquitous constituent of matter. Though its presence in atoms was generally acknowledged, it was also evident that electrons made a negligible contribution to atomic mass. From Thomson's and Millikan's results,

---

2. Thomson's value for the ratio m/e appears to differ from Equation (5.3) by several orders of magnitude because he used a different set of units, cgs units rather than SI units.

3. Millikan was the second American physicist to win the Nobel Prize, Michelson having been so honored in 1907. The 1923 Nobel citation noted the measurement of the electronic charge as well as Millikan's confirmation of Einstein's photoelectric equation.

$$m = \left(\frac{m}{e}\right)e = \frac{1.602 \times 10^{-19}}{1.76 \times 10^{11}} = 9.10 \times 10^{-31} \text{ kg} \quad \textbf{(5.5)}$$

whereas the mass of a hydrogen atom

$$M_{\rm H} = \frac{1}{N_A} = \left(\frac{6.02 \times 10^{26}}{\text{kgmol}}\right)^{-1} = 1.66 \times 10^{-27} \text{ kg}$$

is nearly 2000 times as great.

In those years, speculation about the structure of the atom was a favorite activity, one might almost say a parlor game among physicists. One model, proposed by Thomson about 1910, was dubbed the "plum pudding" model: a spherical object containing N electrons confined in a homogeneous jellylike but relatively massive positive charge distribution whose total charge cancels that of the N electrons.

While Sir Ernest Rutherford was a professor at McGill University in Montreal, he and Royds had shown that α particles emanating from radioactive elements were doubly ionized helium atoms (He^{2+}).[4] Shortly after Rutherford arrived in Manchester, England, he asked two young assistants, Geiger and Marsden, to examine the scattering of these particles by a thin foil of gold. Very little deflection was expected as a result of the smeared-out, neutral charge distribution associated with the Thomson model, but some information might be gained by a careful study of the distribution of the weakly scattered α particles. To everyone's surprise, Geiger and Marsden found that a substantial number of α particles were scattered through angles greater than 90°. As Rutherford put it,

> *It was quite the most incredible event that ever happened to me in my life. It was as incredible as if you fired a 15-inch shell at a piece of tissue paper and it came back and hit you.*

Figure 5.5 Sir Ernest Rutherford (1871–1937). *(The Bettmann Archive/Bettmann Newsphotos.)*

Results of scattering experiments are generally stated in terms of a function called the *scattering cross section*. We digress briefly to define and explain this concept.

5.2 SCATTERING CROSS SECTION

Suppose a uniform flux of I particles per unit area and unit time is incident on a fixed scattering center whose force field deflects the motion of these projectiles. The situation is portrayed in Figure 5.6, page 94, for scattering of charged particles by the Coulomb interaction with a scattering center whose charge is of the same sign. The number of incident particles scattered per unit time into an element of solid angle $d\Omega = \sin\theta \, d\theta \, d\phi$ may be written

$$dn = I\sigma_d(\theta, \phi) \, d\Omega \quad \textbf{(5.6)}$$

---

4. E. Rutherford and T. Royds, *Phil. Mag.* **17,** 281 (1909).

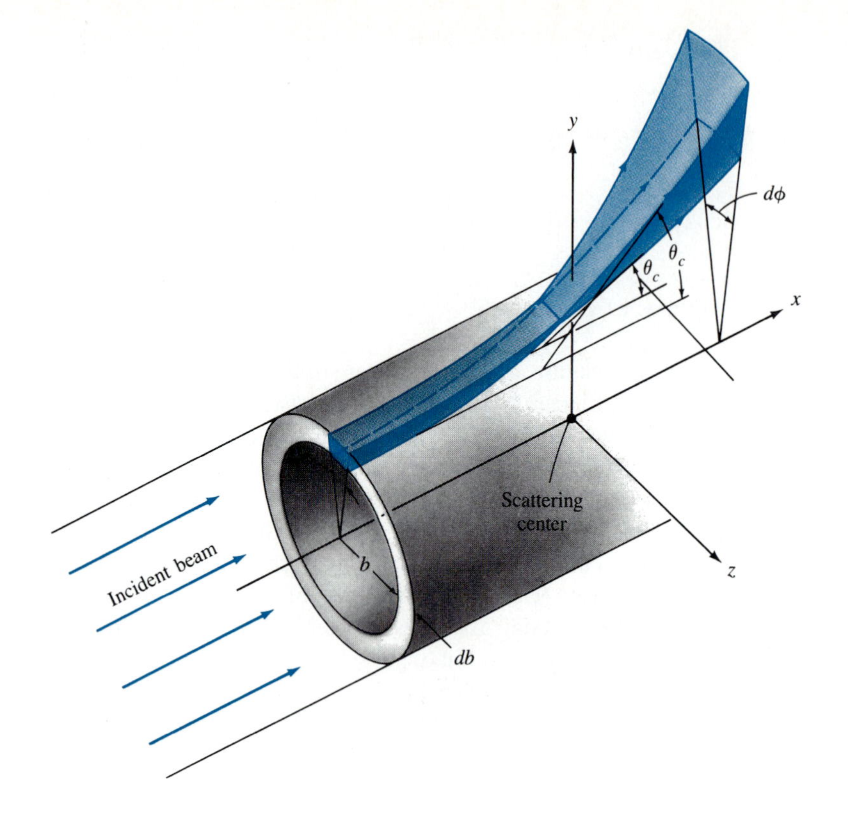

Figure 5.6 Scattering of particles of a beam by the central repulsive force field of a fixed scattering center. Particles that pass through the element of area $b\ d\phi\ db$ are scattered into the solid angle $d\Omega = \sin\theta d\theta d\phi$.

The proportionality factor $\sigma_d(\theta, \phi)$, which has the dimension of area, is called the *differential scattering cross section*. It represents that cross-sectional area of the incident beam whose particles will subsequently be scattered into the elemental solid angle $d\Omega$ (see Figure 5.6).

If, as is the case for the Coulomb interaction, the force between the projectile and the target is a central force, $\sigma_d(\theta, \phi)$, is independent of the azimuthal angle ϕ and

$$\sigma_d(\theta) = 2\pi\sigma_d(\theta, \phi) \tag{5.7}$$

where now $N\sigma_d(\theta) \sin\theta\ d\theta$ is the number of particles scattered per unit time through an angle θ into the element $d\theta$. From Figure 5.6 we see that $\sigma_d(\theta)$ is the cross-sectional area of the annular ring of radius b and width db shown in Figure 5.6. The distance b is known as the *impact parameter*.

The *total scattering cross section*, σ_0, is the integral of $\sigma_d(\theta, \phi)$ over 4π steradians; i.e.,

$$\sigma_0 = \int_0^{4\pi} \sigma_d(\theta, \phi)\ d\Omega = \int_0^{2\pi}\int_0^{\pi} \sigma_d(\theta, \phi) \sin\theta\ d\theta\ d\phi \tag{5.8}$$

$$= 2\pi \int_0^{\pi} \sigma_d(\theta) \sin\theta\ d\theta$$

5.3 COULOMB (RUTHERFORD) SCATTERING

We now derive the expression for $\sigma_d(\theta)$ for Coulomb scattering. We shall assume that the scattering center is fixed; that is, we neglect the recoil of the target particle. This is not a restrictive assumption. As shown in most mechanics texts, it is always possible to reduce a two-body collision between particles of masses m and M to a form equivalent to the collision of a particle of mass

$$\mu = \frac{mM}{m + M} \tag{5.9}$$

with a particle of infinite mass, which therefore does not recoil. The quantity μ is known as the *reduced mass*.

The force acting on the incident particle of charge ze in the vicinity of another particle of charge Ze is

$$F = \frac{zZe^2}{4\pi\epsilon_0 r^2} \tag{5.10}$$

As first demonstrated by Newton, the trajectory of the incident particle is a hyperbola, shown in Figure 5.7. Here we have indicated the scattering angle θ and an angle ψ, the angle between the radius vector from the scattering center to the projectile particle and the line of symmetry, the z axis.

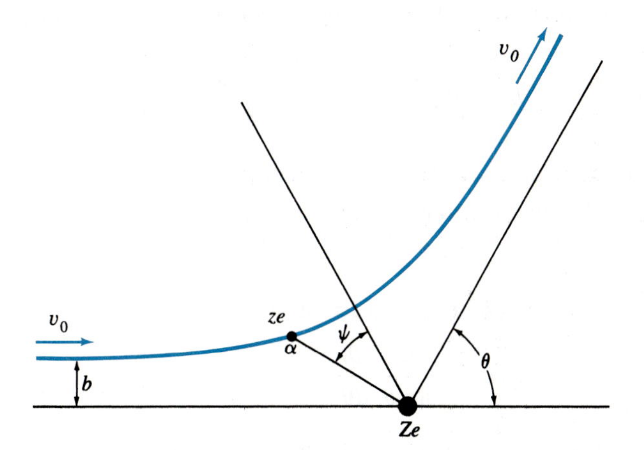

Figure 5.7 The geometry of Coulomb scattering. The particle of charge *ze* is scattered by a fixed nucleus of charge *Ze*. The *z* axis is the line that bisects the complement to the scattering angle θ.

Since the scattering center remains stationary, the kinetic energy of the projectile is the same before and after the collision; consequently, the speed of the particle, v_0, is also the same before and after the collision. However, the particle does suffer a change in momentum given by

$$\Delta p_z = 2mv_0 \cos \frac{\pi - \theta}{2} = 2mv_0 \sin \frac{\theta}{2} \tag{5.11}$$

This change in momentum is equal to the impulse; i.e.,

The potential energy is

$$\mathrm{PE} = -eV(r_n) = -\frac{e^2}{4\pi\epsilon_0 r_n} \tag{5.30}$$

The total energy is

$$E_n = \mathrm{KE} + \mathrm{PE} = -\frac{e^2}{8\pi\epsilon_0 r_n} = -\frac{me^4}{32\pi^2\epsilon_0^2\hbar^2 n^2} = -\frac{E_0}{n^2} \tag{5.31}$$

where $$E_0 = \frac{me^4}{32\pi^2\epsilon_0^2\hbar^2} = 2.2 \times 10^{-18}\ \mathrm{J} = 13.6\ \mathrm{eV} \tag{5.32}$$

The energy of the system is negative because we have adopted the standard convention of assigning zero potential energy to the two charges when they are infinitely far removed from each other. The negative energy simply means that these are *bound states;* to separate the two charges we must supply energy to the system. The *ground state* of the system is that for which $n = 1$; it is the state in which the hydrogen atom will normally exist and to which it will tend to return if excited to a higher energy state. The *first ionization energy* of an atom is the energy that must be supplied to the atom to remove one electron from the neutral atom when in its ground state. It follows that the ionization energy of the hydrogen atom is 13.6 eV.

Using the then-current values for the various physical constants, Bohr obtained $a_0 = 5.15 \times 10^{-11}$ m and $E_0 = 13$ eV. He now felt certain that he was on the right course because the "values are of the same order of magnitude as the linear dimensions of the atom . . . and the ionization potential."

If the system consists of a nucleus of charge Ze and a single electron, the radii and energies of the stationary states are given by

$$r_n = \frac{a_0 n^2}{Z} \qquad E_n = -\frac{Z^2 E_0}{n^2} \tag{5.33}$$

For example, the ground-state energy of singly ionized helium ($Z = 2$) is -54.4 eV, and that of doubly ionized lithium is -122.4 eV.

EXAMPLE 5.3 What are the radius and energy of the first excited state of doubly ionized lithium?

SOLUTION For lithium, $Z = 3$. The first excited state of the ion is that for which $n = 2$. Hence,

$$r_2 = \frac{4a_0}{3} = 0.0705\ \mathrm{nm} \qquad E_2 = -\frac{9E_0}{4} = -30.6\ \mathrm{eV}$$

5.5 THE ENERGY-LEVEL DIAGRAM; EMISSION AND ABSORPTION OF RADIATION

Bohr had postulated that in a transition from one stationary state to another, electromagnetic radiation, whose frequency is given by Planck's formula, is either absorbed or emitted. That is,

$$h\nu = \frac{hc}{\lambda} = |E_f - E_i| \tag{5.34}$$

where E_i and E_f are the energies of the atom in its initial and final states. If $E_i > E_f$, a photon of frequency ν is emitted. To raise an atom from a state of energy E_i to one of energy $E_f > E_i$ by the absorption of a photon, the photon must have a frequency ν given by Equation (5.34). The latter process is often referred to as *resonance absorption*.

These absorption and emission processes are most easily visualized by means of an energy-level diagram, such as the one for the hydrogen atom shown in Figure 5.11, page 104. Here, the vertical axis represents the energy, and the horizontal lines the energies of the stationary states, each identified by the quantum number n. Since all stationary bound states have negative energies, zero energy (the ionized atom) is represented by the line at the top of the diagram.

Energy levels of the hydrogen atom for $n = 1, 2, 3, 4, \ldots$ are shown in Figure 5.11. It is customary to indicate a transition between two states by an arrow. A downward arrow corresponds to a downward transition of the atom, accompanied by the release of energy, and conversely for an upward arrow.

The wavelength of a particular emission or absorption line is now readily calculated from Equations (5.34) and (5.31). One obtains

$$\frac{1}{\lambda} = \frac{\nu}{c} = \frac{E_0}{hc}\left|\frac{1}{n_f^2} - \frac{1}{n_i^2}\right| = R_\infty\left|\frac{1}{n_f^2} - \frac{1}{n_i^2}\right| \tag{5.35}$$

where

$$R_\infty = \frac{E_0}{hc} = 1.097373 \times 10^7\ \mathrm{m}^{-1} \simeq 1.1 \times 10^7\ \mathrm{m}^{-1} \tag{5.36}$$

is the *Rydberg constant*. The subscript ∞ indicates that this is the appropriate value of the Rydberg constant if the nuclear mass is assumed infinite. In fact, of course, the nucleus has a finite mass and, in the Bohr model, the nucleus and electron rotate about their common center of mass. As in the collision problem, the finite mass of the nucleus can be taken into account by replacing the electron mass by the reduced mass $\mu = mM_N/(m + M_N)$, where M_N is the mass of the nucleus. Since the Rydberg constant is E_0/hc and E_0 is proportional to m [see Equation (5.32)], the value of the Rydberg constant is slightly altered when we take the nuclear mass into account. Although quite small, the concomitant change in the wavelengths of the spectral lines is readily measured. It was, in fact, this small shift in spectral wavelengths that led to the discovery of "heavy hydrogen."

When $n_f = 2$, the expression for the wavelengths of emission lines due to

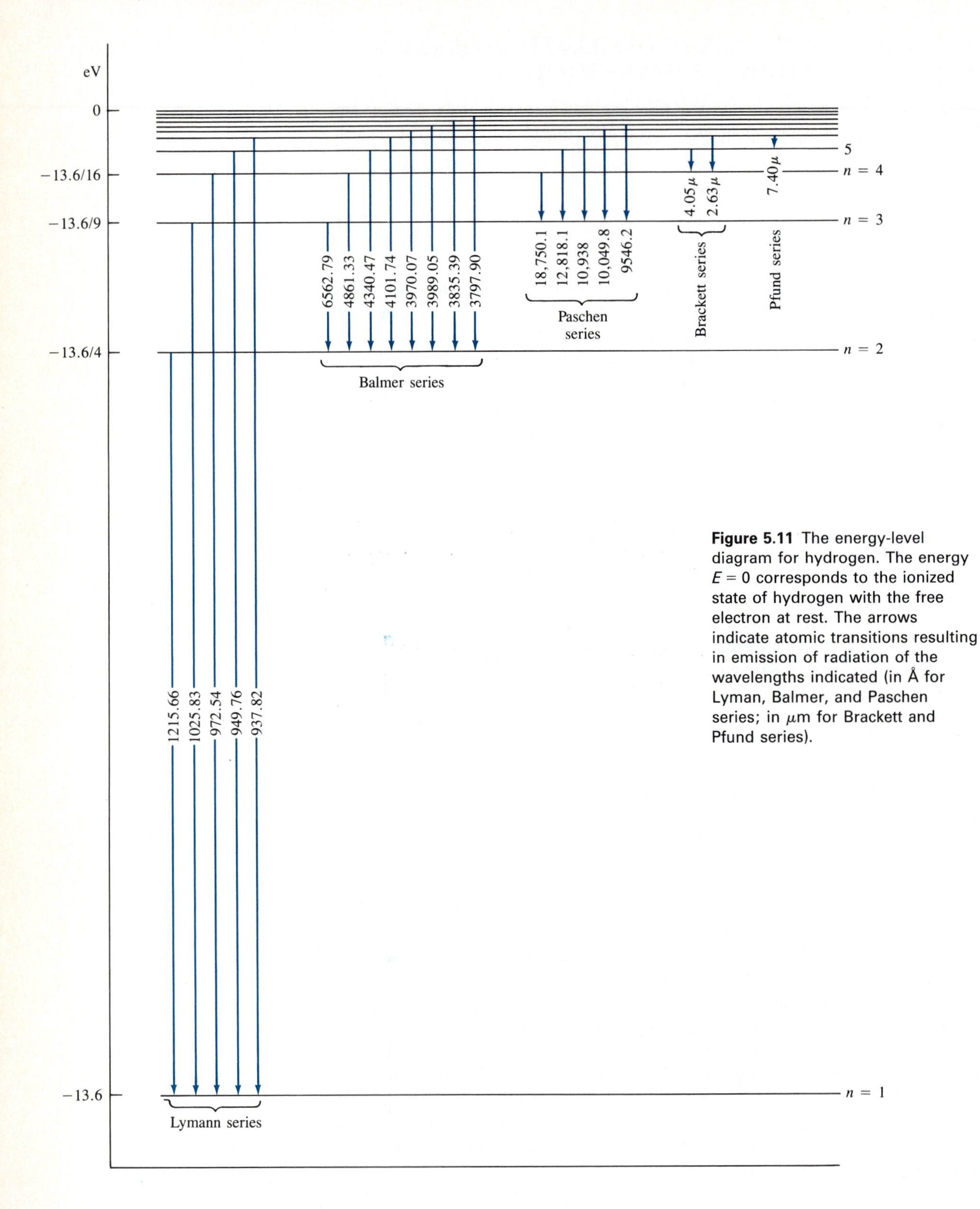

Figure 5.11 The energy-level diagram for hydrogen. The energy $E = 0$ corresponds to the ionized state of hydrogen with the free electron at rest. The arrows indicate atomic transitions resulting in emission of radiation of the wavelengths indicated (in Å for Lyman, Balmer, and Paschen series; in μm for Brackett and Pfund series).

transitions from states $n_i = 3, 4, 5, \ldots$ is identical to that proposed by Balmer some thirty years earlier. This group of lines of the hydrogen spectrum is therefore referred to as the *Balmer series*. These are the only lines of the hydrogen spectrum that fall in the visible region.

Transitions from $n_i = 4, 5, 6, \ldots$ to $n_f = 3$ give rise to infrared radiation. By 1913, these lines had been observed and studied by Paschen, but the remaining spectral series corresponding to $n_f = 1, 4, 5, \ldots$ had not been seen. In his paper, Bohr submitted

> *If we put t_2 [n_f] = 1 and 4, 5, . . . we get series respectively in the extreme ultraviolet and the extreme ultra-red, which are not observed, but the existence of which may be expected.*

Within a few years the spectral series of hydrogen in the ultraviolet and in the far infrared were observed and their wavelengths measured.

EXAMPLE 5.4 Calculate the Rydberg constant for ordinary hydrogen. Use this value to determine the long- and short-wavelength limits of the Lyman ($n_f = 1$) and Balmer series.

SOLUTION The Rydberg constant for a system with nuclear mass M_N may be written

$$R_N = \frac{R_\infty}{1 + m/M_N}$$

The ratio of the electron mass to the proton mass is

$$\frac{m}{M} = \frac{0.511 \text{ MeV}/c^2}{938.3 \text{ MeV}/c^2} = \frac{1}{1836}$$

Therefore,

$$R_\text{H} = \frac{1.097373 \times 10^7}{1 + \frac{1}{1836}} = 1.096776 \times 10^7 \text{ m}^{-1}$$

The long- and short-wavelength limits of a given series correspond to the transitions from $n_i = \infty$ to n_f and from $n_i = n_f + 1$ to n_f, respectively. For the Lyman series, $n_f = 1$, and the two wavelength limits are given by

$$\frac{1}{\lambda_\text{long}} = R_\text{H}\left(1 - \frac{1}{4}\right) = 0.822582 \times 10^7 \text{ m}^{-1} \qquad \lambda_\text{long} = 121.568 \text{ nm}$$

$$\frac{1}{\lambda_\text{short}} = R_\text{H} \qquad \lambda_\text{short} = 91.17632 \text{ nm}$$

For the Balmer series, $n_f = 2$. Hence,

$$\frac{1}{\lambda_{\text{long}}} = R_{\text{H}}\left(\frac{1}{4} - \frac{1}{9}\right) = 0.1523300 \times 10^7 \text{ m}^{-1} \qquad \lambda_{\text{long}} = 656.469 \text{ nm}$$

$$\frac{1}{\lambda_{\text{short}}} = R_{\text{H}}\left(\frac{1}{4}\right) = 0.274194 \times 10^7 \text{ m}^{-1} \qquad \lambda_{\text{short}} = 364.705 \text{ nm}$$

The results are in excellent agreement with the measured wavelengths for these lines.

5.6 CHARACTERISTIC X-RAY LINES

In the preceding chapter we saw that the spectrum from an X-ray tube consists of two parts, a broad background extending to a short-wavelength cutoff and, superimposed on this, a series of intense, sharp lines whose wavelengths depend on the target material. Here we shall be concerned with the origin of these characteristic X-ray lines.

In the second of his papers, Bohr attempted to extend the theory to atomic systems more complex than hydrogen. That proved to be a difficult task, and it soon became apparent that only a few broad generalizations would emerge from the effort. Among them were, however, a number of significant conclusions that were later incorporated into a theory of atomic structure that has withstood the test of time. In particular, Bohr recognized that the extranuclear electrons must be organized into orbital groups, some close and tightly bound to the nucleus, others with larger orbits and less tightly bound. Bohr, in fact, proposed a classification scheme similar to the one subsequently formulated by Pauli and others, and now recognized as the correct model of atomic structure.

According to that scheme, the electrons of a many-electron atom may be thought of as falling into "shells," each shell corresponding to a quantum number

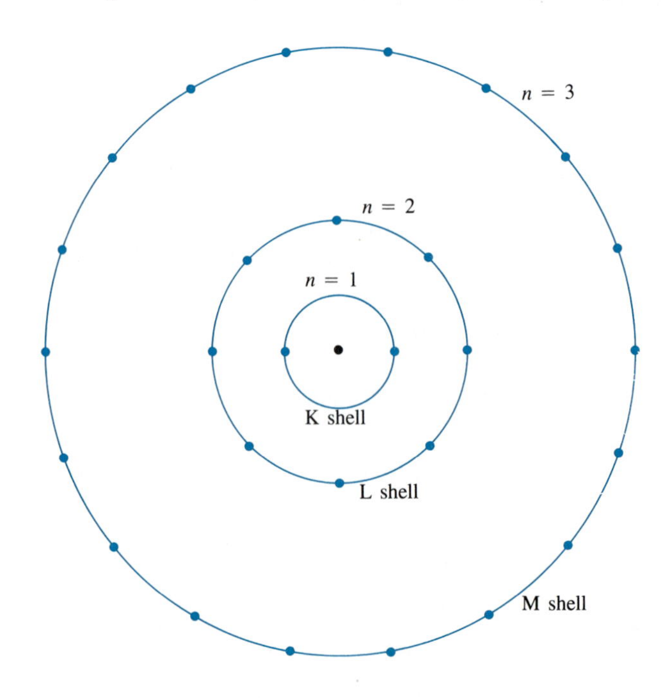

Figure 5.12 Schematic diagram of the electronic structure of the K, L, and M shells. The K ($n = 1$) shell can accommodate no more than 2 electrons, the L ($n = 2$) shell no more than 8 electrons, and the *M* ($n = 3$) shell no more than 18 electrons.

n. The first of these, the so-called K shell with $n = 1$, can hold no more than 2 electrons. The next, the L shell ($n = 2$), can accommodate 8 electrons. The $n = 3$, or M shell, holds a maximum of 18 electrons. Although there are some exceptions for $n > 3$, electrons in shells of low quantum number are the more tightly bound. The total number of electrons equals the atomic number Z.

In 1913, Harry Moseley, whom Rutherford later described as his most talented student, measured the wavelengths of the characteristic X rays of a large number of elements. He discovered a remarkable regularity with respect to atomic number, displayed in Figure 5.13, called *a Moseley plot,* in which the square root of the

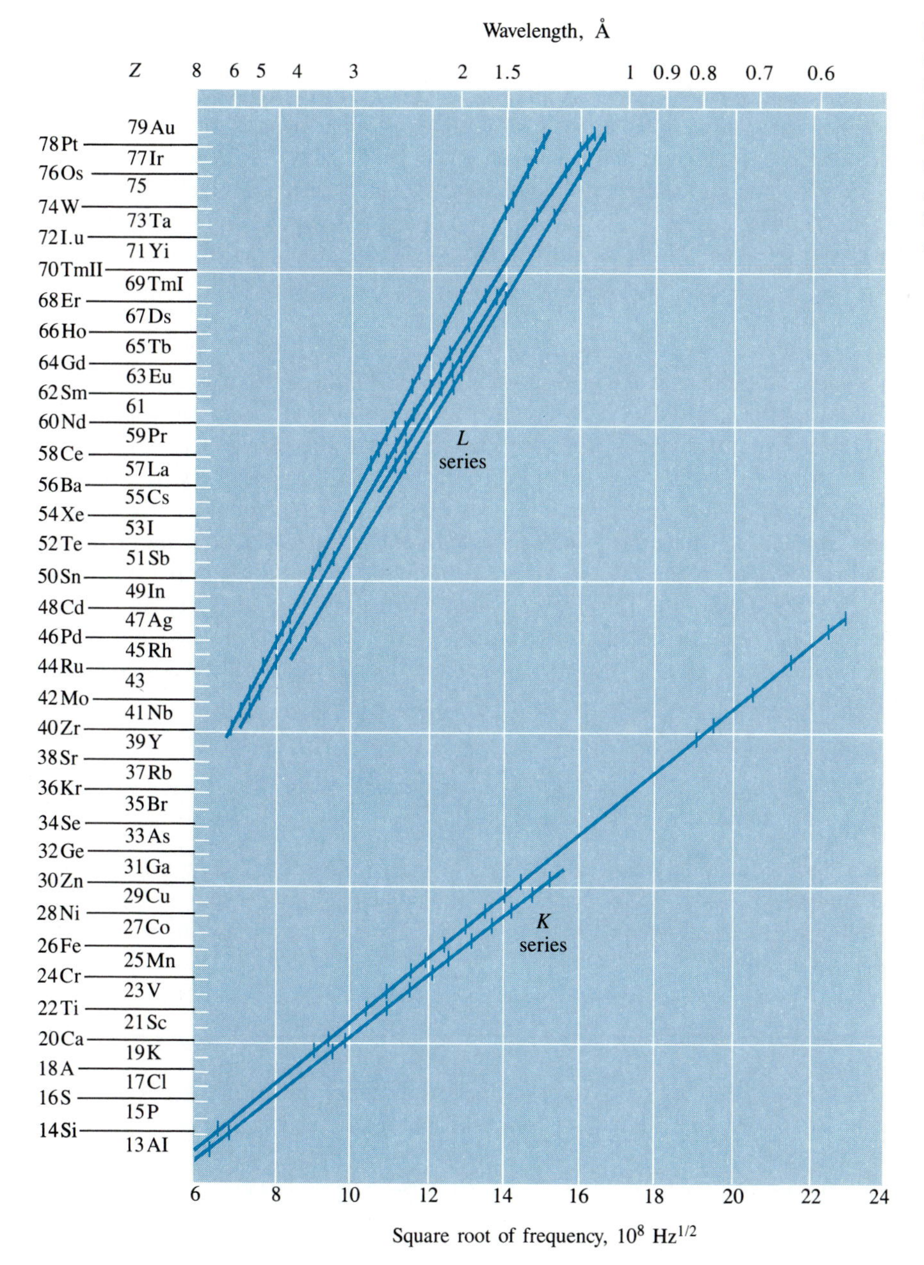

Figure 5.13 Moseley plot of the characteristic X-ray lines of a large number of elements. The *K*-series lines result from transitions of an electron to an empty state in the K shell. The two *K* lines represent the K_α and K_β lines, arising from transitions of an L- and an M-shell electron, respectively.

Figure 5.14 Harry Moseley (1887–1915). *(University of Oxford, Museum of History of Science/ Courtesy AIP, Niels Bohr Library.)*

frequency of the X-ray line is plotted against atomic number. Since such a plot shows a series of straight lines, it follows that, for each series,

$$\sqrt{\nu} = B(Z - a)$$

or, squaring both sides,

$$\nu = \frac{c}{\lambda} = B^2(Z - a)^2 \tag{5.37}$$

Here B and a are constants determined by the slope and intercept of the straight line.

These results find a ready explanation in terms of the Bohr model and the atomic shell scheme just described. In a heavy atom, such as Au, Nb, or W, the two electrons in the K shell are very tightly bound. Each electron feels almost the full Coulomb field of the nucleus of charge Z; part of the nuclear charge is screened by the other $n = 1$ electron, and even electrons in shells with $n \geq 2$ contribute slightly to this screening of the nuclear charge. Consequently, each K-shell electron "sees" a nuclear charge of $(Z - a)e$, where $a \simeq 1$. Each K-shell electron therefore has an energy

$$E_1 = -(Z - a)^2 E_0 \text{ eV}$$

and it will require that much energy to remove one of these electrons from the atom. If that happens, for instance, as a result of bombardment by energetic electrons, an atomic electron from one of the outer shells can drop into the vacated state in the K shell, thereby lowering the energy of the system. This happens very quickly, usually in less than a nanosecond. The atom then emits radiation in the X-ray region of the spectrum, and since the two energy states are well defined, so is the frequency (and wavelength) of the emitted X ray. The binding energy of the initial state of the electron that makes the transition, though small compared with that in the final state, cannot be neglected. For instance, for the K_α line, the X-ray line corresponding to an L shell to K shell transition, the energy of the initial state is approximately given by

$$E_i = -\frac{(Z - 8)^2 E_0}{4}$$

since that L-shell electron prior to the transition felt the nuclear charge screened by the other seven L-shell electrons and the remaining K-shell electron. The energy difference $E_i - E_f$ divided by Planck's constant is then closely approximated by the Moseley relation.[9]

EXAMPLE 5.5 Estimate the wavelength of the characteristic K_α X ray from niobium. The K_α line arises from the transition of an L-shell electron into an empty K-shell state.

---

9. A correct calculation of the energies of characteristic X rays is very difficult and complicated, requiring the application of quantum mechanics and numerical procedures.

SOLUTION We shall assume that $a = 1$ for the K-shell electron and take $a = 8$ for the L-shell electron. The atomic number of niobium is $Z = 41$. The energies E_i and E_f are then

$$E_i = -\frac{(41 - 8)^2(13.6 \text{ eV})}{4} = -3700 \text{ eV}$$

$$E_f = -(41 - 1)^2(13.6 \text{ eV}) = -21{,}760 \text{ eV}$$

The energy difference for the transition is 18,060 eV, and the corresponding wavelength is

$$\lambda = \frac{1240 \text{ eV} \cdot \text{nm}}{18{,}060 \text{ eV}} = 0.0687 \text{ nm}$$

When Moseley did his work, the atomic numbers of the known elements were not really well established. What could be determined by direct measurement was atomic weight; atomic number was believed to be a monotonic function of atomic weight. Among Moseley's achievements was the correct assignment of atomic numbers and the prediction of atomic numbers of some missing elements.[10]

At the outbreak of World War I, Moseley enlisted in the British army. He was one of many young men who lost their lives in the ill-fated attack at Gallipoli in 1915.

5.7 THE FRANCK-HERTZ EXPERIMENT

An atom may be raised from the ground to an excited state by other than electromagnetic radiation. For example, in a collision of an electron with an atom, the latter may be excited provided that the initial energy of the electron is at least equal to the energy difference between the excited and ground states of the atom.

Suppose a gas of atoms is introduced into a tube such as that of Figure 5.15.

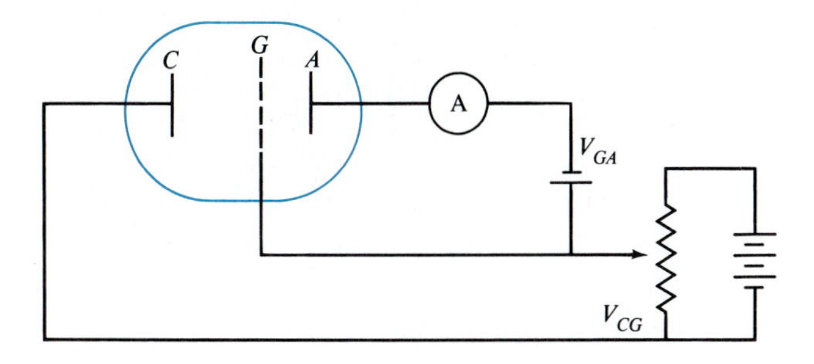

Figure 5.15 Schematic drawing of the circuit for a Franck-Hertz experiment. The current in the external circuit is measured as the cathode-grid potential is slowly varied while maintaining a fixed grid-anode retarding potential.

10. Prior to Moseley's work, nickel, with atomic weight 58.7, was placed ahead of cobalt, atomic weight 58.9, in the periodic table. Moseley's plot showed conclusively that the element of atomic number 27 must be cobalt and that atomic number 28 had to be assigned to nickel. One of the missing elements, hafnium ($Z = 72$), whose existence was predicted by Moseley, was later discovered in Bohr's laboratory (the name *hafnium* derives from its place of discovery, Copenhagen).

Electrons released from the cathode C are accelerated toward the grid G by the electric field established by the potential V_{CG}, and after passing through the grid have a kinetic energy of eV_{CG} eV. They are then subjected to a small retarding potential V_{GA} before being collected at the anode. If, in its progress from cathode to anode, an electron loses no energy, it will arrive at the anode with the energy $e(V_{CG} + V_{GA}) = eV_{CA}$ eV and will contribute to the current in the circuit as measured by the ammeter A. However, if eV_{CG} is greater than the excitation energy of the gas atoms from their ground state to the lowest excited state, some of the electrons will, on colliding with a gas atom, lose much of their kinetic energy and then have insufficient energy left to overcome the retarding potential V_{CA}. Consequently, as the cathode-grid potential is gradually increased, one should observe a drop in current when $eV_{CG} \simeq E_{x1}$, where E_{x1} is the first excitation energy of the gas atoms. If the potential V_{CG} is then increased again, the current should rise, but another trough should appear when $eV_{CG} \simeq 2E_{x1}$ or $eV_{CG} \simeq E_{x2}$, where E_{x2} is the second excitation energy of the gas.

In 1914, James Franck and Gustav Hertz demonstrated such resonance absorption using a beam of electrons projected through mercury vapor.[11] Their results are shown in Figure 5.16. Moreover, as Franck demonstrated in a later experiment, when the electron energy is sufficient to excite the mercury atoms, they return to the ground state with the emission of radiation of wavelength

$$\lambda = \frac{hc}{E_x - E_g}$$

where E_x and E_g are the excited- and ground-state energies of mercury.

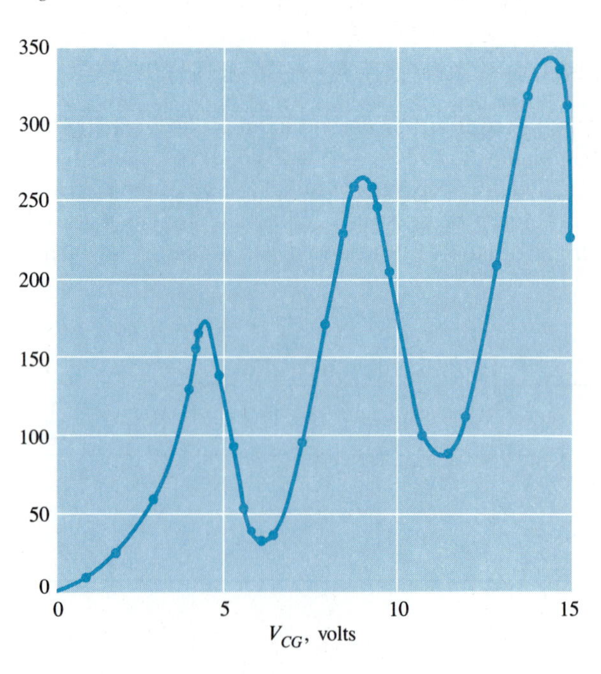

Figure 5.16 Current versus grid potential as measured by James Franck and Gustav Hertz. The first excitation energy of mercury atoms is 4.9 eV. When the accelerating potential between cathode and grid exceeds 4.9 V by a small amount, the anode current drops sharply as electrons that have lost this energy in exciting mercury atoms cannot overcome the grid-to-anode retarding potential.

11. J. Franck and G. Hertz, *Verh. d. D. Phys. Ges.* **16,** 457 (1914). James Franck and Gustav Hertz were awarded the 1925 Nobel Prize in physics for this work.

5.8 THE CORRESPONDENCE PRINCIPLE

In the preceding sections, the energy levels of the hydrogen atom were derived from the quantization rule on angular momentum, Equation (5.25). Though this is by far the most direct path to Equations (5.27) and (5.31), it is not the one followed by Bohr, who stated that quantization rule on the very last page of his article. To forge the connection between radiation due to transitions between stationary states and classical electrodynamics, Bohr postulated a general theorem now known as the *correspondence principle,* which may be phrased as follows:

> *The dynamics of a system, as described by quantum theory, must agree with the classical description of that system in the limit of very large quantum numbers.*

The classical frequency of radiation emitted by a system such as the hydrogen atom is just the frequency of the orbital motion of the electron. For large n, and recalling that $L = n\hbar = mr_n^2\omega$, we have

$$\nu = \frac{\omega}{2\pi} = \frac{n\hbar}{2\pi m r_n^2} = \frac{me^4}{32\pi^3\epsilon_0^2\hbar^3 n^3} = \frac{2E_0}{hn^3} \tag{5.38}$$

where we have used Equations (5.27) and (5.31).

The frequency of the radiation due to transition from the quantum state $n_i = n$ to $n_f = n - 1$ is

$$\nu = \frac{E_0}{h}\left[\frac{n^2 - (n-1)^2}{n^2(n-1)^2}\right] = \frac{E_0}{h}\left[\frac{2n - 1}{n^2(n-1)^2}\right] \simeq \frac{2E_0}{hn^3} \qquad \text{if} \qquad n \gg 1 \tag{5.39}$$

which is identical to the classical result, Equation (5.38).

It is interesting to note that the correspondence principle immediately suggests a generalization of the Bohr model. We saw that the spectral lines emitted by an excited hydrogen atom correspond to transitions for which $\Delta n = n_i - n_f \geq 1$. There should be no reason why for large n the quantum jumps must be limited to $\Delta n = 1$. However, if we do allow larger changes in n, such as $\Delta n = 2$ or 3, the frequencies of the emitted radiation will, according to the quantum model, be integral multiples of the fundamental frequency given by Equation (5.39). Classically, however, a charge in a circular orbit radiates only at the rotational frequency. This discrepancy between the classical and the quantum model in the correspondence limit is removed if elliptical as well as circular orbits are allowed.

The extension of the Bohr model to include elliptical orbits was carried out by Sommerfeld, who demonstrated that one could thereby account for the so-called fine structure of spectral lines. The critical parameter in these calculations is a dimensionless constant, the *fine structure constant*

$$\alpha = \frac{e^2}{2\epsilon_0 hc} = \frac{1}{137} \tag{5.40}$$

which is the ratio v_0/c, where v_0 is the classical orbital speed of the electron in the ground state of hydrogen. Although Sommerfeld's explanation of fine structure proved to be incorrect, the fine structure constant does play a profound role in the quantum theory of atoms, and we shall encounter it again in Chapter 9.

The inclusion of elliptical orbits required the specification of a second quantum number to characterize the eccentricity of the orbit. Again, as the theory developed, this second quantum number assumed a somewhat different but important significance. The need for more than one quantum number was also appreciated by Sommerfeld as an essential link to classical physics. In classical mechanics, the complete description of a trajectory involves three coordinates; the correspondence principle implies that in quantum theory, three independent quantum numbers should make their appearance. As we shall see shortly, these suggestions are supported by the more elaborate quantum mechanics developed by Schrödinger and Heisenberg.

SUMMARY

Experiments by Geiger and Marsden on scattering of α particles by thin metal foils led Rutherford to propose the nuclear model of the atom. The inherent instability of the model was obviated by Bohr's postulate that electrons in atoms can exist only in *stationary states*. In Bohr's theory, these stationary states are prescribed by the condition.

$$L = n\hbar \tag{5.25}$$

where L is the angular momentum, n is an integer, and $\hbar = h/2\pi$.

For the hydrogen atom, application of Equation (5.25) and the classical equations of motion of a negatively charged electron in the Coulomb field of the proton yields the radii of the allowed circular orbits,

$$r_n = n^2 a_0 \qquad a_0 = \frac{4\pi\epsilon_0\hbar^2}{me^2} = 5.29 \times 10^{-11}\ \text{m} \tag{5.28}$$

and the energies of the corresponding stationary states,

$$E_n = -\frac{E_0}{n^2} \qquad E_0 = \frac{me^4}{32\pi^2\epsilon_0^2\hbar^2} = 13.6\ \text{eV} \tag{5.31}$$

Energy-conserving transitions between stationary states account for the observed absorption and emission lines of atomic hydrogen. Their wavelengths are given by

$$\lambda^{-1} = R_{\text{H}}\left|\frac{1}{n_{\text{f}}^2} - \frac{1}{n_i^2}\right| \tag{5.35}$$

where R_{H} is the *Rydberg constant* for hydrogen. For an infinitely massive nucleus the Rydberg constant is $R_\infty = 1.097373 \times 10^7\ \text{m}^{-1}$.

The characteristic X-ray lines emitted by atoms arise when an electron from an outer shell of the atoms drops into an empty state in an inner shell that was created by bombardment with energetic electrons.

Resonance absorption of atoms using energetic electrons was demonstrated by Franck and Hertz.

The connection between quantum theory and classical dynamics is expressed by Bohr's *correspondence principle:*

The dynamics of a system, as described by quantum theory, must agree with the classical description of that system in the limit of very large quantum numbers.

Bibliography

Biographies of Bohr and Rutherford

A. P. French and P. J. Kennedy (eds.), *Niels Bohr; A Centenary Volume,* Harvard University Press, Cambridge, Mass., 1985.

Ruth Moore, *Niels Bohr,* MIT Press, Cambridge, Mass., 1985.

David Wilson, *Rutherford,* MIT Press, Cambridge, Mass., 1983.

The Nuclear Atom

Robert B. Leighton, *Principles of Modern Physics,* McGraw-Hill, New York, 1959.

See also references listed at the end of Chapter 4.

PROBLEMS

5.1 The dimensions for the Thomson apparatus of Figure 5.2 are as follows. The separation between plates P_1 and P_2 is 1.2 cm, and the length of the plates is 3.0 cm. The distance between the center of the deflecting region (the midpoint between the deflecting plates) and the screen at the end of the tube is 36.0 cm.

(a) Suppose the magnetic field in the deflecting region is 0.001 T and the potential difference between cathode and anode is 1800 V. What potential difference should be maintained between the deflecting plates so that the electron beam will traverse the deflecting region undeviated?

(b) If now the magnetic field is turned off and the potential between the plates remains at the value determined in part (a), what will be the deflection of the electron beam at the screen?

5.2 In analyzing the data in Thomson's e/m experiment, the effect of gravity on the trajectory of the electrons is always neglected. Justify that neglect.

5.3 A beam of protons (H^+), α particles (He^{2+}), singly ionized helium atoms, deuterons, and H_2^- ions passes through a velocity selector. All the beam particles that emerge have a speed of 7×10^6 m/s. This heterogeneous beam then enters a region of uniform magnetic field of 0.2 T, directed perpendicular to the particles' velocity. Calculate the radius of curvature of the trajectory for each of the beam's constituents.

5.4 Derive an expression that relates the distance of closest approach between an α particle and a very massive nucleus of charge Ze to the kinetic energy of the incident particle. Calculate the distance of closest approach of a 4.5-MeV and of a 9.0-MeV α particle to a lead nucleus.

5.5 Obtain an expression for the distance of closest approach between an α particle and a nucleus of charge Ze and mass M, taking into account the recoil of the target nucleus.

•5.6 Derive an expression that relates the distance of closest approach to the impact parameter b for scattering of a particle of mass m and charge ze by a very massive nucleus with charge Ze. (Neglect nuclear recoil.)

•5.7 Figure 5.17 shows the results of an experiment in which α particles were scattered through 60° by a thin lead foil. From

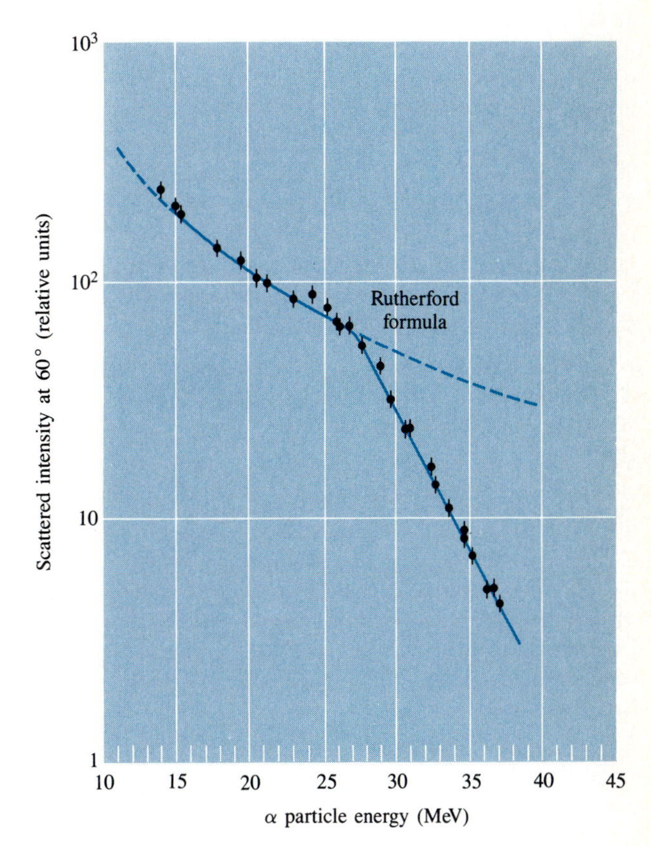

Figure 5.17 Problem 5.7.

those data, estimate the radius of the lead nucleus, using the result of Problem 5.6.

5.8 A beam of 7.0-MeV protons is incident on a 2.0-μm-thick silver foil. What fraction of the beam is scattered through angles greater than (a) 10°? (b) 20°? (c) 90°? What fraction of the incident beam is scattered less than 20°?

5.9 In a scattering experiment, a 2.0-nA proton beam is incident on a 3.0-μm-thick tungsten foil. The number of protons scattered through angles greater than 90° is 4×10^4 per second. Determine the energy of the incident protons.

5.10 In a scattering experiment, the area of the particle detector is 0.6 cm^2 and the distance between the detector and a 1.0-μm-thick tantalum target is 50 cm. The incident beam of 8.0-MeV protons carries a current of 0.5 nA. How many protons per second will be detected for a scattering angle of (a) 60°? (b) 90°?

•5.11 A beam of 6.0-MeV protons is incident on an indium target. Determine the distance of closest approach of the protons to indium nuclei for protons scattered through 60° and through 120°. For these protons, also calculate the kinetic energy at the distance of closest approach to the target nuclei.

5.12 What is the difference in wavelength of the first (long-wavelength) line of the Balmer series for hydrogen and deuterium?

5.13 Suppose that the force of attraction between a proton and an electron were proportional to r rather than $1/r^2$; that is, $F = -kr$. Use the angular momentum quantization condition to show that the stationary-state radii are given by $r_n = (n\hbar/\sqrt{km})^{1/2}$ and that the stationary-state energies are $E_n = n\hbar\omega$, where ω is the angular frequency of revolution of the particle of mass m in this central force field.

5.14 In calculating atomic energy levels, relativistic corrections have been neglected. (a) Show that this is a good approximation for hydrogen; estimate the error due to this approximation. (b) Estimate the magnitude of the relativistic correction for the wavelength of the K_α line of tungsten.

5.15 Show that the speed of an electron in the stationary state of quantum number n in the Bohr model is given by $v = \alpha c/n$, where α is the fine structure constant.

5.16 Express the energy of the stationary states of the hydrogen atom in terms of the fine structure constant α and the rest energy of the electron.

5.17 In the Bohr-Rutherford model of hydrogen, the electron moves in a circular orbit about the proton. (a) Calculate the electric current associated with that orbital motion for the ground state. (b) Determine the magnetic field at the proton due to this current. (c) Repeat these calculations for the first excited state of hydrogen.

5.18 Equation (5.34) assumes that all of the energy of the atomic transition appears as electromagnetic radiation. (a) Show that to conserve both energy and momentum, the atom must recoil. (b) Show that in consequence of this recoil, the energy of the photon that is emitted is, to good approximation, given by $h\nu = \Delta E/(1 + \Delta E/2Mc^2)$, where M is the mass of the atom.

5.19 The He$^+$ ion is essentially identical to the hydrogen atom, except that the nucleus has a mass four times that of the proton and a charge twice that of the proton. (a) What transitions of the He$^+$ ion will give emission lines whose wavelengths are very nearly equal to the first two lines of the Balmer series? Calculate the difference between the wavelength of the first Balmer line and the emission line from He$^+$ that is of nearly the same wavelength.

5.20 The *muonium* atom is the bound state of a μ^- meson (muon) and a proton. The mass of the muon is $207m_e$, where m_e is the electron mass. (a) Calculate the short- and long-wavelength limits of the Balmer series of muonium. (b) Are there any transitions of muonium that give emission lines in the visible region? If so, give a few of these transitions.

5.21 *Positronium* is the bound state of an electron and a positron. Calculate the energies of the three lowest bound states of positronium.

5.22 The atoms of a gas of atomic hydrogen are excited to the $n = 4$ state. What are the wavelengths of the lines emitted if (a) only transitions for which $\Delta n = 1$ are allowed and (b) transitions with $\Delta n \geq 1$ are allowed?

5.23 What is the wavelength of the Balmer line that corresponds to the $n = 5$ to $n = 2$ transitions?

5.24 A line of wavelength $\lambda = 1091$ nm is observed in the hydrogen spectrum. Identify the transition that leads to this line.

5.25 What is the minimum potential that must be applied across an X-ray tube to observe the K_α line (a) of molybdenum? (b) of tungsten?

5.26 Calculate the approximate wavelength of the K_α lines of molybdenum and tungsten.

5.27 The most energetic characteristic X ray emitted by an X-ray tube has a wavelength of 0.212 nm. What is the most likely target material?

5.28 The tube for a Franck-Hertz experiment is filled with a gas whose atoms have excited states 6.0 and 7.6 eV above the ground state. Assume that excitation to other states is negligible. Sketch the curve of current versus grid voltage that one might obtain for cathode-to-grid voltages up to 20 V.

5.29 The following are some of the wavelengths, in nanometers, of lines emitted by singly ionized helium: 121.5, 164, 230.6, 320.3, 541, 656. Identify the transitions that give rise to these emission lines.

•5.30 Emission lines from a hot gas are broadened by several effects, among them random motion of the gas atoms (Doppler broadening). If the gas is atomic hydrogen and its temperature is 700 K, estimate the Doppler broadening of the first line of the Balmer series.

The Wave Nature of Particles

Chapter 6

I thus arrived at the following overall concept which guided my studies: For both matter and radiation, light in particular, it is necessary to introduce the corpuscle concept and the wave concept at the same time. In other words, the existence of corpuscles accompanied by waves has to be assumed in all cases.

Louis Victor duc de Broglie, Nobel Prize address, 1929

6.1 INTRODUCTION: THE DE BROGLIE RELATION

Although by 1924 most physicists had become reconciled with the wave-particle dualism of electromagnetic radiation, the suggestion by Prince Louis de Broglie that particles, such as electrons and protons, should exhibit wavelike properties was greeted with almost universal skepticism, even derision.[1]

That very speculative proposition in de Broglie's doctoral dissertation, for which, at the time, there was not a shred of experimental support, was based on a conviction that nature is inherently symmetric. If one form of energy which usually propagated as a wave could assume corpuscular character, another, which is usually associated with a particle of finite mass, might well display wave properties.

De Broglie deduced the connection between particle and wave properties from the Einstein-Planck expression for the energy of an electromagnetic wave and the classical result for the momentum of such a wave. The two expressions are

$$E = h\nu \tag{6.1}$$

$$p = \frac{E}{c} \tag{6.2}$$

Figure 6.1 Louis Victor duc de Broglie (1892–1987). *(Culver Pictures.)*

If we now replace E by $h\nu = hc/\lambda$ we arrive at the de Broglie relation

$$p = \frac{h}{\lambda} \qquad \lambda = \frac{h}{p} \tag{6.3}$$

Although the conjecture that a wave of wavelength $\lambda = h/p$ is associated with a particle seemed rather farfetched, it had the merit of providing an interpretation of Bohr's angular momentum quantization rule that could be appreciated in

1. L. de Broglie, *Ann. de Phys.* **10,** 22 (1925). Prince Louis Victor de Broglie was awarded the 1929 Nobel Prize "for his discovery of the wave nature of electrons."

terms of classical arguments. In classical mechanics, the wavelengths of natural resonances are simply related to the spatial dimensions of the vibrating system. Standing waves in organ pipes and on strings, and resonant frequencies of electromagnetic waves in cavities, are but a few examples. If, as de Broglie surmised, one should associate some sort of wave with a particle, such as the electron in orbit about the proton of the hydrogen atom, it would be reasonable to impose a similar resonance condition on the wavelength of that electron.

To maintain a standing wave over the circumference of a circular orbit, the wavelength must be an integral fraction of that circumference. That is,

$$2\pi r_n = n\lambda_n \tag{6.4}$$

Such standing waves are shown schematically in Figure 6.2. If we now use the de Broglie relation (6.3), we get

$$2\pi r_n = \frac{nh}{p_n} = \frac{nh}{mv_n}$$

and we immediately obtain the Bohr postulate

$$mv_n r_n = L_n = \frac{nh}{2\pi} = n\hbar$$

Thus de Broglie's hypothesis not only links the Bohr model to classical ideas but confers a new significance on the quantum number n: It is the number of de Broglie wavelengths that fit into the circumference of Bohr's stationary orbits. The concept of a stationary or, at least, long-lived state of a quantum system as a *resonance* has proved very fruitful, especially in nuclear and elementary particle physics. We shall return to these ideas later.

Though suggestive and appealing, the existence of particle waves was by no means confirmed by such heuristic arguments. It remained to establish the presence of truly wavelike properties, namely, interference and diffraction.

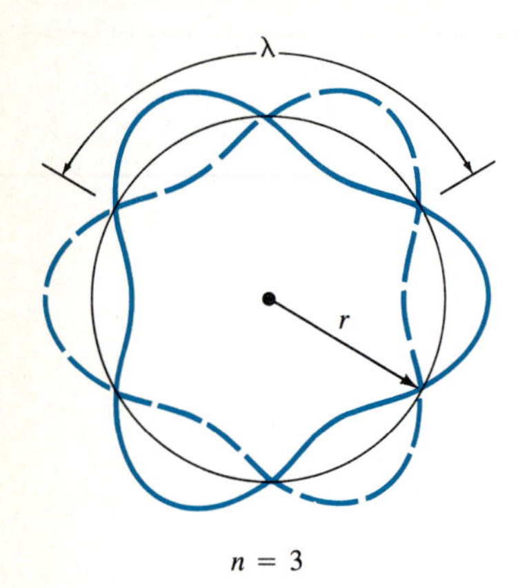

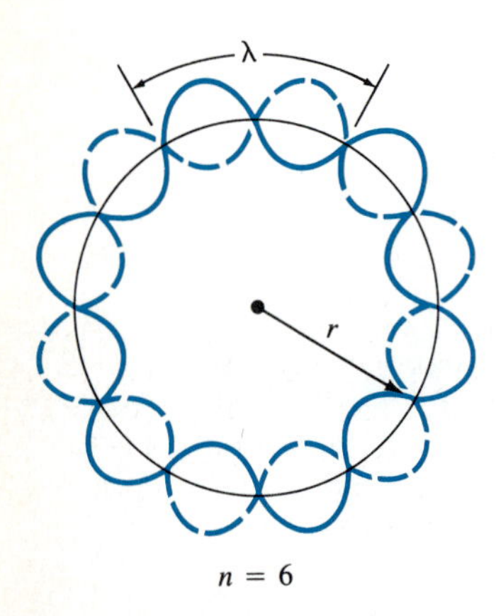

Figure 6.2 Standing de Broglie waves of electrons around the circumference of Bohr orbits. Resonances for $n = 3$ and $n = 6$ are shown.

6.2 EXPERIMENTAL EVIDENCE OF ELECTRON WAVES

It was Elsasser who first called attention to the fact that the de Broglie wavelength of electrons of moderate energy, 10 to 1000 eV, is comparable to the lattice spacing of crystals. Consequently, if electrons did behave like waves, one should observe peaks similar to the Laue spots of X-ray diffraction when electrons are reflected from the surface of a crystal. If electrons are transmitted through a very thin polycrystalline foil, one should see a pattern similar to the Debye-Scherrer rings of X-ray diffraction.

Confirmation of Laue diffraction by electron waves came within a year of Elsasser's suggestion; it was a classic example of serendipity. As Davisson recalled in his Nobel lecture, work had started on the project in 1919, five years before de Broglie's paper, with experiments on secondary emission of electrons from metal surfaces.

> *. . . chance intervened again; it was discovered, purely by accident, that the intensity of elastic scattering varies with the orientation of the scattering crystals. Out of this grew, quite naturally, an investigation of elastic scattering by a single crystal of predetermined orientation. . . . Thus the New York experiment was not, at its inception, a test of the wave theory. Only in the summer of 1926, after I had discussed the investigation in England with Richardson, Born, Franck and others, did it take on this character.*[2]

The experiment involved the measurement of the number of electrons reflected elastically from the surface of a carefully cleaned metal crystal as a function of angle and electron energy. The dependence of the reflected intensity from the [111] face of a single crystal of nickel on electron energy (and, hence, wavelength) is shown in Figure 6.4. From the direction of that peak and the known spacing of the [111] planes in nickel one could calculate the wavelength of the 54-eV electrons, using the Bragg equation

$$2d \sin \theta = m\lambda$$

(see Problem 6.5). Here d is the separation between adjacent crystal lattice planes, θ is the angle between the direction of propagation of the incident waves of wavelength λ and the reflecting planes, and m is an integer. The calculated wavelength agreed with the de Broglie relation within experimental uncertainty.

Debye-Scherrer diffraction rings of electronic origin were observed almost at the same time by G. P. Thomson, son of Sir J. J. Thomson.[3] Subsequently, diffraction effects using atomic beams and beams of neutrons were also observed. The latter, in particular, are especially valuable probes in the study of magnetically ordered structures because, owing to their electrical neutrality, neutrons interact with atoms of a crystal principally via their magnetic fields.

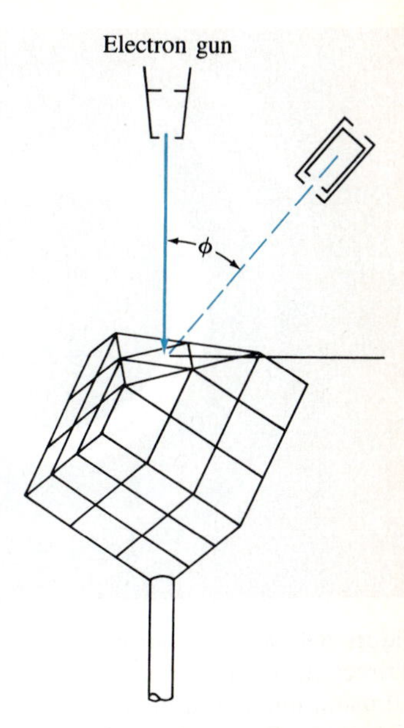

Figure 6.3 Schematic diagram of the Davisson-Germer experiment. Monoenergetic electrons impinge on the [111] face of a cubic crystal (nickel) and are reflected toward the detector. The angle ϕ as well as the energy of the incident electrons is adjustable.

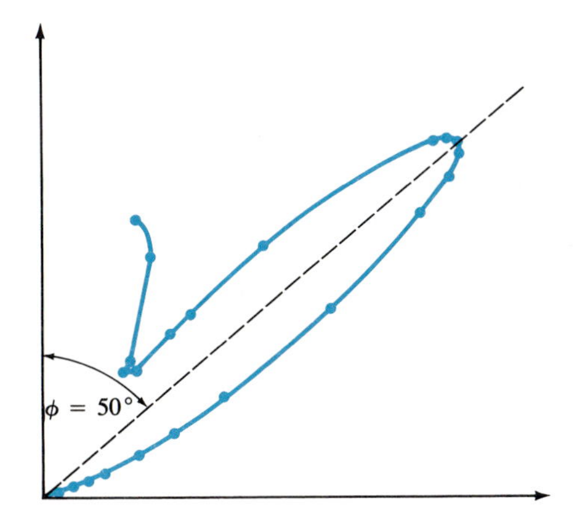

Figure 6.4 In the Davisson-Germer experiment, Bragg reflection, evidenced by a pronounced peak in the reflected intensity, is observed at a scattering angle of 50° when the energy of the incident electrons is 54 eV.

2. C. J. Davisson, Nobel lecture, 1937.

3. C. J. Davisson and G. P. Thomson shared the 1937 Nobel Prize for "their experimental discovery of the diffraction of electrons by crystals."

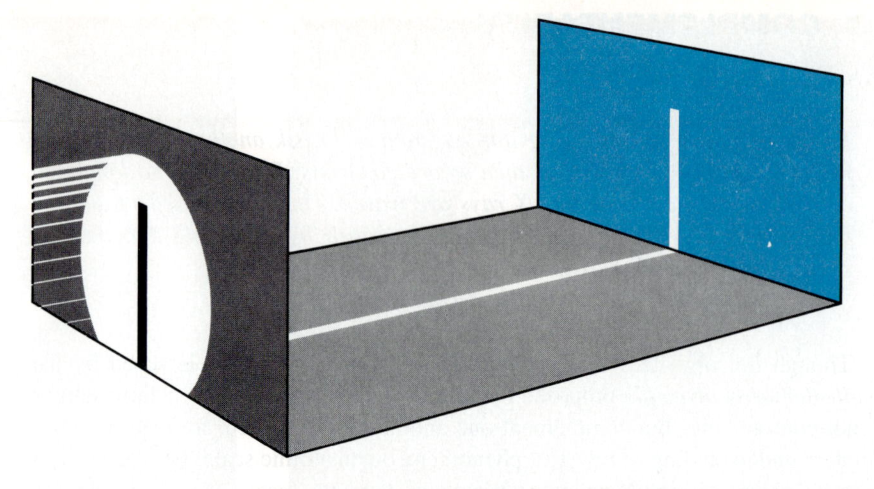

Figure 6.7 According to classical mechanics, the image of a slit upon which a beam of particles is incident is the geometric projection of the slit.

nature of particles. As mentioned earlier, interference and diffraction are the hallmarks of waves, and their observation using electrons gave unequivocal proof of de Broglie's hypothesis. Diffraction is also the phenomenon that best demonstrates the restriction on simultaneous spatial and momentum localization expressed by Equation (6.5).

Suppose we send a beam of particles whose momentum is p_x through a very narrow slit, as indicated in Figure 6.7. The particles that emerge from the slit at $x = 0$ must have been localized in the y direction within an uncertainty $\Delta y = d$, where d is the width of the slit. Classically, if the particles continue to propagate in the x direction, they will strike the screen at $x = D$ within a narrow strip of width d.

If, however, we perform this experiment using a beam of light, we know that the light intensity at the screen will exhibit the typical diffraction pattern shown in Figure 6.8. The width of that pattern, defined by the angular separation between the two minima bordering the central maximum, is 2θ, where θ is given by

$$\sin\theta \simeq \theta = \frac{\lambda}{d} = \frac{\lambda}{\Delta y} \tag{6.6}$$

provided $\theta \ll 1$, or $\lambda \ll d$ (see Figure 6.9). Equation (6.6) demonstrates that the

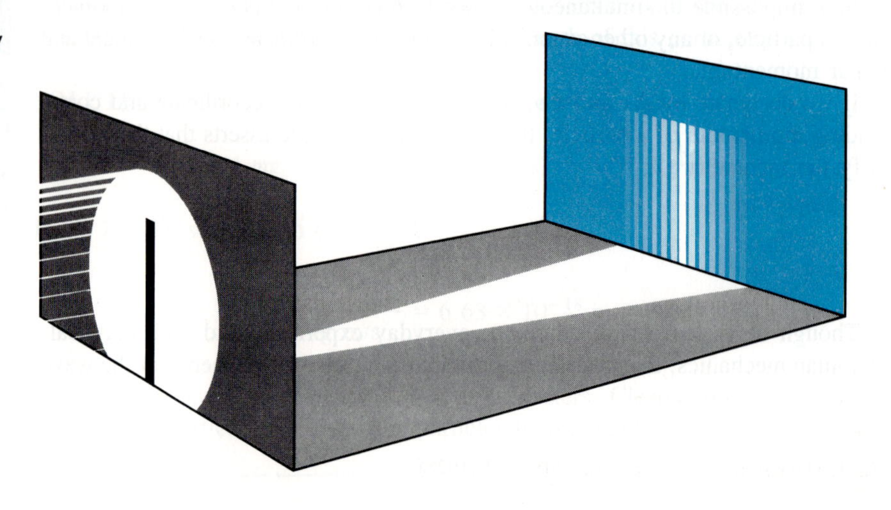

Figure 6.8 If a slit is illuminated by waves (e.g., light), a diffraction pattern is seen on the screen.

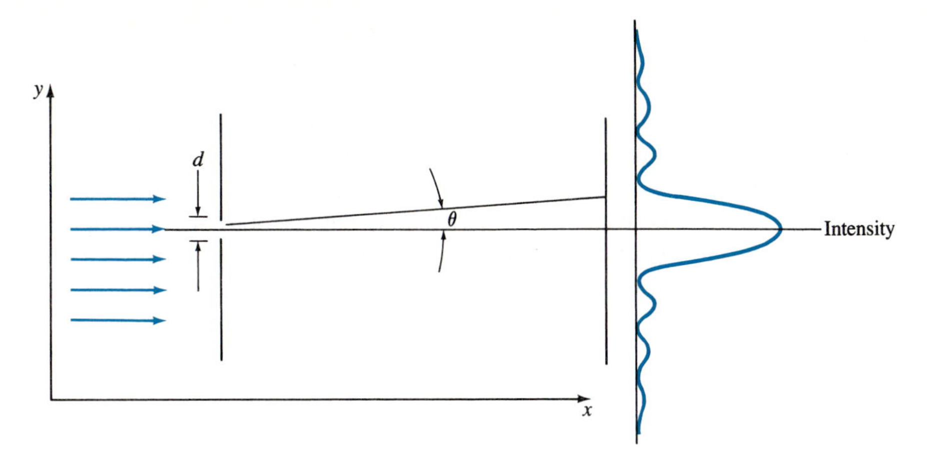

Figure 6.9 The intensity of the diffraction pattern from a thin slit using monochromatic light. The width of the pattern is defined by the angle θ between the central maximum and the first minimum. The angle θ is proportional to the wavelength of the light and inversely proportional to the slit width.

more we try to delimit the position of the wave at $x = 0$ by reducing the width of the slit, the more we broaden the diffraction pattern.

Since particles exhibit wavelike behavior, we can use Equation (6.6) and the de Broglie relation to determine the spread in p_y, the y component of the momentum. Particles that strike the screen at a point δy from the center must have exited the slit moving at an angle $\theta = \tan^{-1}(\delta y/D)$ with respect to the x axis, where D is the distance between slit and screen. Consequently, their x and y components of momentum must have been related by $\tan\theta = \Delta p_y/p_x$, where Δp_y is the y component of momentum of a particle that strikes the screen at the point δy. Combining these two expressions and Equation (6.6), we have

$$\frac{\Delta p_y}{p_x} = \frac{\delta y}{D} = \tan\theta \simeq \theta = \frac{\lambda}{\Delta y} \tag{6.7}$$

provided $\theta \ll 1$. But from the de Broglie relation, the wavelength associated with the particles is $h/p \simeq h/p_x$. Thus,

$$\Delta y \, \Delta p_y \simeq h \tag{6.8}$$

which is consistent with the uncertainty relation, Equation (6.5). We see that the uncertainty principle is a direct and necessary consequence of the wave nature of particles.

The uncertainty principle also places an upper bound on the precision of an energy measurement, and, in light of the mass-energy equivalence, on the measurement of the mass of a particle. If the time interval during which an energy (or mass) measurement can be performed is limited to Δt, the uncertainty in that measurement is related to Δt by

$$\Delta E \, \Delta t \geq \frac{\hbar}{2} \tag{6.9}$$

This limitation comes about as follows. Suppose we wanted to determine the frequency of an oscillator. Frequency measurement is essentially a counting pro-

$$E \geq \tfrac{1}{2}m\omega^2(\Delta x)^2 + \frac{\hbar^2/8m}{(\Delta x)^2} \tag{6.17}$$

The minimum value of E is obtained by differentiating Equation (6.17) with respect to Δx and setting the derivative equal to zero. The result is

$$(\Delta x)_{\text{min}} = \sqrt{\frac{\hbar}{2m\omega}} \tag{6.18}$$

and the corresponding minimum value of the energy is

$$E_{\text{min}} = \tfrac{1}{2}\hbar\omega \tag{6.19}$$

The lowest energy state of the harmonic oscillator is not the state in which the mass is ''at rest'' but that state in which the mass vibrates with an amplitude given by Equation (6.18) and an energy of $\frac{1}{2}\hbar\omega$.

EXAMPLE 6.5 Width of Atomic Spectral Lines. When an atom is raised to an excited state, it usually decays to the ground state in about 10^{-8} s. Consequently, there is an uncertainty in the energy of this short-lived excited state and a corresponding uncertainty in the wavelength of the radiation emitted as the atom returns to the ground state. It is of interest to examine this uncertainty width, and compare it to other line-broadening effects.

SOLUTION The wavelength of the emitted photon is

$$\lambda = \frac{hc}{E_i - E_f} \tag{6.20}$$

where E_i is the energy of the initial, excited state and E_f the energy of the final, ground state. According to Equation (6.9) this energy difference has an uncertainty of

$$\Delta E \geq \frac{\hbar}{2\,\Delta t}$$

If we differentiate Equation (6.20) with respect to E_i we obtain

$$\frac{d\lambda}{dE_i} = -\frac{hc}{(E_i - E_f)^2} = -\frac{\lambda}{E_i - E_f}$$

Consequently,

$$\frac{\Delta\lambda}{\lambda} = -\frac{\Delta E}{E_i - E_f}$$

In this instance, ΔE is in the neighborhood of 5×10^{-8} eV, or about eight orders of magnitude smaller than the energy difference of atomic transitions, which are of the order of a few electrovolts. The uncertainty line width for atomic transitions is, therefore, quite small, and is almost always overshadowed by other sources of line broadening such as the Doppler effect (see Problem 5.30). The lifetimes of excited nuclear states are, however, very much shorter, typically between 10^{-20} and 10^{-18} s, and γ rays from most nuclear transitions have natural line widths of the order of a few keV.

6.5 THE WAVE-PARTICLE DUALITY AND COMPLEMENTARITY: A *GEDANKEN* EXPERIMENT

The conclusive experiment confirming the wave theory of light was Young's demonstration of two-slit interference. Let us imagine repeating this same experiment using, now, a beam of monochromatic (i.e., monoenergetic) electrons as illustrated in Figure 6.10. Instead of a photographic film, we place a dense array of detectors at the screen; each detector signals the arrival of an electron by a pulse. Moreover, the pulse height is proportional to the charge deposited in the detector.

We begin our thought (*Gedanken* is the German word) experiment by blocking one of the slits, S_2. We note, first of all, that the pulse height from every detector is always the same. This simply tells us that electrons do not break up; each electron delivers a charge $-e$ to the detector.

Next, we note that the rate at which the pulses appear depends on the location of the detector, the greatest counting rate being recorded for the detector directly opposite slit S_1. If we divide the counting rate of each detector by that for the complete bank of detectors, we obtain the relative probability that an electron will strike a given detector. That relative probability is indicated by P_1 in Figure 6.10.

We next block off slit S_1 and open S_2, and observe a very similar pattern, now centered on the point opposite S_2.

Last, we open both slits. Contrary to what one might have expected if electrons propagated as classical small particles, the counting rate is not the sum of P_1 and P_2. Instead, the distribution looks like P_{12}, which is the intensity distribution of the interference pattern produced when a double slit is illuminated by coherent monochromatic light.

The direct, experimental observation of electron double-slit interference is conclusive evidence of the wave nature of these particles. It also might appear to suggest that, unlike a "true" particle, an electron does not pass through either S_1 or S_2 but somehow is partitioned, a portion passing through each of the two slits. Yet, each detector records pulses that correspond to one whole electron, Also, one can easily check that the interference pattern is not due to the mutual Coulomb interaction of electrons that pass nearly simultaneously through the two slits by simply reducing the beam intensity so that, on the average, only one electron per second penetrates either slit. If we then start to record pulses at the detectors, they will at first appear almost randomly distributed; but after a while, an obvious pattern emerges, as in Figure 6.11, page 127.

If a whole electron passes through either S_1 or S_2, one should be able to tell through which of the two slits the electron has gone by observing it just after it has penetrated the mask with the double slit. We can imagine doing this by placing a

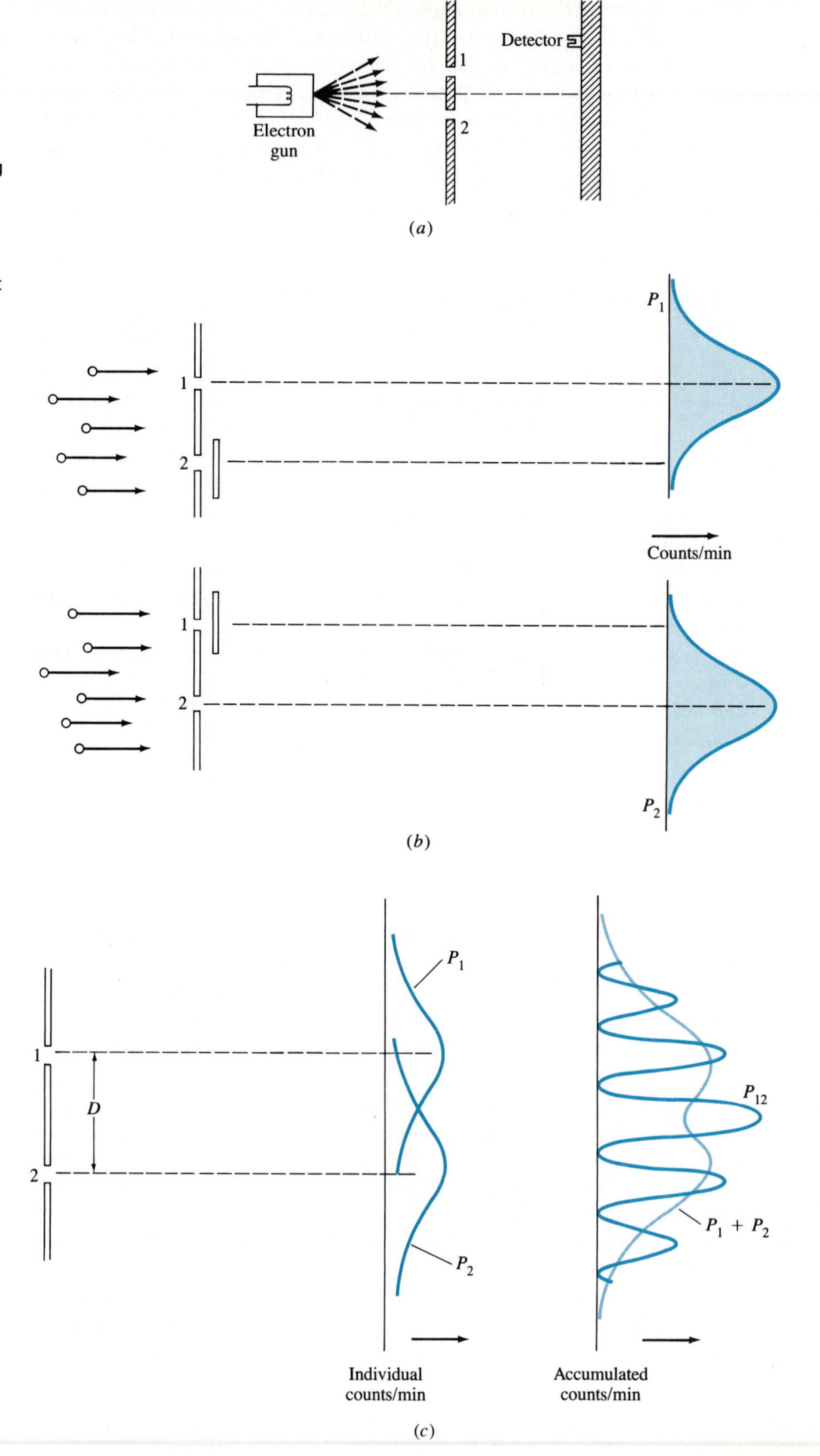

Figure 6.10 The thought *(Gedanken)* experiment in which a double slit is "illuminated" by a beam of monoenergetic electrons. (*a*) The arrangement of source, slits, and detector array (of which only one is shown here). (*b*) The probabilities P_1 and P_2 of observing an electron at a given detector when only slit S_1 or slit S_2, respectively, is open. (*c*) P_{12} is the observed probability distribution when both slits are open. Note that $P_{12} \neq P_1 + P_2$.

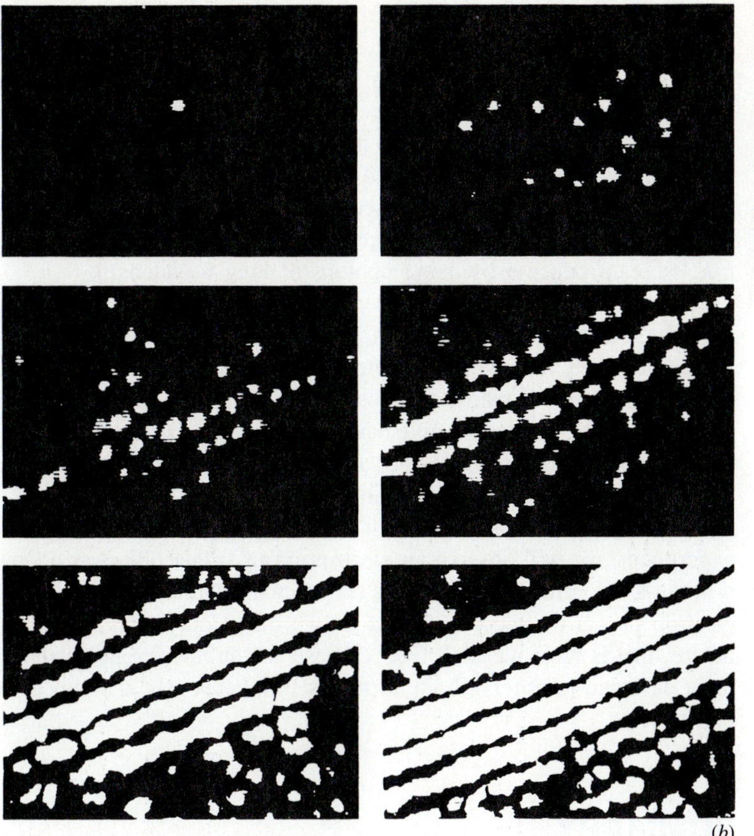

Figure 6.11 (*a*) Development of the double-slit interference pattern when the intensity of the incident electron beam is greatly reduced. (*From G. F. Missiroli and G. Pozzi,* American Journal of Physics, *vol* 44, no. 3, 1976, *p.* 306.) (*b*) An experimentally observed electron interference pattern from a double-slit arrangement (*Stanley Hirschi,* American Journal of Physics, vol. 42, 1974, pp. 4–11.).

light source as shown in Figure 6.12, and looking for flashes of light due to scattering by the electron. The origin of the flash will tell use whether the electron has come through S_1 or S_2.

But when we do this more elaborate experiment, some disaster seems to strike; the interference pattern vanishes just as soon as we turn on the light source. We do see a light flash each time a detector records the arrival of an electron on the screen, and we can now associate every detector signal with an electron's passage through

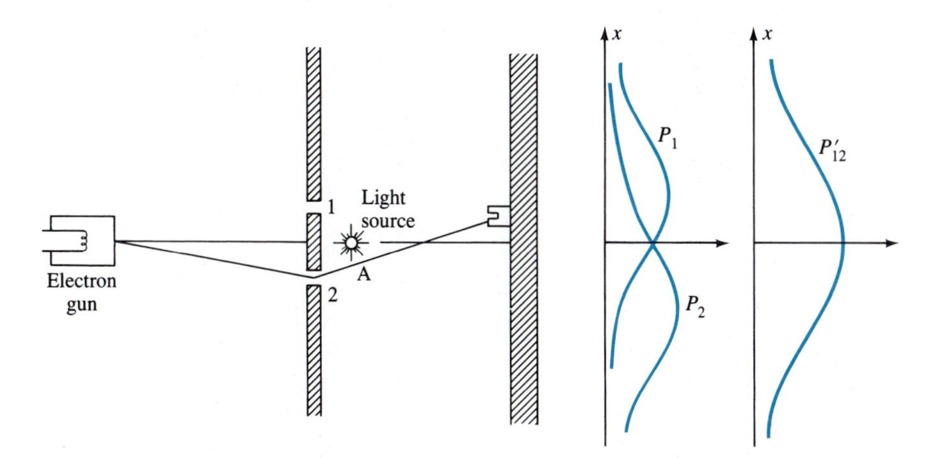

Figure 6.12 A more elaborate *Gedanken* experiment. A light source is placed behind the slit to permit an observer to determine through which of the two slits an electron has passed. If both slits are open and the wavelength of the light from the source is less than the slit separation, permitting sufficient resolution to distinguish between the two slits, then the intensity pattern on the screen is $P'_{12} = P_1 + P_2$. No interference pattern is observed.

either S_1 or S_2. If we plot the distribution of detector pulses, the result is now the sum of P_1 and P_2.

What has happened is that in the very process of scattering a photon, some momentum is transferred to the electron. That is, in the very act of ascertaining the location of the electron, we disturb its momentum and the electron will not strike the screen at the spot where it might have, had no photon disturbed its progress.

The solution to this dilemma seems obvious: use photons of such low momenta that their influence on the electrons is negligible. Since, for photons, $p = E/c = h/\lambda$, we can reduce p by using photons of long wavelength.

When we now try the experiment again, gradually increasing the wavelength of the light source, we find that when $\lambda > d$, the interference pattern becomes evident again. However, because the wavelength of the photon is so large, we can no longer resolve two sources whose separation is d. We will see a flash, but cannot tell whether the electron responsible has gone through S_1 or S_2. The result is consistent with the uncertainty principle. Our attempt to simultaneously determine the position and momentum of the electron cannot succeed.

The complementarity principle of Bohr tells us that in this, as in any other experiment, we will observe an electron or any other object as a ''particle'' if we design the experiment to reveal its corpuscular nature. (Through which slit did the electron pass?) On the other hand, if we direct our effort to observing its wave nature, that is what we shall find. (Do we see an interference pattern?) An experiment designed to observe both attributes simultaneously is doomed to failure from the outset.

SUMMARY

De Broglie proposed that particles manifest wavelike properties with wavelength given by

$$\lambda = \frac{h}{p}$$

The wavelike behavior of electrons was verified experimentally by Davisson and Germer and by G. P. Thomson, who demonstrated electron diffraction by crystals, similar to X-ray diffraction.

The apparent dichotomy of particlelike and wavelike behavior was clarified by Bohr's *complementarity principle*. Whether particlelike or wavelike properties are observed is determined by the nature of the experiment.

The Heisenberg *uncertainty principle* is a direct consequence of wavelike behavior. Accordingly,

$$\Delta x \, \Delta p_x \geq \frac{\hbar}{2} \qquad \Delta E \, \Delta t \geq \frac{\hbar}{2}$$

where Δx, Δp_x, ΔE, and Δt are the uncertainties in position, momentum, energy, and lifetime.

Bibliography

K. K. Darrow, "Davisson and Germer," *Scientific American,* May 1948.

George Gamow, *Thirty Years That Shook Physics,* Anchor/Doubleday, New York, 1966.

See also the bibliography at the end of Chapter 5.

PROBLEMS

6.1 What are the de Broglie wavelengths of (a) an electron, (b) a proton, and (c) an α particle of 2-keV kinetic energy?

6.2 What is the kinetic energy of an electron whose de Broglie wavelength is 50 nm?

6.3 What are the kinetic energies of (a) an electron, (b) a proton, and (c) an α particle whose de Broglie wavelengths are 0.01 nm?

6.4 What is the average de Broglie wavelength of an oxygen molecule at room temperature?

6.5 You are asked to derive the Bragg equation for X-ray or electron-wave diffraction. Consider a set of crystal planes, as in Figure 6.13, and a wave incident on that set of partially reflecting planes. The direction of propagation of the wave, of wavelength λ, makes an angle θ with the crystal planes. Show that the waves reflected from adjacent planes will interfere constructively when $2d \sin\theta = m\lambda$, where d is the separation between adjacent planes.

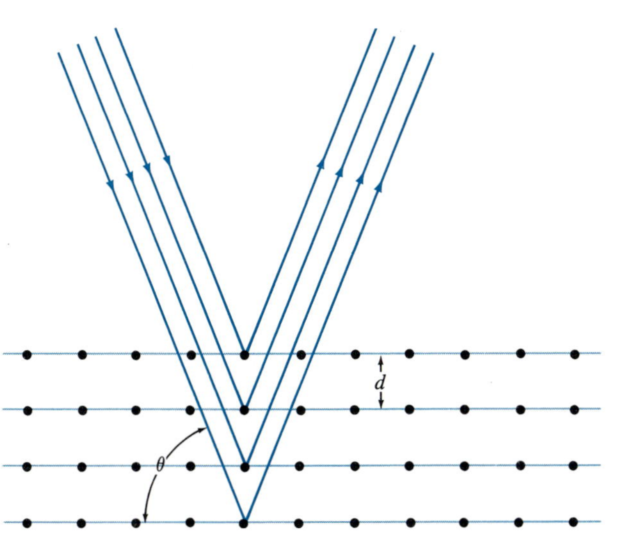

Figure 6.13 Problem 6.5.

6.6 Neutrons are frequently used to study magnetic order in crystals such as iron, nickel, and manganese fluoride. What should be the approximate wavelength of the neutrons used in these experiments? At what energy will neutrons have such a wavelength, and what is the temperature at which neutrons have that thermal energy?

6.7 Determine the approximate size of the hydrogen atom in its ground state by finding the minimum total energy of the orbiting electron consistent with the uncertainty condition.

6.8 Prior to the discovery of the neutron in 1934 it was believed that the nucleus consisted of protons and electrons. For example, ^4He was envisaged as a composite of four protons and two electrons confined within the nuclear radius. What would be the kinetic energy of an electron confined within a sphere of 4×10^{-15} m diameter, the typical diameter of a light nucleus? Is that result consistent with the energies of electrons emitted in β decay, which range from a few keV to about 1 MeV?

•6.9 Show that, in general, the de Broglie wavelength of a particle of rest mass m_0 and kinetic energy E_k is given by

$$\lambda = \frac{hc}{E_k(1 + 2m_0c^2/E_k)^{1/2}}$$

6.10 Show that in the extreme relativistic limit, a particle's de Broglie wavelength approaches that of a photon of the same energy.

6.11 Show that in the classical limit, the expression given in Problem 6.9 reduces to $\lambda = h/m_0v$.

6.12 The nucleus of ^{227}Th is naturally radioactive, decaying to ^{223}Ra with the emission of an α particle. In this process, the daughter nucleus is often left in a short-lived excited state, and then decays to the ground state with the emission of an 80-keV γ ray. The natural line width of this γ ray is 0.6 keV. What is the lifetime of the excited state of the ^{223}Ra nucleus?

•6.13 The Σ^* particle has a rest energy of 1385 MeV. In a series of measurements, the uncertainty width of this rest energy is about 35 MeV. Suppose that when this particle is created in a high-energy collision it travels at a speed of $0.994c$ in the bubble chamber. How long is the track left by this particle?

6.14 In quantum mechanics, apparent violation of energy conservation is possible if the lifetime of the new state of the system is so brief that the uncertainty in its energy is sufficiently large. In 1935 Hideki Yukawa proposed that nucleons are held together by the exchange of a particle, now known as the π-meson, or pion. It was known at the time that nuclear forces are of very short range, about 1.5 fm. Suppose that the pion in traveling between two nucleons moves at nearly the speed of light. What is the maximum rest energy that this particle can have without violating energy conservation? (The pion rest mass is 135 MeV/c^2.)

6.15 Show that the uncertainty relation can also be written $\Delta x\,\Delta\lambda \geq \lambda^2/4\pi$.

•6.16 A beam of electrons with velocity v is normally incident on a slit of width d. On passing through this slit, the electrons strike a screen a distance D from the slit. What should be the value of d so that the image on the screen is of smallest width? Determine this slit size for 10-eV electrons and a slit-to-screen distance of 1 m.

6.17 The nucleus ^{5}He (two protons and three neutrons) is unstable and breaks up in 0.8×10^{-21} s into an α particle and a neutron, releasing a total of 0.9 MeV, which is shared between the α particle and the neutron. Suppose that a ^{5}He nucleus at rest decays in this manner. Calculate the average energy of the neutron and the expected experimental width of the neutron energy spectrum.

The Schrödinger Equation

Chapter 7

Anyone who is not shocked by quantum theory has not understood it.

Niels Bohr

I think I can safely say that nobody understands quantum mechanics.

Richard P. Feynman

7.1 INTRODUCTION

By 1925, quantization of electromagnetic waves, proposed almost apologetically two decades earlier, had become accepted by the physics community. The successes of Planck's radiation law, Einstein's lucid explanation of the photoelectric effect, and Compton's results on inelastic scattering of X rays were irrefutable. Moreover, Einstein's and Debye's theories of the specific heat of solids implied a very broad generalization of the quantum concept to all wave phenomena, not only electromagnetic waves. The previous year, de Broglie had proposed that a wave character should be ascribed to particles, and although his idea was not taken seriously by everyone, it was received with much interest by the more creative and imaginative theorists.

In classical physics, understanding light and its interaction with matter developed slowly and came to full fruition only with the work of Maxwell. It was generally recognized that meaningful progress in understanding particle waves could not be achieved without a mathematical formalism for matter waves analogous to Maxwell's equations for the electromagnetic field. Only then could one hope to answer some of the perplexing questions such as

1. What is the relationship between the particle-wave field—the wave function, as it is now called—and physical observables such as energy, position, momentum, and angular momentum?
2. How do particle fields interact with their environment? Are these waves reflected and refracted as are electromagnetic waves? How should one treat the interaction between particles—for example, the Coulomb interaction between charged particles—in the context of the wave formalism?
3. How can the corpuscular and wave character of electromagnetic waves be reconciled? Presumably, it should be possible to recast Maxwell's equations into a quantum formalism.

In this chapter we shall first present heuristic arguments that lead to a wave equation for particle waves, the famous Schrödinger equation. We shall then consider the first two questions, at least for some of the simplest situations. The formalism alluded to in the last question has been developed and is employed extensively in modern theory. Although we shall refer to it briefly in some later chapters, a detailed discussion is beyond the scope of this text.

7.2 THE ONE-DIMENSIONAL SCHRÖDINGER EQUATION

Figure 7.1 Erwin Schrödinger (1887–1961). Schrödinger and Dirac shared the 1933 Nobel Prize in physics for "the discovery of new productive forms of atomic theory." *(Wolfgang Plaundler/Courtesy AIP, Niels Bohr Library.)*

When embarking on a new and difficult topic it is advisable to simplify the formalism as much as possible, at least at the beginning, so that central issues will not be obscured by mathematical complexities. In this chapter we therefore focus on one-dimensional systems and defer generalization to the real world of three dimensions to Chapter 8.

Let us be clear about one thing from the outset. **We shall not derive the Schrödinger equation.** That equation *cannot* be derived any more than Newton's laws could be. Both are formal, mathematical expressions that prove remarkably successful in describing and predicting natural phenomena. Both ultimately draw their justification from the consistent agreement between theory and experiment, and both must be abandoned or modified in the face of disagreement. Just as Newton's second law is not relativistically invariant, so also is the Schrödinger equation, and it must give way to a relativistic wave equation. But, just as Newtonian mechanics is perfectly adequate in solving a vast complex of nonrelativistic problems, so is the Schrödinger equation when applied to a manifold of problems in atomic physics.

Our aim, then, is to deduce an equation that must be satisfied by some function $\Psi(x,t)$ which shall represent the wave field of a particle. In this undertaking we shall be guided by the classical wave equation

$$\frac{\partial^2\Psi}{\partial x^2} = \frac{1}{v^2}\frac{\partial^2\Psi}{\partial t^2} \tag{7.1}$$

and by the Planck-Einstein equation

$$E = \hbar\omega \tag{7.2}$$

and the de Broglie relation

$$\mathbf{p} = \frac{h}{\lambda} = \hbar\mathbf{k} \qquad k = \frac{2\pi}{\lambda} \tag{7.3}$$

where $\mathbf{k}$ is the wave vector. Since we are restricting ourselves to one dimension, p and k can be treated as scalars.

If Equation (7.1) were used to describe electromagnetic waves, Ψ might represent the vector $\mathbf{E}$ or $\mathbf{B}$; for waves on a string, Ψ would be the transverse displacement; and for sound waves in air, Ψ would be the departure of the local pressure from ambient.

A solution of Equation (7.1) is the familiar plane wave

$$\Psi(x,t) = Ae^{i(kx-\omega t)} \tag{7.4}$$

provided

$$v = \frac{\omega}{k} \qquad \omega = vk \tag{7.5}$$

The velocity v is generally referred to as the *phase velocity*.

If we multiply both sides of Equation (7.5) by $\hbar$ we have

$$\hbar\omega = E = v\hbar k = v(\frac{h}{\lambda}) = vp \tag{7.6}$$

where we have used Equations (7.2) and (7.3).

Equation (7.6) is indeed correct for electromagnetic waves, for which $E = cp$, where c is the velocity of light. For a particle with finite rest mass, however, the nonrelativistic energy is not cp but

$$E = \frac{p^2}{2m} + V(x,t) \tag{7.7}$$

where $V(x,t)$ is the particle's potential energy, which may be a function of both position and time.

To see how we might proceed to construct a differential equation for material particles analogous to the classical wave equation, we note that partial differentiation of the wave function in Equation (7.4) with respect to the coordinate x introduces the factor ik, whereas partial differentiation with respect to t introduces the factor $-i\omega$. Thus

$$i\hbar\frac{\partial\Psi}{\partial t} = \hbar\omega\Psi = E\Psi \tag{7.8}$$

and

$$-i\hbar\frac{\partial\Psi}{\partial x} = \hbar k\Psi = p\Psi \tag{7.9}$$

Equations such as (7.8) and (7.9) that are of the general form

$$(\text{Differential operator})\Psi = (\text{constant})\Psi$$

are called *eigenvalue* equations. The constant is the *eigenvalue* and Ψ the corresponding *eigenfunction*. (The words derive from the German *eigen,* meaning "singular" or "unique.") Since E and p are the energy and momentum, respectively, the differential operators

$$E_{\text{op}} = i\hbar\left(\frac{\partial}{\partial t}\right) \tag{7.10}$$

$$p_{x\text{op}} = -i\hbar\left(\frac{\partial}{\partial x}\right) \qquad \mathbf{p}_{\text{op}} = -i\hbar\nabla \tag{7.11}$$

are identified as the energy and momentum operators in quantum mechanics. Here the symbol ∇ is the *gradient:*

$$\nabla = \mathbf{i}\left(\frac{\partial}{\partial x}\right) + \mathbf{j}\left(\frac{\partial}{\partial y}\right) + \mathbf{k}\left(\frac{\partial}{\partial z}\right) \tag{7.12}$$

where **i, j, k** are the unit vectors along the three Cartesian coordinates.

We now see that the differential equation that corresponds to Equation (7.7) is

$$-\frac{\hbar^2}{2m}\frac{\partial^2\Psi}{\partial x^2} + V(x,t)\Psi = i\hbar\frac{\partial\Psi}{\partial t} \tag{7.13}$$

Equation (7.13) is the one-dimensional *time-dependent Schrödinger equation*. It can be, and has been, applied to a vast number of problems in atomic physics with truly astonishing success. In the realm of that microcosm it has held sway with as much force as have Newton's laws in the macrocosm of projectiles, fluids, and planets. We shall examine a few simple applications presently and find an answer to the meaning and significance of the wave function itself. First, however, let us consider the very common situation where the potential is a function of position only, independent of time.

7.3 THE TIME-INDEPENDENT SCHRÖDINGER EQUATION

If the potential is a function of x only, the differential equation (7.13) is *separable*. That is, it evolves into two different equations, one in the variable x, the other in the variable t. To see how this comes about, recall that the plane-wave solution, Equation (7.4), can be written as a product

$$\Psi(x,t) = Ae^{ikx}e^{-i\omega t}$$

and try a solution for the Schrödinger equation that is the product of a function of x only and a function of t only,

$$\Psi(x,t) = \psi(x)\phi(t) \tag{7.14}$$

Equation (7.13) now takes the form

$$i\hbar\psi(x)\frac{\partial\phi(t)}{\partial t} = -\frac{\hbar^2}{2m}\phi(t)\frac{\partial^2\psi(x)}{\partial x^2} + V(x)\phi(t)\psi(x)$$

and dividing by $\Psi(x,t) = \phi(t)\psi(x)$, we obtain

$$i\hbar\frac{1}{\phi}\frac{\partial\phi}{\partial t} = -\frac{\hbar^2}{2m}\frac{1}{\psi}\frac{\partial^2\psi}{\partial x^2} + V(x) \tag{7.15}$$

The left-hand side of Equation (7.15) is a function of t only, the right-hand side a function of x only. Since Equation (7.15) must be valid for *any* x and t, this can be so only if the two sides of Equation (7.15) are equal to a constant. Recalling that the operator $i\hbar(\partial/\partial t)$ is the energy operator, we set this constant equal to the energy E and arrive at the *time-independent Schrödinger equation*

$$-\frac{\hbar^2}{2m}\frac{\partial^2\psi}{\partial x^2} + V(x)\psi = E\psi \tag{7.16}$$

In three dimensions, the time-independent Schrödinger equation is usually written

$$-\frac{\hbar^2}{2m}\nabla^2\psi + V(\mathbf{r})\psi = E\psi \tag{7.17}$$

where ∇^2 represents the differential operator

$$\nabla^2 = \frac{\partial^2}{\partial x^2} + \frac{\partial^2}{\partial y^2} + \frac{\partial^2}{\partial z^2}$$

Last, since the right-hand side of Equation (7.7) is the classical Hamiltonian, Equation (7.17) is frequently written in the abbreviated form

$$\mathcal{H}\psi = E\psi \qquad \mathcal{H} = -\frac{\hbar^2}{2m}\nabla^2 + V(\mathbf{r}) \tag{7.18}$$

7.4 INTERPRETATION OF THE WAVE FUNCTION: PROBABILITY DENSITY AND EXPECTATION VALUES

Having deduced a wave equation for the wave function ψ, we now turn to the first question asked at the start of this chapter. What is the relationship between ψ and quantities such as position, energy, momentum, or other dynamic variables? What meaning, what physical significance should one attach to the wave function?

To answer these questions we again turn to electrodynamics for guidance. We recall that the energy density of an electromagnetic field is proportional to the square of the electric field, not the field itself. If we expressed the electric field vector in complex form, the energy density would be written $\frac{1}{2}\epsilon_0|\mathrm{E}|^2 = \frac{1}{2}\epsilon_0 E^*E$, where the asterisk means "complex conjugate." Since mass itself is a form of energy and the wave function ψ represents the wave field of a particle of mass m, $\psi^*\psi$ may be thought of as an energy density associated with that mass. Accordingly, the product

$$\psi^*\psi = |\psi|^2 = P(x) \tag{7.19}$$

is to be interpreted as a *probability density*. By this we mean that the probability that the particle whose wave function is $\psi(x)$ will be found in the interval dx is given by

$$P(x)\,dx = \psi^*(x)\psi(x)\,dx \tag{7.20}$$

Note that we can no longer make a precise statement regarding the position of the particle; we can only conclude something about the probability that when we do an experiment to locate the particle, we shall find it in a given interval dx. That is what one might anticipate in light of the uncertainty principle.

The particle, if it exists at all, must be somewhere. In other words, the integral of $P(x)$ over all space should be unity. One therefore imposes a *normalization condition* on the wave function ψ:

$$\int_{-\infty}^{\infty} \psi^*(x)\psi(x)\, dx = 1 \tag{7.21}$$

A wave function that satisfies Equation (7.21) is said to be *normalized*. If $\psi(x)$ is not normalized, that is, if

$$\int_{-\infty}^{\infty} \psi^*(x)\psi(x)\, dx = C \neq 1$$

it can always be normalized by the simple expedient of dividing $\psi(x)$ by $\sqrt{C}$, provided, of course, that the integral does not diverge. Normalization imposes, therefore, a further *boundary condition* on the wave function, namely,

$$\psi(x) \to 0 \qquad \text{as} \qquad x \to \pm\infty \tag{7.22}$$

There is yet another condition that the wave function must meet. Since the energy cannot be infinite, it follows from Equation (7.16) that $\partial^2\psi/\partial x^2$ must also be finite unless $V(x)$ is singular. Consequently, the wave function and its first derivative must be continuous everywhere.

Let us summarize the conditions that the wave function must satisfy:

1. $\psi(x)$ must be a solution of the Schrödinger equation, Equation (7.16).
2. $\psi(x)$ must be normalizable; i.e., $\psi(x) \to 0$ as $x \to \pm\infty$.
3. $\psi(x)$ must be a continuous function of x.
4. $\psi'(x) \equiv \partial\psi/\partial x$ must be a continuous function of x.

As we shall see presently, these conditions generally impose severe limitations on the energy E. It is always possible to integrate the second-order differential equation (7.16) for any nonsingular potential energy and any arbitrary value of the constant E. However, unless E is one of a special class of constants, the solution will diverge rather than approach zero as x approaches infinity. Values of E for which $\psi(x)$ diverges are not physically acceptable solutions of the wave equation. The special values of E for which the four boundary conditions are met are known as the *energy eigenvalues*.

If the potential is not time-dependent, we shall always be able to find one or more eigenvalues of the energy. These are the energies of the stationary states postulated by Bohr. Often, however, we want to know more than just the energy—for example, the momentum or position of the particle. We know from the uncertainty principle that we cannot expect to obtain precise answers to questions such as *What is the position of the particle?* or *What is the momentum of the particle?* Instead, we must be content to know the probability that a measurement of these variables will yield a particular result. The average of a large number of such measurements of a variable is known as the *expectation value* of that variable, and it is this that we can calculate once we know the wave function $\psi(x)$.[1]

For example, since $P(x)\,dx$ is the probability that the particle will be found in the interval dx, the average or expectation value of x, denoted by $\langle x \rangle$, is

$$\langle x \rangle = \int_{-\infty}^{\infty} xP(x)\,dx = \int_{-\infty}^{\infty} \psi^*(x)x\psi(x)\,dx \tag{7.23}$$

and, in general,

$$\langle F(x) \rangle = \int_{-\infty}^{\infty} \psi^*(x)F(x)\psi(x)\,dx \tag{7.24}$$

where $F(x)$ is any function of x.

In classical mechanics, once a problem has been solved, all dynamic variables can be expressed as functions of x (or, in three dimensions, of $\mathbf{r}$). In quantum mechanics, however, it is in principle impossible to write the momentum as a function of x, since that would imply that both p and x can be determined simultaneously, contrary to the uncertainty principle. Thus we cannot expect to use Equation (7.24) to calculate $\langle p_x \rangle$, the expectation value of the momentum along the x axis.

In constructing the wave equation we noted that the differential operator for p_x is $-i\hbar(\partial/\partial x)$. The expectation value of p_x is given by the integral

$$\langle p_x \rangle = \int_{-\infty}^{\infty} \psi^*(x)\left(-i\hbar\frac{\partial}{\partial x}\right)\psi(x)\,dx \tag{7.25}$$

We have now assembled the basic framework of quantum mechanics and are ready to apply these concepts to a few examples. First, however, we digress briefly to consider how a particle that is somewhat localized in space can be represented using the wave approach.

7.5 WAVE PACKETS; GROUP AND PHASE VELOCITIES

A pure sine wave is of infinite extent. Its wavelength and wave vector k are precisely determined. If such a wave represented a particle, we would know its momentum accurately ($p = \hbar k$), but it would be completely nonlocalized. That is, of course, what we expect from the uncertainty principle. However, if we wish to represent a particle that is somewhat localized, we must construct what is called a *wave packet* from a superposition of sine waves of different wavelengths.

---

1. One might well ask why Equation (7.16) yields a precise value for the energy when, according to the uncertainty principle, $\Delta E\,\Delta t \geq \frac{1}{2}\hbar$. The point is that since Equation (7.16) does not involve time (it is the *time-independent* Schrödinger equation), the solution is the wave function for all time; i.e., Δt is infinite. Consequently, a precise value of E does not violate the uncertainty principle. The energy eigenvalue is the energy of the *stationary* state.

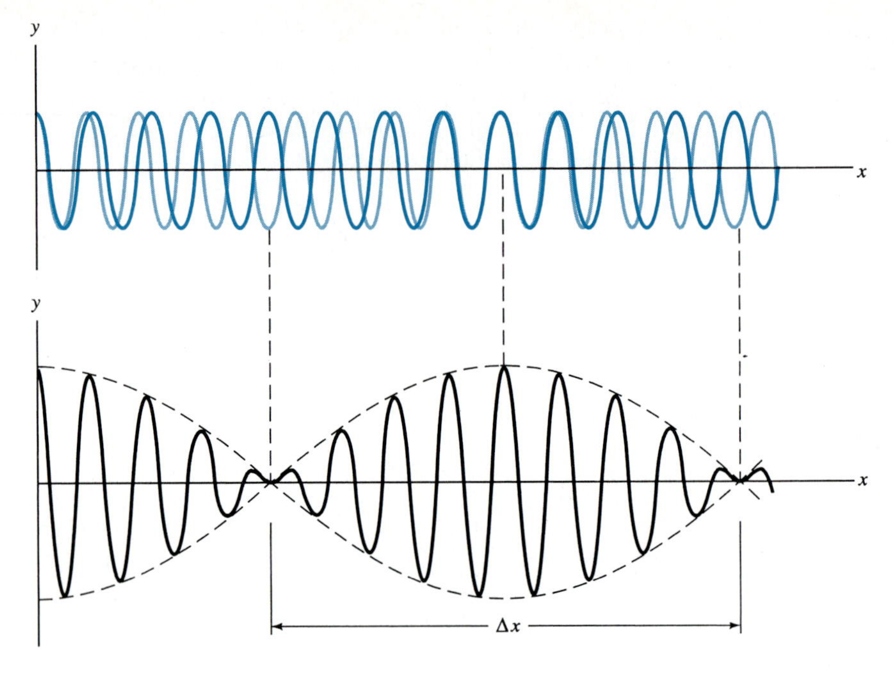

Figure 7.2 Superposition of two waves of nearly equal wavelengths results in a beat pattern, that is, a wave whose wavelength is the average of the wavelengths of the two constituents, and whose amplitude is modulated. The separation between adjacent nodes is inversely proportional to Δk, the difference in the wave numbers.

To see how such a wave packet might come about, consider the superposition of two sine waves of nearly equal wavelengths and frequencies. Their sum is

$$\begin{aligned} y(x,t) &= y_1(x,t) + y_2(x,t) = A \sin (k_1 x - \omega_1 t) + A \sin (k_2 x - \omega_2 t) \\ &= 2A \sin [\tfrac{1}{2}(k_1 + k_2)x - \tfrac{1}{2}(\omega_1 + \omega_2)t] \cos [\tfrac{1}{2}(k_1 - k_2)x - \tfrac{1}{2}(\omega_1 - \omega_2)t] \\ &= 2A \sin (\overline{k}x - \overline{\omega}t) \cos (\tfrac{1}{2}\, \Delta k\, x - \tfrac{1}{2}\, \Delta\omega\, t) \end{aligned}$$

This represents the familiar beat pattern shown in Figure 7.2: a wave whose wave vector and frequency are the averages of those of the two constituent waves and whose amplitude is modulated by the slowly changing cosine function. The rate at which the maximum of a particular beat progresses is $v_g = \Delta\omega/\Delta k$.

Though the superposition of two waves does not form a wave packet, we can construct a true wave packet using a large number of waves of suitable amplitudes; that is, we make a Fourier sum

$$y(x,t) = \sum A_i \sin (k_i x - \omega_i t)$$

and if the interval between adjacent wave vectors and frequencies is very small, we can approximate the sum by a Fourier integral

$$y(x,t) = \int A(k) \sin (kx - \omega t)\, dk$$

An example of such a wave packet (at $t = 0$) is shown in Figure 7.3. This *Gaussian* wave packet

$$y(x) = \sqrt{2\pi}\, \Delta k \exp \left[-\frac{(\Delta k x)^2}{2} \right] \sin k_0 x \tag{7.26}$$

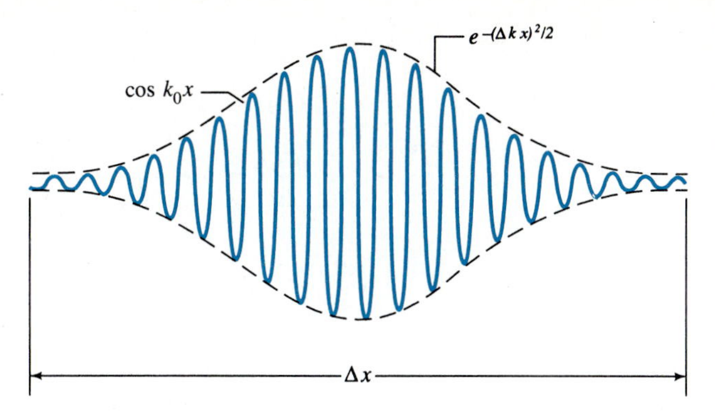

Figure 7.3 The Gaussian wave packet is the result of superposition of a large number of waves whose amplitudes are given by Equation (7.27).

develops if

$$A(k) = \exp\left[-\frac{(k - k_0)^2}{2(\Delta k)^2}\right] \tag{7.27}$$

The packet is centered on k_0; that is to say, the dominant wave has a wave vector k_0, but the wave packet contains a large number of waves whose wave vectors differ from k_0. As $|k - k_0|$ increases, the amplitude of these waves diminishes. The spread in wave vector is characterized by the constant Δk. The larger Δk, the smaller is the spatial spread Δx of the wave packet. Note that these results are consistent with the uncertainty principle: a large value of Δk (large uncertainty in the momentum $\hbar k$) results in a small value of Δx, the uncertainty in position.

The generalization of the argument that gave us the velocity of the beat pattern formed by two waves leads to the result that the wave packet proceeds at the *group velocity*

$$v_g = \frac{d\omega}{dk} \tag{7.28}$$

while the individual waves that make up this wave packet travel at the phase velocity

$$v_p = \frac{\omega}{k} \tag{7.29}$$

Let us now consider two kind of waves, light waves and particle waves. For light waves (in vacuum), $\omega = ck$. Since $d\omega/dk = c$, phase and group velocities are both equal to c. A very short pulse of light will propagate at the same speed as a monochromatic light wave.

Consider next a free particle, that is, one with kinetic energy E moving in a constant potential $V(x) = 0$. For such a particle, the time-dependent Schrödinger equation (7.13) reduces to

$$-\frac{\hbar^2}{2m}\frac{\partial^2\Psi}{\partial x^2} = i\hbar\frac{\partial\Psi}{\partial t} \tag{7.30}$$

whose solution is

$$\Psi(x,t) = Ae^{i(kx-\omega t)} \tag{7.31}$$

where

$$\frac{\hbar^2 k^2}{2m} = \hbar\omega = E \tag{7.32}$$

The group velocity of this particle is

$$v_g = \frac{d\omega}{dk} = \left(\frac{1}{\hbar}\right)\frac{dE}{dk} = \frac{\hbar k}{m} = \frac{p}{m} \tag{7.33}$$

in agreement with the classical result. The velocity of the particle is that of the wave packet, that is, the group velocity v_g, not the phase velocity of the wave function, $\omega/k = \hbar k/2m = p/2m$.

7.6 PARTICLE IN A ONE-DIMENSIONAL SQUARE WELL

Our first example of the application of the Schrödinger equation is for a particle restricted to one-dimensional motion between two potential barriers, as shown in Figure 7.4. Though the situation is already somewhat artificial, we shall, in the first instance, make it even more unrealistic by considering the limit as the height of the potential barrier goes to infinity. We do this because without this further simplification the solution already requires some tedious algebraic manipulation, yielding a transcendental equation that must be solved by numerical methods, thus obscuring some essential features of the problem.

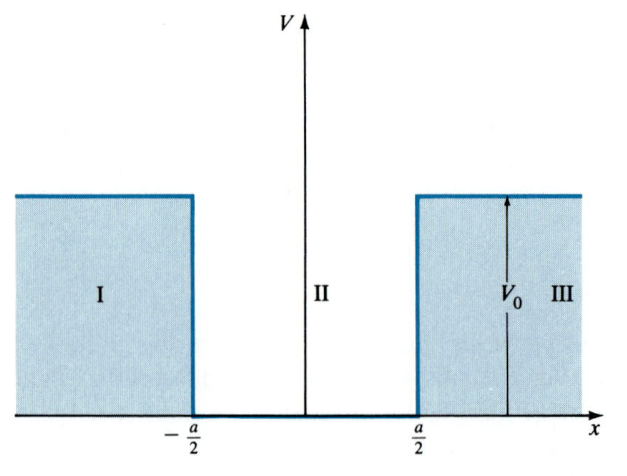

Figure 7.4 The one-dimensional square well. The solutions to the Schrödinger equation in regions I, II, and III must be continuous at the boundaries between these regions.

7.6(a) Infinite Potential Barriers

The one-dimensional region $-\infty < x < \infty$ must now be divided into three parts. For $x < -a/2$ (region I) and $x > a/2$ (region III), $V(x) = \infty$; for $-a/2 < x < a/2$ (region II), $V(x) = 0$. In regions I and III, $\psi(x) = 0$, for otherwise the left-hand side of Equation (7.16) would be infinite, and no meaningful solution could be found.

The fact that $\psi(x) = 0$ for $|x| > a/2$ simply means that outside the potential well the probability density is exactly zero. The particle is truly confined to region II; that is what we would expect from classical physics.

In region II, the time-independent Schrödinger equation is

$$-\frac{\hbar^2}{2m}\frac{\partial^2\psi}{\partial x^2} = E\psi \tag{7.34}$$

For present purposes it is more convenient to write the solution of this simple differential equation as a sum of sines and cosines rather than as a sum of exponentials; that is,

$$\psi(x) = A\sin(kx) + B\cos(kx) \tag{7.35}$$

where A and B are arbitrary constants and

$$k = \sqrt{\frac{2mE}{\hbar^2}} \tag{7.36}$$

Although expression (7.35) is a solution of Equation (7.34) for any value of k, only a selected class will satisfy the boundary conditions that $\psi(x)$ must meet. Since $\psi(x)$ must be continuous,

$$\psi(-a/2) = \psi(a/2) = 0 \tag{7.37}$$

Equation (7.37) can be satisfied if $A = B = 0$, but that is not a meaningful solution because then $\psi(x) = 0$; there is no particle in the well at all. The possible solutions divide into two distinct classes, those for which $A = 0$ and $B \neq 0$, and those for which $A \neq 0$ and $B = 0$.

If $A = 0$, we can satisfy the boundary conditions provided $\cos(ka/2) = 0$, i.e., if

$$k = \frac{(2n+1)\pi}{a} \qquad n \text{ an integer} \geq 0 \tag{7.38}$$

If $B = 0$, the boundary conditions are satisfied if

$$k = \frac{2n\pi}{a} \qquad n \text{ an integer} \geq 1 \tag{7.39}$$

The energy eigenvalues for these allowed solutions are obtained from Equation (7.36). For both groups we obtain

$$E_n = \frac{\hbar^2\pi^2}{2ma^2}n^2 \qquad n \text{ an integer} \geq 1 \tag{7.40}$$

Note that the lowest allowed energy is not zero (which would be the classical

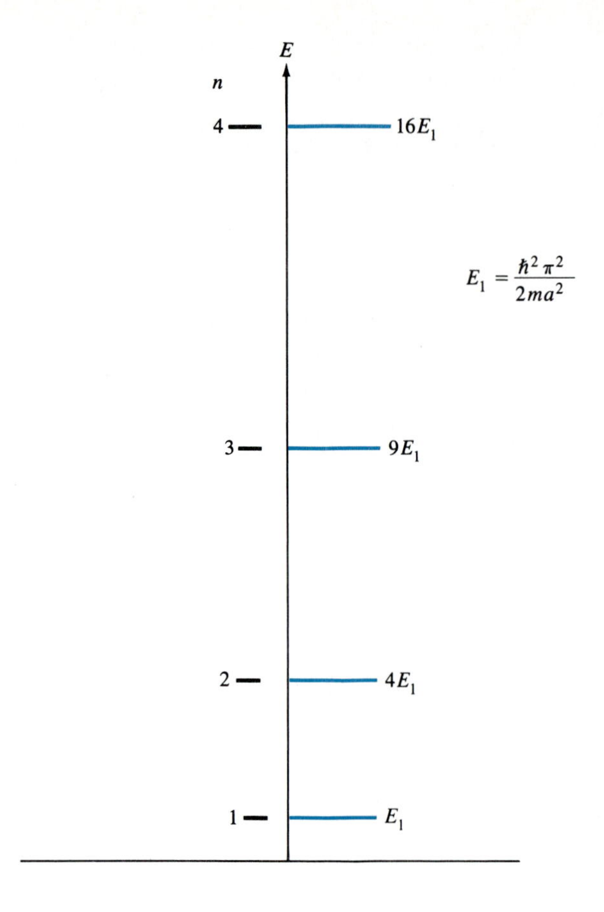

Figure 7.5 Energy-level diagram for a particle in a one-dimensional infinite square well. Note that the spacing between the energy levels increases with increasing quantum number. Note, also, that the successive energy levels alternate between those for even and odd solutions of the wave equation.

result—a particle at rest in this well) but $\hbar^2\pi^2/2ma^2$. The result is consistent with the uncertainty principle in the following sense. Since the particle is confined within the well of width a, its uncertainty in x is $\Delta x = a$. Consequently, its uncertainty in p_x is $\Delta p_x \geq \hbar/2a$, and its energy is therefore $E = (\Delta p_x)^2/2m \geq \hbar^2/8ma^2$.

A few of the wave functions for the two types of solutions are shown in Figure 7.6. These two classes of solution have special character under an inversion operation, a change of x into $-x$. Solutions for which $A = 0$ are even functions of x; that is, $\psi(x) = \psi(-x)$. If $B = 0$, the solutions are odd functions of x: $\psi(x) = -\psi(-x)$. As we shall see presently, this separation of the solutions into either even or odd functions is by no means accidental, nor is it unique to this particular situation.

To complete the problem we must normalize the wave function. It is this procedure that determines the coefficients A and B. Since $\psi(x)$ represents a single particle,

$$\int_{-\infty}^{\infty} \psi^*(x)\psi(x)\,dx = \int_{-a/2}^{a/2} \psi^*(x)\psi(x)\,dx = 1 \tag{7.41}$$

with the result

$$A = B = \sqrt{\frac{2}{a}} \tag{7.42}$$

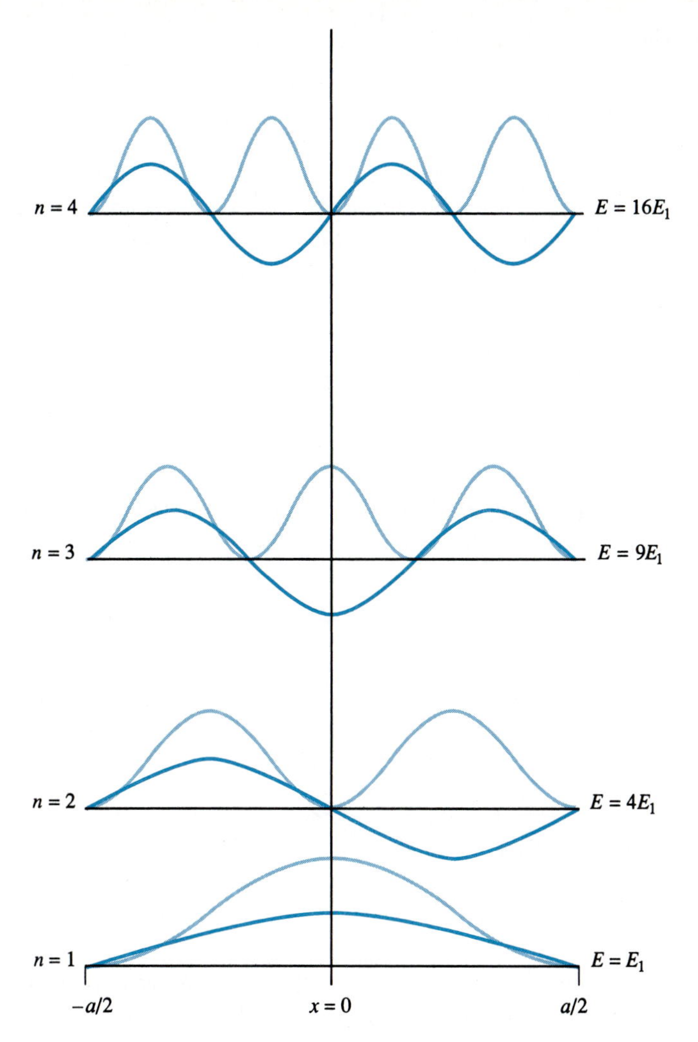

Figure 7.6 The wave functions (solid color) and probability densities (light color) for the ground state ($n = 1$) and first three excited states ($n = 2$, 3, and 4) of a particle confined to an infinite one-dimensional potential well of width a. Note that the wave functions are either odd or even functions of x.

7.6(b) Finite Potential Barriers

We consider next the case where the particle is confined to a potential well of finite depth. The barriers at $x = \pm a/2$ now have a height V_0. The solution in this case proceeds along the same lines as in the preceding case. However, since $V(x)$ is finite, we can no longer assume that $\psi(x) = 0$ for $|x| > a/2$. In region II, the Schrödinger equation is as before, Equation (7.34), with solutions given by Equation (7.35). In regions I and III, however, the Schrödinger equation is

$$-\frac{\hbar^2}{2m}\frac{\partial^2\psi}{\partial x^2} = (E - V_0)\psi \tag{7.43}$$

which we can write as

$$\frac{\partial^2\psi}{\partial x^2} = \alpha^2\psi \qquad \alpha = \sqrt{\frac{2m(V_0 - E)}{\hbar^2}} \tag{7.44}$$

whose solution is

$$\psi(x) = Ce^{\alpha x} + De^{-\alpha x} \tag{7.45}$$

Since $e^{\alpha x}$ diverges as $x \to \infty$ and $e^{-\alpha x}$ diverges as $x \to -\infty$, normalizability of the wave function requires that $D = 0$ in region I and $C = 0$ in region III.

In the following we derive the equation that determines the allowed values of the energy for the even solutions. The derivation of the corresponding equation for the odd solutions is left as Problem 7.7. For the even solutions,

Region I	**Region II**	**Region III**
$\psi(x) = Ce^{\alpha x}$	$\psi(x) = B\cos(kx)$	$\psi(x) = De^{-\alpha x}$

We now apply the boundary conditions on the wave function and its derivative at $x = \pm a/2$. Continuity of $\psi(x)$ gives

$$Ce^{-\alpha a/2} = B\cos(ka/2) \qquad De^{-\alpha a/2} = B\cos(ka/2) \tag{7.46a}$$

Continuity of the first derivative gives

$$\alpha Ce^{-\alpha a/2} = kB\sin(ka/2) \qquad -\alpha De^{-\alpha a/2} = -kB\sin(ka/2) \tag{7.46b}$$

From Equation (7.46a) it follows that $C = D$. If we now divide Equation (7.46b) by Equation (7.46a) we obtain

$$\alpha = k\tan(ka/2) \tag{7.47}$$

Equation (7.47) relates α to k. But α and k are also related through Equations (7.36) and (7.44), with the result

$$k^2 + \alpha^2 = \beta^2 \qquad \beta^2 = \frac{2mV_0}{\hbar^2} \tag{7.48}$$

The allowed values of k, and hence of the energy, are now the solutions of the transcendental equation

$$k\tan(ka/2) = \sqrt{\beta^2 - k^2} \tag{7.49}$$

which can be cast into the dimensionless form

$$X\tan X = \sqrt{Y^2 - X^2} \qquad X^2 = \frac{ma^2E}{2\hbar^2} \qquad Y^2 = \frac{ma^2V_0}{2\hbar^2} \tag{7.50}$$

Equation (7.50) must be solved by numerical methods (or graphically). Note that Equation (7.50) has solutions only if $X < Y$, i.e., if $E < V_0$. That is what we should expect, since we are looking for bound states in this potential, and if the kinetic energy of the particle in the region of zero potential is greater than V_0 the particle is not bound classically. The following is a list of the even-parity solutions for a few choices of Y:

$Y = 2$; $X = 1.02987$. No other bound state.
$Y = 5$; $X = 1.30644$, 3.83747. No other bound state.
$Y = 8$; $X = 1.39547$, 4.16484, 6.83067. No other bound state.
$Y = 15$; $X = 1.47247$, 4.41372, 7.34247, 10.24382, 13.07815. No other bound states.

For the potential well with infinite barriers, the solutions are

$$Y = \infty;\ X = (2n + 1)\pi/2:\ 1.5708,\ 4.7124,\ 7.8540,\ 10.9956,\ 14.1372,\ \ldots$$

The importance of the boundary conditions in determining the energy eigenvalues is illustrated in Figure 7.7. Shown there are the results of integrating the Schrödinger equation for the finite square well potential ($Y = 5.0$) for energies equal to the energy eigenvalues of the two lowest stationary states and for energies slightly less and slightly greater than those eigenvalues. When the energies are not equal to the eigenvalues the wave functions diverge exponentially as x becomes large.

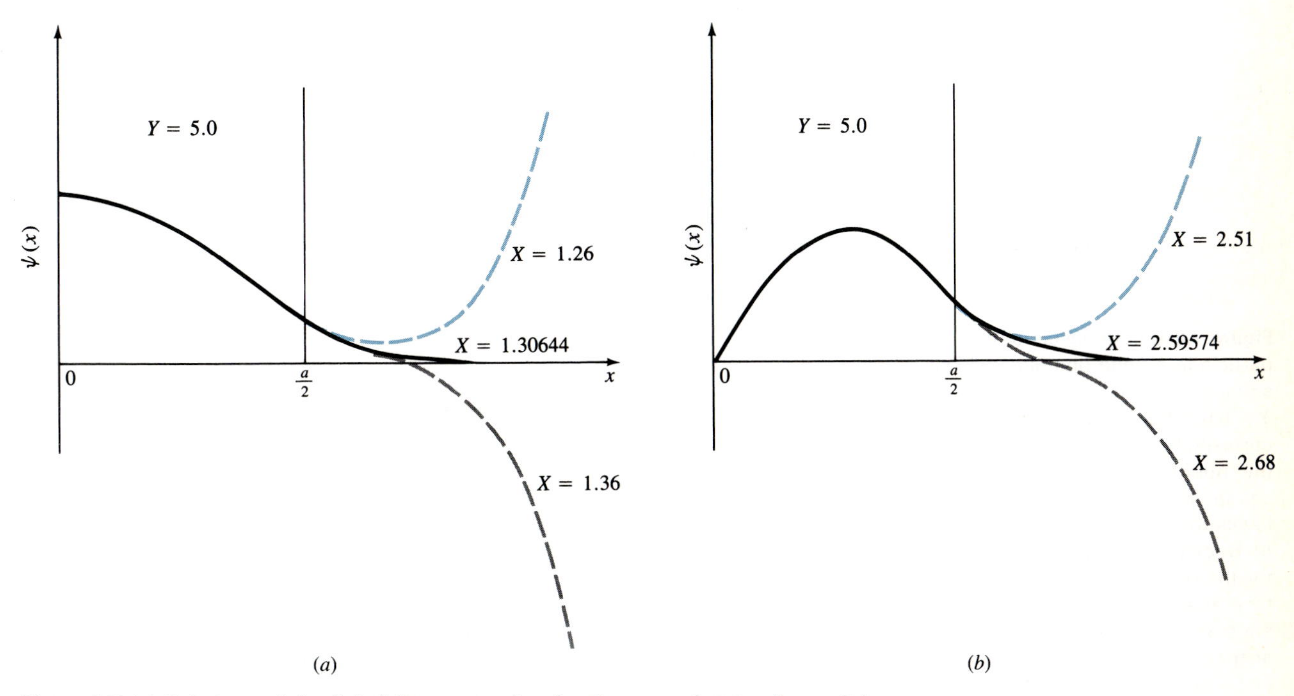

Figure 7.7 (*a*) Solutions of the Schrödinger equation for the ground state of a particle in a finite one-dimensional potential well. The parameter $Y = 5.0$ (see Equation 7.50). The wave function is shown for three choices of the energy parameter X. The boundary of the potential well at $x = a/2$ is indicated. When X is either slightly greater or slightly less than 1.30644, the wave function diverges for large x. (*b*) The wave function for the first excited state of a particle in the potential well with $Y = 5.0$. Again, the wave function diverges unless $X = 2.59574$.

$$\psi(-x) = c\psi(x) \qquad \psi(x) = c\psi(-x)$$

and

$$\psi(x) = c\psi(-x) = c^2\psi(x)$$

It follows then that

$$c = \pm 1$$

If $c = +1$, $\psi(x)$ is even; that is, $\psi(-x) = \psi(x)$. If $c = -1$, $\psi(x)$ is odd; that is, $\psi(-x) = \psi(x)$.

In quantum mechanics, one uses the term *parity* to characterize the symmetry properties of wave functions under inversion of the coordinates, and one speaks of *even-parity* and *odd-parity* solutions. It must be emphasized that the adjectives *even* and *odd* in reference to parity do not necessarily correspond to even or odd quantum numbers. For instance, in the case of the square-well potentials examined in the preceding section, even quantum numbers correspond to odd parity solutions and vice versa.

7.8 TUNNELING

We have already noted that the probability density for a particle confined within a square well of finite depth is nonvanishing on either side of the potential barrier. The penetration of the wave function into classically forbidden regions has very important consequences that have spawned a host of practical devices.

Let us examine the following situation. A particle, with kinetic energy E, approaches a rectangular barrier of height $V > E$ and width a, as shown in Figure 7.9. Classically, such an incident particle is totally reflected by this potential barrier. But, as we have seen, the quantum mechanical solution allows—in fact, requires—that the wave function have a finite amplitude in the region close to the barrier boundary at $x = 0$. If $\psi(x)$ represents a particle incident from the left, the

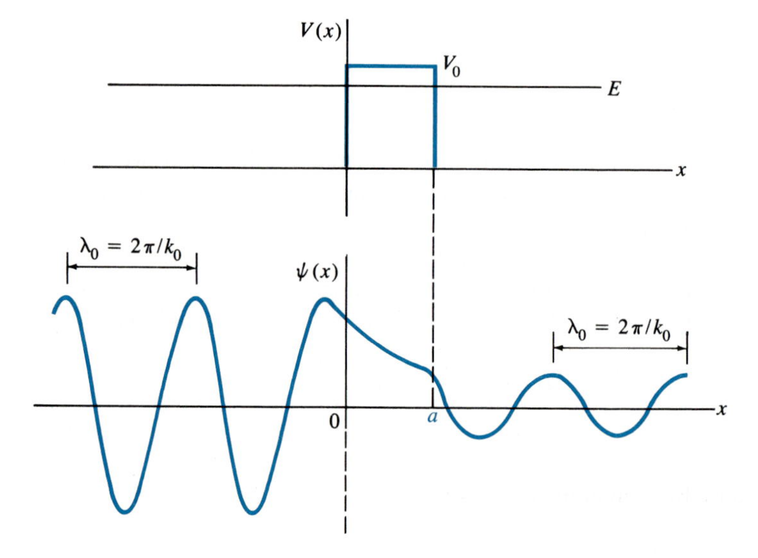

Figure 7.9 A rectangular potential barrier of height V and width a. A particle of energy $E < V$ is incident from the left. A portion of the particle wave is transmitted through the barrier and appears beyond the barrier with reduced amplitude but the same wavelength. Classically, a particle of this energy would be totally reflected by this barrier. There is also a reflected wave not shown in this drawing.

boundary conditions on $\psi(x)$ and $\psi'(x)$ at $x = 0$ lead to an exponentially decaying wave function in the region $0 < x < a$. At $x = a$, that exponential function must again be continuous with the solution of the Schrödinger equation for $x > a$, which is a traveling wave. Since the exponentially diminishing function in the barrier region has a finite, albeit often small amplitude at $x = a$, the wave function for $x > a$ will also be of finite amplitude, and the probability density to the right of the potential barrier will not vanish. The particle, therefore, has a finite probability of penetrating the classically insurmountable barrier and appearing on the opposite side. One refers to this process as *tunneling* through a barrier.

Since in the barrier region the wave function decays as $e^{-\alpha x}$, where α is proportional to $(V - E)^{1/2}$, the probability density $\psi^*\psi$ for $x > a$, that is, the tunneling probability, is proportional to $\exp(-2\sqrt{V - E})$. Furthermore, for a given value of α, the ratio of the wave function at $x = a$ to that at $x = 0$ is proportional to $e^{-\alpha a}$. Hence the tunneling probability for a given barrier height and particle energy diminishes exponentially with the width of the barrier. In practice, tunneling is observed only for very narrow potential barriers. We shall return to these matters when we consider the radioactive decay of nuclei via α particle emission (Chapter 15), and in the description of semiconductor tunnel diodes and a number of superconducting devices (Chapter 13).

7.9 THE HARMONIC OSCILLATOR

The simple harmonic oscillator is one of the most useful model systems in classical as well as quantum mechanics. In classical mechanics, any system in stable equilibrium at rest at the bottom of a potential energy trough, which we take to be at $x = 0$, will tend to return to this position if displaced from equilibrium. If we arbitrarily set $V(0) = 0$, the first nonvanishing term in a power series expansion of $V(x)$ about $x = 0$ will be quadratic in x. Since the potential energy of a simple harmonic oscillator is of the form

$$V(x) = \tfrac{1}{2}kx^2 \qquad \textbf{(7.51)}$$

it follows that a nondissipative system that suffers a small displacement from equilibrium will perform simple harmonic motion. Although the potential function of a real system is generally not given correctly by Equation (7.51), for *small* displacements the harmonic approximation is almost always quite good. For example, although the potential energy of the two atoms of a diatomic molecule such as O_2 is not given by Equation (7.51), the vibrational modes and their energies can nevertheless be calculated to reasonably good approximation using the harmonic oscillator model (see Problem 7.28). For large-amplitude vibrations, departures from the predictions of that model do become pronounced.

Even when the system consists of a large number of masses that are coupled to one another by forces that produce an equilibrium configuration, the response of such a complicated system can often be analyzed through a study of the normal modes of a large number of coupled harmonic oscillators. That is the procedure employed by Debye in deriving an expression for the specific heat of crystals (see Section 4.3).

Although the harmonic oscillator was one of the problems considered by Schrödinger in his initial paper on wave mechanics, its solution already requires mathe-

The expectation value of the momentum p_x is

$$\langle p_x \rangle = \int_{-\infty}^{\infty} \psi^*(x)\left(-i\hbar\frac{\partial}{\partial x}\right)\psi(x)\, dx$$

The wave function $\psi(x)$ must satisfy the following conditions:

1. $\psi(x)$ must be a solution of Equation (7.16).
2. $\psi(x)$ must be normalizable; i.e., $\psi(x) \to 0$ as $x \to \pm\infty$.
3. $\psi(x)$ must be a continuous function of x.
4. $\partial\psi/\partial x$ must be a continuous function of x (except if $V(x)$ is singular).

For a given potential the wave function $\psi(x)$ satisfies conditions 2 through 4 only for certain values of E, the *energy eigenvalues*. For a one-dimensional infinite square well of width a, the energy eigenvalues are

$$E_n = \frac{\hbar^2\pi^2}{2ma^2}n^2 \qquad n \text{ an integer} \geq 1 \tag{7.40}$$

For the one-dimensional harmonic oscillator,

$$E_n = (n + \tfrac{1}{2})\hbar\omega_0 \qquad n \text{ an integer} \geq 0 \tag{7.66}$$

where $\omega_0 = \sqrt{k/m}$ is the angular frequency of the classical oscillator.

If the potential $V(\mathbf{r}) = V(-\mathbf{r})$ (in one-dimension, $V(x) = V(-x)$), the solutions of the Schrödinger equation must be either even or odd functions of $\mathbf{r}$ (or x). The word *parity* is used to characterize the symmetry of the wave functions.

The wave function, and therefore the probability density, may be finite in a region in which the particle would be excluded according to classical mechanics. In particular, there is a finite probability that a particle will *tunnel* through a classically impenetrable barrier.

Bibliography

P. C. W. Davies, *Quantum Mechanics*, Routledge & Kegan Paul, London, 1984.

Steven Gasiorowicz, *Quantum Physics*, Wiley, New York, 1974.

R. L. Liboff, *Introductory Quantum Mechanics*, Addison-Wesley, Reading, MA 1980.

David Park, *Introduction to the Quantum Theory*, 2d ed., McGraw-Hill, New York, 1974.

E. H. Wichmann, *Quantum Physics: Berkeley Physics Course, Vol. 4*, McGraw-Hill, New York, 1967.

PROBLEMS

7.1 A neutron is confined to a one-dimensional infinite potential well whose width is 10^{-14} m. Determine the energies of the ground and first excited states.

7.2 An electron is confined to a one-dimensional infinite potential well. Its first excited stationary state has an energy 1.2 eV above the ground state. Determine the width of the well.

7.3 Evaluate the normalization constant for the wave function of a particle in the first excited state of an infinite one-dimensional potential well.

7.4 Use Equation (7.11) and the result of Problem 7.3 to calculate the expectation values of p_x and p_x^2 for a particle in the ground state and first excited state in an infinite one-dimensional potential well.

7.5 Calculate the uncertainty product $\Delta x\ \Delta p_x$ for a particle in the ground state and first excited state in an infinite one-dimensional potential well.

7.6 Calculate the probability that a particle in a one-dimensional infinite potential well of width a will be found between $x = a/3$ and $x = a/2$ if the particle is (a) in the ground state and (b) in the first excited state. How do these results compare with the classical prediction?

7.7 Complete the solution for the eigenstates of a particle in a finite one-dimensional potential well by deriving an equation similar to Equation (7.50) for the states of odd parity.

7.8 Use numerical or graphical methods to determine the energies of the odd-parity states for electrons in a potential well of width $a = 2$ nm for which the parameter $Y = (ma^2V_0/2\hbar^2)^{1/2}$ is (a) 2.0 and (b) 8.0.

7.9 Suppose that the parameter Y of Equation (7.50) is 5.0 and the particle is an electron. What is the height of the potential barrier if the well width is two Bohr radii?

•7.10 An electron is confined to a one-dimensional potential well whose potential barriers are 6.0 eV. What is the width of the well if the ground-state energy is 0.40 eV?

•7.11 Figure 7.13 shows a one-dimensional potential step of height V_0. Suppose a beam of particles whose energy E is greater than V_0 is incident from the left. Let the wave function for particles of the incident beam be represented by $\psi_i = Ae^{ikx}$, that for the reflected beam be $\psi_r = Be^{-ikx}$, and that for the transmitted beam be $\psi_t = Ce^{ik'x}$. Use the boundary conditions on the wave function to obtain expressions for the ratios $T = (k'/k)(|C|^2/|A|^2)$ and $R = |B|^2/|A|^2$, corresponding to the probabilities for transmission and reflection of a particle, respectively. Show that $T + R = 1$.

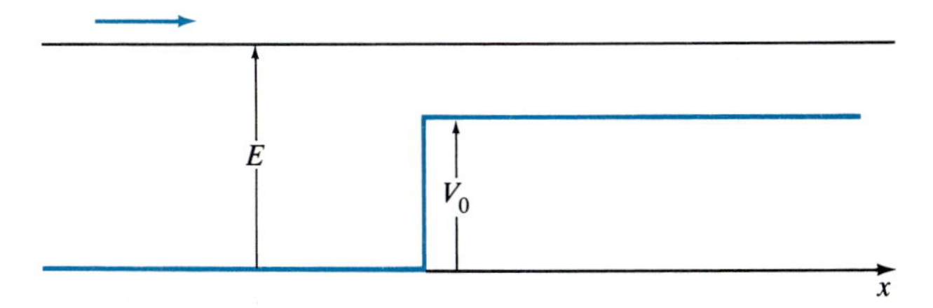

Figure 7.13 A particle of energy $E > V_0$ is incident from the left on a potential step V_0 at $x = 0$.

7.12 Use the result of Problem 7.11 to calculate the transmission and reflection probabilities for a particle of kinetic energy E incident on a potential step of height $V_0 = E/2$.

7.13 Suppose a particle whose kinetic energy is E moves from a region where $V = 0$ to one where $V = -E$, as shown in Figure 7.14. Calculate the reflection and transmission probabilities using the same approach as for Problem 7.11.

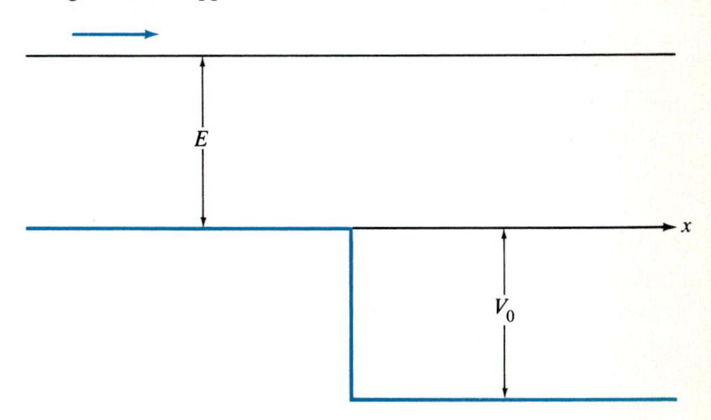

Figure 7.14 A particle of energy E is incident from the left. For $x > 0$, the potential is $V_0 = -E$.

7.14 An electron of kinetic energy $E = 5.0$ eV is incident on a potential barrier of height 5.5 eV and width 10^{-8} m. Estimate the probability that this electron will tunnel through the potential barrier.

7.15 Sketch the wave functions for the ground state and the first two excited states for a particle in each of the one-dimensional potential wells of Figure 7.15. For each function identify the points where $\psi(x) = 0$, $d\psi(x)/dx = 0$, and $d^2\psi(x)/dx^2 = 0$.

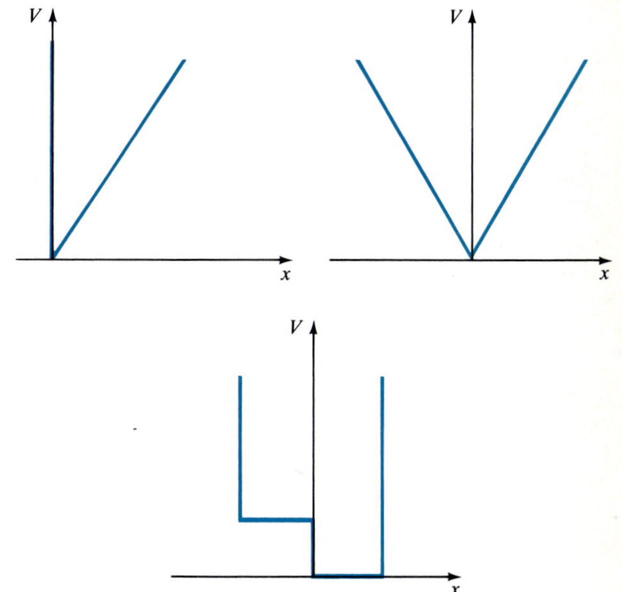

Figure 7.15 Potential wells for Problem 7.15.

7.16 Show that Equation (7.53) is equivalent to Equation (7.52) when α and λ_n are given by Equation (7.54).

7.17 Show that the wave function for the ground state of the harmonic oscillator, Equation (7.68), is properly normalized.

•7.18 Derive the normalized wave function for the first excited state of the harmonic oscillator.

•7.19 Following the same approach as in Example 7.2, use the result of Problem 7.18 to calculate the uncertainty product for the first excited state of the harmonic oscillator. Compare your result with the uncertainty product for the ground state.

7.20 Show that the normalization constant for the wave functions of a particle in an infinite square well is $\sqrt{2/a}$, independent of the quantum number n.

•7.21 Consider the following situation. A particle with kinetic energy E is incident on a square potential well of finite depth V_0 and width a. (a) Describe the motion of the particle in classical terms. What is the relative probability that the particle will be found in an interval Δx near $x = 0$ compared with the probability that it will be found in an equal interval Δx outside the region of the potential well? (b) Sketch the wave function using quantum mechanical arguments. Pay special attention to the wavelength and amplitude for $|x| > a/2$ and $|x| < a/2$. (c) Under what conditions would it be possible to have $\psi(x)$ and $\psi'(x)$ continuous at $x = \pm a/2$ and have the amplitudes equal inside and outside the well? What does this suggest regarding the reflection and transmission probabilities?

7.22 When an electron is incident on a noble-gas atom (e.g., argon, krypton, or xenon) it enters a region of attractive potential that has a fairly well defined boundary at the radius of the atom. The situation is not unlike that discussed in Problem 7.21. Figure 7.16 shows the scattering cross section for electrons incident on noble-gas atoms as a function of their velocity ($\sqrt{E}$). Note that when the electron energy is about 0.7 eV and the scattering atoms are argon, the scattering cross section is very small, indicating that the electron is transmitted almost perfectly in passing through the atomic potential. (The almost total lack of scattering at a critical electron energy is known as the *Ramsauer-Townsend effect*.) The atomic radius of argon is about 0.2 nm. Estimate the depth of the effective potential seen by the incident electron.

7.23 Show that the integral

$$\int_{-\infty}^{\infty} \psi_n^*(x)\psi_m(x)\, dx$$

is identically equal to zero if $n \neq m$ and ψ_n and ψ_m are the wave functions of a particle confined in an infinite square potential well. Wave functions that have this property are said to be *orthogonal*.

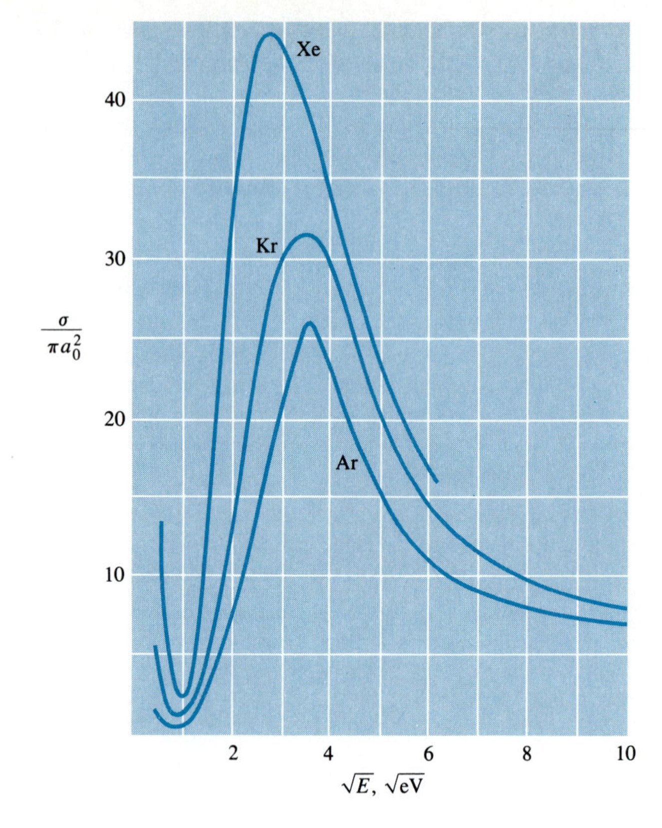

Figure 7.16 Scattering cross section for low-energy electrons incident on noble-gas atoms. Note the sharp minimum when the electron energy is near 1.0 eV, corresponding to a large transmission probability. This is the Ramsauer-Townsend effect.

7.24 Show by direct integration that the wave functions of the harmonic oscillator for $n = 0$ and 1 are orthogonal.

•7.25 Show by direct integration that the wave functions of the harmonic oscillator for $n = 0$ and $n = 2$ are orthogonal.

7.26 Classically, a particle confined to an infinite potential well bounces back and forth between the walls. The probability of finding the particle in an interval Δx is therefore $\Delta x/a$. Show that in the quantum mechanical solution, the probability of finding the particle in an interval Δx is also equal to $\Delta x/a$ if n is so large that $\Delta x \gg a/n$.

7.27 Show that for $n \gg 1$, the expectation value of x^2 for a particle is an infinite one-dimensional square well of width a approaches the classical value $a^2/12$.

The Schrödinger Equation in Three Dimensions: The Hydrogen Atom

Chapter 8

You see that our differential equation automatically selects as the allowed E-values the energy levels of the elliptic orbits quantized according to Bohr's theory. . . . This is very remarkable. It shows that, whatever the waves may mean physically, the theory furnishes a method of quantization which is absolutely free from arbitrary postulates.

Erwin Schrödinger

8.1 INTRODUCTION

The general form of the time-independent Schrödinger equation in three dimensions is Equation (7.17). In Cartesian coordinates, Equation (7.17) is

$$-\frac{\hbar^2}{2m}\left(\frac{\partial^2\psi}{\partial x^2}+\frac{\partial^2\psi}{\partial y^2}+\frac{\partial^2\psi}{\partial z^2}\right)+V(x,y,z)\psi=E\psi \tag{8.1}$$

where $\psi = \psi(x,y,z)$ is a function of the three coordinates. If the potential $V(x,y,z)$ is symmetric in x, y, z, the solution of Equation (8.1) is straightforward. The wave function ψ can then be written as a product

$$\psi(x,y,z) = \psi_x(x)\psi_y(y)\psi_z(z)$$

and Equation (8.1) is separable. That is, one obtains three equations, one for each of the coordinates, and the total energy is the sum of the three energy eigenvalues of those equations. The simplest example of this sort is a particle confined in a cubic "box," the three-dimensional analog of the problem treated in the preceding chapter. We shall, in fact, return to this case in Chapter 11.

Cubic symmetry is, however, the exception rather than the rule. Far more common is spherical symmetry, as for atoms. Here, the potential is a function of r, the distance from the nucleus, and is independent of the polar angle θ and azimuthal angle ϕ. To be sure, one could still write the Schrödinger equation in Cartesian coordinates, but rather than simplify the mathematics, it would complicate matters and obscure some very essential features of the solution. Since the potential has

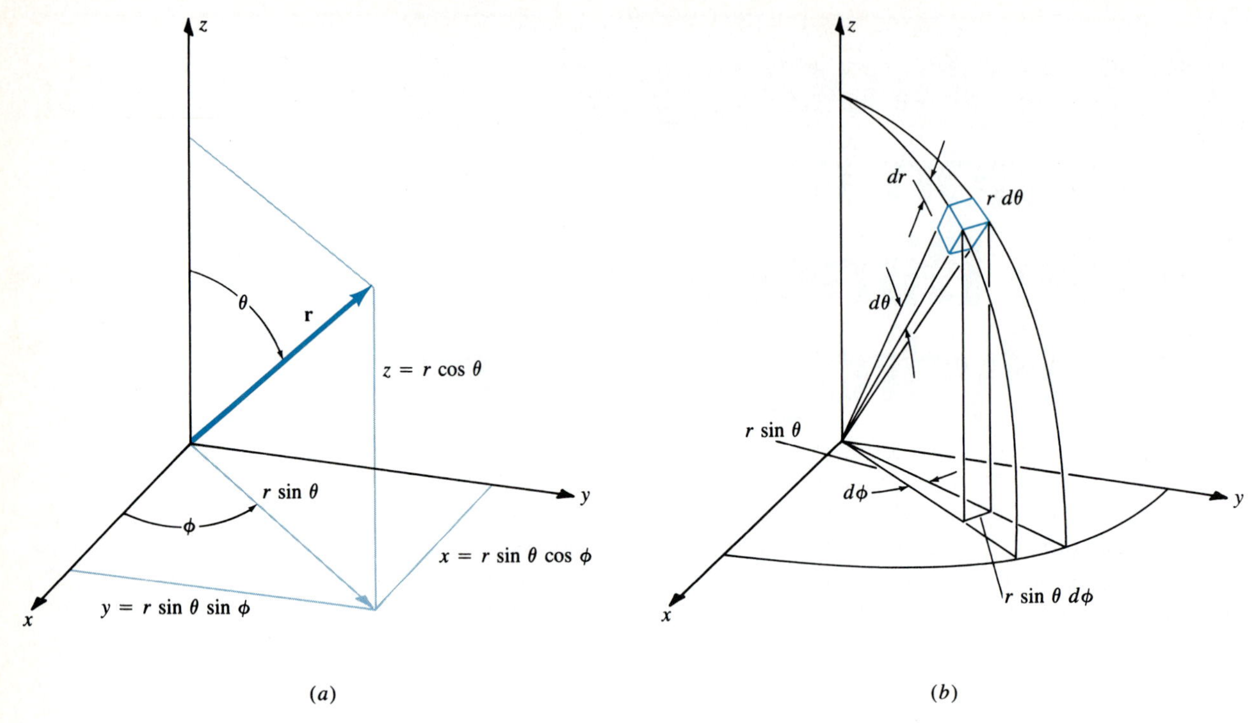

Figure 8.1 (*a*) The relationship between spherical polar coordinates and Cartesian coordinates. (*b*) The element of volume dV in spherical polar coordinates is $r^2 \sin\theta \, dr \, d\theta \, d\phi$.

spherical symmetry, it is sensible to write Equation (8.1) in spherical coordinates (Figure 8.1*a*). If we transform to spherical coordinates using the relations

$$x = r \sin\theta \cos\phi$$

$$y = r \sin\theta \sin\phi$$

$$z = r \cos\theta$$

Equation (8.1) takes the form

$$-\frac{\hbar^2}{2m}\left[\frac{1}{r}\frac{\partial^2(r\psi)}{\partial r^2} + \frac{1}{r^2 \sin\theta}\frac{\partial}{\partial\theta}\left(\sin\theta \frac{\partial\psi}{\partial\theta}\right) + \frac{1}{r^2\sin^2\theta}\frac{\partial^2\psi}{\partial\phi^2}\right] + V(r)\psi = E\psi \quad \textbf{(8.2)}$$

8.2 SOLUTION OF THE SCHRÖDINGER EQUATION IN SPHERICAL COORDINATES

Following the earlier pattern, we look for solutions of the form

$$\psi(r,\theta,\phi) = R(r)\Theta(\theta)\Phi(\phi) \quad \textbf{(8.3)}$$

Before proceeding further, we note that $-(\hbar^2/2m)\,\nabla^2\psi$, the first term of Equation

(8.2), represents the kinetic energy of the mass m. That term is therefore positive definite. In fact, as we shall shortly demonstrate, $-(\hbar^2/2m)$ times the first term within the square brackets corresponds to $p_r^2/2m$, the kinetic energy associated with the radial motion, and the remaining terms to $L^2/2I$, the kinetic energy associated with the angular momentum about the center of coordinates.

If we multiply Equation (8.2) by $(2mr^2 \sin^2\theta)/\hbar^2 R\Theta\Phi$ we obtain

$$-\sin^2\theta\left[\frac{r}{R}\frac{d^2}{dr^2}(rR)\right]+\frac{\sin\theta}{\Theta}\frac{d}{d\theta}\left(\sin\theta\frac{d\Theta}{d\theta}\right)-\frac{1}{\Phi}\frac{d^2\Phi}{d\phi^2}+\frac{2mr^2\sin^2\theta}{\hbar^2}(V-E)=0 \tag{8.4}$$

We have here replaced the partial derivatives by total derivatives, since R, Θ, and Φ are each functions of r, θ, and ϕ only.

Equation (8.4) must hold for all values of the variables r, θ, and ϕ. Since $-(1/\Phi)(d^2\Phi/d\phi^2)$ does not depend on the other two variables, it must be constant, in fact, a positive definite constant, which we denote by m_l^2. The equation that $\Phi(\phi)$ must satisfy is then

$$\frac{d^2\Phi}{d\phi^2}+m_l^2\Phi=0 \tag{8.5}$$

which has the solutions

$$\Phi(\phi)=(2\pi)^{-1/2}e^{im_l\phi} \tag{8.6}$$

where the factor $(2\pi)^{-1/2}$ is the appropriate normalization constant.

Now the wave function must be single-valued. If we rotate our coordinate system through 360° we must surely get back the same solution. Therefore m_l must be zero or a positive or negative integer so that $\Phi(\phi+2\pi)=\Phi(\phi)$.

We next return to Equation (8.4), divide that equation by $\sin^2\theta$, and replace $(1/\Phi)(d^2\Phi/d\phi^2)$ by $-m_l^2$. The result is the equation

$$\frac{r}{R}\frac{d^2}{dr^2}(rR)+\frac{2mr^2}{\hbar^2}[E-V(r)]=-\frac{1}{\Theta\sin\theta}\frac{d}{d\theta}\left(\sin\theta\frac{d\Theta}{d\theta}\right)+\frac{m_l^2}{\sin^2\theta} \tag{8.7}$$

The right-hand side of Equation (8.7) is a function of θ only, the left-hand side a function of r only. Consequently, both must be equal to a constant, which we set equal to $l(l+1)$. The reason for this rather odd choice of the constant is that only if $l=0$ or a positive integer greater than or equal to $|m_l|$ are the solutions of the differential equation for $\Theta(\theta)$ physically acceptable, i.e., they do not diverge. The function $\Theta_{l,m}(\theta)$ is a polynomial in $\cos\theta$, known as a Legendre polynomial.

Finally, we turn to the radial wave equation, the left-hand side of Equation (8.7). That equation now takes the form

$$\frac{d^2}{dr^2}(rR)=-\frac{2m}{\hbar^2}\left[E-V(r)-\frac{\hbar l(l+1)}{2mr^2}\right](rR) \tag{8.8}$$

The solution of Equation (8.8) evidently depends on the potential $V(r)$. If, but only if, the potential $V(r)$ is proportional to $1/r$, as is the Coulomb potential

$$V(r) = -\frac{Ze^2}{4\pi\epsilon_0}\left(\frac{1}{r}\right)$$

are the energy eigenvalues for bound states equal to

$$E_n = -\frac{E_0 Z^2}{n^2}$$

where E_0 is given by Equation (5.32) and n is a positive integer. For all other potentials, the bound-state energies depend on both the quantum numbers n and l. Moreover, the quantum number n must be an integer equal to or greater than $l + 1$. The condition that n be an integer can be traced to the requirement that the probability density vanish as $r \to \infty$.

This completes the solution of the Schrödinger equation in spherical coordinates for the bound states in a spherically symmetric potential. Before we proceed, let us summarize what we have learned so far.

1. The wave function can be written as a product of three functions

$$\psi_{n,l,m}(r,\theta,\phi) = R_{n,l}(r)\Theta_{l,m}(\theta)\Phi_{m_l}(\phi) \tag{8.9}$$

2. The quantum numbers n, l, and m_l must be integers that satisfy the following relations:

$$n \geq 1 \qquad (n = 1, 2, 3, \ldots) \tag{8.10a}$$

$$l \leq n - 1 \qquad (l = 0, 1, 2, \ldots, n - 1) \tag{8.10b}$$

$$|m_l| \leq l \quad (m_l = -l, -l + 1, \ldots, -1, 0, 1, \ldots, l - 1, l) \tag{8.10c}$$

3. The energies of the bound states are quantized. For any negative potential $V(r)$, the bound-state energy levels depend on the strength and shape of the potential and on the quantum numbers n and l.
4. If, but only if, the potential is a Coulomb potential, the bound-state energies are given by

$$E_{n,l} = E_n = -\frac{E_0 Z^2}{n^2} \tag{8.11}$$

where

$$E_0 = \frac{me^4}{32\pi^2\epsilon_0^2\hbar^2} = 13.6 \text{ eV} \tag{8.12}$$

Table 8.1 lists the eigenfunctions $R(r)$, $\Theta(\theta)$, and $\Phi(\phi)$ for an electron in a Coulomb potential for a few values of n, l, and m_l.

TABLE 8.1 Hydrogen Atom Wave Functions for n = 1, 2, and 3

n	l	m_l	$R(r)$*	$\Theta(\theta)$	$\Phi(\phi)$
1	0	0	$\frac{2}{\sqrt{a_0^3}}e^{-\rho}$	$\frac{1}{\sqrt{2}}$	$\frac{1}{\sqrt{2\pi}}$
2	0	0	$\frac{1}{\sqrt{2a_0^3}}\left(1-\frac{\rho}{2}\right)e^{-\rho/2}$	$\frac{1}{\sqrt{2}}$	$\frac{1}{\sqrt{2\pi}}$
2	1	0	$\frac{1}{\sqrt{24a_0^3}}\rho e^{-\rho/2}$	$\sqrt{\frac{3}{2}}\cos\theta$	$\frac{1}{\sqrt{2\pi}}$
2	1	±1	$\frac{1}{\sqrt{24a_0^3}}\rho e^{-\rho/2}$	$\sqrt{\frac{3}{4}}\sin\theta$	$\frac{1}{\sqrt{2\pi}}e^{\pm i\phi}$
3	0	0	$\frac{2}{\sqrt{27a_0^3}}\left(1-\frac{2}{3}\rho+\frac{2}{27}\rho^2\right)e^{-\rho/3}$	$\frac{1}{\sqrt{2}}$	$\frac{1}{\sqrt{2\pi}}$
3	1	0	$\frac{8}{27\sqrt{6a_0^3}}\rho\left(1-\frac{\rho}{6}\right)e^{-\rho/3}$	$\sqrt{\frac{3}{2}}\cos\theta$	$\frac{1}{\sqrt{2\pi}}$
3	1	±1	$\frac{8}{27\sqrt{6a_0^3}}\rho\left(1-\frac{\rho}{6}\right)e^{-\rho/3}$	$\sqrt{\frac{3}{4}}\sin\theta$	$\frac{1}{\sqrt{2\pi}}e^{\pm i\phi}$
3	2	0	$\frac{4}{81\sqrt{30a_0^3}}\rho^2 e^{-\rho/3}$	$\sqrt{\frac{5}{8}}(3\cos^2\theta-1)$	$\frac{1}{\sqrt{2\pi}}$
3	2	±1	$\frac{4}{81\sqrt{30a_0^3}}\rho^2 e^{-\rho/3}$	$\frac{\sqrt{15}}{2}\sin\theta\cos\theta$	$\frac{1}{\sqrt{2\pi}}e^{\pm i\phi}$
3	2	±2	$\frac{4}{81\sqrt{30a_0^3}}\rho^2 e^{-\rho/3}$	$\frac{\sqrt{15}}{4}\sin^2\theta$	$\frac{1}{\sqrt{2\pi}}e^{\pm 2i\phi}$

*The variable $\rho = r/a_0$.

8.2(a) Probability Densities and Expectation Values

In the previous chapter we saw that $|\psi|^2 = \psi^*\psi$ represents the probability density P in the sense that the probability of finding the particle described by the wave function ψ in an element of volume dV is

$$P\,dV = \psi^*\psi\,dV$$

In spherical coordinates, the volume element is $dV = r^2\sin\theta\,dr\,d\theta\,d\phi$ (Figure 8.1(b)). We then have

$$P(r,\theta,\phi)\,dV = |R(r)|^2|\Theta(\theta)|^2|\Phi(\phi)|^2 r^2\sin\theta\,dr\,d\theta\,d\phi$$

Since $\Phi^*\Phi = 1/2\pi$ is independent of the azimuthal angle, the probability density is symmetric about the polar axis. If we integrate over the azimuthal angle ϕ (from $\phi = 0$ to $\phi = 2\pi$), the result is 1. If $l = 0$, the probability density is also independent of the polar angle; i.e., $\psi^*\psi$ is then a function of r only. If $l \geq 1$, the probability density varies with θ, and its form depends on both l and m_l. Representations of the probability densities for some values of n, l, and m_l are shown in Figure 8.2. Note that for $m_l = 0$ the probability density is greatest along the polar axis, whereas for $m_l = l$ it is most concentrated about the azimuthal plane.

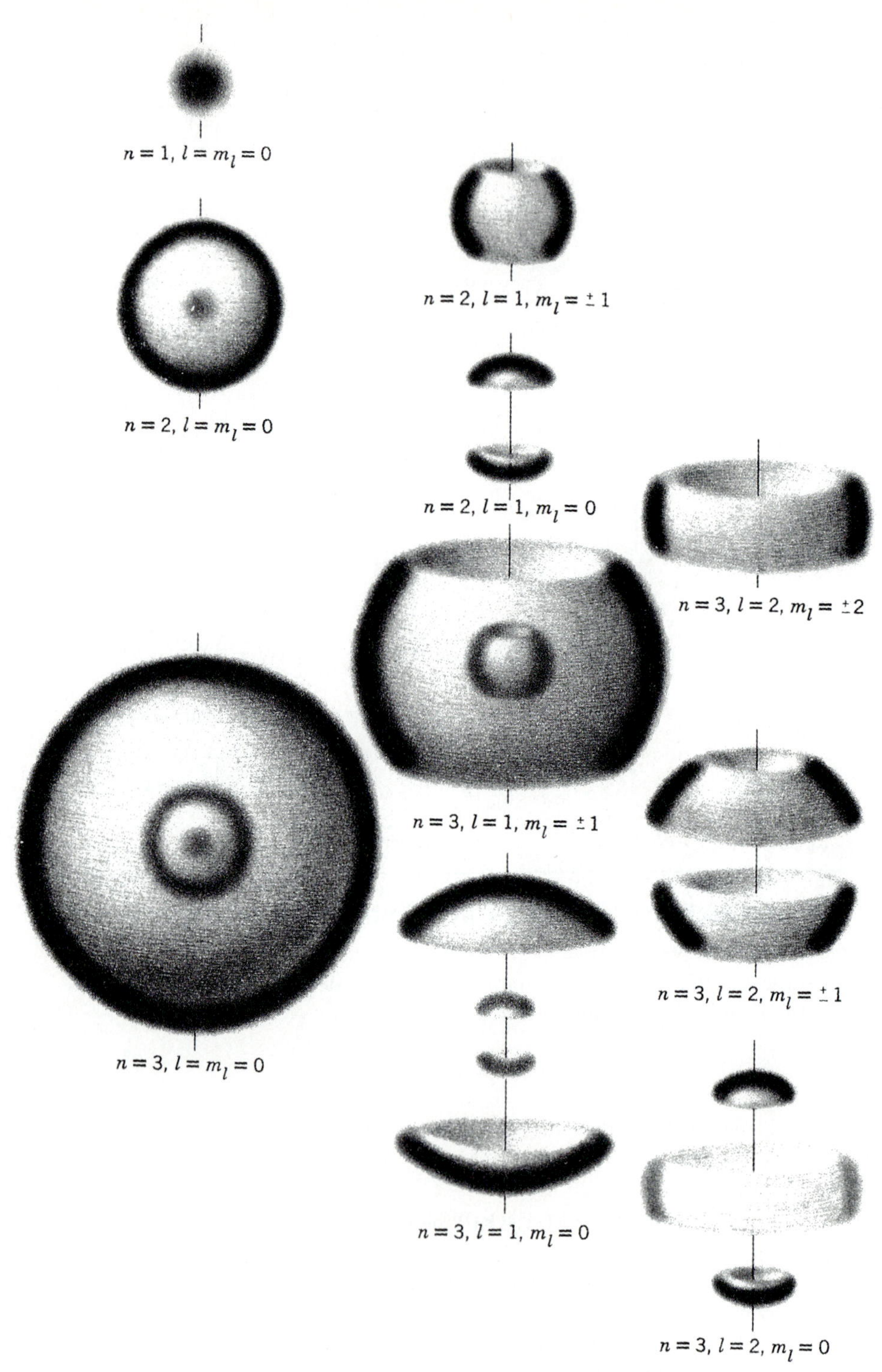

Figure 8.2 Artist's rendition of the probability densities of the hydrogen atom's stationary, states for n = 1, 2, and 3. Note that although the probability densities of the M_l states for $l > 0$ are not spherically symmetric, their sum is spherically symmetric (see Problem 8.10).

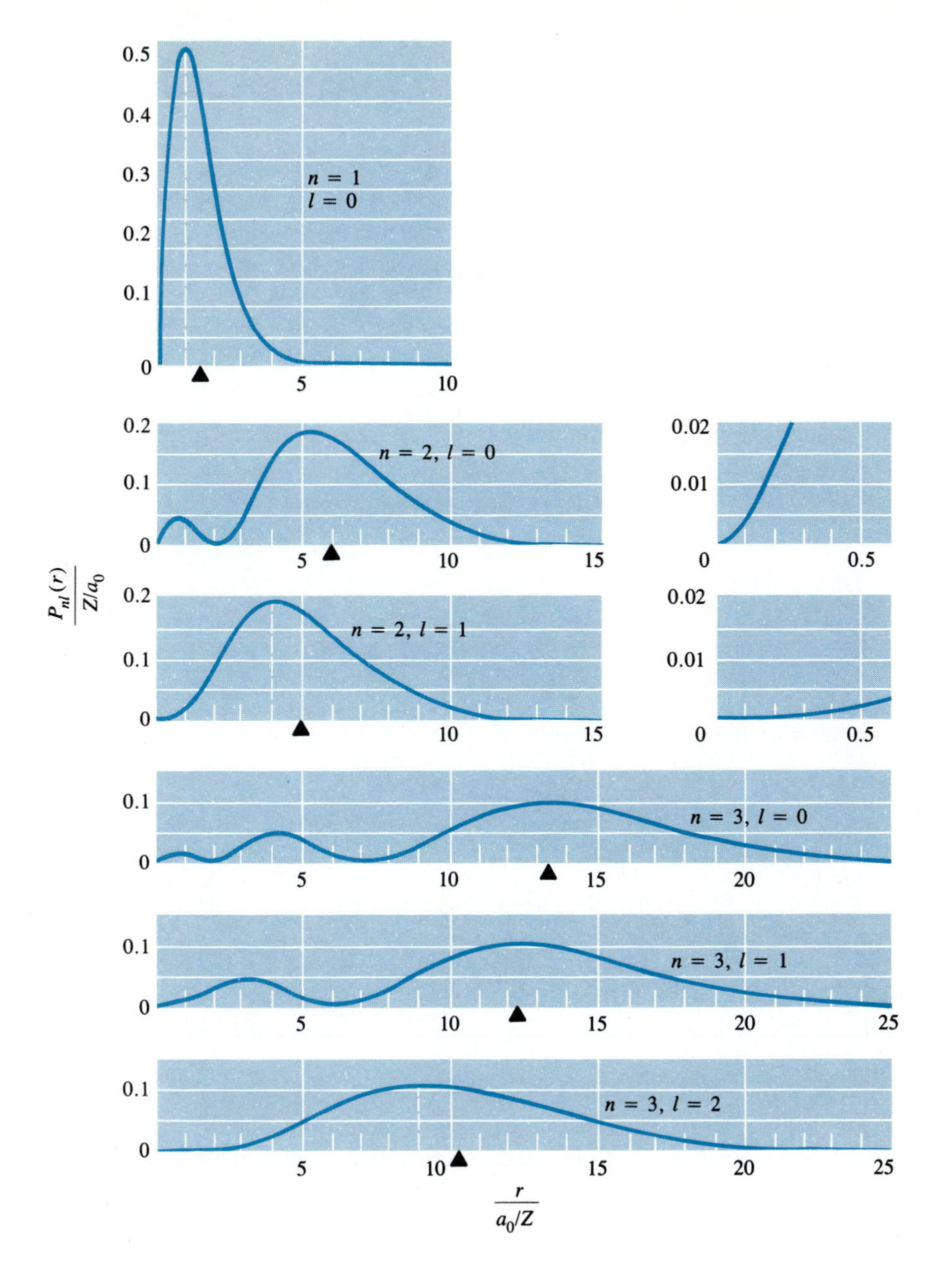

Figure 8.3 The radial probability densities for the states $n = 1$, 2, and 3 of the hydrogen atom. The arrow on the abscissa corresponds to $\langle r_{n,l} \rangle$; the dashed line is the value of r_n according to the Bohr model. Note that $\langle r_{n,l} \rangle$ is always greater than r_n. Also note that when $l = n - 1$, the radial probability density is a maximum at $r = r_n$.

Often, one is interested only in the probability that the electron will be found at a given distance from the nucleus. In that case, one integrates over the polar as well as azimuthal angle, and since $\Theta(\theta)$ is normalized,

$$P_{n,l}(r)\, dr = r^2|R_{n,l}(r)|^2\, dr \tag{8.13}$$

$P_{n,l}(r)$ is referred to as the *radial probability density*. The radial probability densities for a few values of n and l for the hydrogen atom are shown in Figure 8.3. In this figure we have indicated the expectation value of r, $\langle r \rangle$ (see Example 8.2), as well as the radius of the Bohr orbit for the state $l = n - 1$. Note that for states of lower angular momentum, $\langle r \rangle$ is considerably greater than the radius of the Bohr orbit.

EXAMPLE 8.1 Show that the radial wave function $R_{2,0}(r)$ of Table 8.1 is properly normalized.

SOLUTION To demonstrate normalization we must evaluate the integral

$$\int_0^\infty |R_{2,0}(r)|^2 r^2\, dr$$

where
$$R_{2,0}(r) = \frac{1}{(2a_0)^{3/2}}\left(2 - \frac{r}{a_0}\right)e^{-r/2a_0}$$

If we change the variable from r to $x = r/a_0$, the integral takes the form

$$\frac{1}{8a_0^3}\, a_0^3 \int_0^\infty (4x^2 - 4x^3 + x^4)e^{-x}\, dx \tag{8.14}$$

The definite integral

$$\int_0^\infty x^{n-1}e^{-x}\, dx = \Gamma(n) \tag{8.15}$$

where for n an integer

$$\Gamma(n) = (n-1)! \tag{8.16}$$

We obtain

$$\int_0^\infty P_{2,0}(r)\, dr = \frac{a_0^3}{8a_0^3}(4 \times 2 - 4 \times 6 + 24) = 1$$

QED.

EXAMPLE 8.2 Determine the expectation value of r for the states $n = 2$, $l = 0$ and $n = 2$, $l = 1$ for the electron in the hydrogen atom. Also find the value of r for which $P_{2,1}(r)$ is a maximum. Compare the result with the radius of the Bohr orbit for $n = 2$.

SOLUTION We must evaluate the integrals

$$\langle r_{2,0}\rangle = \int_0^\infty rP_{2,0}(r)\, dr \qquad \text{and} \qquad \langle r_{2,1}\rangle = \int_0^\infty rP_{2,1}(r)\, dr$$

where
$$P_{2,0}(r) = r^2|R_{2,0}(r)|^2 \qquad \text{and} \qquad P_{2,1}(r) = r^2|R_{2,1}(r)|^2$$

We again change the variable of integration from r to $x = r/a_0$. The first of the above integrals then differs from that of the previous example only in that the integrand of expression (8.14) is multiplied by the factor a_0x. From Equations (8.15) and (8.16) we then find that

$$\langle r_{2,0}\rangle = \left(\frac{a_0}{8}\right)(4 \times 6 - 4 \times 24 + 120) = 6a_o$$

and $$\langle r_{2,1}\rangle = \frac{a_0}{24}\int_0^\infty x^5 e^{-x}\,dx = 5a_0$$

To determine the radial distance at which $P_{2,1}(r)$ is a maximum we must set its derivative equal to zero and solve for r. Again, we can simplify the numerical work by changing the variable from r to $x = r/a_0$. Then, except for some constants, $P_{2,1}(x) = x^4 e^{-x}$. We now differentiate with respect to x and set the derivative equal to zero:

$$\frac{dP_{2,1}(x)}{dx} = 4x^3e^{-x} - x^4e^{-x} = 0 \qquad x = 4 \qquad r = 4a_0$$

The radial probability density in the state $n = 2$, $l = 1$ is a maximum at the Bohr radius for $n = 2$.

8.3 ANGULAR MOMENTUM IN QUANTUM MECHANICS

In classical mechanics, a particle in a central force field follows a trajectory such that its angular momentum about the center of force is constant. Angular momentum conservation derives from the fact that, since the direction of the force is always along the radius vector, the torque is zero throughout the trajectory. From the correspondence principle, we anticipate that the angular momenta of the stationary-state solutions of Equation (8.2) will also be constants of motion. With that in mind, we now examine the quantum mechanical equations for the angular momentum. We shall see that the quantum numbers l and m_l are intimately related to the angular momentum.

The angular momentum of a particle about the origin is given by

$$\mathbf{L} = \mathbf{r} \times \mathbf{p} \tag{8.17}$$

The quantum mechanical expression for the angular momentum is obtained by replacing the momentum $\mathbf{p}$ in Equation (8.17) by the momentum operator $\mathbf{p}_{\text{op}} = -i\hbar\nabla$ [see Equation (7.11)]. When that is done, the components of the angular momentum are represented by the operators

$$L_x = yp_z - zp_y = -i\hbar\left(y\frac{\partial}{\partial z} - z\frac{\partial}{\partial y}\right) = i\hbar\left(\sin\phi\frac{\partial}{\partial\theta} + \cot\theta\cos\phi\frac{\partial}{\partial\phi}\right) \tag{8.18a}$$

$$L_y = zp_x - xp_z = -i\hbar\left(z\frac{\partial}{\partial x} - x\frac{\partial}{\partial z}\right) = -i\hbar\left(\cos\phi\frac{\partial}{\partial\theta} - \cot\theta\sin\phi\frac{\partial}{\partial\phi}\right) \tag{8.18b}$$

$$L_z = xp_y - yp_x = -i\hbar\left(x\frac{\partial}{\partial y} - y\frac{\partial}{\partial x}\right) = -i\hbar\frac{\partial}{\partial\phi} \tag{8.18c}$$

The operator that represents the total angular momentum is obtained from Equation (8.18) by writing

$$L^2 = L_x^2 + L_y^2 + L_z^2 = -\frac{\hbar^2}{\sin\theta}\frac{\partial}{\partial\theta}\left(\sin\theta\frac{\partial}{\partial\theta}\right) - \frac{\hbar^2}{\sin^2\theta}\frac{\partial^2}{\partial\phi^2} \equiv \Omega \qquad \textbf{(8.19)}$$

In quantum mechanics it is always true that if a dynamical variable is a constant of motion, the wave function is an eigenfunction of the operator that represents that dynamical variable. The eigenvalue of this operator equation then corresponds to the stationary value of that dynamical variable. Since we anticipate from classical arguments that the angular momentum is a constant of motion, we now look for solutions of equations of the form

$$(\text{Operator})\psi = (\text{constant})\psi$$

where (Operator) stands for one of the differential operators representing L_x, L_y, L_z, and L^2. We are interested in solutions of these operator equations for the case in which the wave function is the product $R(r)\Theta(\theta)\Phi(\phi)$.

Let us first consider the operator equation for the z component of the angular momentum,

$$-i\hbar\frac{\partial}{\partial\phi}[R(r)\Theta(\theta)\Phi(\phi)] = \hbar m_l[R(r)\Theta(\theta)\Phi(\phi)]$$

Since the differential operator acts only on the azimuthal angle ϕ, we can cancel the functions $R(r)$ and $\Theta(\theta)$ and obtain

$$\frac{d\Phi}{d\phi} = im_l\Phi$$

which has solutions

$$\Phi(\phi) = (2\pi)^{-1/2}e^{im_l\phi} \qquad \textbf{(8.20)}$$

where $|m_l| = 0, 1, 2, \ldots$ so that $\Phi(\phi)$ is single-valued; we have included here the normalization factor $(2\pi)^{-1/2}$. The eigenfunction in Equation (8.20) is identical to Equation (8.6). In other words, $\Phi(\phi)$ of Equation (8.6) is the eigenfunction of the operator representing the z component of the angular momentum, with the eigenvalue

$$L_z = m_l\hbar \qquad \textbf{(8.21)}$$

The quantum number m_l gives the z component of the angular momentum of the electron in the quantum state n, l, m_l.

Whereas in classical mechanics the z component of the angular momentum can assume any value less than the total angular momentum, in the quantum case it must be zero or an integral multiple of $\hbar = h/2\pi$. The result is reminiscent of the Bohr postulate with, however, one very important difference. The quantum mechanical treatment of the hydrogen atom problem imposes this condition not on the total angular momentum but on the z component of that vector.

Let us next consider the eigenvalue equation for L^2, which we can write

$$\Omega[R(r)\Theta(\theta)\Phi(\phi)] = b[R(r)\Theta(\theta)\Phi(\phi)] \qquad \textbf{(8.22)}$$

where b represents the eigenvalue of that equation. As before, we can cancel $R(r)$, since the operator Ω does not act on the variable r. If we now divide both sides of the resulting differential equation by $\hbar^2\Theta\Phi$ and replace $\partial^2\Phi/\partial\phi^2$ by $-m_l^2\Phi$ according to Equation (8.20), the left-hand side of the resulting equation is identical to the right-hand side of Equation (8.7). As we pointed out earlier, that equation has acceptable solutions only if

$$b = \hbar^2 l(l + 1)$$

where l is a positive integer or zero and

$$l \geq |m_l|$$

Since b is the eigenvalue of the operator representing L^2,

$$L^2 = \hbar^2 l(l + 1) \tag{8.23}$$

and

$$L = \hbar\sqrt{l(l + 1)} \tag{8.24}$$

The quantum number l determines the magnitude of the total angular momentum.

8.3(a) Spatial Quantization

At first, these results relating to the angular momentum of the hydrogen atom (or any spherically symmetric quantum mechanical system) must seem bizarre. According to these results, if, for example, $l = 1$ and we were to measure the component of the angular momentum along some arbitrarily chosen direction in space, we would find that it is either $-\hbar$, 0, or $+\hbar$. Suppose we then measure the component of $\mathbf{L}$ along some other direction in space. Again, our result will be either $-\hbar$, 0, or $+\hbar$! Not only are the possible values of L_z quantized, but no matter what direction of space we select as the polar axis, the result of a measurement of L_z must be one of those three values. How can one understand this apparently paradoxical conclusion?

The crucial point here is that **until we make our measurement,** it makes no sense to even speak of a z axis. It is in the process of performing the experiment that yields L_z that we must do something that establishes a particular spatial direction as the polar axis. For example, one might place the atom between the pole faces of a magnet; the direction of the magnetic field defines a unique direction in space. Only then can one legitimately ask, What is the component of the angular momentum along this direction?

The other curious aspect of these results is that the z component of the angular momentum, $m_l\hbar$, is always less than the magnitude of the total angular momentum, $\sqrt{l(l + 1)}\hbar$. Whatever we select as our direction for spatial quantization, we shall never find that the angular momentum points exactly along that line. That limitation is another consequence of the uncertainty principle. In its most general form, that principle states that the product of any coordinate and its (canonically conjugate) momentum must be greater than $\hbar/2$. In this instance, the coordinate associated with L_z is the azimuthal angle ϕ. The uncertainty principle imposes the condition

$$\Delta L_z\,\Delta\phi \geq \frac{\hbar}{2}$$

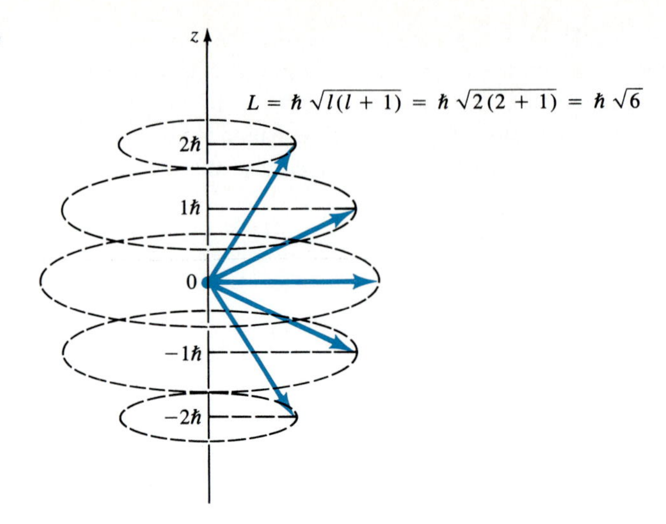

Figure 8.4 The vector model of the angular momentum for $l = 2$.

Since we can determine L_z precisely ($L_z = m_l\hbar$), the azimuthal angle must be completely indeterminate. In other words, L_x and L_y must remain unknown. If, however, L_z were equal to L, we would know that $L_x = L_y = 0$, and thus we would have obtained a result that violates the uncertainty principle.

The vector model, illustrated in Figure 8.4 for $l = 2$, is often helpful in visualizing the angular momentum and its allowed orientations. For $l = 2$, the total angular momentum vector has the magnitude $\sqrt{6}\hbar$; the component of $\mathbf{L}$ along the z direction is restricted to the five values $m_l\hbar$ where $m_l = 2, 1, 0, -1, -2$.

Now that we have seen how the quantum numbers l and m_l are related to the angular momentum of the system, we can make at least a plausible argument for the condition

$$l \leq n - 1$$

If we examine Equation (8.8) we see that it looks just like the one-dimensional Schrödinger equation, Equation (7.16), with $\psi(x)$ replaced by $rR(r)$ and the potential $V(r)$ by an effective potential

$$V_{eff} = V(r) + \frac{\hbar^2 l(l + 1)}{2mr^2} \tag{8.25}$$

We now recognize the second term of Equation (8.25) as the rotational kinetic energy, $L^2/2I = L^2/2mr^2$, sometimes referred to as the *centripetal potential*. In other words, for all values of l except $l = 0$, the effective potential is increased by an amount that depends on l and varies as $1/r^2$.

For bound states, $E_{n,l}$ must be negative. The electron moves in the negative potential due to the attractive Coulomb force of the nucleus, but the effective potential is increased by the electron's orbital motion. A bound state can exist only if on the average V_{eff} is negative. From the radial wave functions of Figure 8.3, and also from the Bohr model, we see that the expectation value of r will be less the smaller the principal quantum number n. Consequently, the centripetal potential is greater the smaller the principal quantum number n, and it is then not altogether unexpected that the limitation on the angular momentum is most severe for small values of n.

8.4 DEGENERACY

As we have seen, the energies of the bound states of the one-electron atom depend on just one quantum number, the principal quantum number n, in the Bohr model as well as in the quantum mechanical treatment of that system. The wave function $\psi(r,\theta,\phi)$, however, is completely specified only if we know the three quantum numbers, n, l, and m_l. Here again, there is a close analogy to the classical situation. The energy of a planet or comet of mass M in orbit about the sun is known once we know the semimajor axis of its elliptical orbit. Yet, to completely trace its trajectory we must also know the direction of the angular momentum vector due to its orbital motion. Three quantities are then needed, two angles as well as the magnitude of the semimajor axis. In general, **the totality of quantum numbers required to completely specify a quantum state equals the number of independent coordinate variables in the Schrödinger equation.**

As in the classical analog of planetary motion, it often happens in quantum mechanics that a number of quite different states have exactly the same energy. That states with the same n but different l and m_l do differ significantly is evident from an examination of Figure 8.2; yet, all of those states have the same energy. These energy states are then said to be *degenerate*. The degree of the degeneracy is the number of different eigenfunctions that yield this energy. For example, the energy state E_2 of the hydrogen atom is fourfold degenerate: that energy state encompasses the quantum states $l = 0$, $m_l = 0$ and $l = 1$, $m_l = +1$, 0, and -1.[1]

EXAMPLE 8.3 What is the degeneracy of the energy state of the hydrogen atom for $n = 4$?

SOLUTION For $n = 4$, l can take on the values 0, 1, 2, and 3. The $n = 4$, $l = 3$ state is sevenfold degenerate ($m_l = -3, -2, -1, 0, 1, 2, 3$), the state $n = 4$, $l = 2$ is fivefold degenerate, the state $l = 1$ is threefold degenerate, and the state $l = 0$ is nondegenerate. Thus, the total number of states with energy $E = E_4$ in the hydrogen atom is $7 + 5 + 3 + 1 = 16$; the $n = 4$ energy level is sixteenfold degenerate.

Degeneracy is almost always a consequence of symmetry. For instance, the fact that all states of given l have the same energy regardless of the value of m_l is a direct consequence of the spherical symmetry of the system. Recall that the quantum number m_l determines the z component of the angular momentum vector. Since under spherical symmetry there is no preferred direction in space, the energy of the quantum system cannot depend on L_z.

The degeneracy with respect to the quantum number l is peculiar to the Coulomb potential. In any other spherically symmetric potential, the energy depends on l as well as on n. Because this degeneracy is not related to an obvious symmetry property, it is sometimes referred to as an "accidental" degeneracy.

---

1. These statements are not strictly correct, because we have neglected the spin of the electron. We shall include the effect of spin on the energy and on the degeneracy of stationary states in the next chapter.

Degeneracy can always be removed or "lifted" by reducing the symmetry that gave rise to it. For instance, as we shall see in the following chapter, the degeneracy associated with the quantum number m_l is lifted if one places the atom in a uniform magnetic field, thereby selecting a particular direction as the polar axis and destroying the spherical symmetry that resulted in the m_l degeneracy.

SUMMARY

Since the Coulomb potential of a point charge is spherically symmetric, the Schrödinger equation for the hydrogen atom is most conveniently solved in spherical coordinates. The wave function $\Psi(r,\theta,\phi)$ is then separable, i.e., $\Psi(r,\theta,\phi) = R(r)\Theta(\theta)\Phi(\phi)$. The boundary conditions that the wave function be single-valued and vanish as $r \to \infty$ lead to the following results:

1. The stationary states are characterized by three integer quantum numbers, n, l, and m_l. These quantum numbers must meet the restrictions

$$n \geq 1$$
$$l \leq n - 1$$
$$|m_l| \leq l$$

2. The quantum numbers l and m_l specify the total orbital angular momentum and its component along a specific direction (z direction) according to the relations

$$L = \sqrt{l(l+1)}\hbar \qquad L_z = m_l\hbar$$

3. The energy eigenvalues are those of the Bohr model, namely,

$$E_n = -\frac{me^4}{32\pi^2\epsilon_0^2\hbar^2}\frac{1}{n^2} = -\frac{13.6}{n^2}\text{ eV}$$

Since the energies of the stationary states do not depend on l or m_l (neglecting electron spin; see Chapter 9), there are n^2 states of the same energy for each n. The states are said to be degenerate.

Degeneracy is generally related to symmetry and can be lifted by destroying the symmetry. The degeneracy associated with the quantum number m_l is related to the spherical symmetry of the Coulomb potential and is lifted if the atom is placed in a magnetic field. The degeneracy associated with the quantum number l is peculiar to the $1/r$ dependence of the Coulomb potential and is referred to as an accidental degeneracy.

Bibliography

R. Eisberg and R. Resnick, *Quantum Physics of Atoms, Molecules, Solids, Nuclei, and Particles,* 2d ed., Wiley, New York, 1985.

D. Kleppner, M. G. Littman, and M. L. Zimmerman, "Highly Excited Atoms," *Scientific American,* May 1981, p. 130.

R. B. Leighton, *Principles of Modern Physics,* McGraw-Hill, New York, 1959.

L. Pauling and E. B. Wilson, *Introduction to Quantum Mechanics,* McGraw-Hill, New York, 1935.

PROBLEMS

8.1 Use the transformation relation between spherical and Cartesian coordinates to show that Equation (8.2) is equivalent to Equation (8.1).

8.2 Show that if $V(r)$ is the Coulomb potential, $R_{1,0}(r)$ is a solution of the radial wave equation, Equation (8.8), when $E = -E_0$.

8.3 Repeat Problem 8.2 for $R_{2,0}(r)$ and $E = -E_0/4$.

8.4 Show that the total number of m_l states for a given value of the principal quantum number n is n^2.

8.5 Sketch the vector model for $l = 3$, showing all possible values of L_z.

•8.6 Show that $R_{2,1}(r)$ is properly normalized.

8.7 Find the expectation value of $V(r)$ for the state $n = 1$. Show that $\langle V(r) \rangle = -2E_0$.

•8.8 Calculate the expectation values of $(x^2 + y^2)$ and of z^2 for the states $n = 2$, $l = 1$, $m_l = 0$ and $n = 2$, $l = 1$, $m_l = 1$.

8.9 Show that $\Theta_{1,0}(\theta)$ is orthogonal to $\Theta_{2,0}(\theta)$. That is, show that

$$\int_0^{\pi} \Theta_{1,0}(\theta)\Theta_{2,0}(\theta) \sin\theta \, d\theta = 0$$

8.10 Show that although $|\psi_{2,1,0}|^2$ and $|\psi_{2,1,\pm1}|^2$ are not spherically symmetric, the sum $\sum_{m_l} |\psi_{2,1,m_l}|^2$ is spherically symmetric.

•8.11 Calculate the probability that the electron in the hydrogen atom will be at a radius less than $\frac{1}{2}r_n$ for $n = 2$, $l = 0$ and for $n = 2$, $l = 1$.

8.12 Show that the radial wave functions $R_{2,0}(r)$ and $R_{1,0}(r)$ are orthogonal.

8.13 In classical mechanics, the kinetic energy of a rotating dumbbell (two point masses M separated a distance a) is $L^2/2I$, where I is the moment of inertia of the dumbbell about its center of mass. Show that the allowed energies of the corresponding system in quantum mechanics are $E = l(l + 1)\hbar^2/2I$, where l is zero or a positive integer.

•8.14 The dipole moment of a charge distribution $\rho(r)$ is given by $\int z\rho(r)\, dV$, and its quadrupole moment is given by $\int (3z^2 - r^2)\rho(r)\, dV$.

(a) Show that the expectation value of the dipole moment of the hydrogen atom is zero for any stationary state.
(b) Show that the expectation value of the quadrupole moment is zero if $l = 0$.
(c) Calculate the expectation value of the quadrupole moment of hydrogen for the state $n = 2$, $l = 1$, $m_l = 1$.

8.15 Calculate the probability that the electron in the $n = 1$ stationary state of hydrogen will be inside the proton, i.e., within a sphere of radius $r = 1.2 \times 10^{-15}$ m. (*Hint:* Use the series expansion for the exponential function.)

8.16 Repeat the calculation of Problem 8.15 for the states $n = 2$, $l = 0$ and $n = 2$, $l = 1$.

In any quantum system, no two electrons can occupy the same quantum state.

In fact, the exclusion principle holds not only for electrons but for all particles whose spin quantum number is an odd half integer ($\frac{1}{2}$, $\frac{3}{2}$, etc.). As we shall show later, the exclusion principle imposes an important symmetry requirement on the wave function of a many-electron system. First, however, we consider the ground-state configuration of atoms of atomic number $Z > 1$.

9.2(a) The Ground State of Atoms

The remarkable regularity of the physical and chemical attributes of elements was first recognized a century before its origin could be explained. In 1817, Johann Dobereiner, a German chemist, pointed out that there were groups of elements with nearly identical chemical properties, such as lithium (Li), sodium (Na), and potassium (K), and also chlorine (Cl), bromine (Br), and iodine (I). As more and more elements were isolated, the triads of Dobereiner grew in number and had to be expanded in size. To the first of the groups above, rubidium (Rb) and cesium (Cs) had to be added, and fluorine (F) joined the second group. The ultimate achievement of this taxonomy was the periodic table of the elements published by Mendeleev in 1869. With amazing insight and unabashed confidence, Mendeleev not merely ordered the known elements in a scheme that has survived to this day, but predicted the existence of elements still unknown. The initial skepticism that greeted Mendeleev's proposal dissipated as, one after another, the blank spaces in his table were filled with the names of newly discovered elements.

The mysterious numerology of the periodic table suddenly became transparent with Pauli's suggestion. Although one cannot obtain an exact analytic solution for the wave function of a many-electron atom, one can, to first approximation, characterize the configuration in terms of single-particle wave functions, each identified by the set of four quantum numbers n, l, m_l, and m_s. The complete specification of these four quantum numbers for every electron of the atomic system is referred to as the *electronic configuration*. Since, to lowest order, the single-particle energies are independent of the quantum numbers m_l and m_s, it suffices to give the values of the principal quantum number n and the orbital quantum number l. The *spectroscopic notation* uses a letter to designate l. The code, which has its origin in the early description of atomic spectra, is as follows:

Orbital quantum number l	0	1	2	3	4	5
Letter designation[4]	s	p	d	f	g	h

For example, the symbol 3p represents an electron in the state $n = 3$, $l = 1$.

According to the Pauli principle, the number of electrons that can be accommodated in a state of given orbital quantum number l equals $2(2l + 1)$; the number of possible choices of m_l is $2l + 1$, and each of these states can hold two electrons, one with $m_s = +\frac{1}{2}$, and one with $m_s = -\frac{1}{2}$. Hence, a p state can hold as many as six electrons. If the electronic configuration includes more than one electron in a given

4. Capital letters are used to identify the orbital angular momentum state of the atom, as distinguished from that of the individual electrons (see Section 9.3).

(n, l) state, this is indicated by a superscript. For example, $3p^4$ means that there are four electrons with $n = 3$ and $l = 1$.

The ground-state configuration of the elements is now constructed by filling the available quantum states in such a manner that the total energy of the system is a minimum consistent with the requirements of the exclusion principle.

Hydrogen (Z = 1). In the ground state, the one electron will be in the $n = 1$, $l = 0$, i.e., the 1s state. In the absence of an external magnetic field, the two spin orientations are degenerate.

Helium (Z = 2). The ground state of He is $1s^2$. The 1s shell is now filled with two electrons of opposite spin orientation.

Lithium (Z = 3). Two of the three electrons occupy the 1s states. The third electron must go into the next state of lowest energy, an $n = 2$ state. In the hydrogen atom, the 2s and 2p states are degenerate, but this degeneracy is a consequence of the inverse-square-law Coulomb potential. The potential in which the $n = 2$ electron of lithium moves is not of that form, and as a result, the s and p states have different energies, with the s state the lower. The reason for this energy difference is apparent from the radial probability densities of the wave functions for $l = 0$ and $l = 1$ shown in Figure 8.3. The electron in the s state ($l = 0$) has a much higher probability density near $r = 0$, i.e., close to the nucleus, than the electron in the p state ($l = 1$). The s electron is said to be in a ''penetrating'' orbit, and feels the strong negative potential due to the large nuclear charge; the potential in which the p electron moves is nearly hydrogenic because the nuclear charge of $3e$ is screened by the two closely bound 1s electrons. Indeed, it is always true in atomic physics that for any n, states of lower l are more tightly bound, i.e., have lower energy, than states of large l. In fact, that energy difference becomes so pronounced in the heavier elements that the 4s states are filled before the 3d states, and the 5s, 5p, and 6s states are filled before any 4f states.

The ground state of lithium is, therefore, $1s^2 2s$.

Beryllium (Z = 4). Following the preceding arguments, one expects and finds that the ground state of Be is $1s^2 2s^2$.

Boron (Z = 5). With the 1s and 2s states fully occupied, the fifth electron goes into a 2p state; the configuration of B is $1s^2 2s^2 2p$.

Carbon (Z = 6) to neon (Z = 10). As more electrons must be accommodated, they fill the remaining 2p states. Since $l \leq n - 1$, the $n = 2$ shell is completed when we reach the neon configuration, $1s^2 2s^2 2p^6$.

Sodium (Z = 11) to argon (Z = 18). The electronic configurations for this group follow the pattern of the preceding eight elements. That of sodium is $1s^2 2s^2 2p^6 3s$; that of magnesium ($Z = 12$) is $1s^2 2s^2 2p^6 3s^2$. With the next element, aluminum, the 3p levels begin to fill, and when argon is reached, the structure is again one of completely filled s and p subshells.[5]

Potassium (Z = 19) to krypton (Z = 36). The valence electron[6] of potassium occupies a 4s state, not a 3d state, for the reason already mentioned. Similarly, the valence electrons of calcium are the two 4s electrons.

---

5. It is customary to refer to the set of states of a given principal quantum number as a *shell* and to the set of states of a given orbital quantum number as a *subshell*.

6. Electrons in the outermost unfilled subshells are referred to as *valence electrons* because they give the atom its chemical valence.

orbital angular momenta of the electron are coupled through the spin-orbit interaction. In classical mechanics, the angular momentum of an isolated system is conserved. This conservation principle applies with undiminished force also in quantum mechanics. However, as in classical mechanics, it is the *total* angular momentum that is conserved, not that of the individual parts of the system. In the case of an electron with orbital angular momentum **L**, it is the total angular momentum, the vector sum of **L** and **S**, that is conserved. By convention, the symbol **J** is used to designate the total angular momentum, and it, too, obeys the now familiar angular momentum quantization rules. That is,

$$J = \sqrt{j(j+1)}\hbar \tag{9.18}$$

where

$$j = |l \pm s| = |l \pm \tfrac{1}{2}| \tag{9.19}$$

and

$$J_z = m_j\hbar \qquad m_j = -j, -j+1, \ldots, j-1, j \tag{9.20}$$

As was mentioned in the preceding section, the ground and excited states of atoms are usually expressed in a notation that specifies the orbital and total angular momenta and in which closed-shell configurations are understood. The spin angular momentum of the state is identified by a superscript whose value is the number of possible spin orientations, known as the *multiplicity* of the state. The total angular momentum quantum number j is identified by a subscript. For example, the ground state of hydrogen is $1^2S_{1/2}$. The 2P state is split by spin-orbit interaction into two levels, the $2^2P_{1/2}$ and $2^2P_{3/2}$ states, with the $2^2P_{1/2}$ the lower of the two. A similar situation prevails in sodium. The ground state, $3^2S_{1/2}$, is not affected by spin-orbit coupling, since $\mathbf{L} = 0$. The first excited state, the 3P state, is split into a $3^2P_{1/2}$ and a $3^2P_{3/2}$ state. The transitions from these two states to the ground state produce the prominent doublet of the D line (see Figure 9.10).

We can now also understand the "anomalous" Zeeman effect, that is, the splitting of a spectral line into an *even* number of lines by a magnetic field. If a sodium atom is placed in a magnetic field, the doubly degenerate $3^2S_{1/2}$ level ($m_j = \pm\frac{1}{2}$) is split by the magnetic field because the atom in that state carries a

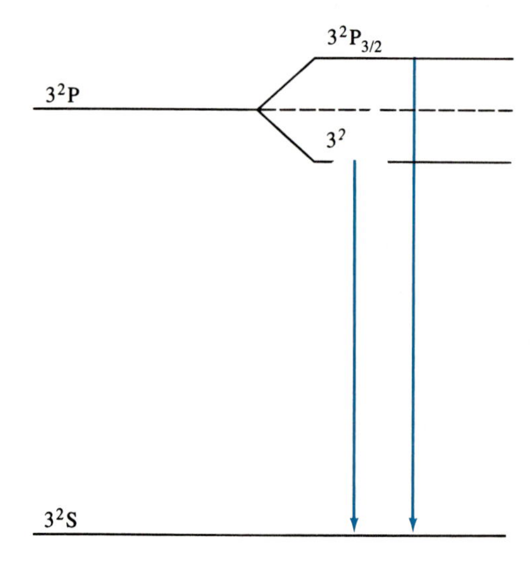

Figure 9.10 Fine structure splitting of the ground and first excited states of sodium. The ground state is not split by spin-orbit interaction since $\mathbf{L} = 0$; the 3^2P state is split by that interaction. The optical transition from the 3^2P states to the ground state results in the doublet structure of the sodium D line.

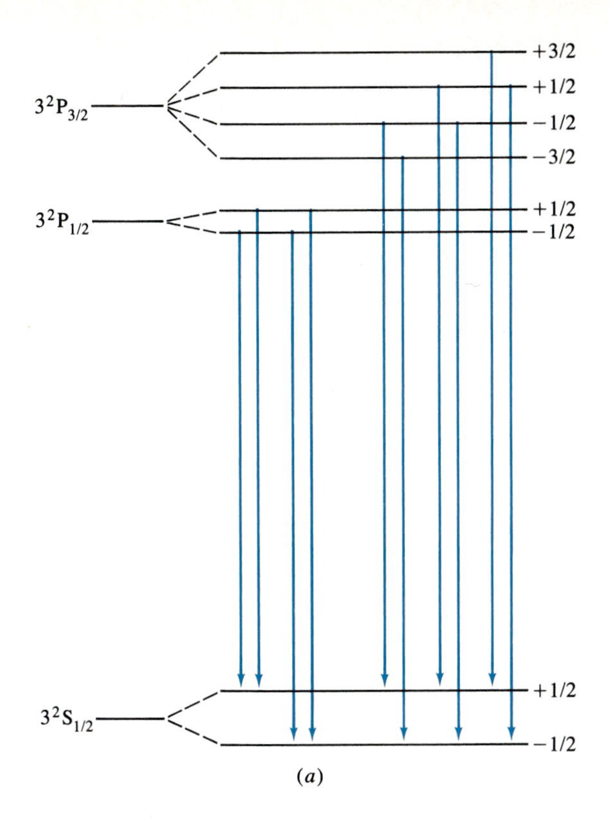

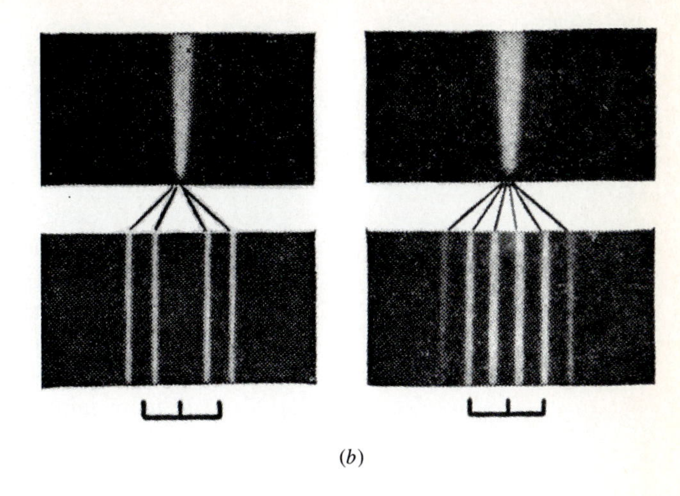

Figure 9.11 (a) Splitting of the $3^2S_{1/2}$, $3^2P_{1/2}$, and $3^2P_{3/2}$ states of sodium by a magnetic field. (b) In a magnetic field the allowed transitions from the $3^2P_{1/2}$ states to the $3^2S_{1/2}$ states result in four lines; the allowed transitions from the $3^2P_{3/2}$ states to the $3^2S_{1/2}$ states result in six lines. *(Harvey Elliot White,* Introduction to Atomic Spectra, *McGraw Hill, New York,* 1934, *p.* 152.*)*

magnetic moment due to the spin of the electron. The degeneracy of the $3^2P_{1/2}$ level is also lifted by the magnetic field. However, the net magnetic moment of the atom in this state is the result of a combination of orbital and spin magnetic moments and differs from that of the $3^2S_{1/2}$ state. The fourfold degenerate $3^2P_{3/2}$ state splits into four separate levels, and the energy separation between those levels differs from that for the $3^2S_{1/2}$ and $3^2P_{1/2}$ states. The energy-level structure for these states is shown in Figure 9.11.[7]

The selection rules for optical transitions are similar to those already discussed. They are

$$\Delta j = 0, \pm 1 \quad \text{(except that } j = 0 \text{ to } j = 0 \text{ is forbidden)}$$

$$\Delta m_j = 0, \pm 1$$

As indicated in Figure 9.11, the $3^2P_{1/2} \rightarrow 3^2S_{1/2}$ transition results in four lines, the $3^2P_{3/2} \rightarrow 3^2S_{1/2}$ transition in six lines. These are shown in Figure 9.11b.

7. The z component of the magnetic moment of an atom in the state j is given by

$$\mu_z = g\mu_B m_j$$

where g is the Landé g factor

$$g = 1 + \frac{j(j+1) + s(s+1) - l(l+1)}{2j(j+1)}$$

9.4 THE EXCLUSION PRINCIPLE AND THE SYMMETRY PROPERTIES OF THE ELECTRONIC WAVE FUNCTION

Prior to the development of quantum mechanics, an atom was portrayed like the solar system, with the electrons representing the planets circulating about the nuclear "sun." Even in the simple Bohr model, that analogy is inappropriate. In the solar system, the gravitational interaction between planets is quite small compared with that between any planet and the very massive sun; interplanetary interactions can, therefore, be treated as small perturbations. In an atom, however, the interaction energy between electrons and between an electron and the nucleus are of almost the same magnitude, and a perturbation approach is inapplicable.

Even more important is the fact that whereas one can identify each planet unambiguously, the electrons of an atom are indistinguishable. This leads to complications that have no classical analog, yet are central in quantum mechanics. To see how the fact that we cannot label electrons influences the character of the wave function of a many-electron system, let us consider the simplest case, the helium atom.

The time-independent Schrödinger equation for the helium atom is

$$-\frac{\hbar^2}{2m}\left[\frac{\partial^2\psi(r_1,r_2)}{\partial r_1^2}+\frac{\partial^2\psi(r_1,r_2)}{\partial r_2^2}\right]+V(r_1,r_2)\psi(r_1,r_2)=E\psi(r_1,r_2) \qquad \textbf{(9.21)}$$

Here r_1 and r_2 denote the coordinates of the two electrons and $\psi(r_1,r_2)$ is the two-particle wave function, a function of both electron coordinates. The potential energy $V(r_1,r_2)$ is, as indicated, also a function of both coordinates, since it includes the mutual Coulomb interaction between the two electrons, and that depends on both r_1 and r_2. Now, although we have attached the labels "1" and "2" to the two electrons, they cannot be distinguished. Consequently, the total probability density of the electron distribution must be unchanged if we arbitrarily interchange the labels "1" and "2." In other words,

$$|\psi(r_1,r_2)|^2 = |\psi(r_2,r_1)|^2$$

Therefore, either

$$\psi(r_1,r_2) = \psi(r_2,r_1) \qquad \text{symmetric wave function}$$

or

$$\psi(r_1,r_2) = -\psi(r_2,r_1) \qquad \text{antisymmetric wave function}$$

Following the procedure adopted in characterizing the electronic configuration of atoms, one might try to solve the Schrödinger equation by writing the wave function as a product of one-electron functions, that is,

$$\psi(r_1,r_2) = \psi_p(r_1)\psi_q(r_2) \qquad \textbf{(9.22)}$$

where p and q are labels that represent the quantum states of the two electrons. That is, p and q each include the four quantum numbers n, l, m_l, and m_s.

The product function in Equation (9.22) is neither symmetric nor antisymmetric with respect to an interchange of the electron coordinates. We can, however, use the product function to construct wave functions of the proper symmetry, namely,

$$\psi_s = \psi_p(r_1)\psi_q(r_2) + \psi_p(r_2)\psi_q(r_1) \qquad \text{symmetric wave function} \qquad \textbf{(9.23a)}$$

$$\psi_a = \psi_p(r_1)\psi_q(r_2) - \psi_p(r_2)\psi_q(r_1) \qquad \text{antisymmetric wave function} \qquad \textbf{(9.23b)}$$

Now according to the Pauli exclusion principle, no two electrons can occupy the same quantum state; that is, they cannot have the same quantum numbers. If we look at the two combinations of one-electron functions, we see that if p and q are the same, the antisymmetric function vanishes identically. In other words, the Pauli principle is equivalent to the requirement that in a many-electron system the total wave function be antisymmetric with respect to the interchange of any two electrons of the system.

The wave function for an electron is the product of the space function $\phi(r)$ and the spin function χ. Similarly, the total wave function of a many-electron system is the product of a total space function and a total spin function. An antisymmetric wave function is then either the product of a symmetric space function and an antisymmetric spin function, or an antisymmetric space function and a symmetric spin function.

In the two-electron helium atom, the spins of the two electrons can be either parallel or antiparallel. If they are antiparallel, the total spin quantum number $S = 0$; if they are parallel, $S = 1$. In the ground state of helium, the quantum numbers n, l, and m_l of the two electrons are the same, and the space function is, therefore, symmetric. Consequently, the spin function must be antisymmetric, and the ground state is a *singlet* state. The total spin quantum number $S = s_1 + s_2 = 0$, and the multiplicity of the state, $2S + 1$, is 1.

If one of the two electrons is raised to a higher energy level, the total wave function of the two-electron system must still be antisymmetric. Now, however, the space function may be either symmetric or antisymmetric, and the spin function of opposite symmetry. Thus one obtains two sets of energy levels, corresponding to singlet and triplet states, i.e., $S = 0$ and $S = 1$. Since in the triplet states, the spins of the two electrons are parallel, the spin wave function for those states is symmetric, the space function antisymmetric.

The energy-level pattern for helium is shown in Figure 9.12. The diagram is divided into two parts, the singlet states and the triplet states. Note that the energies of the triplet states are lower than those of the corresponding singlet states. That energy difference, referred to as the *exchange energy,* derives from the symmetry of the wave function. As we have seen, the space function of the triplet states must be antisymmetric to satisfy the exclusion principle. From Equation (9.23b) it follows that if $r_1 = r_2$, the antisymmetric wave function $\psi_a = 0$, and consequently the probability density vanishes for $r_1 = r_2$. By contrast, the probability density of the symmetric space function for $r_1 = r_2$ may be substantial, as shown schematically in Figure 9.13 for a two-electron wave function constructed from a 1s and a 2s function. **The symmetry property of the wave function ensures that in the triplet state the two electrons are relatively far apart** and their Coulomb repulsion is, therefore, less than in the singlet state. That is a quite general result; the energy of an atomic state with an antisymmetric space function is less than that of the corresponding symmetric space state.

Note also that the energy levels of the triplet system are split by the spin-orbit

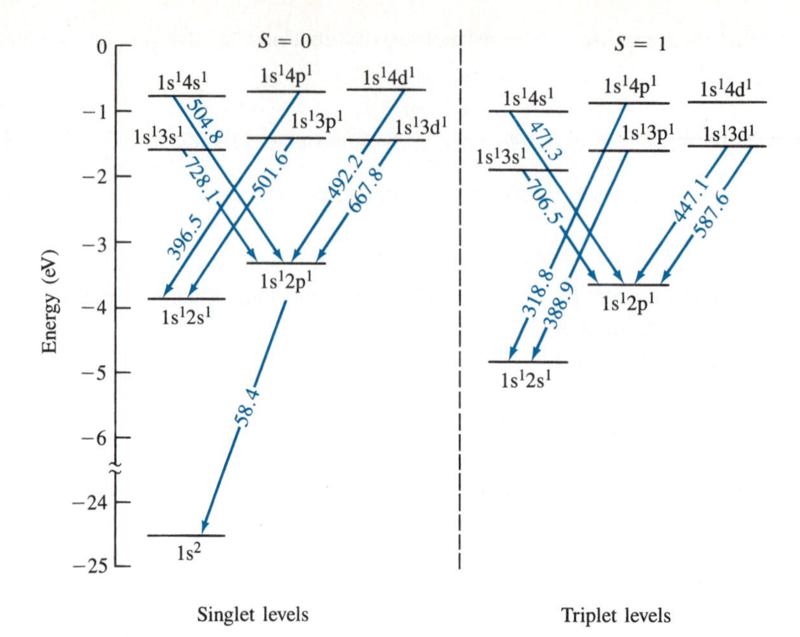

Figure 9.12 Singlet and triplet energy levels of helium. The triplet levels are split by spin-orbiting coupling, but that splitting is not shown in this figure.

Figure 9.13 Probability densities for a two-electron system (schematic). In the symmetric space state, the singlet spin state (*a*), the probability density is large at $x = 0$, i.e., when the two electrons occupy the same region. In the antisymmetric space state, the triplet spin state (*b*), the probability density at $x = 0$ vanishes; the two electrons tend to stay apart from each other. Consequently, the triplet state has the lower energy.

$|\psi_S \chi_A|^2$

(*a*)

$|\psi_A \chi_S|^2$

(*b*)

interaction. This splitting is shown in Figure 9.12 only for a few of the lower triplet levels, and is greatly exaggerated.

The selection rules for optical transitions between energy levels of multielectron atoms are the same as before with one further condition, namely,

$$\Delta S = 0$$

In other words, transitions between triplet and singlet states are ''forbidden.'' Although the selection rule is not absolute, the probability of a forbidden transition is very small compared with that of an allowed transition, and the 2^3S_1 and 2^1S_0 states of helium are both long-lived *metastable* states: the first, because a transition to the ground state must violate the $\Delta S = 0$ selection rule; the second, because the transition must violate the ΔJ selection rule ($J = 0$ to $J = 0$ is ''forbidden''). Such metastable states are central in the design and operation of lasers.

9.5 MASERS AND LASERS

Today, lasers are commonplace: the vast majority of supermarkets use dozens of lasers at their checkout counters; compact disk (CD) players use them; lasers are employed for surveying, in numerous industrial applications, in surgery and dentistry, and in light-fiber communication systems.[8]

The maser, the laser's progenitor, though not so widely used, has also played a crucial role in science as the most sensitive detector and amplifier of microwave radiation known. It was a maser-instrumented radiotelescope that first revealed the ubiquitous blackbody radiation corresponding to a temperature of about 3 K, the residue of the ''Big Bang,'' the traumatic birth of the universe.

The word *laser* is an acronym for *l*ight *a*mplification by *s*timulated *e*mission of *r*adiation; *maser* is the same acronym with *m*icrowave replacing the word *l*ight. In these acronyms, the key word is *stimulated*.

9.5(a) Spontaneous and Stimulated Emission

The concept of stimulated emission, which is basic to maser and laser operation, was first introduced by Einstein in 1916. In his paper, Einstein presented a simple and elegant derivation of Planck's formula, Equation (4.26), using only thermodynamic arguments.[9]

Consider a collection of atoms in thermal equilibrium in a cavity at a temperature T. The distribution of the atoms among the accessible quantum states of energy E_n is given by the Boltzmann factor, i.e.,

$$N_n = Ce^{-E_n/kT} \tag{9.24}$$

where C is a constant. Thus, the numbers of atoms in states E_i and E_j are in the ratio

$$\frac{N_i}{N_j} = e^{(E_j - E_i)/kT} \tag{9.25}$$

---

8. It is ironic that in 1958, when C. H. Townes and A. L. Schawlow submitted an idea for an optical resonant cavity, a critical component of all lasers, to the Bell Laboratory patent office, it was at first rejected on the grounds that ''optical waves had never been of importance to communication and hence the invention had little bearing on Bell System interests.''

9. Einstein always sought very general proofs rather than derivations based on details of specific model systems, and was especially fond of thermodynamics. As he wrote in later years, ''A theory is the more impressive the greater the simplicity of its premises, the more diverse the things it relates, and the more extended its area of applicability. Therefore the deep impression which classical thermodynamics made upon me. It is the only physical theory of universal content which, I am convinced, will never be overthrown.''

The blackbody radiation within the cavity, encompassing the entire electromagnetic spectrum, interacts with the atoms causing transitions between the various energy levels. Einstein recognized that there must be three kinds of transitions. The first is *absorption* of radiation by the atoms as they are excited from a state of low energy to one of higher energy. The second is the inverse process, the downward transition induced by the radiation field; it is this process we now call *induced* or *stimulated emission,* in which radiation is released by the atom as it makes its downward transition. Finally, we know that even in the absence of a radiation field, an atom in an excited state returns to the ground state in a relatively short time, again emitting radiation in the process; the third kind of transition is, therefore, *spontaneous emission*.

The probability for induced transitions between states E_i and E_j is proportional to the intensity $I(\nu)$ of the radiation of frequency

$$\nu = \frac{|E_j - E_i|}{h} \tag{9.26}$$

The probability for spontaneous transitions is, by assumption, independent of $I(\nu)$. Let us designate these probabilities by $B_{ji}I(\nu)$, $B_{ij}I(\nu)$, and A_{ij}, respectively. Here we assume that the state j is at a lower energy than the state i. B_{ij} and A_{ij} refer to emission, B_{ji} to absorption. It is implicit that while the coefficients B and A will depend on the properties of the atom, they do not depend on the temperature of the cavity.

The number of upward transitions from the state j to the state i is then

$$N_j B_{ji} I(\nu)$$

and the number of downward transitions from i to j is

$$N_i[B_{ij}I(\nu) + A_{ij}]$$

In steady state, the number of upward transitions per unit time must equal the number of downward transitions. Consequently, we can write

$$N_j B_{ji} I(\nu) = N_i[B_{ij}I(\nu) + A_{ij}] \tag{9.27}$$

If, as we assumed at the outset, the system is in thermal equilibrium at the temperature T, the ratio N_i/N_j is given by Equation (9.25), and Equation (9.27) reduces to

$$B_{ji}I(\nu)e^{h\nu/kT} = B_{ij}I(\nu) + A_{ij} \tag{9.28}$$

Imagine raising the temperature of the cavity and of the gas of atoms so high that $kT \gg h\nu$; then $h\nu/kT \to 0$ and $e^{h\nu/kT} \simeq 1$. Simultaneously, the intensity of the radiation within the cavity also increases so that, as $T \to \infty$, A_{ij} becomes negligible compared with $B_{ij}I(\nu)$. In the high-temperature limit, therefore, $B_{ij} = B_{ji}$, and since these coefficients are independent of temperature, that equality must hold true in general.

We can now solve Equation (9.28) for the radiation intensity $I(\nu)$ and obtain

$$I(\nu) = \frac{A_{ij}}{B_{ij}} (e^{h\nu/kT} - 1)^{-1} \qquad \textbf{(9.29)}$$

which is equivalent to the Planck distribution, Equation (4.26).

To simplify our discussion, we focus attention on just two energy levels, E_1 and E_2 of our system. Normally, the population of atoms is greatest in the lower state, E_1, and therefore if photons of energy $h\nu = E_2 - E_1$ are incident on the system, the number of upward transitions will be greater than the number of downward transitions; the system will absorb energy from the incident radiation. Suppose, however, that somehow we manage to invert the population so that we have initially more atoms in the excited state E_2 than in the ground state. If we now irradiate the system with even a few photons of the resonant energy, these photons will induce atoms to make transitions from the excited state to the ground state, emitting photons of energy $h\nu = E_2 - E_1$ in the process. Thus instead of removing photons from the incident radiation, the system now augments the number of photons as a result of induced, or stimulated, emission. In other words, the system prepared with a population inversion amplifies the incoming signal. Schematically, we can represent the absorption and stimulated emission processes by

$$A + h\nu \rightarrow A^* \qquad \text{(absorption)} \qquad \textbf{(9.30a)}$$

$$A^* + h\nu \rightarrow A + 2h\nu \qquad \text{(stimulated emission)} \qquad \textbf{(9.30b)}$$

where A^* represents the atom in the excited state.

An important feature of radiation due to stimulated emission is that the phase of the emitted photon is the same as that of the photon which induces the transition; that is, the two photons in Equation (9.30b) are phase-coherent.

By 1920 the theoretical groundwork for lasers had already been prepared.[10] Moreover, much of the technology and information on atomic and molecular energy levels that is needed to design a laser such as the commonly used helium-neon laser was also at hand. There are probably several reasons this development was delayed for some thirty years. First, effective communication between physicists and engineers was not fostered until World War II; physicists inclined to work on problems that appeared to be only very remotely related to practical concerns, and most engineers tended to ignore the work of physicists. Moreover, physicists often regarded involvement in device application as unimportant and unworthy of their talents. Second, the need or potential usefulness of such a device was not recognized at the time.[11]

---

10. In 1924 R. C. Tolman wrote, " . . . molecules in the upper quantum state may return to the lower quantum state in such a way as to reinforce the primary beam by 'negative absorption' " [R. C. Tolman, *Phys. Rev.* 23, 693 (1924)].

11. Recalling his 1947 research on the energy splitting between the $2S_{1/2}$ and $2P_{1/2}$ states of hydrogen, W. Lamb wrote, "The concept of negative absorption was new to us at the time, and we were unaware of the earlier references. I think we understood that the radiation would be coherent, as was the input signal. However, we did not associate negative absorption with self-sustained oscillation. Even if we had done so, at least three factors would have kept us from inventing the maser: (1) our interest was centered on the fine structure of hydrogen, (2) the smallness of the expected absorption (gain), and doubt as to its sign, and (3) the ready availability of oscillators at the frequency used."

9.5(b) Creating a Population Inversion: Pumping

As has been pointed out, an atomic or molecular system will not act as a maser or laser under normal circumstances, because the distribution of the constituents among the available energy levels is such as to result in absorption of incident radiation. "Negative absorption," or net emission, can come about only if the population in the state of higher energy exceeds that in the state of lower energy. Thus, to make a laser or maser work, one must first devise a scheme for achieving the requisite population inversion.

Suppose we could construct an atomic or molecular system that has just three energy levels: the ground state of energy E_1, and two excited states of energies E' and E_m, as shown in the diagram of Figure 9.14.[12] Let us further assume that these levels have the following properties:

1. Atoms are readily excited from the ground state to the state E' on irradiation by light of frequency $\nu_p = (E' - E_1)/h$.
2. The state E' is short-lived, and the probability for spontaneous decay to the state E_m is comparable to the probability for decay back to the ground state.
3. The state E_m is a metastable state; that is, the probability for *spontaneous* decay from E_m to E_1 is extremely small.

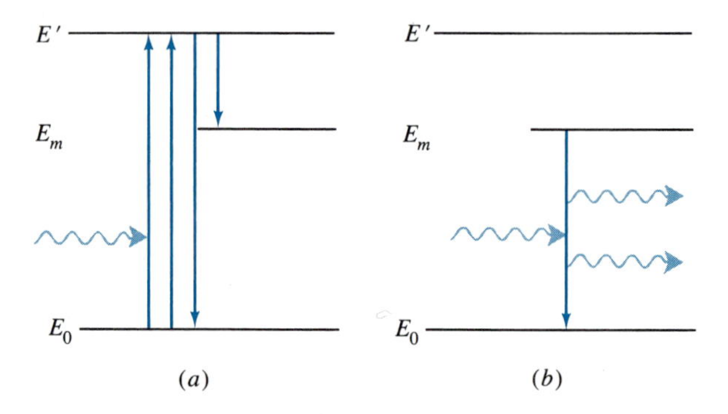

Figure 9.14 Energy levels for a three-level laser. The level E_m is a long-lived metastable state. (*a*) "Pumping" of atoms from the ground state to the excited state E' populates E_m as atoms decay from E' to E_m. (*b*) An incident photon of energy $h\nu = E_m - E_0$ stimulates the transition of atoms in the metastable state to the ground state.

If we now subject the system to radiation of frequency ν_p, the so-called pump frequency, atoms will be excited to the state E' and, of course, the radiation will also induce transitions back down to the ground state. If the pump is very intense, the numbers of atoms in the states E' and E_1 will be equal; as far as these two energy levels are concerned, the effect is the same as if we had immersed the atoms in a cavity of extremely high temperature.

Since there is a fair probability of decay from E' to E_m, not all of the excited atoms return to the ground state; some decay to the state E_m, where they are trapped because this state is metastable. This process partly depletes the state E', but that is made up by further excitation from the ground state. Thus, after a short time there will be an abundance of atoms in the state E_m, and very few in either E' or E_1. We have thus achieved a population inversion with respect to the states E_m and E_1.

12. Of course, such systems do not exist; however, only three or four of the innumerable energy levels of the active molecules directly participate in maser or laser operation. For our purposes, we can ignore the presence of other energy states.

If now photons of energy $h\nu_L = E_m - E_1$ are admitted into the system, they will stimulate transitions to the ground state. If the atoms are in an enclosure that traps most of these photons, each photon induced by the primary beam will cause further stimulated emission before it escapes from the enclosure. The end result is then an avalanche of phase-coherent photons of energy $h\nu_L$.

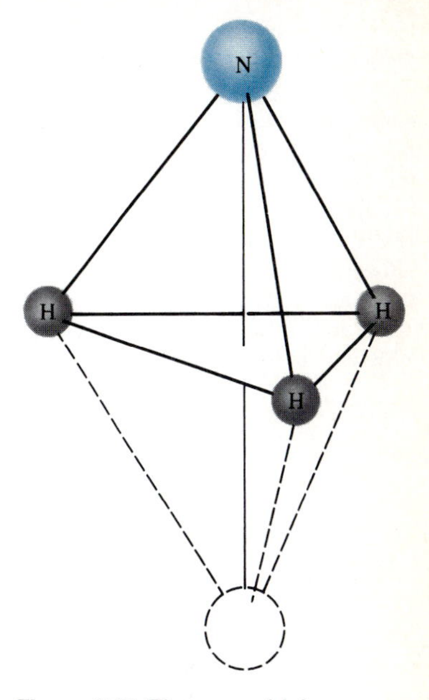

Figure 9.15 The pyramidal structure of the ammonia molecule, NH_3. The system is symmetric with respect to reflection in the plane of the three H atoms.

9.5(c) Maser and Laser Operation

The first successful device was the ammonia maser invented by Gordon, Zeiger, and Townes in 1954. The NH_3 molecule may be envisaged as a triangular pyramid, with the hydrogen atoms at the corners of the base and the nitrogen at the apex (Figure 9.15). The nitrogen atom can be either above or below the plane of the base, the two configurations having exactly the same energy. Thus the potential energy of the nitrogen atom in the field of the three hydrogens may be represented schematically as in Figure 9.16; there are two potential minima located symmetrically about the midplane at $x = 0$. Since the potential is symmetric, the solutions of the Schrödinger equation must have definite parity (see Section 7.7). The wave function describing the nitrogen atom is a linear superposition of the wave functions corresponding to the atom in one or the other of the potential minima; one of these wave functions is symmetric (even parity), the other antisymmetric (odd parity). Both states have nearly the same energy, the energy difference corresponding to a frequency of 23,870 MHz, or a wavelength of 1.256 cm. As this energy difference is only 10^{-4} eV, both states are almost equally populated at ordinary temperatures. However, because the charge distributions differ in the symmetric and antisymmetric states, molecules in the two states can be separated by passing a beam of NH_3 molecules through a strong inhomogeneous electric field.

A schematic diagram of the experimental setup employed by Gordon, Zeiger, and Townes is shown in Figure 9.17. The electric field configuration is such that molecules in the higher of the two energy levels are focused on the opening of a cavity designed to resonate at the frequency of 23,870 MHz. If a small microwave

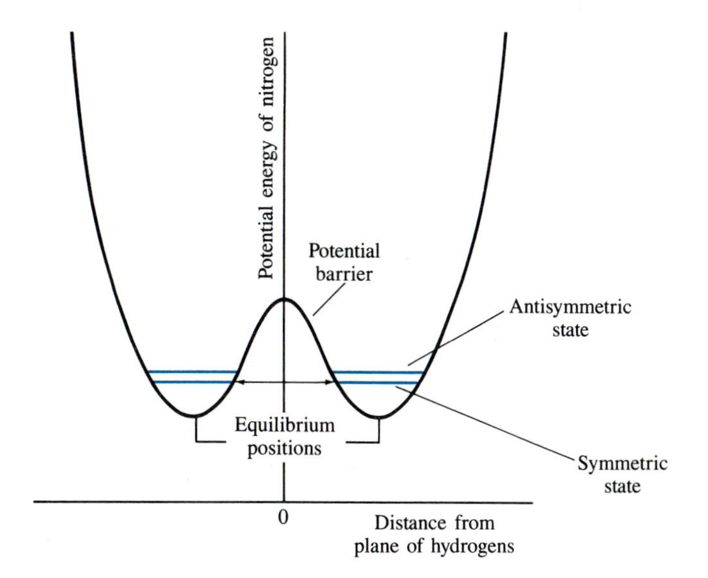

Figure 9.16 The potential energy of the nitrogen atom of NH_3 as a function of distance from the reflection plane. The energy levels of the symmetric and antisymmetric states are shown.

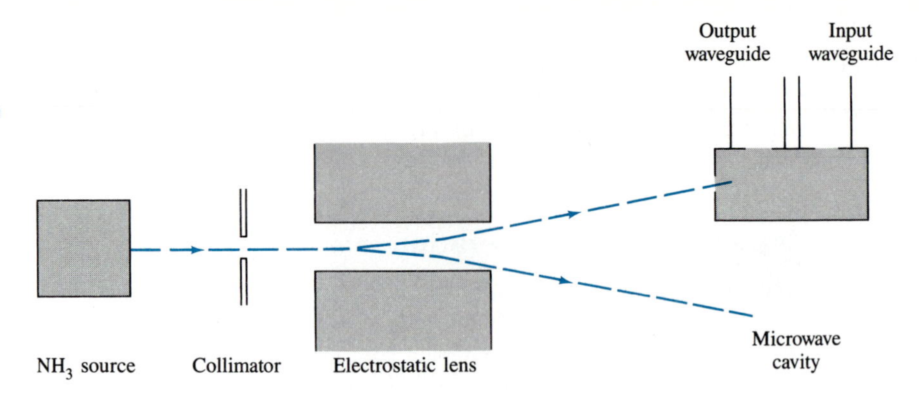

Figure 9.17 Schematic diagram of the ammonia maser. The electrostatic lens focuses only molecules in the upper (*A*) state on the opening of the microwave cavity. Hence, molecules in the cavity are all in the higher antisymmetric state. Transitions to the lower state are induced by admitting microwave radiation of 23,870 MHz frequency.

signal of this frequency is sent into the cavity, it will stimulate transitions of the molecules to the lower state, and the output signal from the cavity will then be much greater than the input; the device acts as a microwave amplifier. Alternatively, one can employ a positive feedback loop and maintain sustained oscillations at the resonant frequency.

The ammonia maser was used primarily as a frequency standard or atomic clock. With careful design of the cavity the width of the resonance could be made as small as 10^{-2} Hz, and ammonia atomic clocks with long-term stabilities of one part in 10^{10} were constructed.[13]

The ammonia maser was never a practical amplifier. Not only is it a cumbersome device, but it operates at only one very sharply defined frequency. A practical maser should be tunable over a reasonable frequency range. A device that meets this requirement is the ruby maser. Here the energy levels between which stimulated emission takes place are two Zeeman levels of the paramagnetic chromium ion in the host alumina crystal. By adjusting the strength of the magnetic field, the Zeeman splitting can be controlled and the maser tuned.

Ruby can also serve as the active material for a laser. When used in this manner, the intense pump radiation is obtained from a xenon flash lamp (see Figure 9.18). The two ends of the cylindrical ruby rod are highly polished and parallel within very close tolerance. One end is silvered, the other partially silvered, so that the stimulated photons are trapped within this optical resonant cavity sufficiently long to induce further transitions. The beam exits from the partially silvered end.

Today there are a great variety of lasers commercially available. Some operate at one or a few frequencies, others can be tuned over a fairly wide frequency band. Their energy outputs range from fractions of a milliwatt to megawatts. Here we shall describe only one other laser system, the helium-neon laser whose bright red beam is familiar to nearly every person today.

---

13. A stability of one part in 10^{10} corresponds to a gain or loss of less than one second per century. Since then, far more precise atomic clocks have been developed. Today, the second is defined as "the duration of 9,192,631,770 periods of the radiation corresponding to the transition between the two hyperfine levels of the ground state of the cesium-133 atom." The stability of a cesium atomic clock is about 4 parts in 10^{12}, corresponding to a gain or loss of no more than one second in six millenia. The recently developed hydrogen maser is much superior still, with an incredible stability of 1 part in 10^{15}; a hydrogen maser atomic clock would not gain or lose more than a second in thirty million years.

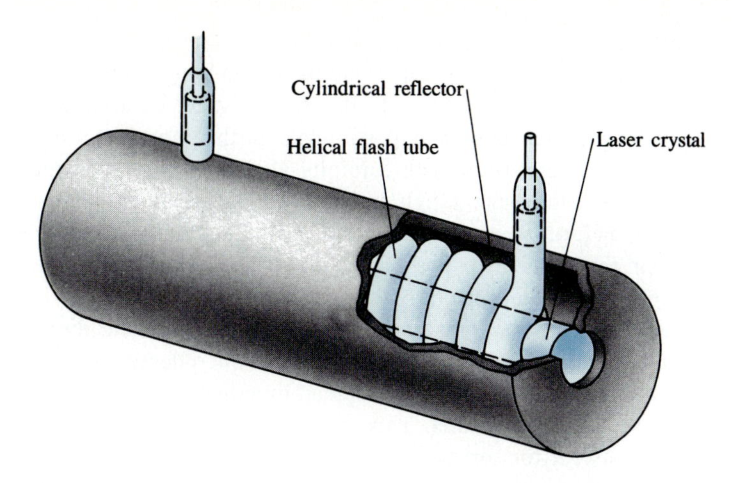

Figure 9.18 Diagram of a ruby laser. Optical pumping is achieved by discharging a xenon flash tube that surrounds the ruby cylinder.

The helium-neon laser consists of a glass tube, 10 to 50 cm long and about 1 cm in diameter, filled with a mixture of ten parts of helium to one part of neon at a pressure of about 1 mmHg. A discharge is maintained by means of a high voltage applied across the tube. Collisions of electrons with helium atoms excite these atoms to the 2^1S_1 state, whose energy is 20.61 eV above the ground state. As we have seen, this is a long-lived metastable state (see page 191). As it happens, neon also has an excited state at almost the same energy (Figure 9.19), and as helium and neon atoms collide within the tube, there is a good chance that the excitation energy of the helium atoms will be transferred to the neon atom, the helium returning to the ground state, and the neon now in the $2p^5 5s^1$ excited state.

Neon atoms in this excited state can decay to the ground state via transitions to intermediate states, one of these being the $2p^5 3p^1$ state shown in Figure 9.19. From there, decay to the ground state proceeds via the $2p^5 3s^1$ state.

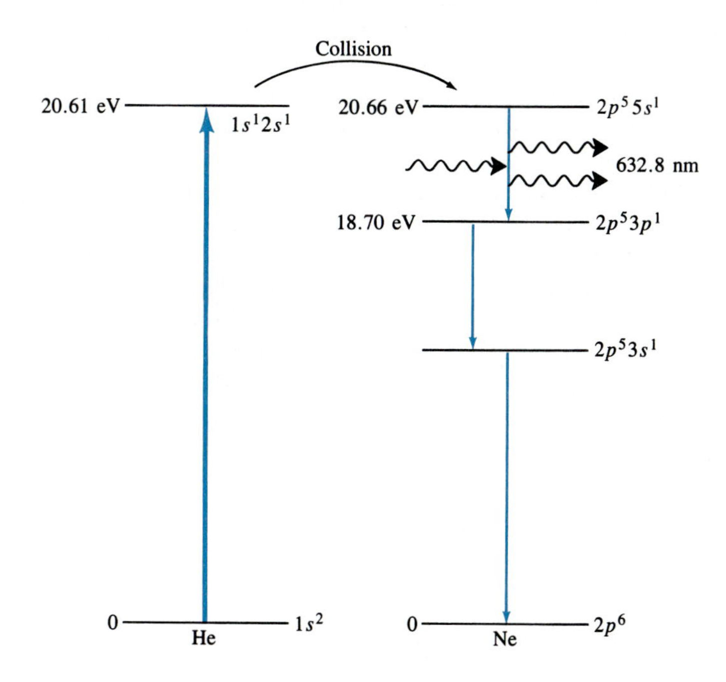

Figure 9.19 Energy levels of a helium-neon laser. Helium atoms are raised to the metastable 2^1S state by collision with electrons in the discharge tube. When a helium atom in this metastable state collides with a neon atom in the ground state, the helium returns to the ground state and the neon atom is excited into the $2p^5 5s$ state at 20.66 eV. The neon atom then decays to the $2p^5 3p$ state with the emission of a 632.8-nm photon. That photon, in turn, stimulates the same transition in other neon atoms that have been raised to the $2p^5 5s$ excited state by collisions with helium atoms.

Since collisions with helium atoms excite only the $2p^5 5s^1$ state of neon, this state will have a population much greater than the lower-lying excited states of this gas. If one of these neon atoms drops to the $2p^5 3p^1$ state, emitting a photon of 632.8-nm wavelength, that photon will stimulate further transitions in the laser tube. The discharge tube is fitted with carefully aligned mirrors, of which one is only partially silvered, to ensure a high probability that a given photon will induce further transitions before it escapes the end of the tube with the partially silvered mirror. The beam that emerges is phase-coherent; spatial coherence of the beam is ensured by alignment of the mirrors.

The coherence of laser beams accounts for their remarkable directionality. Beam divergence of large lasers may be as small as a microdegree, so that a 5-cm-diameter laser beam from Earth will have spread to a diameter of only about 10 m on reaching the surface of the moon. This extreme collimation of the beam makes lasers particularly useful for surveying.

Although the power output of a small laser may be only a few milliwatts, this energy is concentrated in a beam of very small cross section. With lenses or mirrors, the light can be further focused to a very fine point of about one micron diameter, so that the intensity may be many kilowatts per square meter. Lasers are therefore used extensively whenever a precisely controlled amount of energy must be concentrated in some small region. For example, laser tools are used in preparing microelectronic circuitry as well as in medicine. Among medical applications of lasers are their use in eye surgery, particularly attachment of detached retinas; treatment of internal bleeding by coagulation (here the laser beam is directed to the site of hemorrhage by a light pipe); and excision of tumors.

Finally, lasers have made holography possible. Until fairly recently, the compression of three-dimensional objects into two-dimensional images by standard photographic techniques was generally accepted as though it were a fundamental constraint on optical reproduction. In fact, what happens is that much of the information contained in the light that is scattered by the object into the camera lens is not used in photography. The blackening of the photographic emulsion depends on the intensity of the light, not on its phase. It is the phase information that is discarded in ordinary photography.

Holography uses both intensity and phase information to record and reconstruct an image. A schematic diagram of a split-beam holographic setup is shown in Figure 9.20. The beam from a laser is broadened by a beam expander (a diverging lens) and split by a half-silvered mirror. Light from the reference beam as well as light scattered by the object strikes the photographic plate. Since the two sources of light are coherent, an interference pattern is formed at the photographic plate. This interference pattern depends on the intensity of the light scattered by a particular point of the object as well as on the phase difference between the scattered light and the reference beam. The resulting hologram thus appears to be a nearly random arrangement of dark and light regions. However, if the hologram is illuminated using the coherent beam from a laser, the waves scattered from the various points of the photographic plate now interfere in such a way as to reconstruct a three-dimensional image of the object.

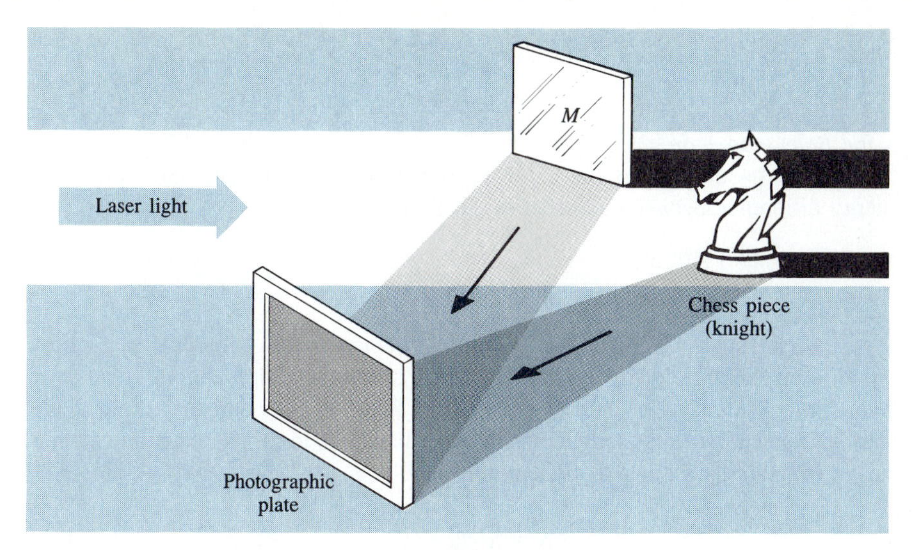

(*a*)

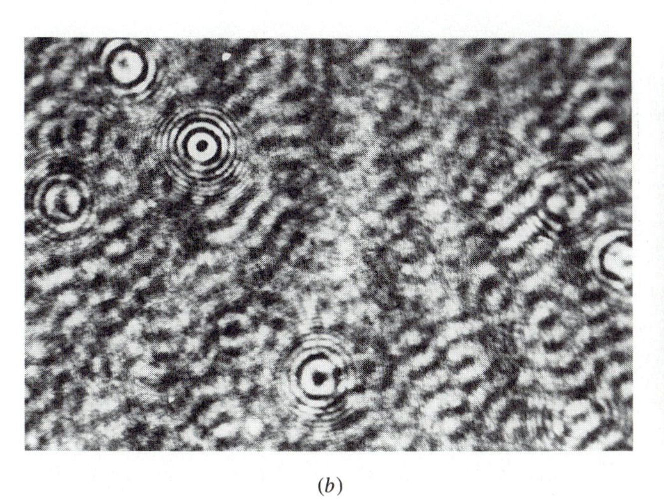

(*b*)

(*c*)

Figure 9.20 (*a*) Schematic diagram of a setup used in split-beam holography. The hologram is produced on the photographic plate by the interference of the primary beam (reflected from mirror *M*) with light scattered to the plate by the object. (*b*) The appearance of a typical hologram does not suggest the wealth of information it contains. *(Ronald R. Erickson.)* (*c*) Photographs of the same holographic image taken from two different angles demonstrate the three-dimensional character of the reconstructed image *(Bell Telephone Laboratories.)*.

SUMMARY

A one-electron (hydrogenlike) atomic system has a magnetic moment

$$\boldsymbol{\mu} = -\frac{e}{2m}\mathbf{L} \tag{9.3}$$

Its component along a specified spatial direction (z direction) is

$$\mu_z = -\frac{e}{2m}L_z = -\mu_B m_l \tag{9.5}$$

where
$$\mu_B = \frac{e\hbar}{2m} \tag{9.6}$$

is the *Bohr magneton*.

In a magnetic field B the $(2l + 1)$-fold degenerate m_l states are split. The energy difference between adjacent m_l levels is

$$\Delta E = \mu_B B$$

This splitting gives rise to the "normal" *Zeeman effect*. Accordingly, application of a magnetic field should split an atomic spectral line into an odd number of lines. The observation of an even number of Zeeman lines and the result of the Stern-Gerlach experiment led Goudsmit and Uhlenbeck to propose that an electron has an *intrinsic* spin angular momentum

$$S = \sqrt{s(s+1)}\hbar \tag{9.11}$$

where
$$s = \tfrac{1}{2}$$

is the electron's *intrinsic spin quantum number*. The component of the spin angular momentum along some specified spatial direction (z direction) is

$$S_z = m_s\hbar \qquad m_s = \pm\tfrac{1}{2} \tag{9.12}$$

The *intrinsic magnetic moment* of the electron is

$$\boldsymbol{\mu}_s = -\frac{2\mu_B \mathbf{S}}{\hbar} \tag{9.13}$$

$$\mu_{sz} = -2\mu_B m_s \tag{9.14}$$

According to the *Pauli exclusion principle* no two electrons can occupy the same quantum state. Consequently, each atomic state characterized by the quantum numbers n, l, m_l can accommodate no more than two electrons, one with $m_s = +\frac{1}{2}$, the other with $m_s = -\frac{1}{2}$. The ground-state electronic configuration of the elements is a direct consequence of the exclusion principle and the condition that the ground state be the state of lowest energy.

The interaction of the intrinsic magnetic moment of the electron with the magnetic field due to its orbital angular momentum results in *spin-orbit coupling* (**L·S** *coupling)* and accounts for the fine structure splitting of spectral lines. The total angular momentum of a given l, s state is

$$J = \sqrt{j(j+1)}\hbar \tag{9.18}$$

where
$$j = |l \pm s| \tag{9.19}$$

and
$$J_z = m_j\hbar \qquad m_j = -j, -j+1, \ldots, j-1, j \tag{9.20}$$

In standard spectroscopic notation the orbital angular momentum of an atomic state is labeled by a letter according to the convention

l	0	1	2	3	4	5
Letter	S	P	D	F	G	H

The *multiplicity* defined as $2S + 1$, where S stands for the total spin quantum number of the atomic system, is indicated by a superscript preceding the letter characterizing the orbital quantum number, and the total angular momentum quantum number j is indicated by a subscript following that letter. The symbol is preceded by the principal quantum number. For example, the excited state of sodium with $n = 3$, $l = 1$, $s = \frac{1}{2}$, $j = l - s = \frac{1}{2}$ is represented by the symbol

$$3^2P_{1/2}$$

To satisfy the requirements of the exclusion principle the total wave function of a many-electron system must be antisymmetric with respect to an interchange of any two electrons of the system.

Masers and lasers operate by the *stimulated emission* of radiation. To achieve lasing, a *population inversion* must be established. This is accomplished by *pumping* the atoms or molecules of the laser into metastable excited states from which they decay to the ground state by stimulated emissions.

Bibliography

T. F. Deutsch, "Medical Applications of Lasers," *Physics Today,* October 1988, p. 56.

Gerhard Herzberg, *Atomic Spectra and Atomic Structure,* 2d ed., J. W. Spinks, 1944.

Robert B. Leighton, *Principles of Modern Physics,* McGraw-Hill, New York, 1959.

D. C. O'Shea, W. R. Callen, and W. T. Rhodes, *Introduction to Lasers and Their Applications,* Addison-Wesley, Reading, Mass., 1977.

A. L. Schawlow, "Laser Light," *Scientific American,* September 1968, p. 120.

Harvey White, *Introduction to Atomic Spectra,* McGraw-Hill, New York, 1934.

PROBLEMS

9.1 Use Table 9.1 to identify three elements that should display a normal Zeeman effect. Indicate, using spectroscopic notation, the transitions involved.

9.2 Determine the normal Zeeman splitting in a magnetic field of 1.2 T. What is the frequency of electromagnetic radiation that might induce transitions between these Zeeman levels?

9.3 A discharge tube containing sodium atoms is placed in a magnetic field of 1.5 T. Calculate the Zeeman splitting of the $3^2S_{1/2}$, $3^2P_{1/2}$, and $3^2P_{3/2}$ states, using the correct values for the Lande g factor (see footnote 7, page 187). How do these splittings compare with the spin-orbit splitting of the 3^2P levels?

9.4 For which of the following atoms would you expect to find spin-orbit splitting of the ground state: Cl, Ar, K, Ca, Sc, Cu, Zn? Justify your conclusions.

9.5 Estimate the spin-orbit splitting of the 4^2P doublet of sodium from the known splitting of the 3^2P doublet. Compare your result with the experimental values of the wavelengths of 330.24 nm and 330.30 nm of the $4^2P \rightarrow 3^2S$ transitions; what might be the reasons for the difference between your estimate and the actual value of the splitting?

9.6 Consider three electrons in states specified by quantum numbers a, b, and c. Let the one-electron states be $\psi_a(r_1)$, $\psi_b(r_2)$, and $\psi_c(r_3)$. Construct a linear combination of product functions $\psi_i(r_j)$ that is antisymmetric with respect to an interchange of any two electrons.

9.7 Show that for a system of N electrons an antisymmetric linear combination of product functions is the determinant

$$\psi(r_1, r_2, \ldots, r_N) = \begin{vmatrix} \psi_1(r_1) & \psi_2(r_1) & \cdots & \psi_N(r_1) \\ \psi_1(r_2) & \psi_2(r_2) & \cdots & \psi_N(r_2) \\ \psi_1(r_3) & \psi_2(r_3) & \cdots & \psi_N(r_3) \\ \cdots & \cdots & \cdots & \cdots \\ \psi_1(r_N) & \psi_2(r_N) & \cdots & \psi_N(r_N) \end{vmatrix}$$

9.8 Give the spectroscopic notation for the ground state and the first excited state of boron. What are the values of n and l of these states? What is the multiplicity of the first excited state? Are either of these states split by spin-orbit coupling? If so, what are their j values?

9.9 Give the ground state of scandium in spectroscopic notation. What is the multiplicity of that state? Is the state split by spin-orbit coupling? If so, what are the j values of the split states?

9.10 Into how many levels is the lowest state of scandium split in a magnetic field? Calculate the energy difference between the Zeeman levels in a field of 0.8 T.

9.11 Which of the following are possible atomic states: $3^3P_{1/2}$, $3^2F_{7/2}$, $4^2F_{5/2}$, $4^2F_{3/2}$, 4^1S_1, 3^1P_1? Explain your answers.

9.12 What is the ground-state configuration of titanium?

•**9.13** A sample of sodium atoms in the ground state is placed in a magnetic field of 2.0 T. What is the Zeeman splitting of the ground state? At what temperature would 48 percent of the atoms of the sample be in the higher energy state? Transitions from the lower to the higher Zeeman level can be induced by means of an alternating magnetic field, a technique known as electron spin resonance (ESR). What is the frequency at which transitions will be induced? Suppose the sample contains one mole of sodium atoms. How much energy can this sample absorb from the alternating magnetic field if it is (a) at room temperature? (b) At $T = 4$ K?

•**9.14** In a Stern-Gerlach apparatus the magnetic field region is 0.8 m long. The field gradient in this region is 14 T/m. The screen on which the atoms emerging from the apparatus are collected is 1.0 m from the end of the field region. The atoms emerge from a cavity that is maintained at a temperature of 1200 K. Calculate the average separation between the spots on the screen if the cavity contains (a) silver and (b) sodium atoms. Assume that all atoms emerging from the cavity are in the ground state.

Molecular Structure and Molecular Spectra

Chapter 10

You may call it the accidental and fortuitous concurrence of atoms.

Lord Palmerston

10.1 INTRODUCTION

In the preceding chapters we focused attention on the structure and spectra of isolated atoms. However, with the exception of the noble gases, atoms appear in nature not as separate entities but as aggregates—molecules, liquids, and crystals. We shall discuss crystalline solids, principally metals and semiconductors, in a later chapter. Here we will be concerned with molecules, mostly molecules of the simplest sort such as the diatomic molecules H_2, NO, and KCl. There are, of course, vastly more complex systems of great interest and importance, notably organic structures such as $C_6H_{16}O_6$ (glucose) or $CH_3(CH_2)_{16}COOH$ (stearic acid). A discussion of their structure and spectra is well beyond the scope of this book. Nevertheless, some of the concepts presented in this chapter apply to these complicated organic molecules.

Figure 10.1 Linus Pauling (b. 1901), recipient of the 1954 Nobel Prize in chemistry for his work on the nature of the chemical bond. *(AP/Wide World Photos.)*

In the century following Dalton's work, chemists studied the chemical properties of numerous compounds, their synthesis, and techniques for dissociating compounds into the constituent elements. Though their efforts shed much light on the subject, the fundamental nature of the forces that hold atoms together to form molecules of such immense diversity remained mysterious until about 1927, when Heitler, London, Born, Oppenheimer, and, later, Pauling and others applied the new techniques of quantum mechanics to this problem.

The very fact that atoms combine to form molecules, that gases condense to liquids and solids, demonstrates that there is a net attractive force between atoms of a particular complex when their separation is greater than the equilibrium distance between them. The basic mechanism responsible for this attraction is largely, though not entirely, electrostatic. Broadly speaking, interatomic bonds in molecules can be classified as either ionic (heteropolar) or covalent (homopolar). A full quantum mechanical calculation of bond energies in both cases is an enormously complex task, requiring lengthy computations that were rarely undertaken prior to the advent of modern computers. Fortunately, a good physical picture can be drawn and much computational drudgery obviated by constructing simplified models that highlight the dominant aspect of the binding forces. It must be remembered, however, that results based on a model depend on the accuracy with which the model reflects the true character of the interatomic forces, and that the real world is far more complicated than the model employed. Frequently, a system will not conform precisely to any model but will exhibit aspects normally associated with two or more different models.

Even a simple molecule, such as Cl_2 or NaCl, consists of a large number of electrons. In principle, the full Hamiltonian of the Schrödinger equation must include the coordinates of all of these electrons as well as those of the nuclei. The potential energy term must include the Coulomb interaction between all of the constituent particles, and the electron wave function must be antisymmetric as required by the Pauli principle. Even writing down the Schrödinger equation, let alone solving it, is a formidable task.

The first step in simplifying the problem is based on our empirical knowledge of the energies involved in chemical reactions, typically between 1 and 10 eV. These are also the excitation and ionization energies of most atoms. By contrast, energies required to remove one of the electrons within the closed-shell structure of an atom, the so-called core electrons, are at least an order of magnitude greater. Consequently, when considering chemical bonding one may, as a first approximation, regard the core electrons as firmly attached to their respective nuclei.

The Hamiltonian of the simplest diatomic molecule, the H_2^+ molecule ion, includes the kinetic and potential energies of the two protons, as well as the energy of the electron. But because the protons are much more massive than the electron, their motion is very slow compared with that of the electron. Therefore one may regard the protons as being at rest when writing the electron's Hamiltonian. In other words, on a time scale appropriate to nuclear motion, the electron accommodates practically instantaneously to changes in the protons' position. In short, when calculating electronic wave functions and energy levels we can disregard the dynamics of the nuclei. This is known as the *adiabatic approximation*.

These two approximations, treating the molecule as composed of atomic cores and valence electrons, and the adiabatic approximation, are generally quite good and form the starting point of practically all molecular calculations.

10.2 IONIC BONDING

The prototype ionic molecule is sodium chloride, NaCl. In this and other alkali-metal halides, the alkali metal is singly ionized (Na^+) and the halogen is in an ionic state as a negative ion (Cl^-). The molecule as a whole is electrically neutral but has a large permanent dipole moment.

As we saw in Chapter 9, the ionization energies of the alkali metals are quite small, 5.14 eV for sodium, and even less for the heavier alkali metals. The neutral halogen Cl lacks just one electron to meet the highly stable noble-gas electron configuration. In fact, that structure is so stable that the energy of the Cl^- ion is actually *less* than that of the neutral atom by 3.62 eV; that is, 3.62 eV must be expended to remove the extra electron from Cl^- and regain the neutral halogen atom. Chlorine is said to have an *electron affinity* of 3.62 eV.

The transfer of the sodium valence electron to the chlorine atom requires only $(5.14 - 3.62)$ eV $= 1.52$ eV of energy. A net energy reduction, the requisite for molecular stability, derives from the electrostatic interaction between these oppositely charged ions. Since both ions have closed-shell configurations, their charge distributions are spherical provided they do not come too close together. Hence, the Coulomb energy is

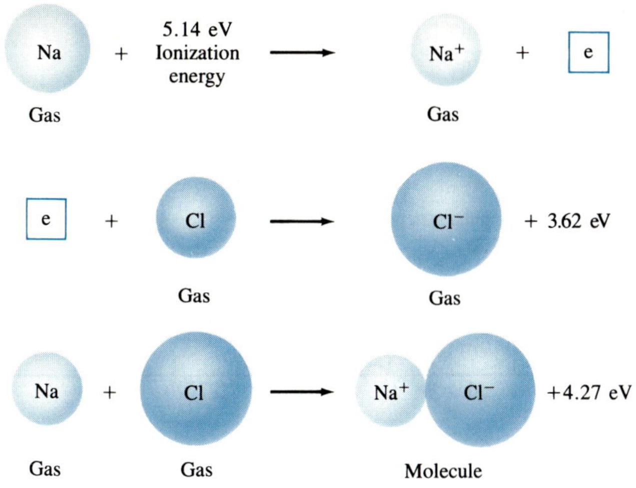

Na + 5.14 eV Ionization energy ⟶ Na^+ + e
Gas Gas

e + Cl ⟶ Cl^- + 3.62 eV
Gas Gas

Na + Cl ⟶ Na^+ Cl^- +4.27 eV
Gas Gas Molecule

Figure 10.2 The energetics of formation of the NaCl molecule.

$$V = \frac{-e^2}{4\pi\epsilon_0 r} \qquad \frac{1}{4\pi\epsilon_0} = 9 \times 10^9 \text{ J}\cdot\text{m/C}^2 \tag{10.1}$$

The energy of the ion relative to that of the two isolated neutral atoms is then

$$E = E_{\text{ion}}(\text{Na}) - E_{\text{el.aff}}(\text{Cl}) - \frac{e^2}{4\pi\epsilon_0 r} \tag{10.2}$$

and is negative for $r < 0.95$ nm. If Equation (10.2) were the complete expression for the molecular energy, the two ions should overlap, which of course does not happen. It is the so-called *core repulsion,* which has its origin in the exclusion principle, that keeps the ions apart. When the ions approach so closely that the core wave functions overlap, they are modified and, in effect, the core electrons are forced into higher energy states. The new core wave functions may be thought of as linear combinations of atomic wave functions with the admixture of eigenfunctions of higher atomic energy levels. That admixture becomes more pronounced the greater the overlap and, as a result, the core state energies increase dramatically with diminishing internuclear separation. The sum of the short-range but rapidly increasing positive core repulsion energy and the long-range attractive electrostatic energy results in a potential energy curve of the form shown in Figure 10.3. For NaCl, the potential energy minimum occurs at a nuclear separation of 0.236 nm, where the energy is 4.27 eV below that of the isolated neutral atomic constituents. The *molecular dissociation energy* of NaCl is 4.27 eV.

We have here considered only the ground state. As in atoms, the valence electrons of a molecule can occupy various excited states whose wave functions satisfy the Schrödinger equation and the boundary condition that $\psi(r) \rightarrow 0$ as $r \rightarrow \infty$. Since the electronic wave functions of excited states are more spread out than the ground-state wave function, the equilibrium internuclear separation in the excited electronic states is somewhat greater than in the ground state, and the potential energy curve generally exhibits a broader and shallower minimum.

the space function must be symmetric (even parity). As with the H_2^+ ion, only the even-parity state is stable. The ground state of H_2 is therefore the $S = 0$ singlet state. Detailed calculations yield $r_0 = 0.074$ nm, $E = -31.7$ eV, or 4.5 eV less than the energy ($E = -2 \times 13.6 = -27.2$ eV) of two isolated H atoms, in agreement with experimental results. In chemical parlance, the singlet state is the *bonding* state, the *triplet* ($S = 1$) state the *antibonding* state.

10.4 MOLECULAR ENERGY LEVELS AND MOLECULAR SPECTRA

As might be expected, the complex structure of polyatomic molecules results in complex electronic energy-level schemes and correspondingly complex spectra. However, though the electronic configuration is the dominant factor in determining molecular energies, vibration of the nuclei about their equilibrium separation and rotation of the molecule about its center of mass also contribute to the total energy. These energies are also quantized; vibrational energies follow approximately the harmonic oscillator pattern, while the rotational energies are given by $E_L = L(L + 1)\hbar^2/2I$, where I is the moment of inertia of the molecule about the rotation axis. In the following we focus attention on diatomic molecules, because these provide the simplest examples of the principal features of molecular energy levels. We shall see how a study of molecular spectra can be used to gain information on bond lengths and force constants for these simple molecules.

10.4(a) Rotational Energy Levels

We picture the diatomic molecule as a dumbbell, two unequal masses M_1 and M_2 a distance r_0 apart, where r_0 is the equilibrium separation. Since, however, the form of the potential function is determined in large measure by the electronic configuration, r_0 for the excited states of the molecule is generally greater than for the ground state.

The moment of inertia of the molecule is

$$I = \mu r_0^2 \tag{10.6}$$

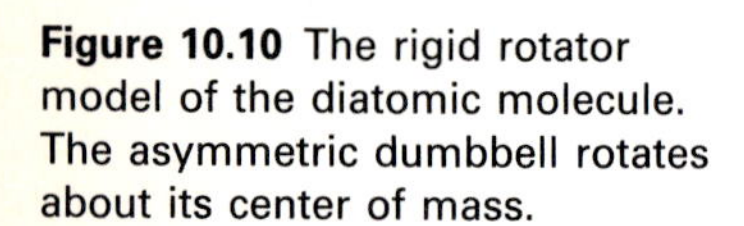
Figure 10.10 The rigid rotator model of the diatomic molecule. The asymmetric dumbbell rotates about its center of mass.

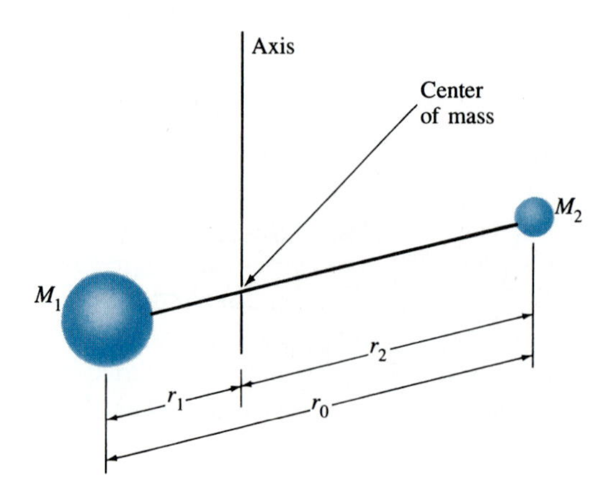

where μ is the reduced mass

$$\mu = \frac{M_1 M_2}{M_1 + M_2} \tag{10.7}$$

The rotational energy of the molecule is quantized by virtue of the quantization of the angular momentum. Accordingly,

$$E_L = \frac{L(L+1)\hbar^2}{2I} \tag{10.8}$$

where the angular momentum quantum number L is a positive integer or zero.

This energy-level scheme is shown in Figure 10.11. Transitions between these levels produce the *pure rotation spectrum* of the molecule. Although all diatomic molecules have rotational energy levels, not all exhibit pure rotation spectra. Molecules such as H_2, and other symmetric molecules—for example, CO_2 (O═C═O)—that carry no permanent dipole moment, cannot absorb or emit electric dipole radiation by changing only the rotational quantum state. For molecules

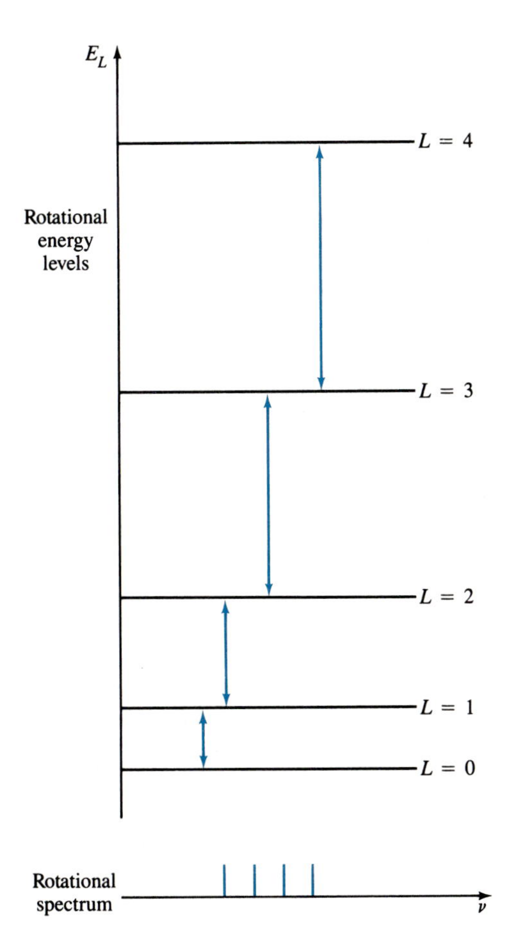

Figure 10.11 Energy levels and allowed transitions (absorption) of of rigid rotator.

region. The separation between vibrational states is usually two to three orders of magnitude greater than that between rotational levels.

The true potential function of the diatomic molecule is not parabolic, increasing more steeply for $r < r_0$ and more gradually for $r > r_0$. These departures from the ideal harmonic oscillator, which become more pronounced with increasing vibrational amplitude, have several important consequences. First, the energy-level spacing diminishes with increasing n (see Problem 7.28). Second, the expectation value of r, which is r_0 for small values of n, also increases as n increases. Third, the usual selection rule $\Delta n = \pm 1$ is not strictly obeyed.

We have treated rotational and vibrational motions of the molecule as independent modes. However, when departures from the simple model (rigid rotator for rotations, harmonic oscillator for vibrations) are taken into account, these modes are coupled. Centrifugal distortion for higher values of L increases the equilibrium separation r_0 and changes not only the moment of inertia but also the effective force constant and, thereby, the vibrational frequency. Conversely, the increase in $\langle r \rangle$ with increasing n due to the anharmonicity of the potential not only alters the vibrational energy levels but increases the moment of inertia of the molecule, thus reducing the spacing between adjacent rotational levels. These effects are all reflected in molecular spectra. A careful analysis of such spectra provides information not only on bond lengths and force constants but on higher-order terms in the Taylor series expansion of $V(r)$, Equation (10.11).

10.4(c) Molecular Spectra

The complete energy-level scheme for the electronic ground state and first excited state of a diatomic molecule is shown in Figure 10.13. Note that, as indicated in this figure, the potential function in the excited electronic state is shallower and its minimum occurs at a greater internuclear separation than in the ground state. The

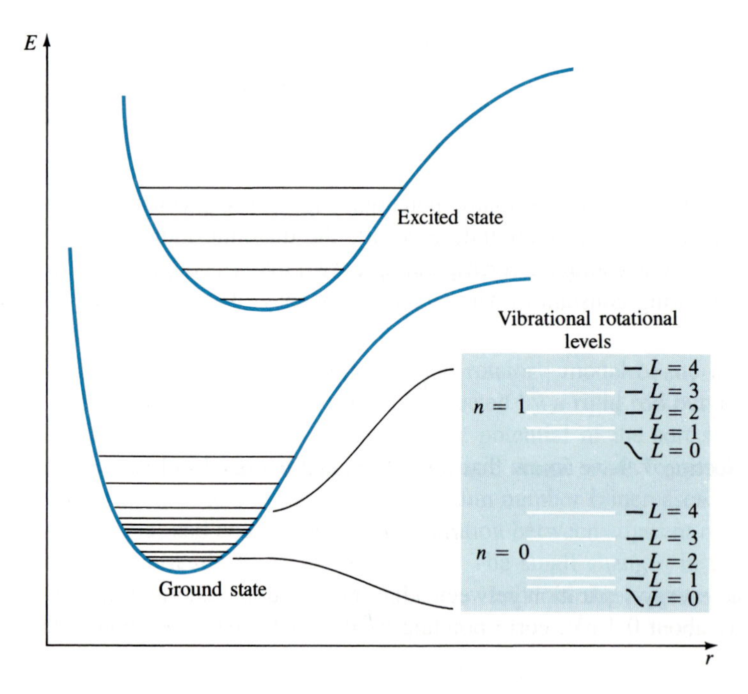

Figure 10.13 Schematic diagram of the energy levels of a diatomic molecule, showing the ground state and first excited electronic state, the vibrational levels of those states, and (in the inset) the rotational levels of the lowest two vibrational states.

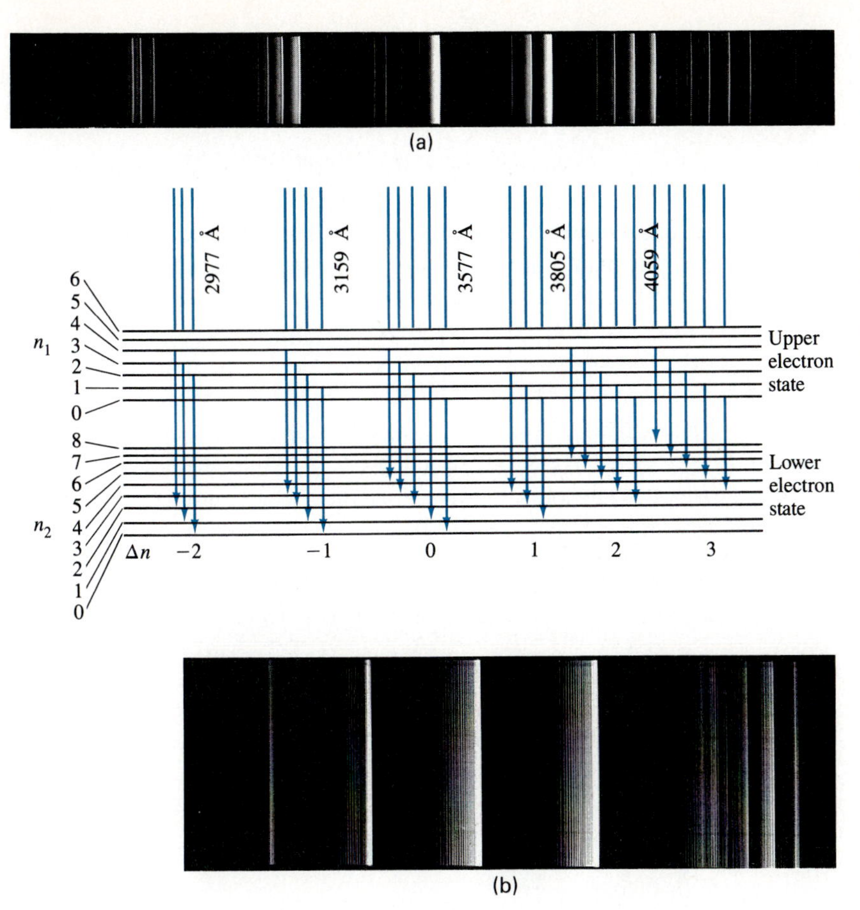

Figure 10.14 A portion of the emission spectrum of N_2 under (*a*) low and (*b*) high resolution. Under low resolution, transitions between various vibrational levels appear as bands (the various vibrational transitions are indicated in the energy-level diagram); at higher resolution, the bands reveal themselves as a series of closely spaced lines, corresponding to transitions between the rotational states. *(J. A. Marquisce.)*

vibrational levels are not uniformly spaced; the level spacing decreases with increasing n, and because the potential is shallower in the excited electronic state, the spacing of vibrational levels in this state is smaller than in the ground state. The rotational levels are shown in the inset.

Under low to moderate resolution, the emission spectrum due to a given electronic transition of a diatomic gas appears to consist of a series of bands rather than sharp lines. Each band corresponds to a transition between two vibrational levels that satisfies the selection rule $\Delta n = \pm 1$. The bands are unevenly spaced because the separation between vibrational levels depends on the electronic state and on the vibrational quantum number n. When these bands are examined under high resolution, they reveal themselves as a series of closely spaced lines, each line corresponding to a transition between two rotational states for which $L' = L \pm 1$.

Molecular spectroscopy is usually done by infrared absorption, whereby only the vibration-rotation levels of the ground electronic state are excited. Since at room temperature $kT \simeq 0.025$ eV, molecules in a gaseous absorption cell are normally in their ground electronic and vibrational states. The most densely populated rotational level at room temperature is, however, not the ground state $L = 0$. We recall that the degeneracy of an angular momentum state is $2L + 1$, and this increase in the number of accessible states with increasing L compensates to some extent for the decrease in the probability of occupancy given by the Boltzmann factor, $e^{-E_L/kT}$. The population of the level of quantum number L in the ground electronic and vibrational state is

$$N_L = N_0(2L + 1)e^{-L(L+1)\hbar^2/2IkT} \tag{10.14}$$

where N_0 is the total number of molecules present. The angular momentum quantum number of the most highly populated level is the integer nearest to

$$L_m = \frac{1}{2}\left(\sqrt{\frac{4IkT}{\hbar^2}} - 1\right) \tag{10.15}$$

The possible rotation-vibration transitions from the lowest vibrational state are shown in Figure 10.15. The selection rule $\Delta n = \pm 1$ limits the possible transitions to the state $n = 1$. Transitions between rotational states must satisfy the angular momentum selection rule $\Delta L = \pm 1$.

The energy of the initial state is

$$E_i = E_e + \frac{1}{2}\hbar\omega_v + \frac{L(L + 1)\hbar^2}{2I}$$

and that of the final state is

$$E_f = E_e + \frac{3}{2}\hbar\omega_v + \frac{(L + 1)(L + 2)\hbar^2}{2I} \qquad \Delta L = +1$$

$$E_f = E_e + \frac{3}{2}\hbar\omega_v + \frac{(L - 1)L\hbar^2}{2I} \qquad \Delta L = -1$$

where E_e is the energy of the ground electronic state and corrections due to anharmonicity and centrifugal distortion have been neglected. The energy differences between initial and final states are, therefore,

$$\Delta E_R = \hbar\omega_v + \frac{(L + 1)\hbar^2}{I} \qquad \Delta L = +1,\ L = 0, 1, 2, \ldots$$

$$\Delta E_P = \hbar\omega_v - \frac{L\hbar^2}{I} \qquad \Delta L = -1,\ L = 1, 2, 3, \ldots$$

We have here used the subscripts R and P because infrared spectroscopists refer to the two branches as the *R branch* and the *P branch*. For the P branch, L cannot be equal to zero, as that would imply an $L = 0$ to $L = -1$ transition, and L must be zero or a positive integer.

A typical infrared absorption spectrum of a diatomic molecule is that of HCl, shown in Figure 10.16, page 220. That spectrum displays several noteworthy features.

First, there is no absorption line at the center of the spectrum, at the energy $h\nu = \hbar\omega_v$. The "missing line" is that for which $\Delta L = 0$, forbidden by the angular momentum selection rule. The frequency of this missing line is, however, readily determined, and from it the interatomic force constant k can be calculated.

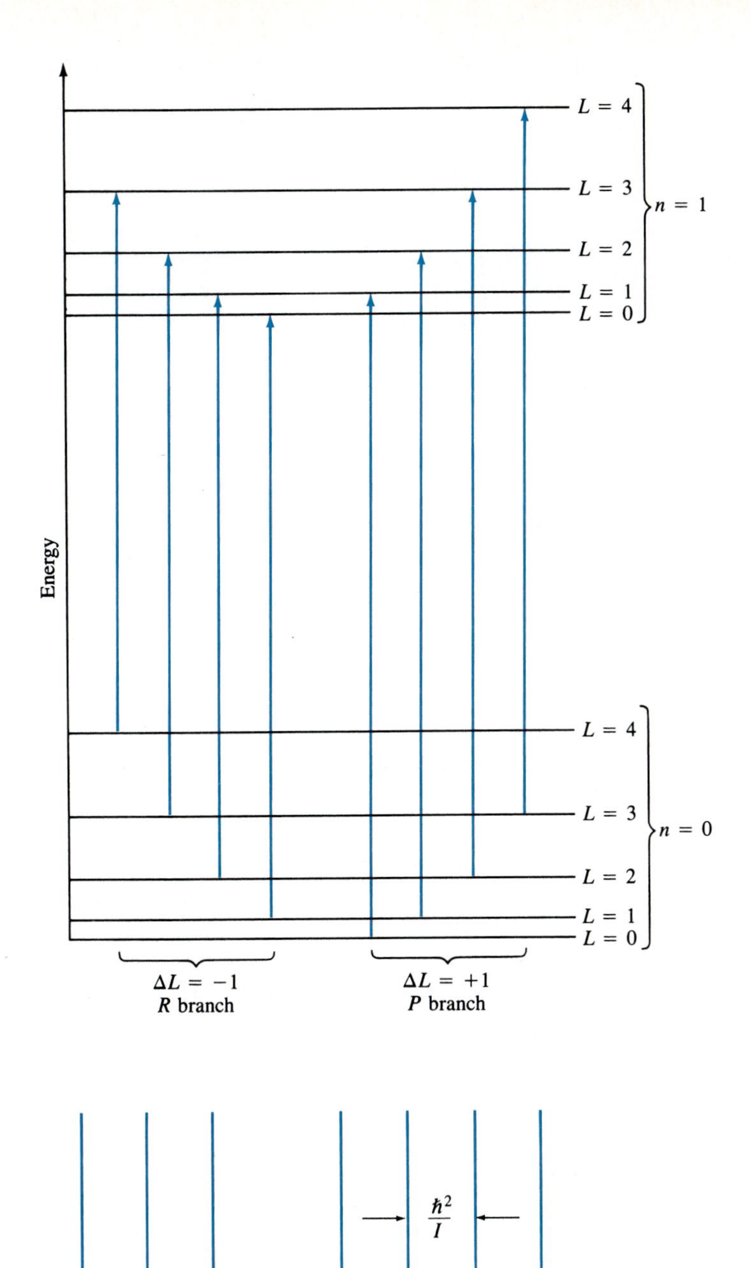

Figure 10.15 Allowed optical transitions of the vibration-rotation absorption spectrum of a diatomic molecule. The corresponding photon energies are shown below. Note that the central line, corresponding to the harmonic oscillator energy $\hbar\omega_v$, is missing because transitions for which $\Delta L = 0$ are forbidden.

Second, the intensities of the lines in both branches vary slowly with L, reaching a maximum for $L = 3$. The intensities of the absorption lines reflect the relative populations of the initial angular momentum levels, given by Equation (10.14).

Third, the spacing between adjacent lines in the P branch increases with increasing L, while that in the R branch decreases as L increases. If the rigid rotator

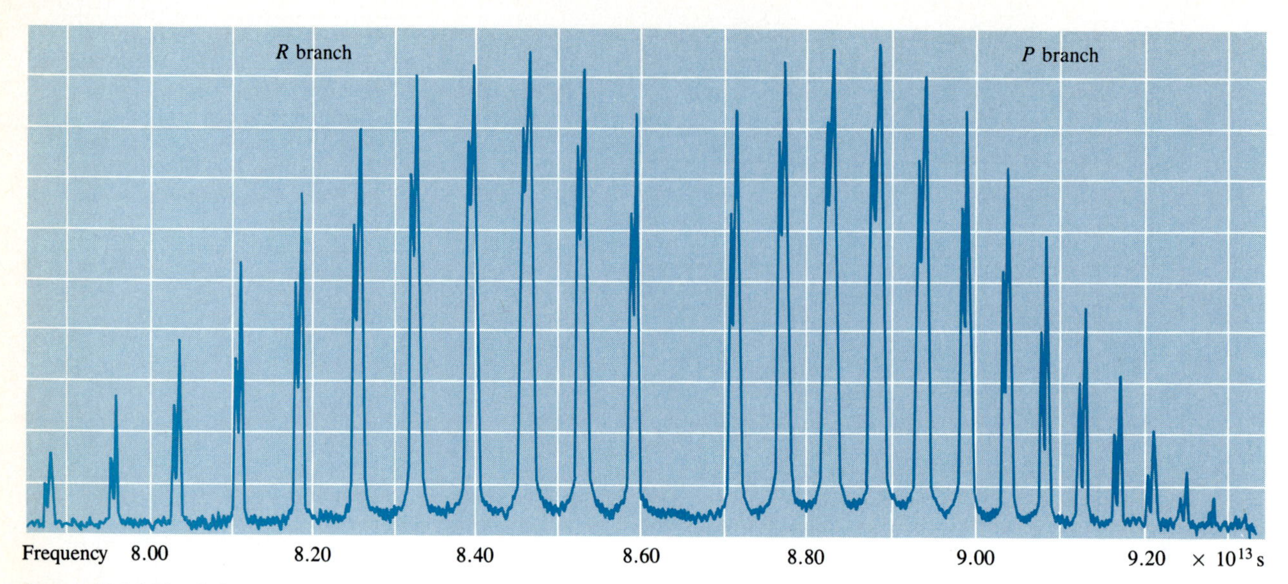

Figure 10.16 The infrared absorption spectrum of HCl.

model of the diatomic molecule were valid, the spacing in both branches should be independent of L and equal to $\hbar^2/I$. The monotonic change in level spacing is due to centrifugal distortion and the consequent increase in the moment of inertia with increasing L.

Last, each "line" is in fact a doublet. This doublet structure has nothing to do with spin-orbit coupling, discussed in Section 9.3, but arises from the presence of two naturally occurring isotopes of chlorine, ^{35}Cl (75.5 percent abundance) and ^{37}Cl (24.5 percent abundance). The vibrational frequencies and moments of inertia of H^{35}Cl and H^{37}Cl differ slightly, and the energy differences between the vibration-rotation levels of H^{37}Cl are somewhat less than for H^{35}Cl.

EXAMPLE 10.3 Use the results shown in Figure 10.16 to calculate the interatomic force constant k and the equilibrium separation between the proton and chlorine nucleus in HCl. Show that, within the available accuracy limit, k is independent of isotopic composition. Estimate the temperature of the gas in the absorption cell.

SOLUTION From Figure 10.16, the vibration frequency of the more abundant isotopic species appears to be 8.65×10^{13} Hz. The reduced mass of the H^{35}Cl molecule is $\mu = \frac{35}{36}$ u $= 1.614 \times 10^{-27}$ kg. The force constant k is given by

$$k = \omega^2\mu = 4\pi^2(8.65 \times 10^{13}\ \text{Hz})^2(1.614 \times 10^{-27}\ \text{kg}) = 477\ \text{N/m}$$

If k does not depend on isotopic composition, the vibrational frequency of the H^{37}Cl molecule should be

$$\nu_{37} = \nu_{35}\left(\frac{\mu_{35}}{\mu_{37}}\right)^{1/2} = \nu_{35}(0.9992) \simeq 8.64 \times 10^{13}\ \text{Hz}$$

From Figure 10.16 it appears that the isotopic doublet splitting is about 0.01×10^{13} Hz.

The energy separation between the first lines of the P and R branches is $2\hbar^2/I$. The moment of inertia is therefore

$$I = \frac{\hbar}{\pi\,\Delta\nu} = \mu r_0^2$$

From Figure 10.16, $\Delta\nu \simeq 0.12 \times 10^{13}$ Hz for either isotopic species. Hence

$$I = \frac{1.05 \times 10^{-34}\ \text{J·s}}{\pi(0.12 \times 10^{13}\ \text{Hz})} = 2.78 \times 10^{-47}\ \text{kg·m}^2$$

The equilibrium internuclear separation is

$$r_0 = \left(\frac{\hbar}{\mu\pi\,\Delta\nu}\right)^{1/2} = \left[\frac{1.05 \times 10^{-34}\ \text{J·s}}{(1.614 \times 10^{-27}\ \text{kg})\pi(0.12 \times 10^{13}\ \text{Hz})}\right]^{1/2} = 0.131\ \text{nm}$$

From Equation (10.15),

$$(2L_m + 1)^2 = \frac{4kTI}{\hbar^2} \qquad kT = \frac{[(2L_m + 1)\hbar]^2}{4I}$$

where L_m is the angular momentum quantum number for maximum intensity. From Figure 10.16 it appears that $L_m \simeq 3$. Hence

$$kT \simeq \frac{[7(1.05 \times 10^{-34})]^2}{4(2.78 \times 10^{-47})} = 4.87 \times 10^{-21}\ \text{J}$$

Solving for T, one obtains $T = 353$ K $= 80$ °C.

Polyatomic molecules—for example, complicated organic molecules—have numerous vibrational modes of diverse frequencies. In many cases, such molecules are built up from combinations of smaller molecular units whose internal bonding forces are substantially stronger than those that bind the units to each other to form the larger complex. For instance, butyl alcohol contains three CH_2 groups, a CH_3 group, and an OH group. The bond strength of the OH group is much greater than that which holds this radical to the CH_3 group, and consequently the vibration frequency of the OH group is nearly the same as that of a free OH radical. The presence of various components in a large organic molecule can often be deduced and the molecular structure deciphered from the infrared absorption spectrum.

SUMMARY

The force that bonds the atoms of a molecule is predominantly electrostatic. The binding energy of an ionic molecule such as NaCl is

$$E = E_{\mathrm{el.aff}}(\mathrm{halide}) - E_{\mathrm{ion}}(\mathrm{alkali}) + \frac{e^2}{4\pi\epsilon_0 r_0} - V_c$$

where $E_{\mathrm{el.aff}}$(halide) is the electron affinity of the halogen atom, E_{ion}(alkali) is the ionization energy of the alkali-metal atom, the third term is the magnitude of the electrostatic potential energy of the two ions at their equilibrium separation r_0, and V_c is the core-repulsion energy.

The electrostatic energy that binds homopolar molecules arises from a purely quantum mechanical effect, the exclusion principle, which requires that the total electronic wave function of the molecule be antisymmetric with respect to an interchange of any two electrons. In H_2, for example, the bonding state is that for which the total spin state is antisymmetric and the spatial part of the two-electron wave function is symmetric. For this *singlet,* or *bonding,* state of the molecule, the electronic probability density is large in the region between the two protons and the bonding derives from the electrostatic attraction of the two protons to this negative charge distribution.

Molecular energy levels are determined by the electronic state and the vibrational and rotational states of the molecule. Typically, the energy separation between electronic energy levels is of the order of 1 eV, that between adjacent vibrational levels is of the order of 0.1 eV, and that between low-lying vibrational levels is of the order of 0.001 eV. The rotational energy levels are given by

$$E_L = \frac{L(L+1)\hbar^2}{2I} \tag{10.8}$$

where L is the angular momentum quantum number of the molecule and I its moment of inertia in the particular electronic, vibrational, and rotational state. The vibrational energy is approximately

$$E_n = (n + \tfrac{1}{2})\hbar\omega_v \tag{10.13}$$

where ω_v is the angular frequency of the classical harmonic oscillator.

The infrared absorption spectra of molecules exhibit bands rather than sharp lines. Each band corresponds to a $\Delta n = \pm 1$ transition between vibrational energy levels. High resolution reveals that each band is, in fact, a series of closely spaced lines corresponding to transitions for which $\Delta L = \pm 1$. Infrared absorption data on diatomic molecules provide information on the equilibrium separation between the two nuclei and the force constant in the harmonic oscillator approximation, as well as on departures from the harmonic oscillator approximation and on the coupling between vibrational and rotational energy state.

Bibliography

Gordon M. Barrow, *Introduction to Molecular Spectroscopy,* McGraw-Hill, New York, 1962.

Gerhard Herzberg, *Molecular Spectra and Molecular Structure,* 2d ed., Krieger, Melbourne, FL, 1989.

Linus Pauling, *The Nature of the Chemical Bond,* Cornell Univ. Press, Ithaca, NY, 1960.

PROBLEMS

10.1 Calculate the force due to core repulsion and the core-repulsion potential energy in KBr at the equilibrium separation. If the core-repulsion potential is of the form $V_c(r) = A/r^n$, what are the values of the coefficient A and exponent n?

10.2 Repeat Problem 10.1 for LiCl and RbCl.

10.3 The equilibrium separations in KBr and RbCl are nearly the same, about 0.28 nm. Determine the ratio of the core-repulsion potentials at equilibrium for these two ionic molecules.

10.4 The equilibrium separation in KI is 0.32 nm. Estimate the dissociation energy of KI. (The experimental value is 3.37 eV.)

10.5 Use the results of Problem 10.1 to estimate the vibrational frequency of the KBr molecule. Do this by expanding the potential near r_0 in a Taylor series, retaining only the term proportional to $(r - r_0)^2$.

10.6 Repeat Problem 10.5 for LiCl using the results of Problem 10.2.

10.7 Derive Equation (10.6) for the moment of inertia of a diatomic molecule.

•**10.8** Derive Equation (10.15) for the angular momentum state with the highest occupancy.

•**10.9** Figure 10.17 shows the vibration-rotation absorption spectrum of the diatomic molecule HBr. From the data determine the force constant in the harmonic approximation and the equilibrium separation between the H^+ and Br^- ions. Also estimate the temperature of the absorption cell that was used in obtaining the data.

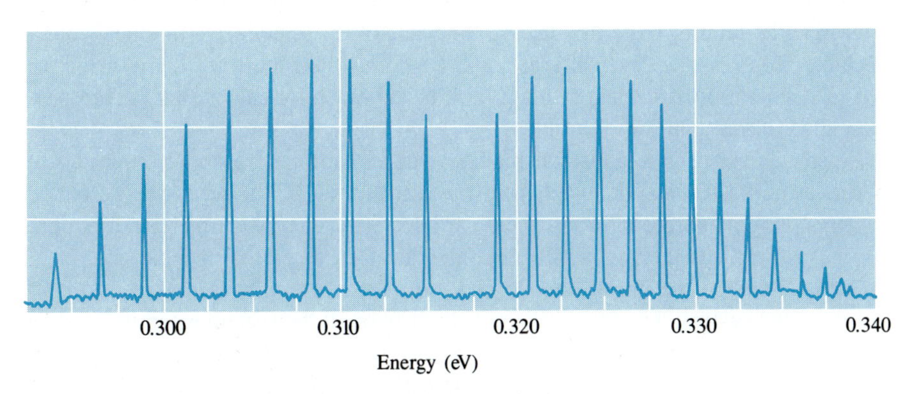

Figure 10.17 The infrared absorption spectrum of HBr.

10.10 The equilibrium separation between the nuclei of the N_2 molecule is 0.11 nm. Calculate the energy separation between the $L = 0$, 1, 2, and 3 rotational levels. What are the wavelengths of electromagnetic radiation corresponding to these energy differences?

10.11 The force constant of the H_2 bond is 575 N/m. Determine the energies of the three lowest vibrational levels of H_2, HD, and D_2 (D is the symbol for deuterium, "heavy" hydrogen).

10.12 Suppose an absorption cell contains a mixture of $H^{35}Cl$ and $D^{35}Cl$. Sketch the absorption spectrum that you would expect, indicating a few of the frequencies of the absorption lines.

10.13 At what temperature might one expect to observe significant absorption due to $n_v = 1$ to $n_v = 2$ transitions in a gas of CO?

10.14 Calculate the wavelengths of the three absorption lines of the pure rotation spectrum of KBr corresponding to transitions from the $L = 0$, 1, and 2 rotational levels.

10.15 What is the angular momentum quantum number of the most highly populated rotational level of KBr at a temperature of 180 °C?

•**10.16** The effective potential energy of a diatomic molecule in the angular momentum state L can be written as

$$V_{\text{eff}} = \frac{1}{2}\mu\omega^2(r - r_0)^2 + \frac{L(L+1)\hbar^2}{2\mu r^2}$$

where r is the internuclear separation and r_0 the internuclear equilibrium separation when $L = 0$. (a) Obtain an expression for the average internuclear separation r_L when $L \neq 0$. (b) Express the effective potential energy in terms of r and r_L. (c) Expand the effective potential obtained in part (b) about $r = r_L$ and determine the vibrational frequency in the harmonic approximation for $L > 0$.

Chapter 11 Statistical Physics

The principal means of ascertaining truth are based on probabilities.

Pierre Simon de Laplace

11.1 INTRODUCTION

All experiments in physics fall into one of two broad categories, those that probe properties of the individual constituents of a system and those that reveal average properties of a large system. Examples of the first are measurements of emission and absorption spectra of a gas and scattering experiments of the sort performed by Geiger and Marsden. Although a large number of atoms or nuclei participate in the experiment, the results can be understood in terms of the properties of the constituent particles. Experiments of the second category are measurements of pressure, heat capacities, electric and magnetic susceptibilities, or thermal expansion, to mention just a few. In such experiments it is the response of the system as a whole that is of interest. Indeed, some basic concepts such as temperature, pressure, and electrical and thermal conductivities are meaningful only for large systems because they are average properties of the constituents.

11.2 CLASSICAL STATISTICAL PHYSICS

In Newtonian physics one could conceivably calculate the trajectory of every molecule in a liter of a gas, given the initial conditions and knowing, as we do, the forces that molecules exert on each other. One could then follow these trajectories and determine the total average impulse on the walls of the container—in other words, calculate the pressure. Even if that could be done—and the uncertainty principle tells us it cannot—that is clearly not a sensible approach for calculating the pressure exerted by more than 10^{20} particles.

Instead, one uses a statistical method whereby the macroscopic properties at thermal equilibrium are deduced from the distribution of the total energy of the system among its numerous constituent particles. This energy distribution function, $f(E_i)$, is defined as the probability that a state of energy E_i is occupied.

It is expedient to introduce two new concepts, the *microstate* and the *macrostate* of the system. The microstate is a description of the system wherein the state (e.g., position and momentum in classical physics; quantum numbers in quantum physics) of every particle is identified. The macrostate is a less detailed characterization of the system; it specifies only the number of particles that occupy a given state.

Classical statistical physics is based on the following fundamental postulates:

1. The particles of the system are identical but distinguishable.
2. There is no restriction on the number of particles that may occupy a particular energy state.

3. At thermal equilibrium, the distribution of particles among the accessible energy states is the most probable distribution consistent with prescribed constraints such as total energy and total number of particles.
4. Every microstate of the system has equal a priori probability.

Although the statistical approach is appropriate only to systems consisting of a large number of identical particles, we shall illustrate the formalism by considering a "gas" of only 6 particles. We further assume that each particle can occupy states of energy $E = n\epsilon$, where n is an integer, and impose the constraint that the total energy of the system is 9ϵ. Our aim is to determine the equilibrium distribution, that is, the probability of occupancy of every accessible energy state. To do this, we calculate the average number of particles in each of the allowed energy states using the four postulates of statistical physics just given; the distribution function is then obtained by dividing the average number of particles in each energy state by 6, the total number of particles of the system.

We begin by listing all possible macrostates. Table 11.1 lists the 26 different

TABLE 11.1 Macrostates and Microstates of Six Identical Particles with a Total Energy of 9ϵ

Macrostate	Number of Particles with Energy $n\epsilon$										Number of Microstates
	0ϵ	1ϵ	2ϵ	3ϵ	4ϵ	5ϵ	6ϵ	7ϵ	8ϵ	9ϵ	
A	5	0	0	0	0	0	0	0	0	1	6
B	4	1	0	0	0	0	0	0	1	0	30
C	4	0	1	0	0	0	0	1	0	0	30
D	3	2	0	0	0	0	0	1	0	0	60
E	4	0	0	1	0	0	1	0	0	0	30
F	3	1	1	0	0	0	1	0	0	0	120
G	2	3	0	0	0	0	1	0	0	0	60
H	4	0	0	0	1	1	0	0	0	0	30
I	3	1	0	1	0	1	0	0	0	0	120
J	3	0	2	0	0	1	0	0	0	0	60
K (F)	2	2	1	0	0	1	0	0	0	0	180
L	1	4	0	0	0	1	0	0	0	0	30
M	3	1	0	0	2	0	0	0	0	0	60
N	3	0	1	1	1	0	0	0	0	0	120
O (F)	2	1	2	0	1	0	0	0	0	0	180
P (F)	2	2	0	1	1	0	0	0	0	0	180
Q	1	3	1	0	1	0	0	0	0	0	120
R	0	5	0	0	1	0	0	0	0	0	6
S	3	0	0	3	0	0	0	0	0	0	20
T (F)	2	1	1	2	0	0	0	0	0	0	180
U	1	3	0	2	0	0	0	0	0	0	60
V	2	0	3	1	0	0	0	0	0	0	60
W (F)	1	2	2	1	0	0	0	0	0	0	180
X	0	4	1	1	0	0	0	0	0	0	30
Y	1	1	4	0	0	0	0	0	0	0	30
Z	0	3	3	0	0	0	0	0	0	0	20
Total Number of Microstates											2002

macrostates of the system of six particles. Each macrostate encompasses a number of microstates; for example, macrostate B contains 30 microstates because there are 6 different particles that could be in the energy state 8ϵ; and having chosen the particle in that energy state, there are then 5 particles that can have the energy ϵ, giving a total of $6 \times 5 = 30$ different and distinct microstates. In general, the number of microstates contained within a macrostate is given by

$$\mathcal{N} = \frac{N!}{n_1!n_2!n_3! \cdots} \tag{11.1}$$

where N is the total number of particles of the system and n_j is the number of particles in the jth occupied level. The last column of Table 11.1 lists the number of microstates of each of the 26 macrostates. The total number of microstates is 2002.

The average number of particles in a state of energy E_i is

$$\mathrm{n}_i = \sum_j n_{ij} P_j \tag{11.2}$$

Here n_{ij} is the number of particles of energy E_i in the macrostate j and P_j is the probability of that macrostate. Using the basic postulate of equal a priori probability, P_j is just the number of microstates contained in macrostate j divided by the total number of microstates, 2002. Thus the average number of particles in the energy state $E = 0\epsilon$ is

$$\begin{aligned}\mathrm{n}_0 = (&5 \times 6 + 4 \times 30 + 4 \times 30 + 3 \times 60 + 4 \times 30 + 3 \times 120 + 2 \times 60 \\ &+ 4 \times 30 + 3 \times 120 + 3 \times 60 + 2 \times 180 + 1 \times 30 + 3 \times 60 + 3 \times 120 \\ &+ 2 \times 180 + 2 \times 180 + 1 \times 120 + 0 \times 6 + 3 \times 20 + 2 \times 180 + 1 \times 60 \\ &+ 2 \times 60 + 1 \times 180 + 0 \times 30 + 1 \times 30 + 0 \times 20)/2002 \\ = &\ 2.143\end{aligned}$$

Similarly,

$$\mathrm{n}_1 = 1.484 \quad \mathrm{n}_2 = 0.989 \quad \mathrm{n}_3 = 0.629 \quad \mathrm{n}_4 = 0.378 \quad \mathrm{n}_5 = 0.210$$
$$\mathrm{n}_6 = 0.105 \quad \mathrm{n}_7 = 0.045 \quad \mathrm{n}_8 = 0.015 \quad \mathrm{n}_9 = 0.003$$

Figure 11.1 shows n_i as a function of E_i. Also shown is the function $Ae^{-\beta E}$, and it is apparent that n_i closely approximates this exponential function.

It often happens that there are several distinct quantum states with the same energy. The number of such *degenerate* states is the *statistical weight* g_i of energy E_i. The average number of particles with energy E_i is then given by

$$\mathrm{n}(E_i) = g_i f(E_i) \tag{11.3}$$

The classical Maxwell-Boltzmann distribution function, derived by a mathematical technique known as the calculus of variations, is the exponential function

$$f_{\mathrm{MB}}(E) = Ae^{-E/kT} \tag{11.4}$$

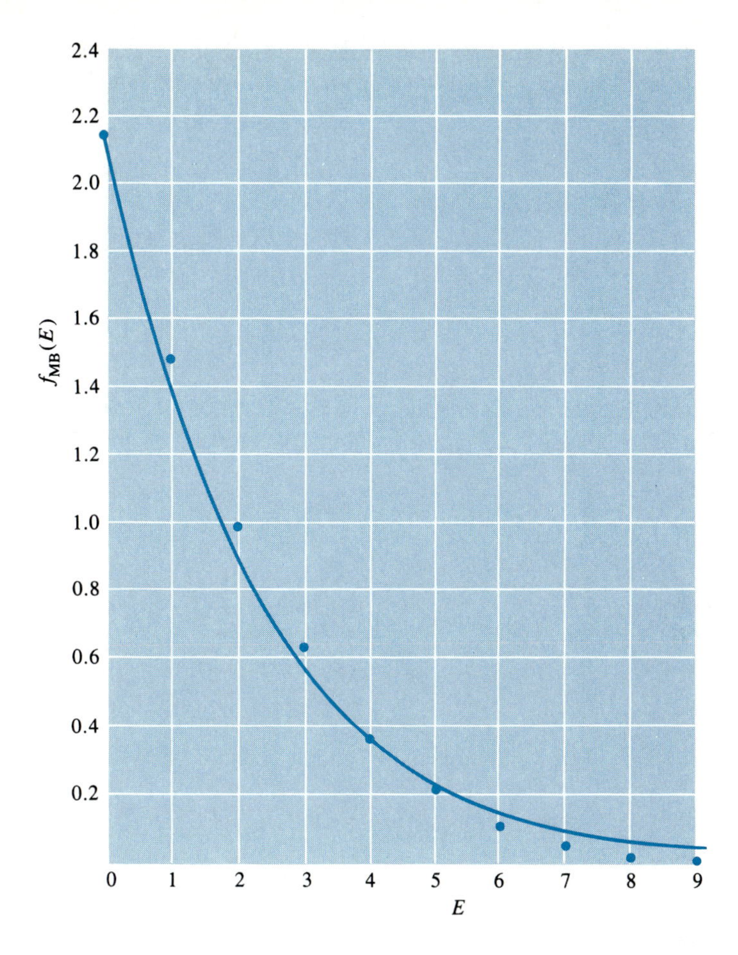

Figure 11.1 The distribution function for a classical gas of six noninteracting identical particles of total energy 9ϵ. Also shown is the exponential function $f_{MB}(E) = Ae^{-E/kT}$.

Here T is the absolute temperature, $k = 1.38 \times 10^{-23}$ J/K is the Boltzmann constant, and A is a normalization constant obtained from the condition

$$N = \sum n(E_i) = \sum g_i f_{MB}(E_i) \tag{11.5}$$

In classical mechanics, the energy is a continuous function; the statistical weight and the distribution function are also continuous functions of the energy. The statistical weight is then referred to as the *density of states* $g(E)$; that is, $g(E)\, dE$ is the number of states in the energy interval between E and $E + dE$.

The normalization summation, Equation (11.5), now becomes an integral

$$N = \int_0^\infty g(E) f_{MB}(E)\, dE = \int_0^\infty n(E)\, dE \tag{11.6}$$

where $n(E)\, dE$ is the number of particles of the system with energy between E and $E + dE$.

Consider a classical gas of N point particles per unit volume. The energy of a particle is its translational kinetic energy:

$$E = \tfrac{1}{2}mv^2 = \tfrac{1}{2}m(v_x^2 + v_y^2 + v_z^2) \tag{11.7}$$

The number of states with energy between E and $E + dE$ equals the number of states in "velocity space" within the spherical shell between v and $v + dv$ whose volume is $4\pi v^2\, dv$. That is,

$$g(E)\, dE = 4\pi v^2\, dv \tag{11.8}$$

From Equation (11.7), $v^2 = 2E/m$ and $dv = (2mE)^{-1/2}\, dE$. Hence,

$$g(E) = \frac{4\pi}{m}\left(\frac{2E}{m}\right)^{1/2} \tag{11.9}$$

We can now determine the normalization constant A of the Maxwell-Boltzmann distribution for this gas. From Equation (11.6) we have

$$N = \frac{4\pi}{m^{3/2}}\int_0^\infty (2E)^{1/2} A e^{-E/kT}\, dE = \frac{4\pi A\sqrt{2}}{m^{3/2}}\int_0^\infty E^{1/2} e^{-E/kT}\, dE \tag{11.10}$$

The definite integral

$$\int_0^\infty x^n e^{-ax}\, dx = \frac{\Gamma(n+1)}{a^{n+1}} = \frac{n!}{a^{n+1}} \tag{11.11}$$

where $$\Gamma(n+1) = n\Gamma(n) \quad \text{and} \quad \Gamma(\tfrac{1}{2}) = \sqrt{\pi} \tag{11.12}$$

Using Equations (11.11) and (11.12) we find

$$N = \left(\frac{2\pi kT}{m}\right)^{3/2} A \qquad A = \left(\frac{m}{2\pi kT}\right)^{3/2} N \tag{11.13}$$

The average value of any function of the energy is now obtained by integrating that function over the distribution, weighted by the density of states, and dividing by the total number of particles.

EXAMPLE 11.1 Determine the average energy and the average speed of the classical ideal gas.

SOLUTION The average energy of the system is given by

$$\begin{aligned}\langle E\rangle &= \frac{1}{N}\int_0^\infty E g(E) f_{\mathrm{MB}}(E)\, dE \\ &= \left(\frac{m}{2\pi kT}\right)^{3/2}\left(\frac{4\pi}{m}\right)\left(\frac{2}{m}\right)^{1/2}\int_0^\infty E^{3/2} e^{-E/kT}\, dE = \frac{3}{2}kT\end{aligned}$$

The average speed is

$$\langle v\rangle = \frac{1}{N}\int_0^\infty vg(E)f_{\text{MB}}(E)\,dE = \frac{1}{N}\left(\frac{2}{m}\right)^{1/2}\int_0^\infty E^{1/2}g(E)f_{\text{MB}}(E)\,dE$$

$$= \left(\frac{8kT}{\pi m}\right)^{1/2}$$

EXAMPLE 11.2 A gas of atomic hydrogen is placed in a magnetic field of 1.0 T. At what temperature would 51 percent of the hydrogen atoms be in the lower Zeeman level?

SOLUTION The degeneracy of each of the two Zeeman levels is 1 (see Chapter 9). The ratio of probability of occupancy of the two levels is then

$$\frac{N_+}{N_-} = \frac{e^{-E_+/kT}}{e^{-E_-/kT}} = e^{\mu B/kT} = \frac{51}{49} = 1.0408$$

Hence,

$$\frac{\mu B}{kT} = \ln 1.0408 \simeq 0.0408$$

and

$$T = \frac{\mu B}{0.0408k} = \frac{(9.27\times 10^{-24}\text{ J/T})(1.0\text{ T})}{0.0408(1.38\times 10^{-23}\text{ J/K})} = 16.5\text{ K}$$

At that temperature and at atmospheric pressure, hydrogen gas has already condensed to a liquid.

EXAMPLE 11.3 According to the simple Bohr model, what fraction of the atoms of atomic hydrogen are in the first excited state at room temperature?

SOLUTION The energies and degeneracies of the ground and first excited states of hydrogen are

$$E_0 = -13.6\text{ eV},\ g(0) = 2 \qquad E_1 = -3.40\text{ eV},\ g(1) = 8$$

Note that in assigning a statistical weight of 8 to the first excited state we have neglected the spin-orbit splitting of the excited state.

The ratio of the number of atoms in the excited state to that in the ground state is then

$$\frac{N_1}{N_0} = \frac{8}{2}\exp\left[\frac{(3.40 - 13.6)\text{ eV}}{(8.62\times 10^{-5}\text{ eV/K})(300\text{ K})}\right]$$

$$= 4e^{-394} = 4\times 10^{-171} \simeq 0$$

The probability of finding even a single atom of hydrogen in the first excited state at room temperature is effectively zero.

11.3 QUANTUM STATISTICAL PHYSICS

The principal difference between classical and quantum statistics is that in quantum statistics one acknowledges that identical particles in a system are, in fact, indistinguishable. There is no label that one can attach to a particular atom in a liter of liquid helium that sets it apart from all other helium atoms, assuming all are in the same quantum state.

It follows that in our example of six identical particles it is impossible to identify a particular microstate of the 120 possible microstates that comprise the macrostate F. Since these individual microstates cannot be observed by any conceivable experiment, they must not be counted as separate, identifiable arrangements in a statistical treatment of the system.

There is yet a further restriction that applies to systems of indistinguishable particles of half-integral spin, such as electrons, protons, and neutrons. These particles obey the Pauli exclusion principle, and the assumption that any number of particles of the system can occupy a particular energy state is no longer valid. If the particles in our example were electrons no more than two (spin up and spin down) could occupy a particular energy state.

We now consider the consequences of the fact that the particles of our system are indistinguishable and of the restriction imposed by the Pauli principle.

11.3(a) Bose-Einstein and Fermi-Dirac Distributions

Again, we begin by considering the small system of six particles whose total energy is 9ϵ. We assume that these are particles of zero or integral spin, for example, helium atoms, so that the second postulate of classical statistics—no restriction on the number of particles in a given energy state—remains valid. Because the particles are indistinguishable, we must not count the various microstates of a given macrostate as distinguishable states of the system. Instead of 2002 states that can occur with equal a priori probability, we now have only 26 states. These are again presumed to have equal a priori probability, and we proceed as before in calculating the average number of particles in a particular energy state. For instance, the average number of particles in the energy state $E = 0\epsilon$ is now

$$\begin{aligned} \mathsf{n}_0 &= (5 + 4 + 4 + 3 + 4 + 3 + 2 + 4 + 3 + 3 + 2 + 1 + 3 + 3 + 2 + 2 + 1 \\ &\quad + 0 + 3 + 2 + 1 + 2 + 1 + 0 + 1 + 0)/26 \\ &= 59/26 = 2.269 \end{aligned}$$

Similarly,

$$\mathsf{n}_1 = 1.538 \quad \mathsf{n}_2 = 0.885 \quad \mathsf{n}_3 = 0.538 \quad \mathsf{n}_4 = 0.269 \quad \mathsf{n}_5 = 0.192$$
$$\mathsf{n}_6 = 0.115 \quad \mathsf{n}_7 = 0.077 \quad \mathsf{n}_8 = 0.038 \quad \mathsf{n}_9 = 0.038$$

In Figure 11.2 we show n_i for this case. Note that n_i for these six indistinguishable particles of integral spin is quite similar to the classical distribution, but differs from it in one crucial aspect. The average number of particles in the lowest energy state is significantly greater than for the classical case where the particles are presumed to be distinguishable. As we shall see, that pattern becomes all the more pronounced

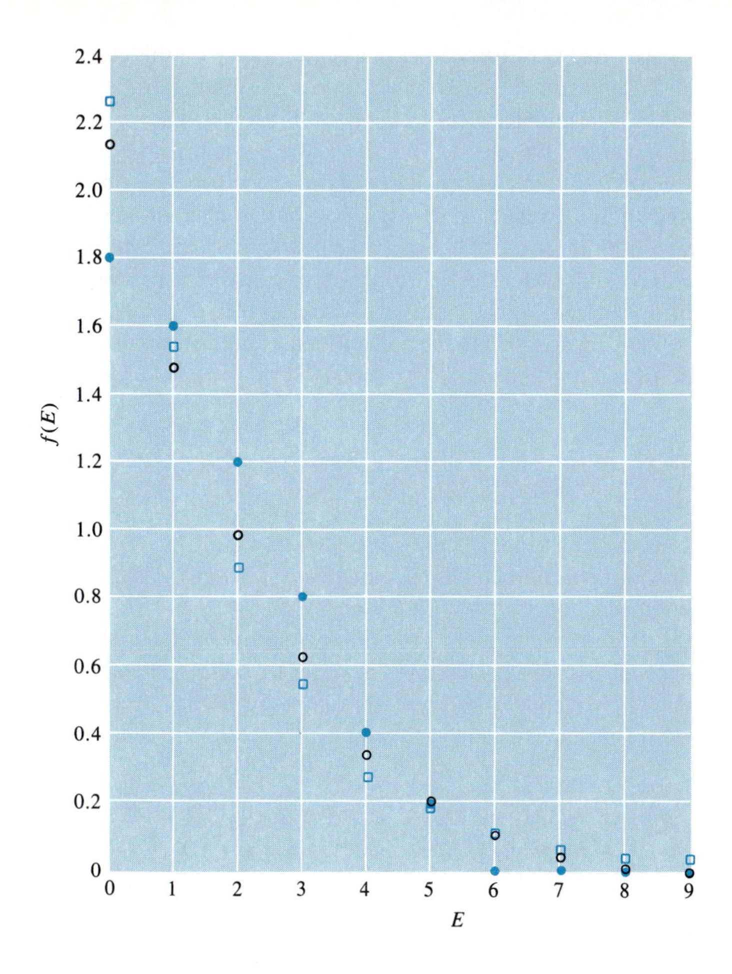

Figure 11.2 The distribution functions for six noninteracting identical particles of total energy 9ϵ. Open circles, classical particles; open squares, bosons; solid circles, fermions.

the lower the temperature (i.e., the smaller the total energy of the system), and can lead to some quite remarkable physical consequences.

As in the classical case, the treatment can be extended to a system of a large number of indistinguishable particles, with the result that the distribution function is given by

$$f_{\mathrm{BE}}(E) = \frac{1}{Be^{E/kT} - 1} \qquad \textbf{(11.14)}$$

Equation (11.14) is the *Bose-Einstein* distribution; particles that obey this distribution function are called *bosons*.

Let us next impose on our system of six indistinguishable particles the further restriction that no more than two particles may occupy an energy state. Those macrostates that meet that condition are labeled "(F)" in Table 11.1, and there are just 5 out of the 26 that satisfy the Pauli principle. As before, we calculate the average number of particles with energies $n\epsilon$, with the result:

$n_0 = 1.8 \quad n_1 = 1.6 \quad n_2 = 1.2 \quad n_3 = 0.8 \quad n_4 = 0.4 \quad n_5 = 0.2$

$n_6 = n_7 = n_8 = n_9 = 0$

The average number of particles in each energy state for this case is also shown in Figure 11.2. As required, the average number in even the lowest energy level is less than 2.0 (1.8 in this instance); except for this lowest energy level, n_i is now greater than for either the classical or the Bose-Einstein distributions for $i < 5$, while $n_i = 0$ for $i \geq 6$. As we shall see shortly, for particles obeying the Pauli principle, known as *fermions,* the energy distribution function is such that at low temperatures and/or large particle density the distribution function is almost exactly 1 for energies less than a critical energy, the so-called *Fermi energy,* and almost exactly zero for energies greater than the Fermi energy.

For a system of a large number of fermions, the distribution function is the *Fermi-Dirac* distribution given by

$$f_{\mathrm{FD}}(E) = \frac{1}{Ce^{E/kT} + 1} \tag{11.15}$$

As before, the constants B and C in Equations (11.14) and (11.15) are determined by the normalization condition, Equation (11.5) or (11.6), except that now the distribution function is either $f_{\mathrm{BE}}(E)$ or $f_{\mathrm{FD}}(E)$.

Both the Bose-Einstein and Fermi-Dirac distribution functions resemble the classical Maxwell-Boltzmann distribution, differing only by the addition of ± 1 in the denominator. Moreover, we know that in many instances classical statistical physics yields results in excellent agreement with experiment. Since integrals of the form

$$\int_0^\infty F(E) f_{\mathrm{BE}}(E)\, dE \qquad \text{and} \qquad \int_0^\infty F(E) f_{\mathrm{FD}}(E)\, dE$$

must be evaluated numerically, whereas integrals of the form

$$\int_0^\infty F(E) f_{\mathrm{MB}}(E)\, dE$$

have analytic solutions, it is useful to know under what conditions one may approximate f_{BE} and f_{FD} by f_{MB}.

Equations (11.14) and (11.15) approach f_{MB} when B or C is much greater than $e^{-E/kT}$. In that limit,

$$f_{\mathrm{BE}}(E) \simeq \frac{1}{B} e^{-E/kT} = Ae^{-E/kT} \ll 1 \qquad \text{and} \qquad f_{\mathrm{FD}} \simeq \frac{1}{C} e^{-E/kT} = Ae^{-E/kT} \ll 1$$

The quantum distributions approach the classical distribution when the probability of occupancy of a state of energy E is much less than 1. That is what we should expect. If the probability of occupancy of a given quantum state is very small, the likelihood that two or more particles would be in that state is negligible. Generally, this condition prevails when the particle density is small and/or the temperature is high. To make the criterion for applicability of classical statistics to a quantum gas more precise, we must first obtain the proper form of the density of states so that we can evaluate the coefficient A.

We consider a gas of N particles confined in a cube of side length a. We recall that the energy levels of a particle confined in a one-dimensional square well of width a are given by

$$E_n = \frac{\hbar^2\pi^2}{2ma^2} n^2 \tag{7.40}$$

For a particle confined in a three-dimensional square well, the energy eigenvalues are given by

$$E_n = \frac{\hbar^2\pi^2}{2ma^2}(n_1^2 + n_2^2 + n_3^2) = E_0(n_1^2 + n_2^2 + n_3^2) \tag{11.16}$$

where the three positive integers n_1, n_2, and n_3 are the quantum numbers that characterize a particular quantum state. To calculate the density of states, that is, the number of quantum states between E and $E + dE$, we first determine the number of eigenstates with energy less than E and then differentiate with respect to E.

The number of states of energy less than E is equal to the number of points within the spherical octant of radius $R = (E/E_0)^{1/2}$ (see Figure 11.3). The number of these points is

$$N = \left(\frac{1}{8}\right)\left(\frac{4}{3}\pi R^3\right) = \frac{1}{6}\pi\left(\frac{E}{E_0}\right)^{3/2} \tag{11.17}$$

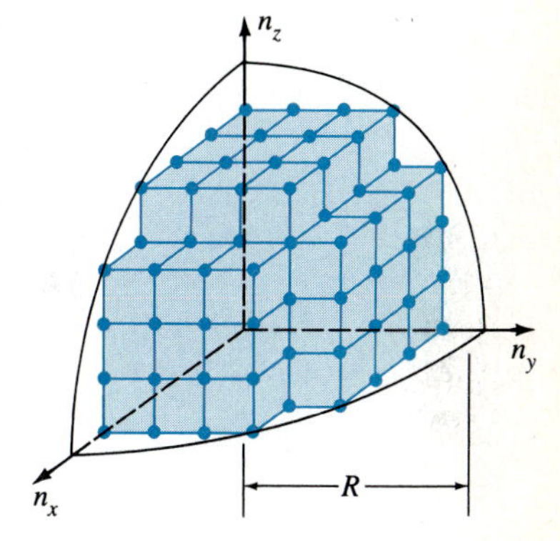

Figure 11.3 The allowed quantum states of a three-dimensional gas of noninteracting particles.

Consequently, the density of states is

$$\frac{dN}{dE} = \frac{\pi}{4}E_0^{-3/2}E^{1/2} = \frac{2\pi}{h^3}(2ma^2)^{3/2}E^{1/2}$$

and the density of states per unit volume is

$$g(E) = \frac{2\pi}{h^3}(2m)^{3/2}E^{1/2} \tag{11.18}$$

If the gas particles are electrons, each energy level can accommodate two electrons (spin up and spin down), and the energy density of states per unit volume is then

$$g_e(E) = \frac{4\pi}{h^3}(2m)^{3/2}E^{1/2} \tag{11.19}$$

To see if, in a particular instance, the classical approximation to the quantum distribution is applicable, we *assume* that we can indeed replace Equation (11.14) or (11.15) by Equation (11.4) and evaluate the constant A. For a gas of N electrons per unit volume,

$$N = \int_0^\infty g_e(E)Ae^{-E/kT}\,dE = A\frac{4\pi}{h^3}(2m)^{3/2}\int_0^\infty E^{1/2}e^{-E/kT}\,dE$$

$$= \frac{2A}{h^3}(2\pi mkT)^{3/2}$$

and $$A = \frac{Nh^3}{2}(2\pi mkT)^{-3/2} \tag{11.20}$$

As mentioned previously, the coefficient A will be small if N is small or T large.

The simplest, yet often adequate model of a metal assumes that the valence electrons of the atoms constitute a "gas" of electrons that are free to move about within the metal. For a semiconductor, such as germanium or silicon, a free-electron model can also be applied; the density of "free" electrons is, however, much smaller than for a metal, typically between 10^{21} and 10^{24} m^{-3}.

The density of free electrons in the monovalent metal silver is

$$N = \frac{(6.02 \times 10^{23}\ \text{mol}^{-1})(10.5\ \text{g/cm}^3)}{108\ \text{g/mol}} = 5.85 \times 10^{22}\ \text{cm}^{-3} = 5.85 \times 10^{28}\ \text{m}^{-3}$$

Hence at room temperature ($T = 300$ K),

$$A = \frac{(5.85 \times 10^{28})(6.63 \times 10^{-34})^3}{2[2\pi(9.11 \times 10^{-31})(1.38 \times 10^{-23})(300)]^{3/2}} = 2.34 \times 10^3$$

which is much greater than 1; the classical approximation is evidently not valid in that case.

Since $A \gg 1$ for $T = 300$ K, we anticipate that the distribution function at that temperature will approximate that at $T = 0$ K. We therefore consider the situation at $T = 0$.

As $T \rightarrow 0$ K, the N electrons of the system will occupy the lowest energy states consistent with the restriction imposed by the Pauli principle, i.e., in accordance with the Fermi-Dirac distribution function. It is convenient and instructive to replace the constant C by $e^{-E_F/kT}$, where E_F is the *Fermi energy* of the system. The physical significance of E_F will become apparent shortly. Equation (11.15) then takes the form

$$f_{FD}(E) = \frac{1}{e^{-E_F/kT}e^{E/kT} + 1} = \frac{1}{e^{(E-E_F)/kT} + 1} \tag{11.21}$$

It follows from Equation (11.21) that at any temperature the probability of occupancy of the state of energy E_F is $\frac{1}{2}$. Moreover, if $|E - E_F| \gg kT$, $f_{FD}(E) \simeq 1$ for states of energy $E < E_F$, and $f_{FD}(E) \simeq 0$ for $E > E_F$. In the limit as $T \rightarrow 0$, all states up to $E = E_F$ are fully occupied and states with $E > E_F$ are empty. The distribution function for $T = 0$ K is as shown in Figure 11.4. The Fermi energy at $T = 0$ K is now readily calculated from the normalization condition

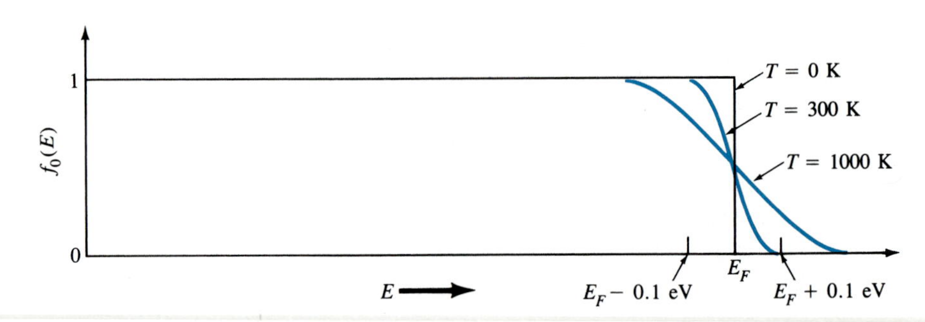

Figure 11.4 The Fermi-Dirac distribution function at $T = 0$ K (black) and finite temperature (color).

$$N = \int_0^\infty g_e(E) f_{\text{FD}}(E)\, dE$$

Since $f_{\text{FD}}(E) = 1$ for $0 < E \leq E_\text{F}$ and $f_{\text{FD}}(E) = 0$ for $E > E_\text{F}$, the normalization integral becomes

$$N = \int_0^{E_\text{F}} g_e(E)\, dE = \frac{4\pi}{h^3}(2m)^{3/2} \int_0^{E_\text{F}} E^{1/2}\, dE = \frac{8\pi}{3h^3}(2mE_\text{F})^{3/2} \quad \textbf{(11.22)}$$

and

$$E_\text{F} = \frac{h^2}{8m}\left(\frac{3N}{\pi}\right)^{2/3} = \frac{\hbar^2}{2m}(3\pi^2 N)^{2/3} \quad \textbf{(11.23)}$$

For silver, with $N = 5.85 \times 10^{28}$ m^{-3}, the Fermi energy at $T = 0$ K is 8.81×10^{-19} J = 5.51 eV. At room temperature, $kT \simeq 0.025$ eV $\ll E_\text{F}$. Therefore, even at room temperature, the distribution function does not differ significantly from that shown in Figure 11.4 except in the small energy range of width $kT \simeq 0.025$ eV about $E = E_\text{F} = 5.51$ eV; the Fermi distribution at that temperature is also shown schematically in Figure 11.4, where the region over which $f_{\text{FD}}(E)$ differs from 1 or 0 has been greatly exaggerated.

When the system is a free-electron gas with $N \simeq 10^{22}$ m^{-3}, as in a typical semiconductor, $A(300\text{ K}) \simeq 4 \times 10^{-4}$, and the Maxwell-Boltzmann distribution is then an excellent approximation to the Fermi-Dirac distribution. We shall return to these matters in Chapter 13.

In the case of a gas of bosons, such as helium at NTP, the coefficient A is approximately 3×10^{-6}. (The density is only 2.7×10^{25} m^{-3}, and the mass of the helium atom is about 8000 times that of an electron.) Still, the tendency of bosons to occupy states already heavily populated leads to a phenomenon known as *Bose-Einstein condensation*. It is a different sort of condensation from the gas-liquid transition that one normally thinks of; the particles "condense" in the sense that an enormous number (all of them as $T \to 0$ K) occupy the lowest quantum state.

11.4 LIQUID HELIUM AND BOSE-EINSTEIN CONDENSATION

The experiment began at 5:45 a.m. on July 10, 1908. The previous day, 75 liters of liquid air had been prepared. Now, under the supervision of Heike Kamerlingh Onnes, who had been appointed to the chair of physics at the University of Leiden 26 years earlier, technicians started the hydrogen liquefier. At 1:30 p.m. the 20 liters of liquid hydrogen required to precool the helium liquefier was at last ready for transfer. This exacting task took considerable time, but by mid-afternoon everything was ready for the assault on liquefaction of helium. This was the moment for which Kamerlingh Onnes had planned with painstaking, meticulous care and worked with single-minded determination for a quarter century.

Figure 11.5 Heike Kamerlingh Onnes (seated) and J. D. van der Walls in the cryogenic laboratory at Leiden. *(Loek, Zuyderduin/Huygens Lab., Leiden.)*

The central cryostat held a constant-volume helium gas thermometer, and as helium gas commenced circulating through the liquefier, Kamerlingh Onnes and his assistants watched the pressure gauge of the thermometer with increasing apprehension. For quite some time the pressure remained steady; valves were adjusted, different pressures were tried, and at last the temperature in the central chamber began to drop, slowly and erratically. But before long the temperature again refused to diminish further. At 7:30 p.m. the 20 liters of liquid hydrogen had been used up, and still no liquid could be seen in the central vessel. Apparently the great experiment had failed.

All afternoon other faculty had come into the laboratory to watch the activity, and it was then, when success seemed to have slipped from Kamerlingh Onnes's grasp, that Professor Schreinemakers suggested that the thermometer might be immersed in the boiling liquid and, therefore, give a steady reading. Perhaps the liquid is so transparent that it is difficult to see. Schreinemakers proposed illuminating the vessel from below, and no sooner was this done than the liquid-gas interface became clearly visible. The chamber held no less than 60 cm^3 of the precious liquid. The last element that had heretofore steadfastly remained in gaseous form had finally condensed at 4.2 K. The door to a broad new vista of experimental research beyond the wildest imaginings of any of the participants in the laboratory had at last been opened. Indeed, Kamerlingh Onnes, who made one of the crucial discoveries (superconductivity), at first failed to understand and appreciate its momentous nature.

We tend to think of experiments as primarily aimed at verification or refutation of theory. That is, indeed, an essential function. But equally important is the discovery of new facts, of entirely unsuspected new phenomena as distinguished from anticipated regularities. Such new facts cannot be revealed until the appropriate techniques have been developed. In the case of superconductivity and superfluidity the discoveries had to await the achievement of very low temperatures, the liquefaction of helium. Rarely has a Nobel Prize been more merited than the one awarded to Kamerlingh Onnes in 1913.[1]

Superfluid liquid helium can serve as an object lesson on how very difficult it is to make a truly new discovery. With hindsight, it is incredible that thirty years could have passed since that memorable day in 1908 before the spectacular phenomenon of superfluidity was observed, or, at any rate, recorded in the literature, for there can be little doubt that it must have been observed innumerable times by researchers in many cryogenic laboratories.

Already on the day of the first liquefaction of helium, Kamerlingh Onnes, in his attempt to solidify this substance by a further reduction in temperature, lowered the temperature through the critical point of transition to the superfluid phase. In the following years, as he investigated the thermal and other properties of liquid helium, he must certainly have seen the sudden cessation of violent boiling that is the most dramatic manifestation of that transition. In fact, he did record one less spectacular effect associated with that phase change, the discontinuity in the thermal expansion coefficient.[2] He also observed a much more dramatic anomalous behavior of the specific heat, but suspecting that the trouble lay with some inexplicable malfunction of equipment, failed to publish this result. He was not the only one who did so. Nicholas Kurti recalled that on one occasion about 1930, while performing a measurement on the magnetic properties of some salts at low temperatures, he was greatly distressed that the vacuum seal around his sample chamber developed a substantial leak just as the temperature of the surrounding helium bath was reduced below about 2.2 K. The vacuum held perfectly at 4.2 K, and continued to hold to even lower temperatures, 3.5 K, 3.0 K, 2.5 K; but then suddenly helium entered

---

1. The Nobel citation reads, ". . . for his investigations on the properties of matter at low temperatures which led, *inter alia,* to the production of liquid helium."

2. H. Kamerlingh-Onnes, *Proc. Konink. Akad. Wetensch.* **13,** 1113 (1911).

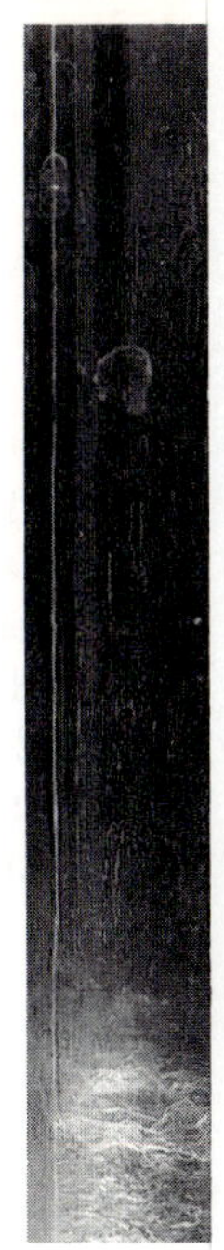

(a)

the chamber from the surrounding liquid. He repeated the cycling to higher and lower temperatures a few times, with the same result, and then, in desperation, allowed the system to warm to room temperature and resoldered the seal—having been, perhaps, the first to observe the superfluidity of Helium II.[3]

In 1932, McLennan, Smith, and Wilhelm[4] at last described the sudden, dramatic change in the liquid's appearance as the temperature passed through that of the thermal expansion discontinuity: ". . . the liquid underwent a marked change, and the rapid ebullition ceased instantly. The liquid became very quiet." In doing so they did no more than record an observation that had certainly been made previously. They failed to draw the "obvious" conclusion. But who, except perhaps a pedantic undergraduate student, would conclude that the thermal conductivity of the liquid had suddenly increased a millionfold or more? Certainly no mature, experienced physicist would be fool enough to seriously suggest this explanation for the cessation of ebullition. Yet that is exactly what occurred; not until 1936 was the thermal conductivity of helium measured and this fact confirmed.

What then are the experimental manifestations of the exceptional new phase of liquid helium? They are displayed in Figures 11.6 through 11.8. First, historically, there is the cusp in the density of the liquid at 2.17 K. Second, as shown in Figure 11.7, there is the enormous anomaly in the specific heat first recorded by Keesom and Clusius[5] and later measured with incredible precision by Buckingham and Fairbank,[6] which was a clear signal of a phase transition at that temperature. Third, the

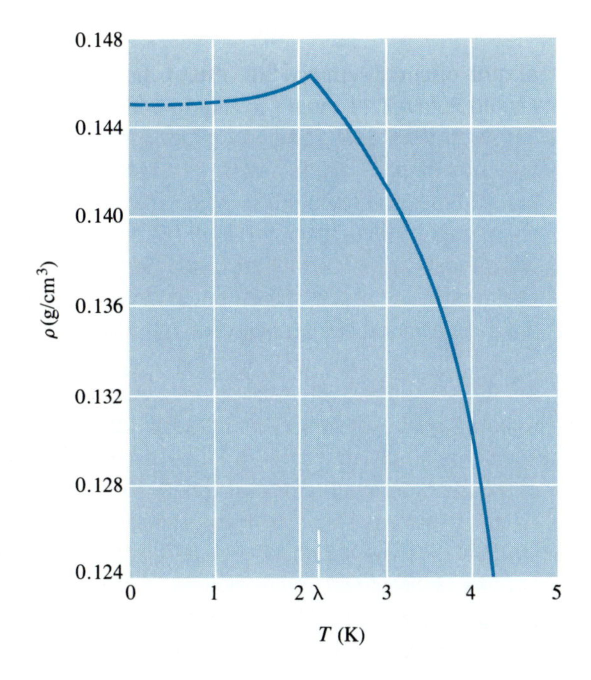

Figure 11.6 The density of liquid helium as a function of temperature. Note the discontinuity at T_λ = 2.17 K.

3. It is common practice to refer to the superfluid phase of liquid helium as Helium II.

4. J. C. McLennan, H. D. Smith, and J. O. Wilhelm, *Phil. Mag.* **14,** 161 (1932).

5. H. Keesom and K. Clusius, *Proc. Konink. Akad. Wetensch.* **35,** 307 (1932).

6. M. J. Buckingham and W. M. Fairbank, "The Nature of the λ-Transition," in *Progress in Low Temperature Physics* III (1961), p. 138.

Figure 11.7
liquid heliur
to the super
of early mea
obtained by
Fairbank in 1
data points r
by only a fe

trons in a metal, one must employ the Fermi-Dirac distribution; in a semiconductor at room temperature the electron density is generally so small that the Maxwell-Boltzmann distribution may be used.

At very low temperatures the particles of a Bose system condense into the ground state. The remarkable properties of superfluid helium, Helium II, are consequences of *Bose-Einstein condensation.*

Bibliography

C. Kittel, *Thermal Physics,* Wiley, New York, 1969.

C. Lane, *Superfluid Physics,* McGraw-Hill, New York, 1962.

F. London, *Superfluids,* vol. II, Dover, New York, 1964.

D. McLachlan, Jr., *Statistical Mechanical Analogies,* Prentice-Hall, Englewood Cliffs, N.J., 1968.

K. Mendelssohn, *The Quest for Absolute Zero: The Meaning of Low Temperature Physics,* McGraw-Hill, New York, 1966.

F. Reif, *Fundamentals of Statistical and Thermal Physics,* McGraw-Hill, New York, 1965.

PROBLEMS

11.1 Use Table 11.1 to show that for the system of six identical but distinguishable particles with total energy 9ϵ, the average number with energy $n_i\epsilon$ is as stated on page 226.

11.2 Consider a "gas" of six identical particles, each of which can occupy a state of energy $E_i = n_i\epsilon$, where n_i is an integer $(0, 1, 2, \ldots)$. The total energy of the system is $E = 8\epsilon$. Construct a table, like Table 11.1, listing all possible macrostates, and determine the number of microstates associated with each macrostate.

•**11.3** Calculate the distribution function for the gas of Problem 11.2 if (a) the particles are distinguishable, (b) the particles are indistinguishable bosons, and (c) the particles are indistinguishable fermions. (d) Fit the points of part (a) to the exponential function $f_{\text{MB}}(E_i) = Ae^{-\beta E_i}$. Determine the values of A and β. {*Hint:* Plot $\ln n_i$ versus E_i; $\ln [f_{\text{MB}}(E_i)] = \ln A - \beta E_i =$ constant $- \beta E_i$.}

11.4 Consider a gas of seven fermions, each of which can occupy a state of energy $n\epsilon$. (a) What is the minimum total energy of this gas? (b) What is the Fermi energy of this gas when the total energy is a minimum? (c) What is the minimum average energy of this gas?

11.5 Suppose the gas of seven fermions of Problem 11.4 has a total energy of 13ϵ. Identify the possible macrostates of the system and calculate the distribution function for this gas. Estimate the Fermi energy and compare the result with that of Problem 11.4.

11.6 Repeat Problem 11.5 if the total energy is 12ϵ.

11.7 What are the most probable values of the speed and of v^2 of a classical gas? How do these compare with $\langle v \rangle$ and $\langle v^2 \rangle$?

11.8 The statistical variation of a quantity F is $[\langle F^2 \rangle - \langle F \rangle^2]^{1/2}$. Calculate the statistical variation of the energy for a classical gas.

•**11.9** Photons are "particles" with spin 1 and, therefore, obey Bose-Einstein statistics. In an enclosure at a temperature T, the total number of photons is not fixed; photons are constantly absorbed and emitted at the walls of the container. In this case, the normalization constant of the Bose-Einstein distribution must be set equal to 1, i.e., $B = 1$.

(a) Show that for a photon gas the probability of occupancy of states in the energy interval dE is given by

$$P(E)\, dE = \frac{4V}{\hbar c \lambda^2} \frac{1}{e^{E/kT} - 1}\, dE$$

where V is the volume of the enclosure and λ is the wavelength of a photon of energy $E = \hbar\omega$.

(b) Use the result of part (a) to derive the Planck blackbody radiation formula.

11.10 Obtain an expression for the number of photons per unit volume and unit energy range in a blackbody and for the total number of photons per unit volume. Determine the number of photons per cm^3 at $T = 1.0$ K and $T = 300$ K. *Note:*

$$\int_0^\infty \frac{x^2}{e^x - 1}\, dx \simeq 2.4$$

11.11 Calculate the heat capacity per unit volume of vacuum.

11.12 Calculate the Fermi energy of copper at $T = 0$ K.

11.13 What is the average energy of an electron gas at $T = 0$ K? Express your result in terms of its Fermi energy.

•11.14 Consider a gas of fermions confined to a two-dimensional space.

(a) Obtain an expression for the density of states per unit area.
(b) Derive an expression for the Fermi energy of this gas as a function of the number of fermions per unit area.

11.15 At what temperature is the amount of superfluid helium (Helium II) equal to that of the normal fluid?

11.16 Calculate the probability of finding an electron in silver with an energy of (5.90 ± 0.1) eV at $T = 300$ K and at $T = 1000$ K.

11.17 When a star whose mass is two solar masses has exhausted its nuclear fuel it is likely to collapse to a *neutron star*. The density of such a star is so great that its radius would be only about 10 km. Since neutrons are spin-$\frac{1}{2}$ particles, they obey Fermi-Dirac statistics. Determine the Fermi energy of the neutrons in such a neutron star.

•11.18 Show that the condition for the validity of the classical approximation to a Fermi gas or a boson gas is equivalent to the requirement that the average separation between the particles be much greater than the de Broglie wavelength of particles of thermal energy.

Chapter 12 Solid-State Physics I: Structure of Crystalline Solids and Electron Energy Bands

Whatever is most beautiful and regular is also found to be most useful and excellent.

Sir D'Arcy Wentworth Thompson

All things are literally better, lovelier and more beloved for the imperfections that have been divinely appointed.

John Ruskin

12.1 INTRODUCTION

What is a "solid"? Although the answer seems obvious, a precise definition is not so readily phrased. One, based on mechanical properties, could be:

A solid is a substance that responds elastically to a shear stress.

Fluids, when sheared, flow, with the rate of flow determined by their viscosity. Fluids do not return to their initial configuration upon removal of the shear stress; they do not deform elastically.

The trouble with this definition is that some solids, such as glass and plastics, flow like liquids at elevated temperatures but gradually become increasingly viscous as the temperature is lowered, and their response to a shear stress is ultimately indistinguishable from that of solids such as copper.

What distinguishes copper from glass is that the former exhibits a characteristic *crystal structure*. The positions of the atoms of the crystal are correlated over long distances; that is, a crystal has *long-range order*. Glass is more like a liquid than a crystal as regards the arrangement of the atoms. There is *short-range order:* the placement of atoms immediately surrounding a given one is *nearly* the same for all atoms of the glass. But over distances of perhaps ten or more interatomic spacings, the correlation that persists over many thousand interatomic distances in crystals cannot be discerned in glass.

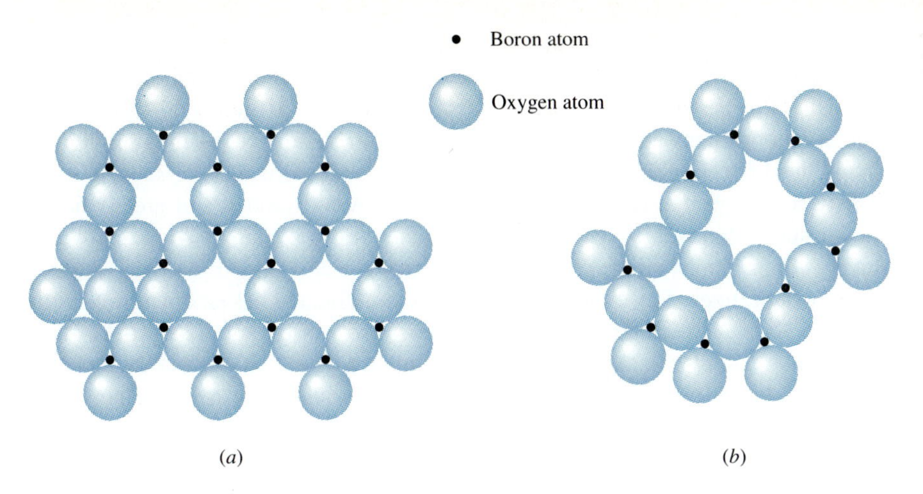

Figure 12.1 Two-dimensional representation of (*a*) crystalline and (*b*) glassy B_2O_3. In the amorphous, glassy state, short-range order persists, but there is no long-range order.

Along with this liquidlike absence of long-range order goes the lack of a well-defined phase transition that separates the solid and liquid phases. For crystalline solids, the transition from the solid to the liquid phase occurs abruptly at a critical pressure and temperature, the melting point, with the concomitant absorption of the latent heat of melting.

In the following, we restrict the term *solid* to mean

A substance in which the average positions of the constituent atoms form a lattice structure

In other words, by *solid* we shall mean a crystalline solid.

We begin this chapter with a brief account of the cohesive forces that hold the atoms in a crystal together. Section 12.3 considers some crystal defects. Every crystal, no matter how carefully prepared, contains some imperfections; indeed, many important properties of solids depend critically on the number and type of defects that are present. The final portion of the chapter concerns itself with the electronic energy levels in crystals. It is in the solution of the Schrödinger equation that we shall find the answer to the question why some solids are good conductors, others insulators, and yet others semiconductors. A detailed treatment of electronic properties of solids—specific heat, conductivity, superconductivity—and a discussion of a few device applications are deferred to the next chapter.

12.2 COHESIVE FORCES IN CRYSTALS

A crystalline solid is distinguished from other atomic aggregates by the three-dimensional periodicity of the positions of the atoms. Selecting a small polyhedron as the basic building block, one can construct a crystal that fills all space by repeated displacement of this *unit cell* along three noncoplanar directions. Suitable geometric shapes for unit cells are, for example, a cube, a regular dodecahedron, a hexagonal cylinder, or any parallelepiped. Not all polyhedra, however, are acceptable unit cells, because the requirement of translational symmetry restricts the rotational symmetry of the unit cell. Consider generating a two-dimensional lattice: we could use parallelograms, triangles, and hexagons as unit cells, but we cannot match up regular pentagons so as to fill all space.

Figure 12.2 Not all regular polyhedra can serve as unit cells. In two dimensions, rectangles and hexagons can be translated along two directions and fill all space, but a regular pentagon is not an acceptable unit cell.

Figure 12.3 (Top) W. H. Bragg *(Weber Collection Courtesy/AIP, Niels Bohr Library.)* and (bottom) W. L. Bragg *(Physics Today Collection/Courtesy AIP, Niels Bohr Library.).*

The structure of crystals could be surmised, as was done by early mineralogists, from their external morphology and lattice spacings determined from density measurement. Precise structure analysis, however, had its beginning in 1912 when, following a brilliant suggestion of Max von Laue, Friedrich and Knipping demonstrated the diffraction of X rays by the regular array of atoms in a crystal.[1] The use of X rays as probes in studying crystal structures was exploited and greatly developed in succeeding years, especially by the father-and-son team of W. H. and W. L. Bragg.[2] Their impact on the complexion of British science is perhaps without parallel in modern times. Since the days of the Braggs, England has emerged as a world leader in X-ray crystallography. In recent years, X-ray structure analysis has concentrated on complex organic systems, and it is surely no accident that the structure of DNA was unraveled by two biologists in Cambridge, one a young American Ph.D. who had gone there for the express purpose of working in the renowned crystallographic laboratory of Sir Lawrence Bragg.

A simple explanation of X-ray diffraction by a crystal was given by W. L. Bragg. Suppose we consider a particular set of atomic planes in a crystal. If each plane reflects a fraction of the incident X-ray beam, these partially reflected waves will interfere constructively provided the path difference of waves reflected from adjacent planes is an integral number of wavelengths. If the distance between atomic planes is d, the path difference between successive reflections is $2d \sin \theta$, where θ is the angle between the incident beam and the atomic planes (see Figure 12.4). Thus one obtains the famous Bragg law

$$2d \sin \theta = n\lambda \tag{12.1}$$

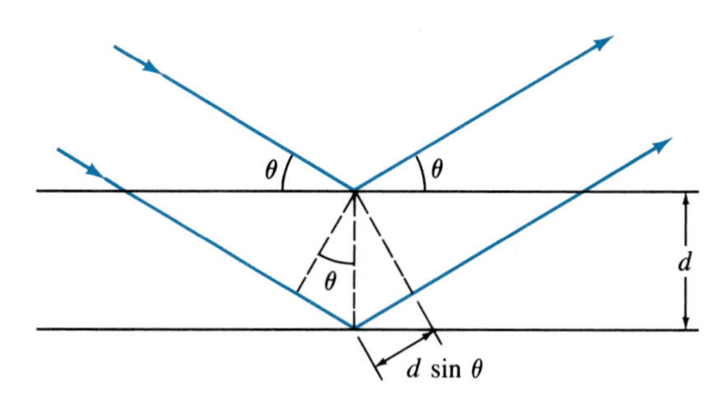

Figure 12.4 The condition for Bragg reflection. The angle between the incident beam and the crystal planes is θ. Waves reflected by successive planes interfere constructively if the path difference is an integral number of wavelengths, i.e., if $2d \sin \theta = n\lambda$.

A typical X-ray diffraction pattern is shown in Figure 12.5, which clearly displays the fourfold symmetry of the crystal planes.

1. W. Friedrich, P. Knipping, and M. von Laue, *Ber. K. Bayer. Akad. Munchen* **17,** 303 (1912).

2. W. H. Bragg and W. L. Bragg were recipients of the 1915 Nobel Prize "for their services in the analysis of crystal structure by means of X-rays."

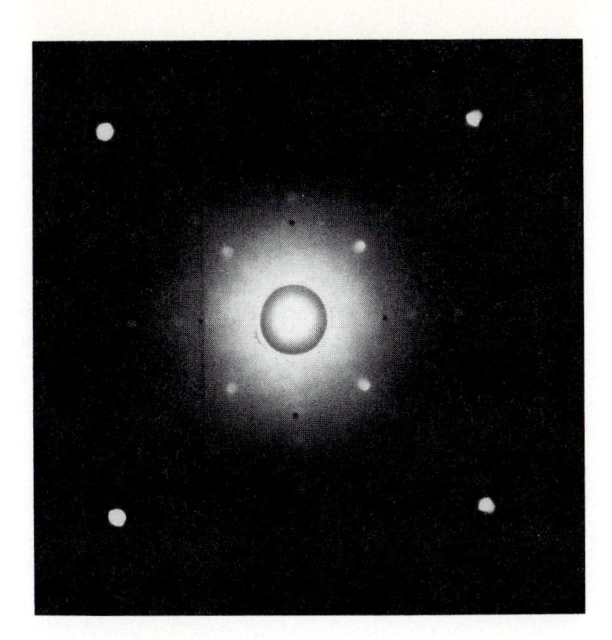

Figure 12.5 A Laue back-reflection X-ray photograph of a sodium chloride crystal. Note the fourfold symmetry of the spots, characteristic of the fourfold symmetry of a cube for rotation about a cube edge. If the crystal were rotated so that the beam was directed along a body diagonal, the picture would show threefold rotation symmetry. Back-reflection X-ray diffraction is commonly used to orient single crystals, that is, to determine the crystal axes relative to the sample's external morphology. A "white" (polychromatic) beam of X rays must be used to obtain such patterns.

EXAMPLE 12.1 An X-ray beam is incident on a cubic crystal along one of the cube axes. Reflection spots are observed at an angle of 90° with respect to the incident beam. The wavelength of the X rays is 0.26 nm. Determine the lattice parameter of the cubic crystal.

SOLUTION Since the angle between the incident and reflected beam is 90°, the angle between the reflecting crystal planes and the incident beam is 45°. Consequently, the reflecting planes are the (110) crystal planes, perpendicular to a face diagonal of the crystal (see Figure 12.6). The distance separating these planes is $a/\sqrt{2}$.

From the Bragg law we have

$$2\left(\frac{a}{\sqrt{2}}\right)\sin 45^\circ = 0.26 \text{ nm} \qquad a = 0.26 \text{ nm}$$

But what determines the structure that is assumed when a liquid slowly cools and solidifies? In its broadest terms, the answer is quickly stated: The atoms or molecules will arrange themselves in such a way that the free energy is minimized. The trouble lies, of course, with the enormously complex task of calculating the free energy for various possible lattice structures from empirical data. For some crystals, for example, the ionic crystals such as NaCl and CsCl and also for covalent crystals such as diamond or silicon, the very nature of the binding forces severely limits the possible crystal structures. Metals and molecular crystals, however, display a vast variety of lattice structures and often undergo phase changes in the solid state in which atoms rearrange themselves in a different periodic pattern.

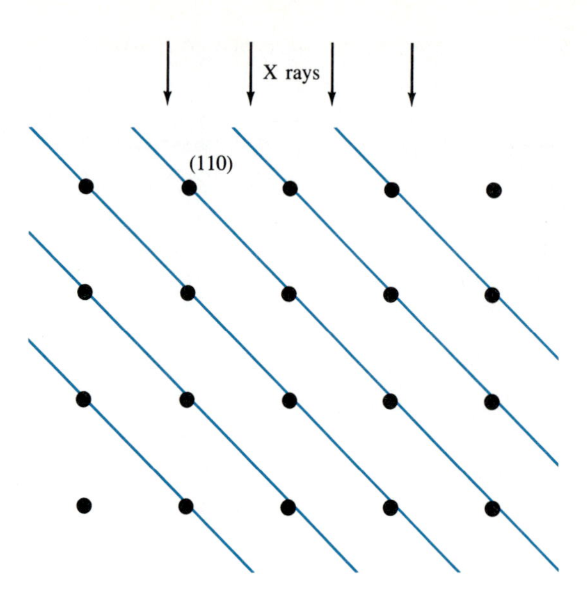

Figure 12.6 Example 12.1.

12.2(a) Ionic Crystals

The dominant cohesive force in ionic crystals is the Coulomb interaction between charge distributions of almost spherical symmetry. Since opposite charges attract and like charges repel, the most stable arrangement is one in which every positive ion is surrounded by nearest-neighbor negative ions, and vice versa. The face-centered cubic (fcc) lattice of NaCl, the body-centered cubic (bcc) lattice of ZnS,

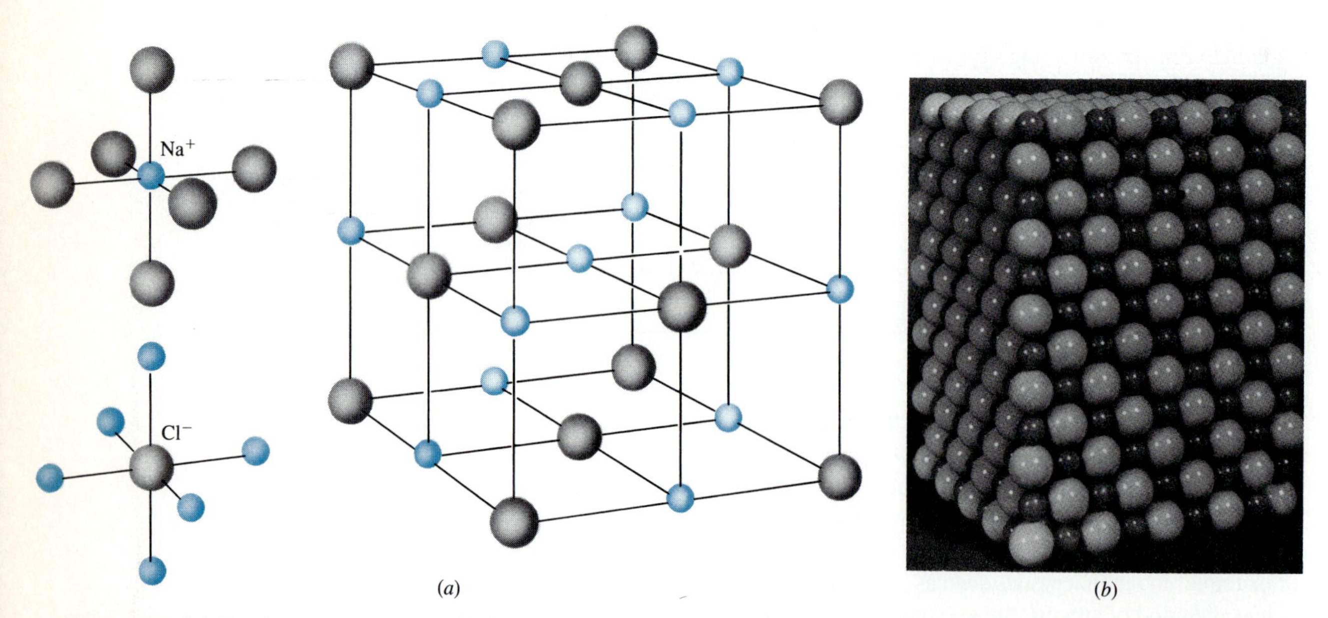

Figure 12.7 (*a*) The face-centered cubic (fcc) crystal structure of NaCl. Each positive ion is surrounded by six nearest-neighbor ions, and vice versa. (*b*) Model of a NaCl crystal; note the close packing of the spheres representing the Na^+ and Cl^- ions *(Courtesy A. N. Holden and P. Singer. From Charles Kittel,* Introduction to Solid State Physics, *6th ed., Wiley, New York, 1986.).*

and the simple cubic (sc) lattice of CsCl meet this requirement. Which of these is energetically favored depends largely on the relative sizes of the two kind of ions. In CsCl the coordination number (the number of equidistant nearest neighbors) is 8, whereas in NaCl the coordination number is only 6; it would seem that the CsCl structure should be favored. However, in the fcc lattice the ionic spheres are more closely packed than in the CsCl lattice, and this close packing tends to reduce the overall energy by bringing oppositely charged ions closer together.

If all but nearest-neighbor interactions could be neglected, the potential energy of one ion in NaCl would be

$$V = -6 \frac{e^2}{4\pi\epsilon_0 a} \tag{12.2}$$

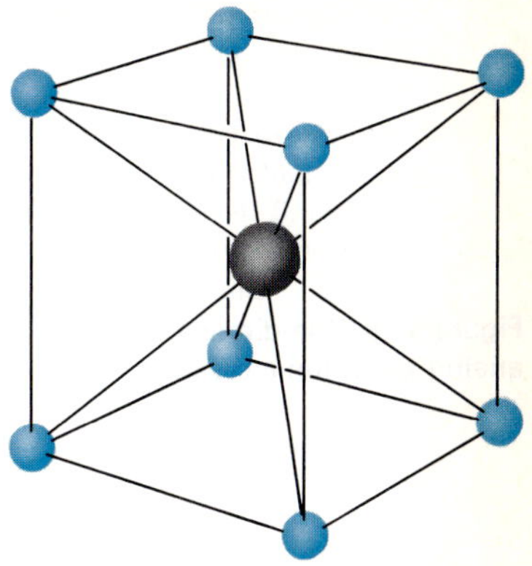

Figure 12.8 The simple cubic (sc) crystal structure of CsCl. The coordination number (number of equidistant nearest neighbors) is 8.

where a is the nearest-neighbor distance. However, surrounding the six nearest-neighbor ions are twelve next-nearest-neighbor ions of the same sign, and beyond those are eight ions of opposite sign, and so on. For the NaCl crystal, therefore, the electrostatic potential energy of an ion of either sign is a sum of terms:

$$V = -\sum_j \frac{e^2}{4\pi\epsilon_0 r_j} = -\frac{e^2}{4\pi\epsilon_0 a}\left(6 - \frac{12}{\sqrt{2}} + \frac{8}{\sqrt{3}} - \cdots\right) = -\alpha \frac{e^2}{4\pi\epsilon_0 a} \tag{12.3}$$

where α is the so-called *Madelung constant*.

Here the first term in the summation comes from the six nearest Cl^- neighbors at a distance a from the central Na^+ ion; the second, from the twelve next-nearest-neighbor Na^+ ions at a distance of $\sqrt{2}a$; the third, from the eight next Cl^- ions at a distance of $\sqrt{3}a$; and so forth. Evidently, the convergence of this series is very slow, and calculation of the Madelung constant entails subtle computational techniques. Ewald and, more recently, Frank developed methods of arranging and counting neighboring ions of opposite sign so that the resulting series converges fairly rapidly.

The Madelung constant, which clearly depends only on crystal structure and not on the kind of ions involved, is given in Table 12.1 for some of the more common structures.

TABLE 12.1

Structure	Madelung Constant	Coordination Number
NaCl (fcc)	1.747565	6
CsCl (sc)	1.762675	8
ZnS (bcc)	1.6381	4

EXAMPLE 12.2 Determine the Madelung constant of a one-dimensional ionic crystal.

SOLUTION The "crystal" consists of alternate positive and negative charges, separated by a distance a. The potential energy of a given charge is then

12.3(a) Dislocations; Plastic Deformation and Crystal Growth

All solids, especially metals, deform plastically if stressed beyond their elastic limit. Indeed, since ancient times, the malleability of metals has been one of their most valuable attributes.

It is a curious fact that a carefully prepared single crystal of zinc, copper, or gold is much softer and more readily elongated or bent than one that has already suffered some deformation. Moreover, when a crystal is pulled and extended, even cursory visual examination reveals that this elongation is not the result of homogeneous deformation but is due to large shear strains, called *slip,* on widely separated slip planes (see Figure 12.14).

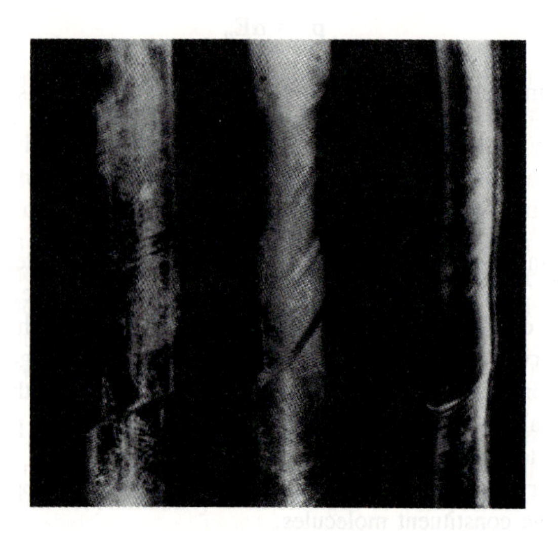

Figure 12.14 Photograph of single zinc crystals that have been plastically deformed under tension; the slip bands are clearly visible *(E. R. Parker. From Charles Kittel,* Introduction to Solid State Physics, *6th ed., New York, Wiley, 1986.).*

The *critical shear stress* is that at which the material fails to behave elastically but, instead, suffers plastic (permanent) deformation. When the critical shear stress of a pure single crystal is compared with what one would calculate on the basis of the known interatomic forces in the crystal, one finds that the experimental value is generally two to three orders of magnitude smaller.

The dilemma posed by this low critical shear stress was resolved in 1934 by Taylor and Orowan,[3] who suggested that crystals include linear defects which they named *dislocations*. An *edge dislocation* is the mismatch at the end of a crystal plane that terminates not on an external boundary but within the crystal, as shown in Figure 12.15. The important features to note are (1) that when a dislocation moves from one side of a crystal to the opposite, the result is a displacement of the upper portion of the crystal relative to the lower by one lattice parameter, and (2) that the shear stress required to initiate and maintain dislocation propagation is very small compared to the elastic shear modulus; a rough estimate suggests about 10^4 N/m^2 for most close-packed metallic crystals. Repeated motion of dislocations over the same plane results in macroscopic slip.

Another type of dislocation is the *screw dislocation,* which corresponds to a partial cut in a crystal and the relative displacement of the two portions by one

3. G. I. Taylor, *Proc. Roy. Soc.* **A 145,** 362 (1934). E. Orowan, *Zeit. f. Physik* **89,** 605 (1934).

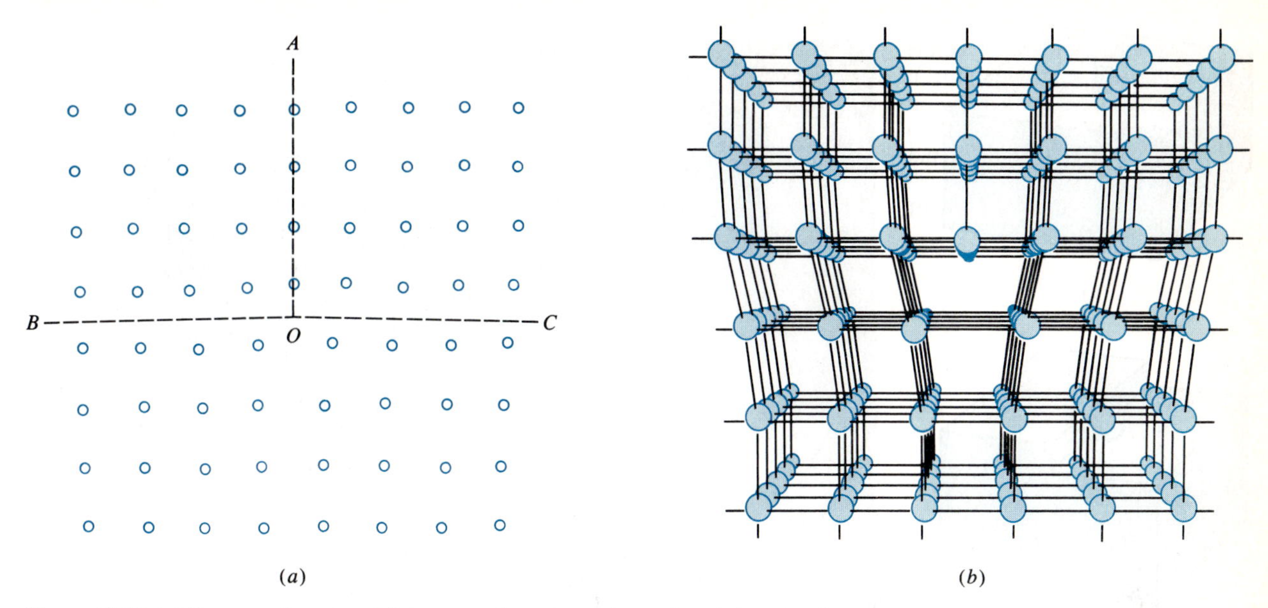

Figure 12.15 (*a*) Cross-sectional and (*b*) perspective view of an edge dislocation in a cubic crystal.

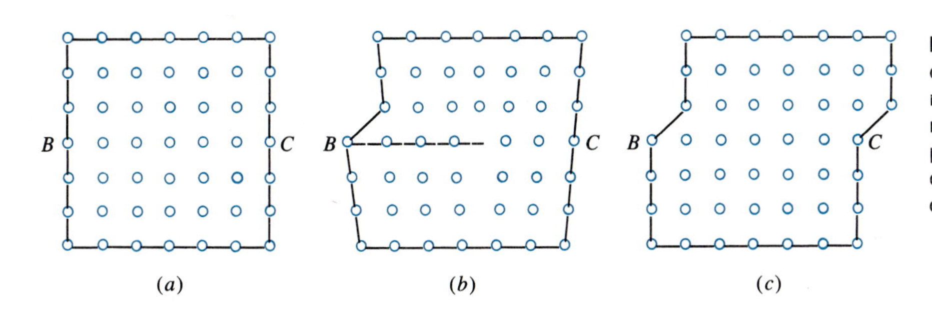

Figure 12.16 Motion of an edge dislocation in the direction *B-C* results in slip of the upper portion relative to the lower by one lattice parameter. Note the edge dislocation near the center of the crystal in part *b*.

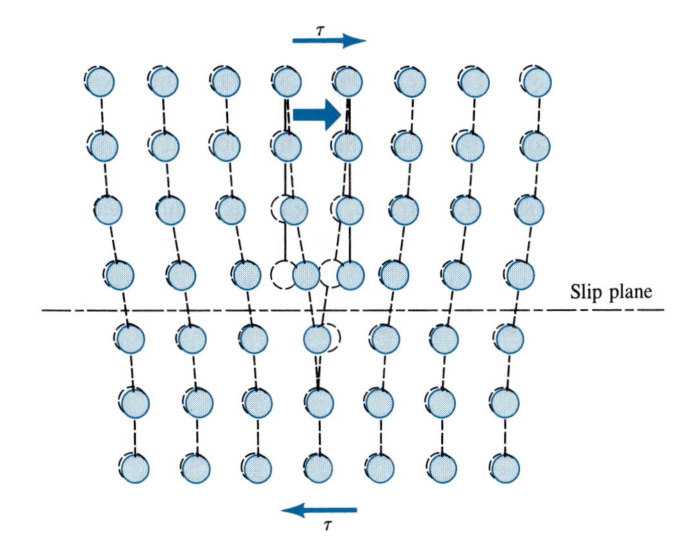

Figure 12.17 Rearrangement of atoms in a crystal as an edge dislocation moves one lattice parameter under the action of a shear stress τ.

identical and, therefore, correspond to the same eigenstate, the physically significant k values are limited to the range 0 to $2\pi/a$. In practice, it is more convenient to let k run from $k = -\pi/a$ to $k = +\pi/a$. This range

$$-\frac{\pi}{a} < k \leq +\frac{\pi}{a} \tag{12.17}$$

is known as the *first Brillouin zone* of the one-dimensional crystal. In three dimensions, the first Brillouin zone is a polyhedron in the three-dimensional space of the wave vectors **k**.

12.4(b) Energy Bands and Effective Masses

To lowest order, the solution of the one-dimensional Schrödinger equation in the tight-binding approximation leads to the result (see Section 12.B)

$$E(k) = E_0 - 2E_1 \cos ka \tag{12.18}$$

where E_0 is the energy of the electron in the free atom and E_1 is an *overlap integral*, a constant that depends on details of the periodic potential and on the amount of overlap of atomic wave functions at neighboring lattice sites.

The energy $E(k)$ as a function of k is shown in Figure 12.21. As k varies from $k = 0$ to $k = \pm\pi/a$, $E(k)$ goes from

$$E(0) = E_0 - 2E_1$$

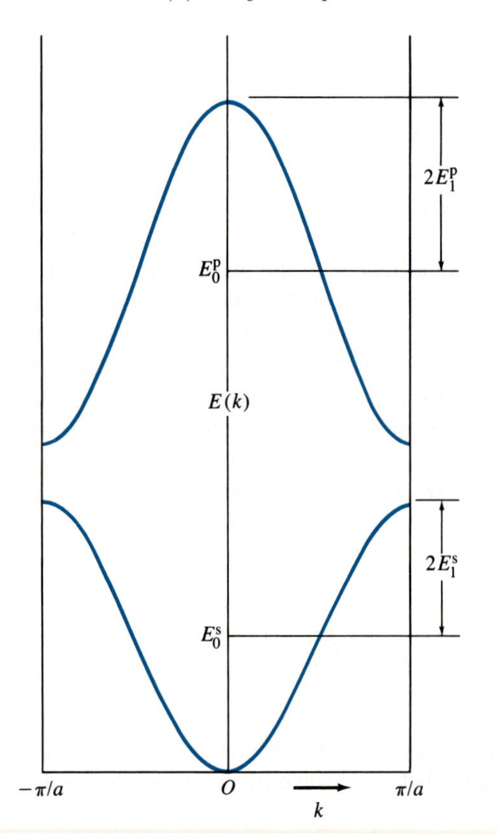

Figure 12.21 Energy bands in the tight-binding approximation. Bands derived from s and p atomic wave functions are shown. The widths of the bands are determined by the overlap integrals E_1^s and E_1^p and are centered on E_0^s and E_0^p, where the energies E_0 are those of the atomic states.

to

$$E(\pm\pi/a) = E_0 + 2E_1$$

The N atomic energy levels have spread into an *energy band* of width $4E_1$. The bandwidth is proportional to the overlap integral E_1 and increases as the separation between atoms is reduced and the overlap of atomic wave functions becomes more pronounced. In the limit of large a, $E_1 \rightarrow 0$ and the "band" reduces to an N-fold degenerate set of levels with $E = E_0$.

The sign as well as the magnitude of E_1 depends on the character of the atomic wave functions. For the 3s functions of sodium, for example, E_1 is positive. But bands are also formed from the superposition of excited-state wave functions. For the 3p ($l = 1$) wave functions, E_1 is negative, so that the 3p band has a *maximum* at $k = 0$ and minima at $k = \pm\pi/a$. Depending on the atomic energy levels E_0^s and E_0^p and on the magnitudes of the overlap integrals E_1^s and E_1^p, the s and p bands may overlap or one may find that there is a forbidden energy region, an *energy gap* between the bands.

The widths of the 2p, 3s, 3p, and 4s bands of sodium as functions of interatomic separation are shown in Figure 12.22. At the observed nearest-neighbor distance of 0.367 nm, the 3s and 3p bands overlap. However, if that distance were greater than 0.7 nm, an energy gap would open between those bands.

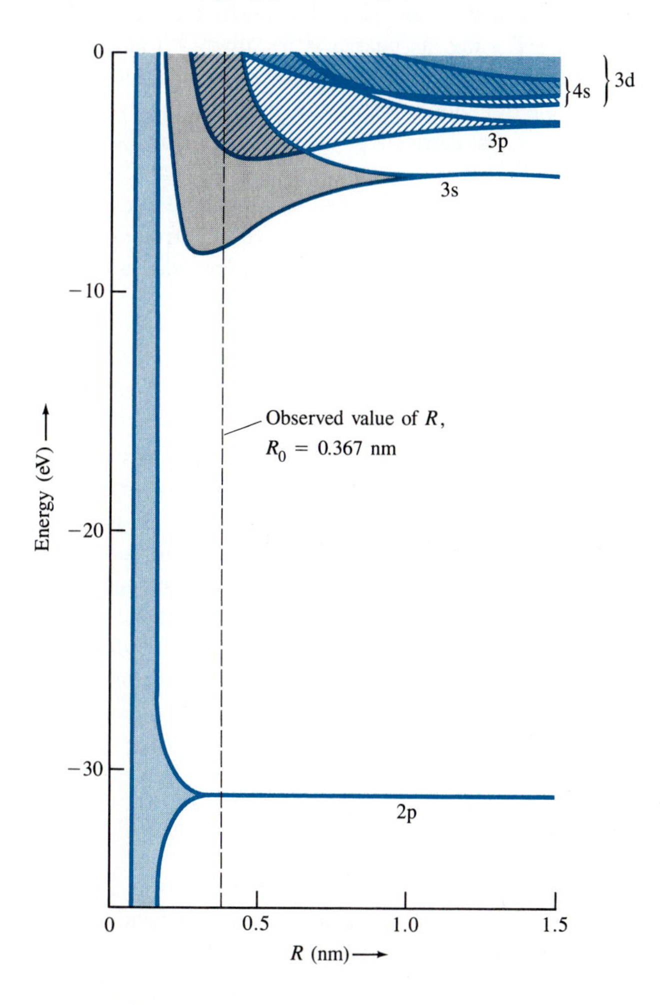

Figure 12.22 Energy bands of sodium as functions of interatomic distance. At the interatomic distance of the sodium crystal, the 3s band is already very broad and overlaps the 3p, 4s, and 3d bands.

Near $k = 0$ we can expand the cosine function and write

$$E(k) = E_0 - 2E_1\left(1 - \frac{k^2a^2}{2}\right) = E_0 - 2E_1 + (E_1a^2)k^2 \qquad \textbf{(12.19)}$$

We see that near $k = 0$, that is, near the energy minimum, $E(k)$ is proportional to k^2. We can write Equation (12.19) as

$$E(k) = \text{constant} + \frac{\hbar^2k^2}{2m^*} \qquad \textbf{(12.20)}$$

where m^* is the *effective mass* of the electrons of that band in the crystal and is given by

$$m^* = \frac{\hbar^2}{2E_1a^2}$$

in this tight-binding approximation. The behavior of electrons in states of small k will be much the same as that of free electrons provided we replace the free-electron mass by m^*. Near the top of the band, a similar expansion about $k = \pi/a$ yields a **negative** effective mass of the same magnitude.

If an electric field **E** acts on an electron of positive mass, the particle accelerates in the direction $-\mathbf{E}$ since its charge is negative. An electron with negative effective mass would, according to Newton's second law, accelerate in the opposite direction, that is, in the direction of **E**. That is precisely the dynamic response of a particle of positive mass that carries a positive charge e. In analyzing the response of an electron distribution that occupies nearly a full band, it is then far more convenient, and common practice, to do so in terms of the response of positively charged ''holes'' to which one ascribes a positive effective mass.

In the nearly free-electron approximation the dependence of energy on wave vector is again

$$E(k) = \frac{\hbar^2k^2}{2m^*} \qquad \textbf{(12.21)}$$

where the effective mass depends on the strength of the periodic lattice potential. As in the tight-binding approximation, the energy is not a continuous function of k. Energy gaps appear whenever $k = \pm n(\pi/a)$, where n is an integer (see Figure 12.23). Since the wave functions are of the Bloch form, $\psi_k(x) = \psi_{k\pm 2\pi/a}(x)$ and, consequently, $E(k) = E(k \pm 2\pi/a)$; one can therefore map the so-called extended-zone curve of Figure 12.22 into the first Brillouin zone as in Figure 12.24. The result looks now very much like Figure 12.21. In this ''reduced-zone'' scheme there are several energy bands whose minima appear alternately at the center and edge of the Brillouin zone.

12.5 METALS, INSULATORS, AND SEMICONDUCTORS

In the tight-binding approximation, each energy band is formed from atomic states whose degeneracy is $2(2l + 1)$. Consequently, an energy band that derives from N atoms will be able to accommodate $2(2l + 1)N$ electrons.

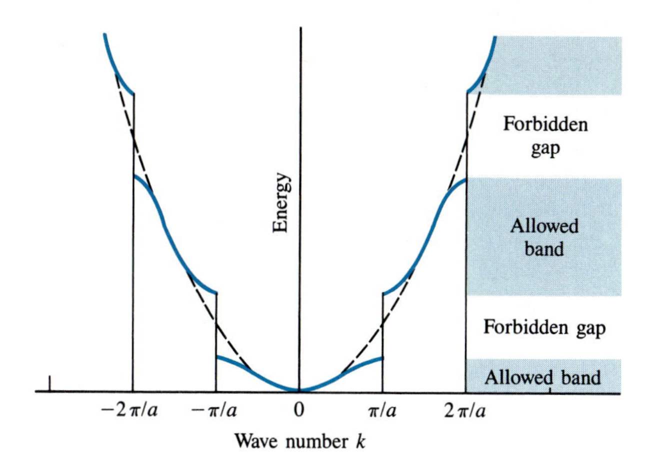

Figure 12.23 Energy bands in the nearly free-electron approximation. Forbidden energy gaps appear for $k = \pm n\pi/a$.

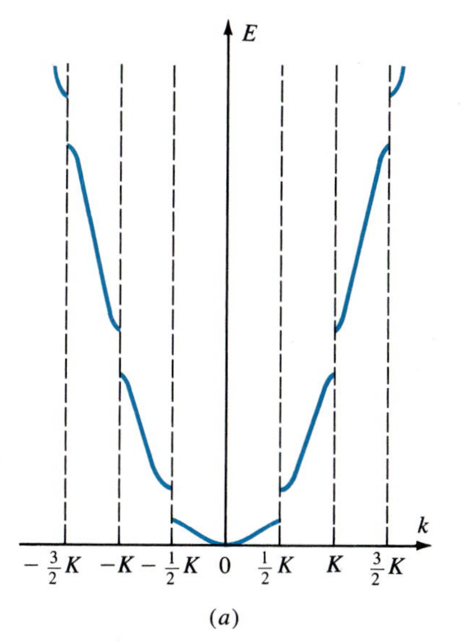

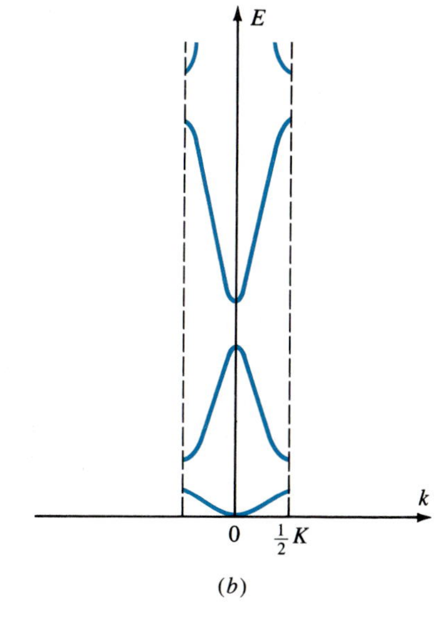

Figure 12.24 (*a*) The energy bands of Figure 12.23 can be mapped into the first Brillouin zone. (*b*) In this reduced-zone scheme, the energy bands have a form very similar to that obtained with the tight-binding approximation.

If an energy band is only partially filled, electrons in that band can readily make transitions to other **k** states of nearly the same energy by absorbing a small amount of energy from an electric field or from thermal fluctuations. In particular, an electron distribution for which the total momentum is zero and which, therefore, carries no net current is easily modified by an electric field E_x to one that does carry current by filling some states with negative k_x and removing electrons from states with positive k_x. We anticipate, then, that crystals with partially filled bands will be good conductors, i.e., metals. For instance, since an s band can hold $2N$ electrons and the alkali metals contribute only one valence electron per atom, alkali-metal crystals should be metallic as, indeed, they are. By the same token, however, the alkaline earths, Be, Mg, Ca, should be good insulators, since their valence bands should be completely filled and readjustment of electrons within these bands is prohibited by the exclusion principle. Yet, these materials are also metallic. The reason for their metallic properties is overlap of the s and p bands (see

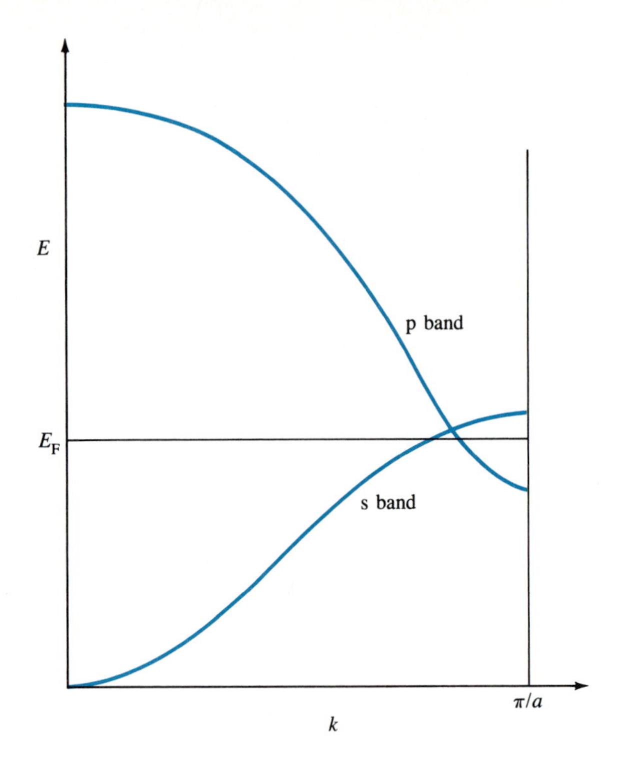

Figure 12.25 If the s band and the p band overlap, as shown here, a divalent metal is a good conductor because neither band is fully occupied even though the total number of electrons would just fill the s band if there were no overlap. The Fermi energy is indicated.

Figure 12.25). In these substances, the valence electrons derived from atomic s states occupy, in the crystal, states from both s and p bands.

In insulators the highest occupied energy band is truly filled and is separated from the next higher band by a substantial energy gap (3 eV or more). The large energy gap makes it most unlikely that an electron could be thermally excited across this gap at temperatures below the melting point.

Semiconductors are really insulators, at least as far as their band structure is concerned. However, the energy gaps between the highest occupied band, the *valence band,* and the next higher band, the so-called *conduction band,* are small (0 to 1.5 eV). In germanium, for example, the band gap is about 0.7 eV; in silicon it is about 1.1 eV.

Because the energy gap is relatively small, electrons from the filled valence band can be thermally excited to unfilled states in the conduction band. Both electrons in the conduction band and holes in the now partly filled valence band contribute to the electrical conductivity of the sample. Since the number of mobile charge carriers depends sensitively on temperature, increasing exponentially with T, the conductivity of pure semiconductors also increases nearly exponentially with temperature, in sharp contrast to metals, whose conductivity diminishes with increasing temperature. A semiconductor in which the majority of charge carriers derive from thermal excitation of electrons from the valence to the conduction band is said to be *intrinsic*.

What makes semiconductors so useful in electronic circuitry is the fact that the number and type of charge carriers, electrons, or holes, can be controlled by the addition of certain impurities. Suppose we replace a germanium atom of the crystal with an arsenic atom. Since arsenic is pentavalent, one electron will be left over, so to speak, after the four covalent bonds with the nearest-neighbor germanium atoms

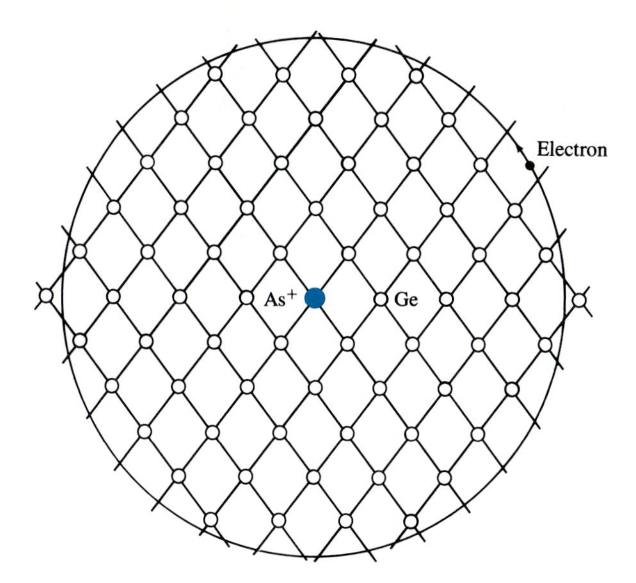

Figure 12.26 A neutral arsenic donor atom in the ground state in the germanium lattice (schematic diagram).

have been formed. This extra electron is only weakly bound to the arsenic atom and can be thermally excited into the conduction band even well below room temperature. Thus, even at room temperature, arsenic-''doped'' germanium conducts moderately well by virtue of electrons provided by the arsenic *donors*. Because the charge carriers are negatively charged electrons, a donor-doped semiconductor is said to be *n-type*.

If, on the other hand, a trivalent gallium atom replaces a germanium atom in the crystal, one of the covalent bonds about the gallium impurity will be unsaturated. Relatively little energy is now required to shift one of the valence electrons from a neighboring saturated germanium bond to the gallium site. The unsaturated bond at that germanium atom readily migrates through the crystal. In the band picture, the unsaturated bond corresponds to an unfilled state in the valence band, i.e., a hole. Gallium and similar impurity atoms that can accept an extra electron in the semiconducting crystal are called *acceptors*. Since conduction is now via positively charged holes, acceptor-doped semiconductors are said to be *p-type*.

Any semiconductor that conducts by virtue of carriers derived from donors or acceptors is said to be *extrinsic*. Of course, at sufficiently elevated temperatures, the number of thermally excited electrons and holes will exceed the number of carriers due to donors or acceptors, and an extrinsic semiconductor will become *intrinsic* at high temperature.

Although a precise calculation of the ionization energies of donors and acceptors in a semiconductor is beset with difficulties, one can obtain a good order-of-magnitude estimate as follows.

When a pentavalent arsenic atom replaces a lattice atom in a germanium crystal, four of the five valence electrons are committed to the four homopolar bonds with the nearest-neighbor germanium atoms. This then leaves one valence electron and a singly charged arsenic ion as a bound system in the germanium crystal. That system is essentially like a hydrogen atom, and we can determine the ionization energy using the Bohr model, provided we replace the free-electron mass by the effective mass of a conduction electron in germanium and replace the permittivity of free space by $\kappa\epsilon_0$, where κ is the dielectric constant of germanium. From Equation (5.32) we then have

$$E_i \approx \frac{m^* e^4}{32\pi^2\kappa^2\epsilon_0^2\hbar^2} = (13.6 \text{ eV})\left(\frac{m^*}{m}\right)\left(\frac{1}{\kappa^2}\right) \tag{12.22}$$

For germanium (see Table 13.2), $m^*/m = 0.20$ and $\kappa = 16$; thus $E_i \approx 0.011$ eV, an energy about one-fourth kT at room temperature. This rough extimate is in fair agreement with the experimental value of 0.0127 eV (see Table 13.4).

Our use of the effective mass and dielectric constant in Equation (12.22) is predicted on the presumption that the orbit radius of the valence electron bound to the arsenic atom is so large that this electron senses the periodic crystal potential as does an electron in the conduction band of germanium. That assumption can be justified, a posteriori, by calculating the orbit radius, using again the Bohr model. From Equation (5.28) we have, in this case,

$$r_0 \approx \frac{4\pi\kappa\epsilon_0}{m^* e^2} = \left(\frac{m}{m^*}\right)\kappa a_0 = 80a_0$$

Since the nearest-neighbor separation in germanium is slightly less than $5a_0$, the orbit of the bound electron encompasses about a thousand lattice sites, and the assumption that this electron senses the periodic lattice potential is well justified.

*12.A APPENDIX: THE BLOCH-FLOQUET THEOREM

We present here a proof of the Bloch-Floquet theorem for a one-dimensional periodic system. As before, we use the so-called periodic boundary condition; that is, we imagine the one-dimensional chain of N atoms as a large circle that closes upon itself.

Since $\mathcal{H}(x + a) = \mathcal{H}(x)$, it follows that

$$\psi(x + a) = C\psi(x) \tag{12.A1}$$

where C is some constant. Then $\psi(x + 2a) = C\psi(x + a) = C^2\psi(x)$ and

$$\psi(x + Na) = C^N\psi(x) \tag{12.A2}$$

The requirement that the eigenfunction be single-valued imposes the condition

$$C^N = 1 \qquad \text{or} \qquad C = e^{i2\pi n/N} = e^{ika} \tag{12.A3}$$

where $k = 2\pi n/Na$ and n is an integer. If we now write

$$\psi(x) = u_k(x)e^{ikx} \tag{12.A4}$$

then

$$\psi(x + a) = u_k(x + a)e^{ik(x+a)} = Cu_k(x + a)e^{ikx}$$

and we see that Equation (12.A1) can be satisfied only if $u_k(x + a) = u_k(x)$, that is, if $u_k(x)$ is periodic with the periodicity of the lattice.

*12.B APPENDIX: SOLUTION OF THE ONE-DIMENSIONAL SCHRÖDINGER EQUATION IN THE TIGHT-BINDING APPROXIMATION

We recall that when two hydrogen atoms are brought together to form the H_2 molecule, the wave functions of the electrons can be constructed from linear combinations of atomic wave functions. The molecular wave functions, which must satisfy the Pauli exclusion principle, result in four eigenstates: the bonding state, for which $S = 0$, and three degenerate antibonding states with $S = 1$. In the tight-binding approximation one also constructs the wave function of the valence electrons of the giant molecule (the crystal) from a linear combination of atomic wave functions. Accordingly, we write

$$\psi_k(x) = \sum_n C_k^n \phi(x - x_n) \tag{12.B1}$$

Here k is the quantum number of the crystal eigenfunction, the C_k^n are the coefficients of the expansion, $\phi(x - x_n)$ is an atomic wave function centered on an atom at lattice point x_n, and the summation extends over all N lattice points.

It can be shown that the function in Equation (12.B1) is of the Bloch form only if the coefficients C_k^n are

$$C_k^n = e^{ikx_n} \tag{12.B2}$$

Thus

$$\psi_k(x) = \sum_n \phi(x - x_n)e^{ikx_n} \tag{12.B3}$$

The energy $E(k)$ of the state k is given by the solution of Equation (12.19), where $V(x)$ is roughly of the form shown in Figure 12.27. We now substitute Equation (12.B3) for $\psi_k(x)$ and write

$$\mathcal{H}\sum_n \phi(x - x_n)e^{ikx_n} = E(k)\psi_k(x) \tag{12.B4}$$

To solve Equation (12.B4) one multiplies both sides by $\psi_k^*(x)$ and integrates over x. The right-hand side is just

$$E(k)\int \psi_k^*(x)\psi_k(x) = NE(k) \tag{12.B5}$$

provided that the atomic wave functions $\phi(x - x_n)$ are normalized and do not overlap significantly; that is, we assume that

$$\int \phi^*(x - x_n)\phi(x - x_m) = \begin{cases} 0 & \text{if } n \neq m \\ 1 & \text{if } n = m \end{cases} \tag{12.B6}$$

The left-hand side of Equation (12.B4) becomes

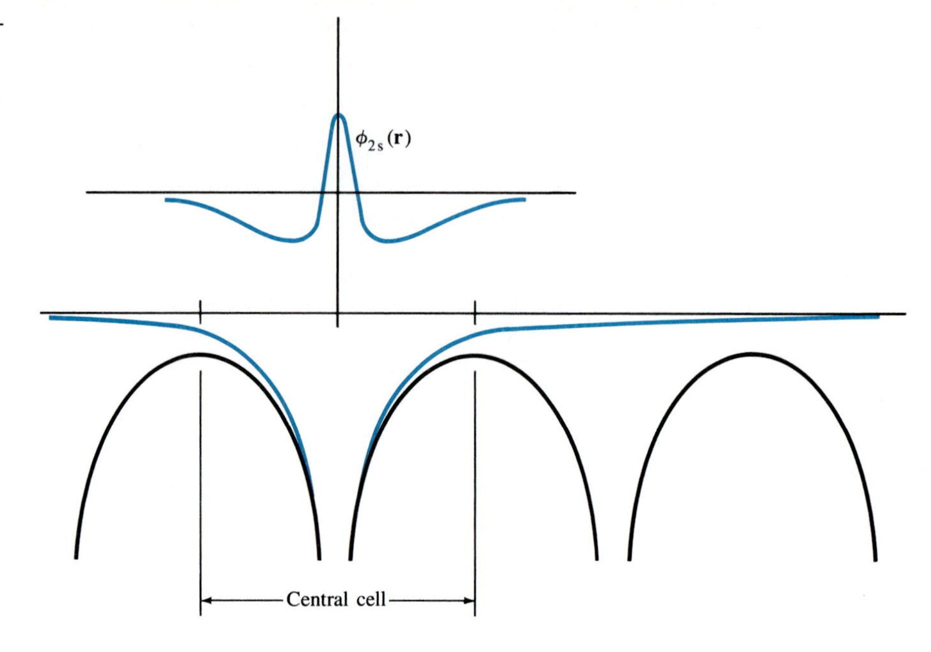

Figure 12.27 Potentials in the tight-binding approximation. *Black curve,* crystal potential; *colored curve,* atomic potential. Also shown, schematically, a 2s-type wave function centered at the origin.

$$\sum_n \sum_m e^{ik(x_n - x_m)} \int \phi^*(x - x_m)\mathcal{H}\phi(x - x_n)\, dx = \sum_n \sum_m \mathcal{H}_{nm} e^{ik(x_n - x_m)} \quad \textbf{(12.B7)}$$

We have taken the exponentials outside the integral since the integration extends over the variable x, not over the lattice points x_n, x_m, and have replaced the integral itself by the symbol $\mathcal{H}_{nm}$. Note that $\mathcal{H}_{nm}$ involves *atomic* wave functions centered at x_n and x_m. These wave functions decay to zero exponentially beyond the atomic radius, which is about the same as the nearest-neighbor distance a. Consequently, the only terms in the summation on the right-hand side of Equation (12.B7) that are likely to be of significant magnitude are those for which $x_n = x_m$ and $x_n = x_m \pm a$. For $x_n = x_m$,

$$\mathcal{H}_{mm} = \int \phi^*(x - x_m)\mathcal{H}\phi(x - x_m)\, dx = E_0 \quad \textbf{(12.B8)}$$

where E_0 is very nearly equal to the eigenvalue of the atomic energy level for which $\phi(x)$ is the eigenfunction, because in the vicinity of x_m the periodic lattice potential is almost equal to the potential of the ion core.

For $x_n = x_m \pm a$, we write

$$\int \phi^*(x - x_m \pm a)\mathcal{H}\phi(x - x_m) = -E_1 \quad \textbf{(12.B9)}$$

Thus the summation in Equation (12.B7) reduces to

$$\sum_n E_0 - E_1(e^{ika} + e^{-ika}) = N(E_0 - 2E_1 \cos ka)$$

and we obtain

$$E(k) = E_0 - 2E_1 \cos ka \quad \textbf{(12.B10)}$$

where E_0 is the energy of the electron in the free atom and E_1 is an *overlap integral,* a constant that depends on details of the periodic potential and on the amount of overlap of atomic wave functions at neighboring lattice sites.

SUMMARY

Broadly speaking, crystalline solids fall into one of four categories. In order of diminishing cohesive energy, these are ionic, covalent, metallic, and molecular crystals. Crystal structure is usually determined by means of X-ray diffraction. Diffraction peaks appear when the *Bragg condition*

$$2d \sin \theta = n\lambda$$

is satisfied.

All crystals contain some imperfections, classified according to their geometry as point defects (impurities, vacancies, interstitials), line defects (edge and screw dislocations), and planar defects (grain boundaries, stacking faults, external surfaces).

In a crystalline solid the stationary states of atomic electrons broaden into energy bands. In the tight-binding approximation the bandwidth is a function of the crystal potential and the degree of overlap between atomic wave functions of electrons on neighboring lattice sites. For most metallic crystals the band structure is often calculated using the nearly free-electron approximation, in which the crystal potential is treated as a small perturbation. Whatever approximate method may be employed, the electron wave function in a crystal is always of the *Bloch form*

$$\psi_{\mathbf{k}}(\mathbf{r}) = u_{\mathbf{k}}(\mathbf{r})e^{i\mathbf{k}\cdot\mathbf{r}}$$

where $u_{\mathbf{k}}(\mathbf{r})$ is a function with the periodicity of the lattice. Near $\mathbf{k} = 0$ and also near an extremum of the energy, the electron energy is given by

$$E(k) = \frac{\hbar^2 k^2}{2m^*}$$

where m^* is the *effective mass* of the electron. Near the top of an energy band m^* is negative, and it is then conventional practice to employ the concept of *holes,* particles that carry a positive charge e and whose effective mass is positive.

Metallic crystals have partially filled energy bands; insulators have filled energy bands. Pure semiconductors are essentially insulators with relatively narrow energy gaps between the highest state of the filled (valence) band and the next higher (conduction) band. At moderate temperatures thermal agitation can induce transitions of electrons from the top of the valence band to the bottom of the conduction band. Impurities in semiconductors can also serve as donors or acceptors, donors contributing electrons to the conduction band and acceptors removing electrons from the valence band and thereby creating holes in the valence band. Electronic properties (electronic specific heat, conductivity, Hall effect) of metals and semiconductors are the subject of the following chapter.

Bibliography

F. Agullo-Lopez, C. R. A. Catlow, and P. D. Townsend, *Point Defects in Materials,* Academic Press, New York, 1988.

F. C. Brown, *The Physics of Solids,* W. A. Benjamin, New York, 1967.

J. Christman, *Solid State Physics,* Wiley, New York, 1986.

A. H. Cottrell, *Dislocations and Plastic Flow in Crystals,* Oxford University Press, New York, 1953.

C. Kittel, *Introduction to Solid State Physics,* 6th ed., Wiley, New York, 1984.

W. T. Read, *Dislocations in Crystals,* McGraw-Hill, New York, 1953.

PROBLEMS

12.1 Lithium fluoride, LiF, crystallizes in the NaCl structure. The nearest-neighbor distance in LiF is 0.2014 nm. Calculate the cohesive energy of LiF. Assume that the exponent $n = 9$.

12.2 The experimental value for the cohesive energy of LiF is 242 kcal/mol. Use the data of Problem 12.1 to determine the exponent n.

12.3 The lattice parameter of the simple cubic cell of CsCl is 0.411 nm. Calculate the cohesive energy of this crystal.

12.4 Derive Equations (12.5) and (12.6).

12.5 Consider the one-dimensional ionic lattice of Figure 12.28 consisting of doubly charged positive ions and twice as many singly charged negative ions. Determine the Madelung constant for this chain to within 1 percent.

Figure 12.28 Problem 12.5.

12.6 A beam of white X rays is incident on a cubic crystal as shown in Figure 12.29. At what angles would one observe reflections from the planes labeled (110), (120), (130), and (310)? If the lattice parameter of the crystal is 0.3 nm, what are the wavelengths of the reflected beams?

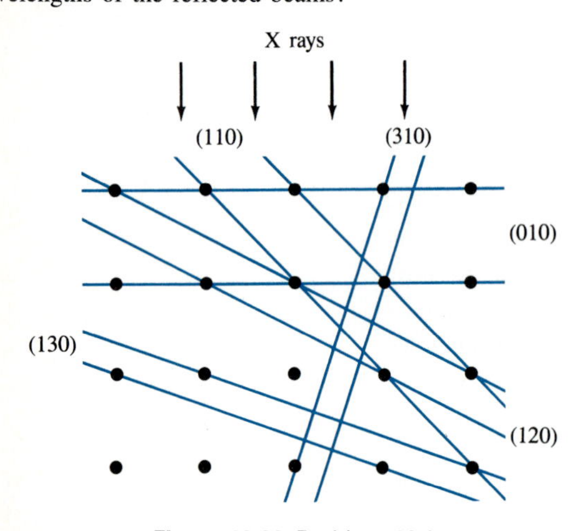

Figure 12.29 Problem 12.6.

12.7 A beam of white X rays is incident on a cubic crystal along a face diagonal. Strong reflections are observed at 90° to the incident beam, and the wavelength of the reflected X rays is 0.2 nm. Determine the lattice parameter of the cubic crystal.

12.8 The formation energies of vacancies and interstitials in copper are 1.5 eV and 4.5 eV, respectively. Calculate the concentration of these defects at room temperature and at 1350 K (a few degrees below the melting point).

12.9 Derive an expression for the molar specific heat of a crystal due to vacancy formation. How large is this contribution compared with the Dulong-Petit value of $3R$ in copper at room temperature and near the melting point?

12.10 Derive an expression for the force that acts on an ion in a monovalent ionic crystal. Express your answer in terms of the Madelung constant α, the equilibrium nearest-neighbor separation a, and the exponent n.

12.11 Show that the restoring force that acts on an ion that is displaced a small distance x from its equilibrium position is given by $F = -kx$, where

$$k = \frac{\alpha e^2}{4\pi\epsilon_0 a^3}(n + 1)$$

12.12 Use the result of Problem 12.11 to calculate the vibration frequencies of Li^+ in LiF and of Na^+ in NaCl. What are the wavelengths of the electromagnetic radiation that correspond to these frequencies?

12.13 Show that the wave function given by Equation (12.B3) is indeed of the Bloch form.

12.14 Consider a two-dimensional square lattice. Follow the procedure employed in Section 12.B to show that the energy is given by

$$E(\mathbf{k}) = E_0 - 2E_1(\cos k_x a + \cos k_y a)$$

and that in this case m^* is isotropic.

12.15 Consider a two-dimensional rectangular lattice with lattice parameters a and b in the x and y directions, respectively. Show that the energy is given by

$$E(\mathbf{k}) = E_0 - 2E_{1x} \cos k_x a - 2E_{1y} \cos k_y b$$

and that the effective mass m^* is now anisotropic. Suppose that the overlap integrals in the x and y directions are related by $(E_{1x}/E_{1y}) = (b/a)^m$. For what value of m is the effective mass isotropic? Is this likely to occur in a real crystal? Explain your answer.

12.16 Use the hydrogenic model to calculate approximate values of the ionization energies of donors in silicon. The effective mass of electrons in silicon is $0.33m$, and the dielectric constant is 12. (For comparison, see Table 13.4)

Chapter 13 Solid-State Physics II: Electronic Properties

In Ogg's theory it was his intent
That the current keep flowing, once sent;
So to save himself trouble,
He put them in double,
And instead of stopping, it went.[1]

George Gamow

13.1 INTRODUCTION

Now that we have seen how the band theory of solids leads to an energy versus wave vector relation similar to that of free electrons, at least over a portion of the $E(k)$ curve, we can examine the electronic properties of crystals using a free-electron approach, provided we replace the free-electron mass by the effective mass m^*. We shall begin by considering the properties of metals. We then turn our attention to semiconductors and semiconductor devices, and in the last section summarize the salient features of superconductors.

13.2 METALS

We imagine the metal to be a box filled with N electrons per unit volume and a homogeneous distribution of positive charge such that the entire system is electrically neutral. We shall account for the periodic lattice potential by assigning to the electron an effective mass m^* that may differ from the free-electron mass m. The earliest theory of metallic conduction, the Drude theory, was based on this model (though it did not allow for an effective mass). However, prior to the formulation of the Pauli principle, it was thought that the electron gas was like an ordinary classical gas of particles, and this presumption led to a puzzling discrepancy between theory and experiment.

13.2(a) The Electronic Specific Heat

If the conduction electrons of a metal constitute a classical gas, their contribution to the specific heat should be $\frac{1}{2}k$ J/K per degree of freedom per electron, or $3(\frac{1}{2}N_A k) = \frac{3}{2}R$ per mole. Since the molar heat capacity of nonconducting solids is $3R$ at room

---

1. R. A. Ogg, an experimental chemist, was the first to suggest that electron pairs might exhibit Bose-Einstein condensation, and that this could explain the phenomenon of superconductivity.

temperature, that of monovalent metals should be $3R + \frac{3}{2}R = \frac{9}{2}R$, and polyvalent metals should exhibit an even larger molar specific heat. Yet, the measured heat capacity of metals gave no evidence of this additional electronic contribution but agreed with the Dulong-Petit value of $3R$. It appeared that, at least insofar as their participation in thermal excitation was concerned, the supposedly free electrons were not free at all.

The reason the electrons of a metal do not contribute significantly to the heat capacity is that they obey the exclusion principle and their distribution is not the classical Maxwell-Boltzmann but the Fermi-Dirac distribution. As we have seen (Section 11.3), at metallic electron densities the Fermi energy E_F at $T = 0$ K is of the order of 5 eV. At $T = 0$ K, the Fermi distribution is a step function, $f_{FD}(E) = 1$ for $E < E_F$ and $f_{FD}(E) = 0$ for $E > E_F$ (see Figure 11.4).

What happens, then, as we raise the temperature? For $T > 0$ K, with the electron gas in thermal equilibrium with the lattice, some electrons are indeed excited into higher energy states. However, since the thermal excitation energy is of order kT, only those electrons within an energy range of about kT of the Fermi energy can gain enough energy from the lattice to make a transition to an unoccupied state of higher energy. Electrons well below E_F cannot make transitions to higher energy states because states accessible to them are already occupied. For $T > 0$ K the Fermi distribution is no longer a step function, but the energy range over which the probability of occupancy falls from 1 to 0 is only of the order of $2kT \ll E_F$ (see Figure 13.1).

We can now understand why the quasi-free electrons of the metal do not contribute significantly to the specific heat, and can estimate the contribution they do make. The number of electrons that can be thermally excited is the integral of the density of states over the energy interval in which the distribution function changes from 1 to 0. Since that interval is small compared to E_F, we may approximate the integral by the product of the density of states at E_F and the energy interval $2kT$, that is, by $2kT\mathcal{N}(E_F)$. If all N electrons contributed to the specific heat, as in a classical gas, the heat molar capacity would be $\frac{3}{2}R$. However, only the fraction $2kT\mathcal{N}(F_F)/N$

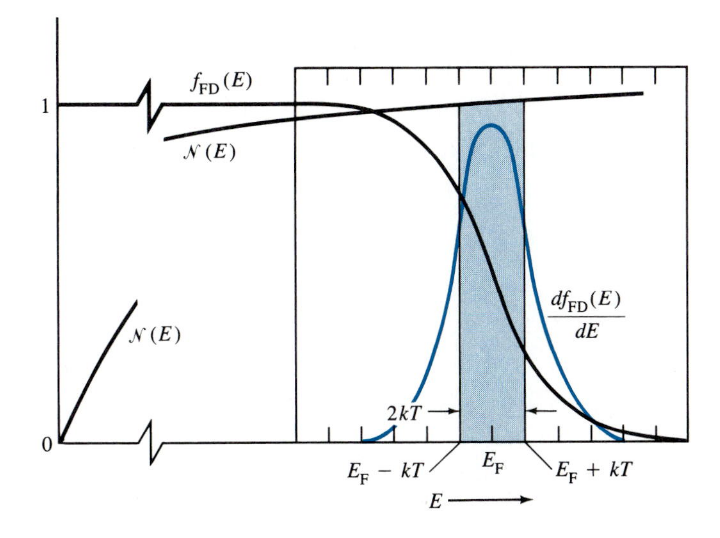

Figure 13.1 The electrons that can contribute to the heat capacity are in that energy interval in which the distribution function $f_{FD}(E)$ changes from 1 to 0, i.e., in the energy region in which $df_{FD}(E)/dE$ is significant. Since the density of states, $\mathcal{N}(E)$, is nearly constant at $E = E_F$, the number of contributing electrons is the product of $\mathcal{N}(E_F)$ and the integral under the curve $df(E)/dE$. This integral may be approximated by the area of the rectangle of height 1 and width $2kT$.

contribute, and therefore the electronic specific heat is approximately

$$C_e = \frac{2kT\mathcal{N}(E_F)}{N}\left(\frac{3}{2}\right)R = \frac{9kT}{2E_F}R$$

The correct expression for C_e is almost the same as this approximate result, namely,

$$C_e = \frac{\pi^2 kT}{2E_F}R \qquad \textbf{(13.1)}$$

Near room temperature, $kT/E_F \simeq \frac{1}{200}$ and the electronic specific heat is an almost negligible fraction of the total heat capacity of the metal. At very low temperatures, however, it can be the dominant term. We recall that well below the Debye temperature the heat capacity due to lattice vibrations, the ''lattice'' heat capacity C_g, varies as T^3 (see Section 4.3). The electronic contribution is proportional to T. It follows that at sufficiently low temperatures, usually below about 1 K, the electronic term will dominate. At such low temperatures the total heat capacity is

$$C = C_e + C_g = \gamma T + BT^3 \qquad \textbf{(13.2)}$$

$$\gamma = \frac{\pi^2 kR}{2E_F} \qquad B = \frac{12\pi^4 R}{5\Theta_D^3}$$

Therefore, if one plots C/T versus T^2, one should obtain a straight line whose slope is a measure of the Debye temperature Θ_D and whose intercept is $\pi^2 kR/2E_F$. Specific heat measurements are frequently employed to determine E_F and, thereby, the effective mass m^*. A typical plot of C/T versus T^2 is shown in Figure 13.2.

EXAMPLE 13.1 From the data of Figure 13.2 determine the effective mass of electrons in potassium. At what temperature are the electronic and lattice contributions equal in this metal?

SOLUTION According to Figure 13.2,

$$\frac{C}{T} = 2.08 \times 10^{-3} + 2.57 \times 10^{-3}T^2 \text{ J/mol·K}$$

The Fermi energy is, therefore,

$$E_F = \frac{\pi^2 Rk}{2(2.08 \times 10^{-3})} = \frac{\pi^2(8.31)(1.38 \times 10^{-23})}{2(2.08 \times 10^{-3})} = 2.72 \times 10^{-19} \text{ J} = 1.7 \text{ eV}$$

The density of conduction electrons in this monovalent metal is

$$N = \frac{(6.02 \times 10^{23} \text{ mol}^{-1})(0.87 \text{ g/cm}^3)(10^6 \text{ cm}^3/\text{m}^3)}{39.1 \text{ g/mol}} = 1.34 \times 10^{28} \text{ m}^{-3}$$

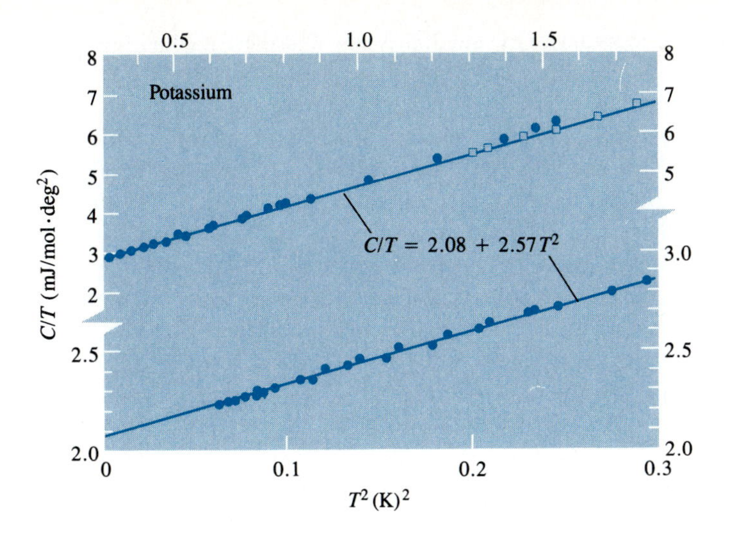

Figure 13.2 The heat capacity of potassium at low temperatures, 0.25 K $< T <$ 1.35 K. C/T is plotted against T^2. The result is a straight line whose intercept is a measure of the electronic contribution and whose slope is proportional to the lattice contribution to the heat capacity. (After W. H. Lien and N. E. Phillips, *Phys. Rev.* **133,** A1370 (1964).)

From Equation (11.23) we have

$$m^* = \frac{\hbar^2}{2E_F}(3\pi^2N)^{2/3} = \frac{(1.05 \times 10^{-34})^2}{2(2.72 \times 10^{-19})}(3\pi^2 \times 1.34 \times 10^{28})^{2/3}$$
$$= 1.09 \times 10^{-30}\text{ kg} = 1.20m$$

The two contributions to the specific heat are equal when

$$T = \left(\frac{2.08}{2.57}\right)^{1/2} = 0.9\text{ K}$$

13.2(b) Electrical Conductivity

The state of a free electron is characterized by its wave vector $\mathbf{k}$. Its plane-wave eigenfunction is also an eigenfunction of the momentum operator with the eigenvalue

$$\mathbf{p} = \hbar\mathbf{k} \tag{13.3}$$

One can show that if a force $\mathbf{F}$ acts on a free electron, its wave vector changes with time according to Newton's second law; that is,

$$\mathbf{F} = \frac{d\mathbf{p}}{dt} = \hbar\frac{d\mathbf{k}}{dt} \tag{13.4}$$

To good approximation, Equation (13.4) remains valid also for electrons in a crystal whose wave function is of the Bloch form.

If we subject a Fermi distribution of electrons in a metal to an electric field $\mathbf{E}$, the wave vector of each electron will develop in time according to Equation (13.4), with $\mathbf{F} = -e\mathbf{E}$. After some short time Δt has elapsed, the initial Fermi distribution f_0, which is spherically symmetric in $\mathbf{k}$-space and centered at $\mathbf{k} = 0$, will have shifted bodily in the direction of $\mathbf{F}$ and will now be centered at $\mathbf{k}' = -(e\mathbf{E}/\hbar)\,\Delta t$. If there were no mechanism by which this new distribution could lose the momentum gained from the electric field in time Δt, that distribution would retain its form indefinitely even if at $t = \Delta t$ we were to remove the field **E**. In a normal metal, however, there are various defects that scatter electrons from an initial state $\mathbf{k}_i$ to some other final state $\mathbf{k}_f$. It is such scattering processes that allow a disturbed distribution to relax toward the equilibrium distribution f_0. The rate at which this relaxation takes place is generally denoted by $w = 1/\tau$, where τ is the *relaxation time*. The various scattering processes establish a balance between the disturbing effect of an applied electric field and the restoring action of relaxation so that in a constant electric field, the steady-state distribution has the form shown in Figure 13.3. That is, the distribution is a spherically symmetric Fermi distribution centered, not at $\mathbf{k} = 0$, but at $\mathbf{k} = -e\mathbf{E}\tau/\hbar$.

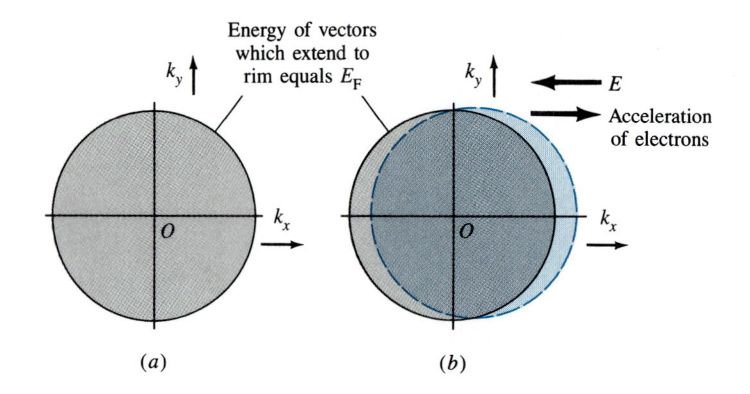

Figure 13.3 Schematic representation of (*a*) the equilibrium and (*b*) the steady-state distribution in a constant electrical field of a degenerate electron gas in a metal. The steady-state distribution is displaced in **k**-space by $-e\mathbf{E}\tau/\hbar$, where τ is the relaxation time.

Suppose an external electric field $\mathbf{E} = (E_x, 0,0)$ is applied to a metal. The resulting current density in the metal is now readily calculated. The current carried by one electron is

$$j_x = -ev_x = -\frac{e\hbar k_x}{m^*} \tag{13.5}$$

and the total current density for the entire distribution of N electrons per unit volume is just the sum of these individual contributions. At equilibrium, the total current vanishes because the distribution f_0 is spherically symmetric about $\mathbf{k} = 0$; for every electron with positive k_x there is one with equal negative k_x. The disturbed distribution is such that every electron has its wave vector augmented by $k_x = -eE_x\tau/\hbar$. Consequently, the total current density is just

$$J_x = N\left(-\frac{e\hbar}{m^*}\right)\left(-\frac{eE_x\tau}{\hbar}\right) = \frac{Ne^2\tau}{m^*}E_x \tag{13.6}$$

The electrical conductivity and resistivity are defined by

$$\sigma = \frac{J}{E} \quad \text{and} \quad \rho = \frac{1}{\sigma} \tag{13.7}$$

Thus,

$$\sigma = \frac{Ne^2\tau}{m^*} \qquad \rho = \frac{m^*}{Ne^2\tau} \tag{13.8}$$

Since N, e, and m^* are constant for a given substance, the observed dependence of the resistivity on temperature and crystal perfection must be contained in the parameter τ, the relaxation time. The reason scattering takes place at all is that the metal is never an ideal crystal, devoid of defects. Stationary defects, such as impurity atoms at substitutional or interstitial sites, scatter electrons because their ionic potentials differ from that of the host ions. The greater the defect concentration, the greater will be the scattering rate and the shorter the relaxation time τ. For dilute alloys we would expect that the scattering rate w will be proportional to solute concentration c, and that the resistivity will increase linearly with c. That is indeed so.

Stationary defects are, however, not the only kind that scatter electrons. The perfect periodic pattern of the ideal crystal lattice is also destroyed by the thermal vibrations of the ions about their normal lattice sites. The relaxation rate due to scattering by *phonons,* the quanta of lattice vibrations, is proportional to the number of phonons present, and that, in turn, depends on temperature. Since the energy of each phonon is $h\nu_{\text{ph}}$, where ν_{ph} is the frequency of the lattice vibration, the number of phonons will be proportional to the total lattice thermal energy, that is, to the integral of the lattice specific heat. At very low temperatures, $C_g = BT^3$, so that $w = 1/\tau$ would vary at T^4; a more detailed calculation suggests that at low tempera-

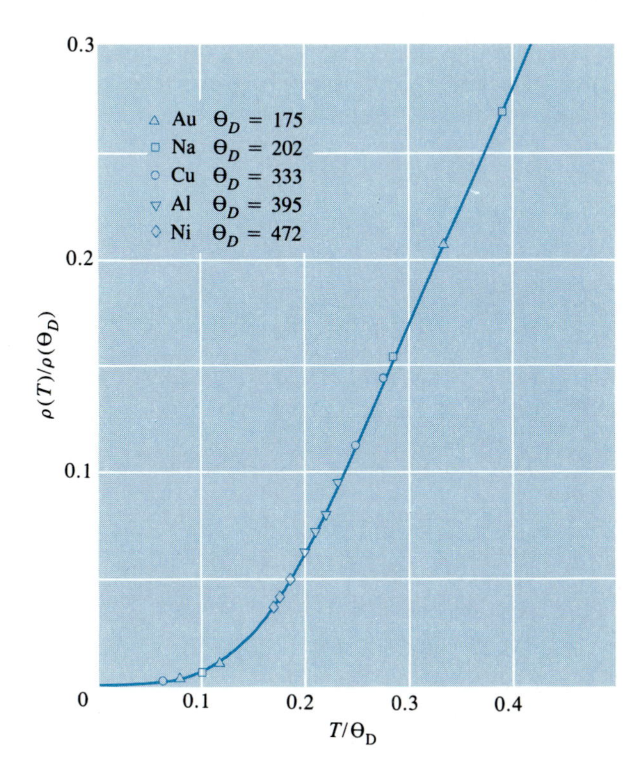

Figure 13.4 The resistivities of several pure metals as functions of temperature. Here the ratio $\rho(T)/\rho(\Theta_D)$ is plotted against T/Θ_D. At low temperature the resistivity increases rapidly with temperature but at higher temperatures becomes a linear function of T/Θ_D.

tures ($T \ll \Theta_D$), the *ideal resistivity* attributable to lattice vibrations is

$$\rho_i(T) \simeq 500\rho_i(\Theta_D)\left(\frac{T}{\Theta_D}\right)^5$$

where $\rho_i(\Theta_D)$ is the ideal resistivity at the Debye temperature.

At room temperature and above, $C_g = 3R$, a constant. It follows that the ideal resistivity should then increase linearly with temperature, and that is indeed what one observes.

Generally it is not just one scattering mechanism that results in relaxation, although a particular one may be dominant. To good approximation the various scattering processes are independent; the presence of a few impurities does not alter the lattice vibration spectrum significantly, and the fact that an impurity participates in the vibrational motion of the lattice does not greatly affect its effectiveness in scattering conduction electrons. The total relaxation rate is then

$$w = \frac{1}{\tau} = \frac{1}{\tau_{\text{ph}}} + \frac{1}{\tau_{\text{im}}} + \frac{1}{\tau_{\text{def}}} \tag{13.9}$$

where τ_{ph}, τ_{im}, and τ_{def} are the relaxation times due to scattering of electrons by phonons, impurities, and lattice defects (vacancies, dislocations, etc.) acting independently. It now follows from Equation (13.8) that the total resistivity of a sample can be expressed as a sum of three separate contributions, i.e.,

$$\rho = \rho_i + \rho_{\text{im}} + \rho_{\text{def}} = \rho_i + \rho_{\text{res}} \tag{13.10}$$

Here ρ_i is the ideal resistivity of the material attributable to electron-phonon scattering only, and ρ_{res} is the *residual resistivity* of the sample that persists even as $T \to 0$ K. The residual resistivity of a sample is an excellent indicator of its crystalline perfection, and one finds "residual resistivity ratios" [$\rho(273\text{ K})/\rho(4.2\text{ K})$]

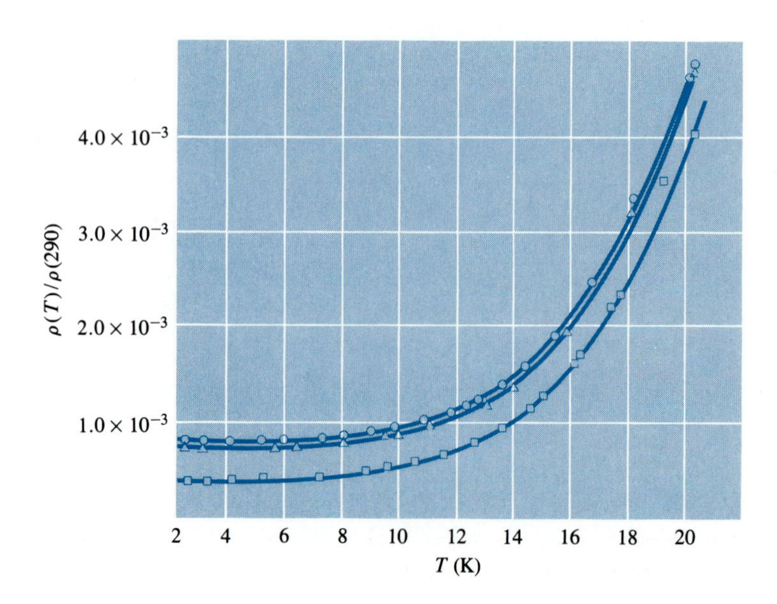

Figure 13.5 The resistivity of three samples of sodium of different purity at low temperature. The three curves are identical but displaced along the ordinate. In other words, $\rho(T) = \rho_i(T) + \rho_{res}$, where ρ_{res} is the temperature-independent contribution due to stationary defects.

quoted in the literature. For quite pure single crystals, residual resistivity ratios in excess of 10^4 can be attained.

13.2(c) Thermal Conductivity; the Wiedemann-Franz Law

The thermal conductivity κ is defined by the relation

$$\mathbf{J}_Q = -\kappa\, \nabla T \tag{13.11}$$

where $\mathbf{J}_Q$ is the thermal energy flux. For a classical gas, the result of kinetic theory is

$$\kappa = \tfrac{1}{3} C_v \langle v \rangle \ell \tag{13.12}$$

Here C_v is the heat capacity per unit volume, $\langle v \rangle$ is the average speed of the particles of the gas, and $\ell = \langle v \rangle \tau$ is their *mean free path,* the average distance traversed by a gas particle between collisions.

In metals, lattice vibrations (phonons) and conduction electrons contribute to thermal conductivity. In pure metals, the phonon contribution, the only effective mechanism in insulators, is usually negligible compared with the electronic contribution.

We can make use of Equation (13.12) to determine the thermal conductivity due to electrons in a metal provided we use the electronic specific heat as obtained in Section 13.2(a) and replace the average speed $\langle v \rangle$ by the speed of those electrons of the Fermi distribution that contribute to the heat capacity. We then obtain

$$\kappa = \frac{N\pi^2 k^2 T \tau}{3m^*} \tag{13.13}$$

If we form the ratio $L = \kappa/\sigma T$, known as the *Wiedemann-Franz* ratio, we find

$$L = \frac{\pi^2 k^2}{3e^2} = 2.45 \times 10^{-8}\ \text{W}\cdot\Omega/\text{K}^2 \tag{13.14}$$

Equation (13.14) predicts that the Wiedemann-Franz ratio is independent of temperature and is the same for all metals. Table 13.1 lists the Wiedemann-Franz ratio for a number of metals, and we see that it is indeed very nearly equal to 2.45×10^{-8} W·Ω/K^2. Since in any solid some thermal energy is carried by lattice vibrations, we see that in a metal this contribution must be quite small, as indeed it is. If phonons did contribute significantly to the thermal conductivity, the Wiedemann-Franz ratio would be greater than predicted by Equation (13.14).

TABLE 13.1 The Ratio $\kappa/\sigma T$ for a Number of Metals at 273 K

Metal	$\kappa/\sigma T$ (10^{-8} W·Ω/K^2)
Cu	2.23
Ag	2.31
Au	2.35
Zn	2.31
Cd	2.42
Sn	2.52
Mo	2.61
Pb	2.47
Pt	2.51

13.3 SEMICONDUCTORS

In semiconductors, as in metals, the departure of the electron distribution from equilibrium under the action of a steady electric field is limited by the relaxation time τ. Conduction in semiconductors, however, differs from that in metals in two important respects. In a metal, the number of conduction electrons is constant, independent of temperature and, to good approximation, also independent of impurity concentration. Second, the number of conduction electrons in a metal is so large that even at elevated temperatures the Fermi distribution function departs only very slightly from the step function at $T = 0$ K.

In an intrinsic semiconductor, on the other hand, the density of charge carriers, electrons and holes, is a sensitive function of temperature. Moreover, the number of electrons thermally excited across the energy gap is so small compared to the density of states in the valence and conduction bands that the probability that a given band state is occupied by more than one electron or hole is insignificant. Consequently, the Pauli principle no longer plays a crucial role, and the distribution function is now practically indistinguishable from the classical Maxwell-Boltzmann distribution (see Section 11.3). Even in extrinsic semiconductors, the doping levels are generally so small that the classical distribution remains an excellent approximation.

To support these qualitative statements, let us calculate the electron and hole equilibrium concentrations for a typical intrinsic semiconductor. We assume that the valence and conduction bands are of standard form, that is,

$$E(k) = \frac{\hbar^2 k^2}{2m^*} \tag{13.15}$$

where m^* is positive for electrons in the conduction band and negative for holes in the valence band.

The electron concentration is given by

$$n = \int_{E_c}^{\infty} \mathcal{N}_c(E) f_0(E)\, dE \tag{13.16}$$

where $f_0(E)$ is the Fermi distribution, $\mathcal{N}_c(E)$ is the density of states in the conduction band, and E_c is the energy at the conduction band edge (see Figure 13.6). The number of holes in the valence band is given by a similar expression, except that we must now multiply the appropriate density of valence band states by the probability that a particular state is unoccupied. Hence,

$$p = \int_{0}^{-\infty} \mathcal{N}_v(E)[1 - f_0(E)]\, dE \tag{13.17}$$

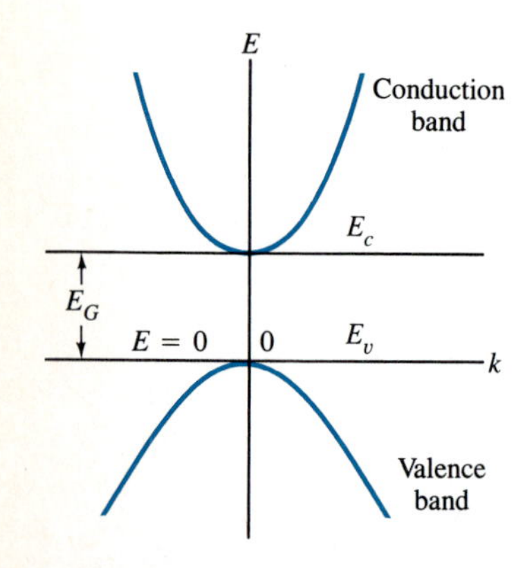

Figure 13.6 The idealized band structure of a semiconductor.

In writing the expressions for n and p we have chosen as our zero of energy the top of the valence band so that $E_c = E_G$, where E_G is the *energy gap*.

Since in the intrinsic semiconductor the excitation of an electron to the conduction band leaves a hole in the valence band,

$$n = p \tag{13.18}$$

In general, it is not possible to obtain a closed analytic solution to these equations. However, as we shall see presently, in nearly all situations of practical interest the Fermi distribution can be replaced by the Maxwell-Boltzmann distribution, and the integrals in Equations (13.16) and (13.17) are then easily evaluated.

In intrinsic semiconductors, the Fermi energy E_F falls roughly midway between the valence and conduction band edges. Since the energy gap E_G is generally between 0.2 and 1.5 eV whereas $kT \simeq 0.025$ eV (room temperature), it follows that for $E \geq E_c$, $e^{(E-E_F)/kT} \geq e^{10} \gg 1$ and

$$f_0(E) \simeq e^{-(E-E_F)/kT} \quad \textbf{(13.19)}$$

for all states in the conduction band. Similarly, for all states in the valence band,

$$1 - f_0(E) \simeq e^{-(E_F-E)/kT} \quad \textbf{(13.20)}$$

With these approximate expressions we can now evaluate n and p, with the results

$$n = 2\left(\frac{m_n kT}{2\pi\hbar^2}\right)^{3/2} e^{(E_F-E_c)/kT} \quad \textbf{(13.21)}$$

$$p = 2\left(\frac{m_p kT}{2\pi\hbar^2}\right)^{3/2} e^{-E_F/kT} \quad \textbf{(13.22)}$$

where m_n and m_p are the electron and hole effective masses. Setting $n = p$, one can now solve for E_F and n (or p) and finds

$$E_F = \frac{1}{2}E_G + \frac{3}{4}kT \ln \frac{m_p}{m_n} \quad \textbf{(13.23)}$$

$$n = p = 2\left(\frac{kT}{2\pi\hbar^2}\right)^{3/2} (m_n m_p)^{3/4} e^{-E_G/2kT} \quad \textbf{(13.24)}$$

Unless $m_p m_n \gg 1$ or $\ll 1$ *and* kT is comparable to E_G, the Fermi energy will indeed be at or very near the center of the band gap. Therefore, provided $E_G \gg kT$, as is usually true, the classical approximation is well justified and Equations (13.21) through (13.24) are valid. Note that in the intrinsic semiconductor the carrier density is an exponential function of the temperature, increasing with increasing temperature at a rate determined by the energy gap E_G.

In the extrinsic semiconductor, the number of electrons or holes is equal to $|N_D - N_A|$, where N_D and N_A are the number of donor and acceptor impurities, and

TABLE 13.2 Energy Gaps, Effective Masses, and Dielectric Constants for Semiconductors

Semiconductor	E_G at 273 K (eV)	Effective Mass m^*/m		Dielectric Constant
		Electrons	Holes	
Ge	0.67	0.2	0.3	16
Si	1.14	0.33	0.5	12
InSb	0.16	0.013	0.6	18
InAs	0.33	0.02	0.4	14.5
InP	1.29	0.07	0.4	14
GaSb	0.67	0.047	0.5	15
GaAs	1.39	0.072	0.5	13

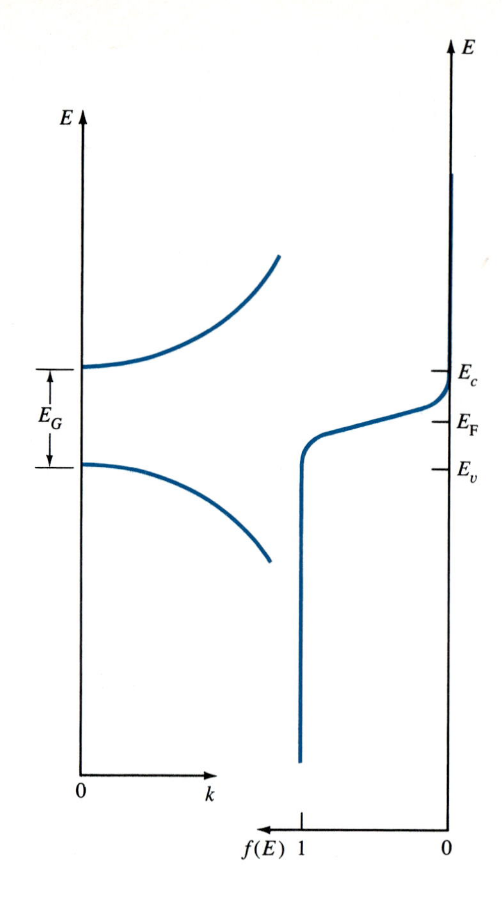

Figure 13.7 The energy bands and Fermi distribution function for an intrinsic semiconductor. Note that since $E_G \gg kT$, $f_{FD}(E) \ll 1$ for $E \geq E_c$ and $1 - f_{FD}(E) \ll 1$ for $E \leq E_v$. Consequently, the classical approximation, $f_{FD}(E) \simeq f_{MB}(E)$, can be used for electrons and holes in the conduction and valence bands, respectively.

provided that $kT \geq E_i$, where E_i is the ionization energy of the majority impurity atoms. Donor and acceptor ionization energies are typically of order 0.01 eV (see Chapter 12). Near room temperature, the carrier concentration is practically constant. At quite low temperatures, generally below that of liquid nitrogen (77 K), the carriers "freeze out" on donor or acceptor impurities and the number of charge carriers falls exponentially with temperature. The exponential factor controlling the temperature dependence is now proportional not to the energy gap but to the ionization energy of the majority doping impurity in the host lattice.

The temperature range over which the carrier concentration is nearly independent of temperature is referred to as the *saturation region*. A curve of carrier concentration as a function of temperature looks roughly as shown in Figure 13.8. Note that the abscissa is not T but $1/T$ and that the ordinate scale is logarithmic; on such a plot the exponential dependence, $e^{-E/kT}$, appears as a straight line.

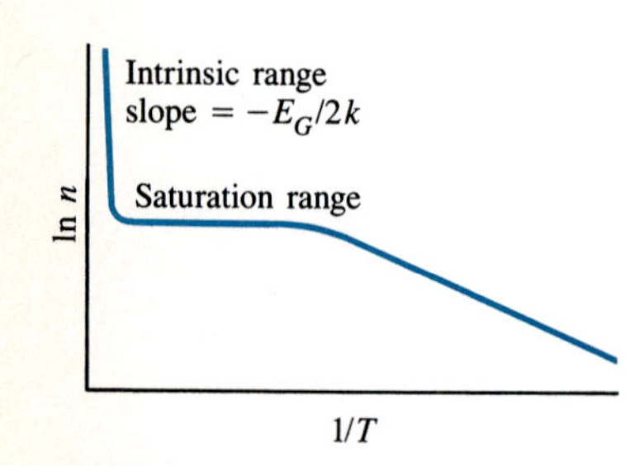

Figure 13.8 The density of electrons in an n-type semiconductor as a function of temperature (idealized schematic representation). At high temperatures ($1/T$ small), the logarithm of n is proportional to $-E_G/2kT$; that is, $n \propto e^{-E_G/kT}$. At very low temperatures n decreases exponentially with diminishing temperature as carriers "freeze out" on donor impurities. In the intermediate region, n is constant.

13.3(a) Conductivity

Because in a semiconductor the number of charge carriers depends critically on temperature and/or impurity concentration, it is more instructive to focus attention on the behavior of a single carrier, electron or hole, in the presence of an applied field than on the conductivity due to the ensemble of carriers. Under the combined influence of the field and relaxation processes, a carrier will attain an average *drift velocity*

$$\mathbf{v}_d = \frac{e\tau}{m^*}\mathbf{E} = \mu\mathbf{E} \tag{13.25}$$

where μ is the *mobility* of the charge carrier. The conductivity can then be expressed as

$$\sigma = ne\mu_n + pe\mu_p \tag{13.26}$$

where μ_n and μ_p are the electron and hole mobilities.

As in a metal, the relaxation time and, hence, the mobility of electrons and holes are temperature-dependent. However, this temperature dependence is very weak compared to the exponential variation of the carrier concentration in the low- and high-temperature regions of Figure 13.8. It is only in the saturation region that the temperature dependence of the mobility determines that of the conductivity. A typical plot of resistivity versus temperature for a semiconductor is shown in Figure 13.9. Note that only in the saturation region does the resistivity increase with increasing temperature (as in a metal).

13.3(b) The Hall Effect

Suppose we place a conducting slab in a uniform magnetic field as shown in Figure 13.10 and pass a current through the slab. The charge carriers may be electrons or holes; for the moment we shall assume that they are electrons. They will move with an average drift velocity $\mathbf{v}_d = -\mathbf{J}/ne$, where $\mathbf{J}$ is the current density and the negative sign appears because negatively charged electrons drift opposite to the direction of the current. In the presence of the magnetic field $\mathbf{B}$ these moving charges experience a Lorentz force

$$\mathbf{F}_B = -e(\mathbf{v} \times \mathbf{B}) \tag{13.27}$$

directed as shown, and will be deflected toward the lower edge of the slab. Since there is no external conducting path connecting the top and bottom surfaces, electrons accumulate on the lower surface, leaving the upper surface with a net positive charge. This charge separation creates an internal electric field, the so-called *Hall field* $\mathbf{E}_\mathrm{H}$, which exerts an electrostatic force on the electrons that opposes the Lorentz force. When

$$\mathbf{F}_B + \mathbf{F}_\mathrm{H} = -e(\mathbf{v} \times \mathbf{B}) - e\mathbf{E}_\mathrm{H} = 0$$

no further charge separation takes place. The magnitude of the Hall field is, therefore,

$$E_\mathrm{H} = v_d B$$

and the potential difference between top and bottom surfaces is

$$V_\mathrm{H} = E_\mathrm{H} d = v_d B d = \left(\frac{J}{ne}\right) Bd$$

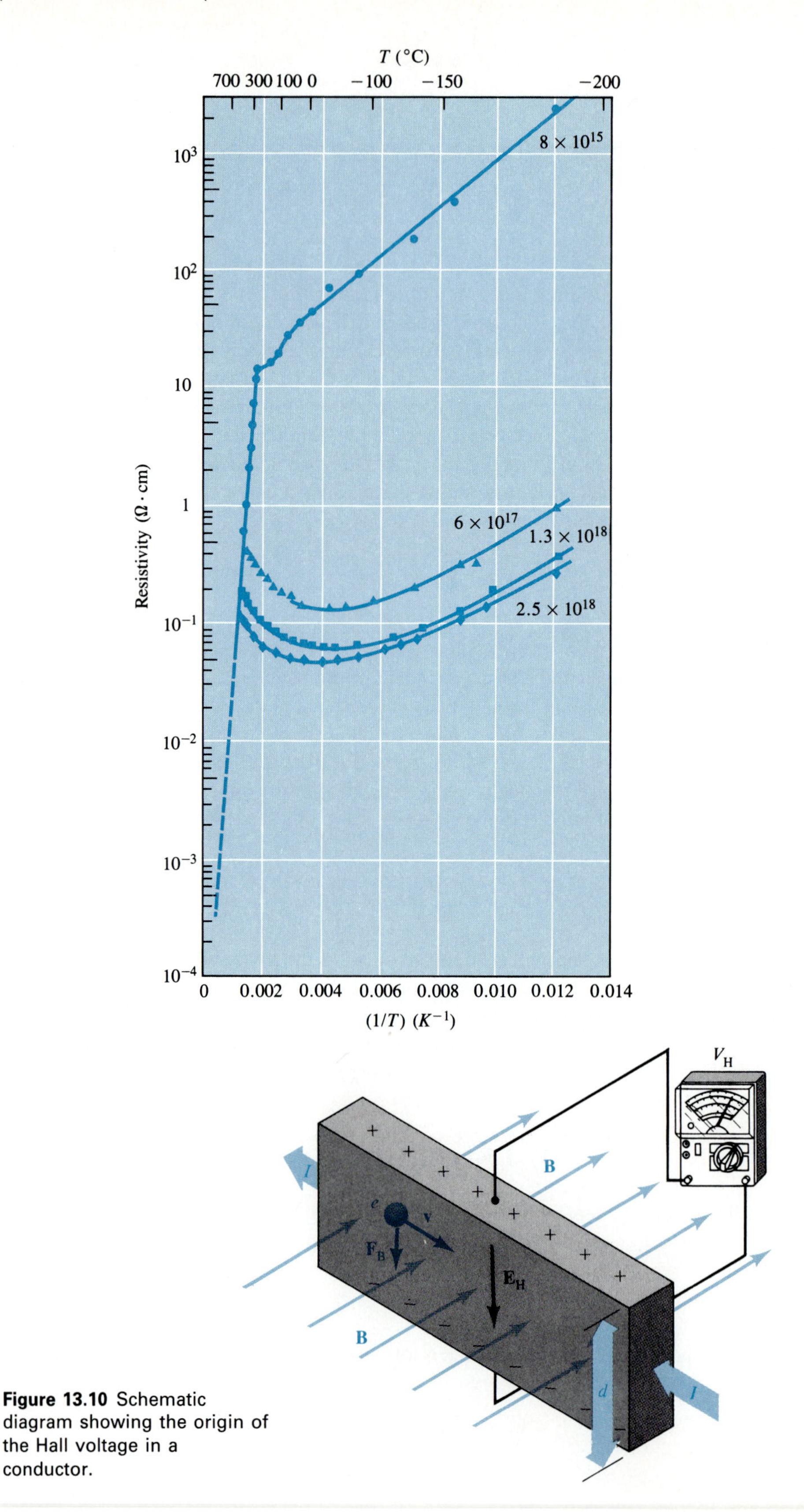

Figure 13.9 Resistivity of four *p*-type silicon samples as functions of temperature. The doping levels—numbers of acceptors per cm³—are indicated. For the purest sample, the intrinsic region commences at about 300 °C. In the other three samples, the saturation region extends from about 0 °C to about 400 °C, and in this temperature range ρ increases with increasing temperature.

Figure 13.10 Schematic diagram showing the origin of the Hall voltage in a conductor.

The *Hall coefficient* is defined by

$$R = \frac{E_H}{JB} \tag{13.28}$$

and is equal to

$$R = \frac{1}{ne} \tag{13.29}$$

Hall measurements are useful in determining not only the density of charge carriers but also their type, electrons or holes. As is readily confirmed, the direction of the Hall field $\mathbf{E}_H$ for a given direction of current and magnetic field is opposite for negative electrons and positive holes.

*13.4 SEMICONDUCTOR DEVICES

It is probably fair to say that the one invention of the latter half of the twentieth century that has had the most profound impact on society is the transistor. Subsequent developments, leading to integrated circuitry, have completely revolutionized communications, banking, and manufacturing industries; computers, faster and far more powerful than any available in the 1960s, are now accessible to most college students.

Here is also one of the most revealing examples of the value of basic research. The earliest radio receivers were so-called crystal sets, which used a metal-semiconductor point contact to demodulate the signal. The rectification property of this metal-semiconductor contact was known at the beginning of the century; however, no one really understood how it worked—and often it did not work all that well. It was only after the fundamental properties of semiconductors were understood that an explanation of the point-contact diode was at hand, and the invention of more reliable and sophisticated devices followed almost immediately. These "solid-state" devices are much more robust, compact, reliable, and efficient than the vacuum tubes in use until about 1955.

In the following we present what is, of necessity, a rather brief survey of some of the more common devices based on semiconductor technology, such as junction diodes, transistors, and various optical and infrared detectors. In the preceding section we discussed equilibrium and steady-state properties of homogeneous semiconductors. Almost all practical devices, however, depend on the presence of inhomogeneities. The first of these to be considered is the *p-n* junction diode.

13.4(a) *p-n* Junction Diodes

p-n junction rectifiers are generally single crystals of germanium, silicon, or some other semiconductor containing *n*- and *p*-type regions separated by a thin interface. The differential doping can be achieved by diffusing different impurities into the substrate or by epitaxial growth.

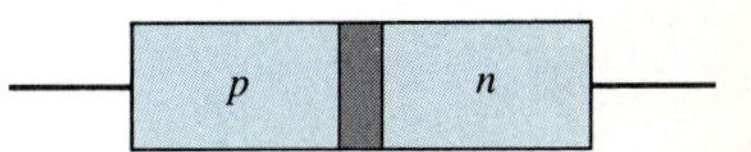

Figure 13.11 A *p-n* junction diode (schematic diagram).

In the absence of an externally applied potential, the Fermi level must be constant throughout the system. Since in the *p*-type region E_F is close to the valence band edge, whereas in the *n*-type region it is near the conduction band edge, the band configuration as a function of position must be as shown in Figure 13.12. How does that strange band shape arise physically?

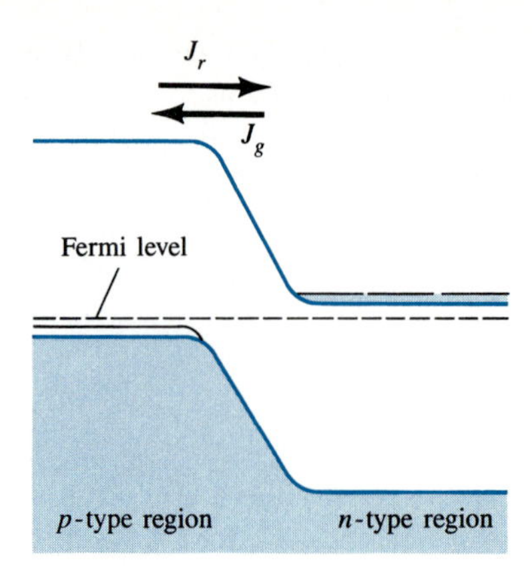

Figure 13.12 Energy bands and Fermi level in a *p-n* junction diode at equilibrium. The recombination and generation currents are indicated.

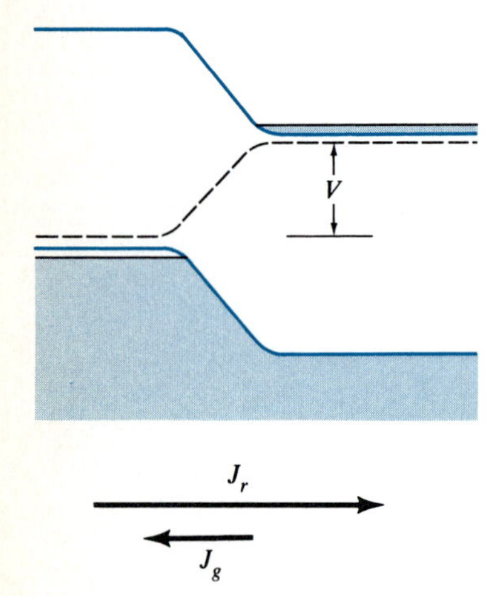

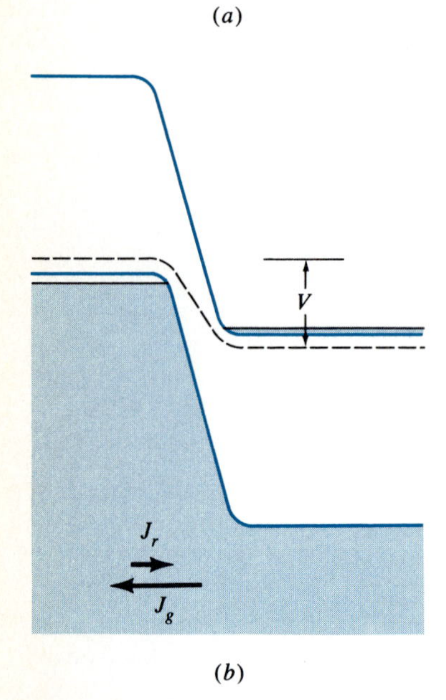

Figure 13.13 Band configuration and Fermi level in a *p-n* junction diode under (*a*) forward bias and (*b*) reverse bias.

Suppose we imagine joining a *p*- and *n*-type semiconductor. Let us assume that initially the band edges are at the same energy. Since there now exists a concentration gradient of electrons and holes in the vicinity of the interface, electrons will diffuse into the *p*-type region and holes will diffuse into the *n*-type region. Consequently, we shall be left with an excess positive charge on the *n* side of the interface and an excess negative charge on the *p* side. The dipole layer established in this manner will grow until the resulting electric field is large enough that the diffusion current due to the concentration gradient is compensated by the conduction current due to this internal electrostatic field.

Even under equilibrium conditions, shown in Figure 13.12, electrons will continue to diffuse into the *p* region and ultimately recombine with holes; similarly, holes will diffuse into the *n* region and recombine there. The charge transport resulting thereby is called the *recombination current* J_r. Simultaneously, electrons thermally excited into the conduction band in the *p* region in the vicinity of the junction "fall down" the potential energy hill into the *n* region; similarly, thermally generated holes in the *n* region near the junction are pushed by the electric field of the junction into the *p* region. The current due to thermally generated electrons and holes is the *generation current* J_g. At equilibrium, that is, in the absence of an externally applied potential, the net current, $J_r + J_g$, is zero, and $J_r = -J_g$.

Suppose we now apply a potential across the device. Since the junction region is nearly denuded of carriers, its resistance is very high compared with that of the *n*- or *p*-type region. Consequently, almost the entire applied voltage appears across this thin junction layer. If the potential is in the "forward" direction, that is, as shown in Figure 13.13*a*, diffusion of electrons into the *p* region and of holes into the *n* region is enhanced because the potential hill which they must surmount is now reduced by the applied potential. Hence J_r is increased. The generation current, on the other hand, remains practically the same as before. It depends only on the rate of generation of electrons and holes in the *p* and *n* regions, respectively, and their diffusion lengths L_n and L_p, the distance they are likely to travel before recombining. If they are generated within a distance of about one diffusion length from the interface, they stand a good chance of reaching the junction. There the dipole field pushes them across the junction.

If the externally applied potential is reversed, the situation is as shown in Figure 13.13*b*. Now the recombination current is reduced, since the charge carriers must overcome a larger potential barrier to diffuse across the junction. However, as before, the generation current remains essentially unaffected.

The probability of surmounting a potential barrier is proportional to the Boltzmann factor $e^{-e\phi/kT}$, where ϕ is the height of the potential barrier. Thus,

$$\frac{J_r}{J_{r0}} = e^{eV/kT} \tag{13.30}$$

where J_{r0} is the recombination current when $V = 0$, which is equal to $-J_g$. Since the net current across the junction is $J = J_r + J_g$, we obtain

$$J = J_g(e^{eV/kT} - 1) \tag{13.31}$$

The actual performance of a *p-n* junction diode conforms almost precisely to these theoretical predictions, as shown in Figure 13.14. If the reverse voltage exceeds a few $kT/e \simeq 0.025$ V, the reverse current saturates to J_g. In the forward direction, however, the current increases roughly exponentially. Note that both ordinate and abscissa scales of Figure 13.14 are logarithmic.

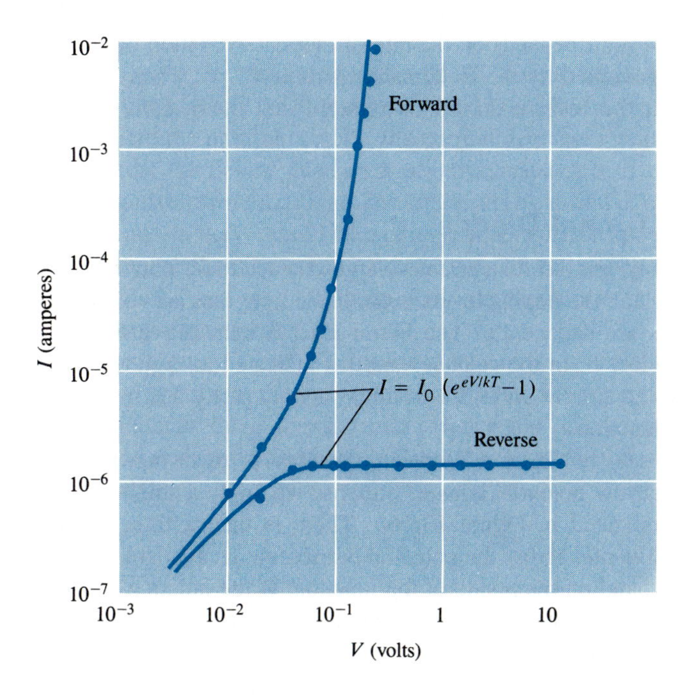

Figure 13.14 Current voltage characteristic of a *p-n* junction diode. The solid line represents Equation (13.31); the circles are experimental values. (From L. V. Azaroff and J. J. Brophy, *Electronic Processes in Materials,* McGraw-Hill, New York, 1963.)

If the reverse voltage across the diode is increased, one reaches a critical voltage at which the reverse current rises dramatically. This sharp increase is due to *avalanche breakdown* of the junction. The few electron-hole pairs thermally excited near the junction are accelerated across the junction by the electric field. In a very strong field they gain sufficient energy to create new electron-hole pairs by impact ionization, and these, in turn, create yet more such pairs. This breakdown current must be limited by an external circuit resistance to prevent damaging the device. The critical reverse-bias voltage at which breakdown occurs can be controlled by

The greater wavelength response of extrinsic photoconductive devices is achieved only at a cost. At room temperature, extrinsic silicon and germanium are in the saturation regime; that is, essentially all donor or acceptor impurities are already thermally ionized. To observe an increase in carrier concentration due to photoionization of donor or acceptor impurities it is necessary to maintain the detector at low temperature, typically at liquid air temperature, 77 K. In some cases, the temperature is reduced to that of liquid helium, 4.2 K. Although a decided inconvenience, there is also a benefit to the low operating temperature, namely, a significant increase in the signal-to-noise ratio.

Photovoltaic Effect

Although a photovoltaic effect can be observed in a homogeneous semiconductor, this is of no technological importance and will not be considered here. The effect is quite small, and, moreover, its response time is long compared to that which occurs at a *p-n* junction.

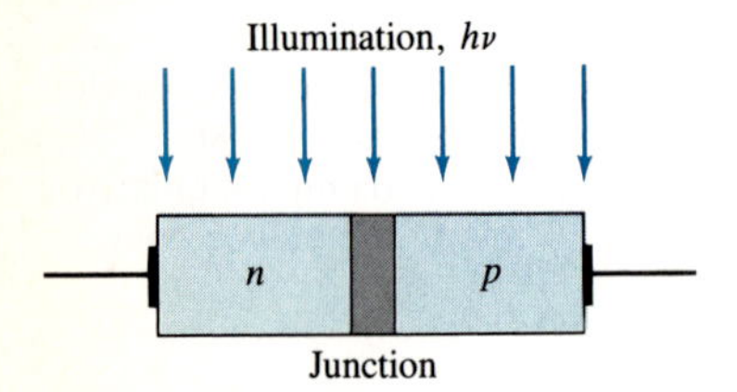

Figure 13.25 A *p-n* junction photodiode under uniform illumination (schematic diagram). The width of the junction region is here much enlarged.

Figure 13.25 is a schematic representation of a *p-n* junction diode showing the junction region much enlarged. Suppose we now illuminate this diode with light whose frequency is above the threshold for electron-hole pair creation. As is apparent from Figure 13.26, electrons and holes produced in the junction itself are driven toward the *n*- and *p*-type region, respectively, by the electrostatic field of the junction. In addition, holes created in the *n* region within about one *diffusion length* L_p will reach the junction before recombining and be driven into the *p* region by the field of the junction. Similarly, electrons created by illumination in the *p* region within a distance of about L_n of the junction will also contribute to charge transport across the junction. In other words, the net effect of illumination is to increase the generation current across the junction. This additional current is given by

$$J_{\text{ph}} = -eR(L_n + L_p) \tag{13.32}$$

where R is the rate at which pairs are created by the incident photons; the negative sign is used to indicate that this current flows in the reverse direction with respect to the standard polarity of the diode. The total current is then the sum of Equations (13.31) and (13.32):

$$J = J_g(e^{eV/kT} - 1) - eR(L_n + L_p) \tag{13.33}$$

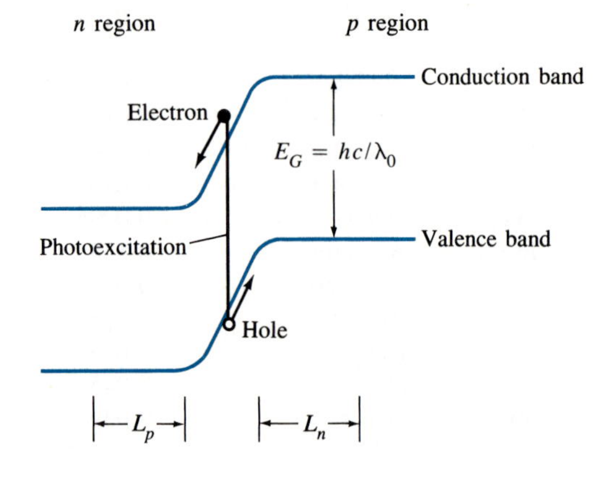

Figure 13.26 Energy bands of a photodiode under illumination when used in the photoconductive mode. Electrons and holes created by absorption of photons at the junction are driven into the *n* and *p* regions, respectively, by the electrostatic field of the junction. Holes within a distance L_p of the junction in the *n* region are also likely to diffuse to the junction before recombining, as are electrons within a distance L_n on the *p* side.

If the junction is reverse-biased so that, in the dark, the current is the reverse saturation current J_g, the change in current is proportional to the incident light intensity. The signal is then a photocurrent, not a photovoltage, and the device is said to operate in the photoconductive mode. It is, however, far more common to operate the diode under open-circuit conditions. In that case, $J = 0$, and from Equation (13.33) it follows that a voltage

$$V = \frac{kT}{e} \ln \left[1 + \frac{eR(L_n + L_p)}{J_g} \right] \tag{13.34}$$

is developed between the terminals of the diode.

Equation (13.34) is, however, valid only for low light intensities. If R is increased, the voltage across the diode does not increase without limit but saturates to the barrier potential. Under very strong illumination the junction behaves as though the temperature were increased to the point where the material becomes intrinsic; an essentially flat band condition then prevails and the differential doping plays no role. Thus, the maximum voltage that can be produced is about equal to E_G/e, typically about 1 volt. Often photodiodes are connected in series to generate a larger output voltage.

One application of photodiodes is in conversion of solar radiation to electrical energy, i.e., as solar cells. The incident radiation is then not monochromatic but covers the entire electromagnetic spectrum corresponding to blackbody radiation at about 6000 K. To utilize as much of this radiation as possible it might seem that a narrow band gap material would be most advantageous. However, as we have just seen, the maximum voltage generated by a photodiode is no greater than E_G/e. Thus a compromise must be effected between maximum utilization of the incident radiation and high output voltage. It turns out that the optimum band gap is about 1.4 eV, roughly that of gallium arsenide. Although under idealized conditions conversion efficiencies of 30 percent should be possible, the efficiency of most solar cells is rarely much better than about 15 percent.

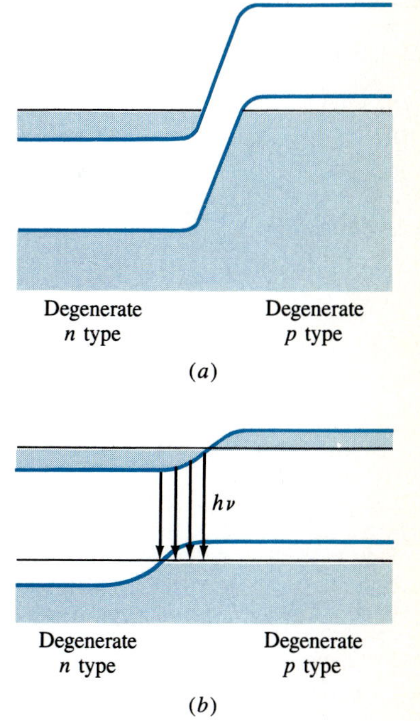

Figure 13.27 A heavily doped *p-n* junction diode (a) with no applied bias and (b) with a large forward bias. Radiative recombination in the junction region results in the emission of light of frequency $\nu \simeq E_G/h$.

Light-Emitting Diodes (LEDs) and Junction Lasers

In the preceding pages we referred briefly to the inverse of electron-hole generation, namely, electron-hole recombination. Whenever such recombination takes place, energy must be conserved. One way, though by no means the only way, this can happen is by the emission of a photon of energy

$$h\nu = E_G + E_n + E_p \tag{13.35}$$

where E_n and E_p are the energies of the electron and hole relative to the conduction and valence band edges; generally, these energies are small compared with E_G. Although in many semiconductors, for example, germanium and silicon, radiative recombination is unimportant compared with nonradiative recombination, there are other materials, for instance, gallium arsenide and cadmium selenide, in which radiative recombination is dominant.

Figure 13.27*a* shows the band configuration of a heavily doped *p-n* junction diode; there is a high concentration of electrons on the *n* side, and of holes on the *p* side. If we now apply a forward bias to this diode, as shown in Figure 13.27*b*, many electrons and holes diffuse into the junction region, where they recombine. If this is a radiative process, the junction is a light-emitting diode (LED).

In the junction of the forward-biased LED the electron and hole concentrations are inverted in the sense in which this term was used to characterize the population inversion of energy levels in a laser (see Section 9.5). The photons emitted by the LED have random phase and polarization, although at fairly high injection levels (large forward bias) one condition for laser action is already achieved: photons that propagate in the plane of the junction trigger further radiative recombinations elsewhere in the junction. Coherence can be enforced by preparing the crystal so that back and front surfaces are parallel and optically flat. Since the dielectric constant of the semiconductor is much greater than 1, the light suffers numerous internal reflections before it escapes, and only waves that interfere constructively survive. The emerging radiation is now coherent and polarized perpendicular to the plane of the junction. Since junction lasers are compact and have fast response times (about one nanosecond), they are commonly used for transmission of information through optical fibers.

13.5 SUPERCONDUCTIVITY

13.5(a) Fundamental Properties of Superconductors

After 1908, the accessibility of previously unattainable low temperatures opened an enormous vista of experimental studies, quite apart from investigations into the properties of helium itself. For example, it was known that the resistivity of pure metals diminished monotonically with decreasing temperature, but the behavior near absolute zero was still a matter of controversy. According to one theory, the resistance should decrease continuously and vanish at $T = 0$ K. On the other hand, it was argued that as very low temperatures are approached, the valence electrons in a metal would be bound to the atoms and the resistance should therefore increase exponentially as T approaches 0 K. The measurement of resistance is a relatively simple matter and was soon initiated at Leiden.

The first metal studied by Kamerlingh Onnes was platinum, but the results were not illuminating: the resistivity reached a plateau and neither fell further nor increased below about 6 K even in the purest specimen. But Kamerlingh Onnes did note that the resistivity at the plateau depended on sample purity, being lowest for the purest sample. The one metal which could be purified better that any other was mercury. Because of its low melting and boiling points it can be distilled and redistilled and made almost entirely free of impurities.

The results of the measurements on mercury were surprising.[2] At 4.2 K, the temperature of liquid helium at atmospheric pressure, the resistivity of pure mercury was fairly high, greater than that of gold. However, as the temperature was reduced slightly, the resistivity dropped so low that it could not be measured. Kamerlingh Onnes now believed that he had, at last, resolved the old controversy. Yet, something was not quite right; instead of falling to zero gradually, the resistance seemed to drop by a factor of about a million within a fraction of a degree near 4.17 K. Even so, it was several years before Kamerlingh Onnes recognized that superconductivity, a term he first used in 1913, was not the manifestation of the expected reduction in resistivity but an entirely new phenomenon, one which defied theoretical explanation for nearly half a century. Today we know that far from being the

2. H. Kamerlingh Onnes, *Commun. Phys. Lab. Univ. Leiden* **12,** 119, 120 (1911).

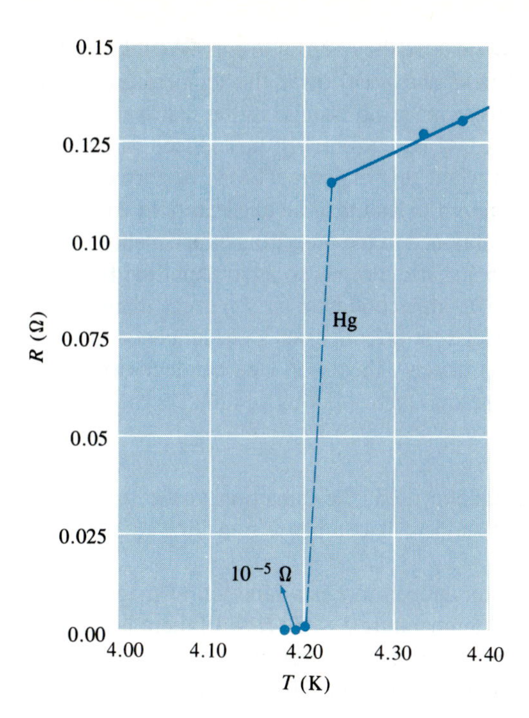

Figure 13.28 The resistance of a sample of mercury at low temperature as reported by H. Kamerlingh Onnes. (*Commun. Phys. Lab. Univ. Leiden* **12,** 120, 1911.)

exception, superconductivity is a common occurrence. The vast majority of metals become superconductors at sufficiently low temperatures, and there are literally hundreds of alloys as well as some semiconductors and ceramics that become superconducting.

Is the resistivity of a superconductor really zero, or only very, very small? Kamerlingh Onnes himself devised the most sensitive method for measuring this resistance. He made a ring of lead (which he also found to be superconducting) and lowered its temperature while the ring was in a magnetic field. He then removed the field, inducing a current in the ring which would tend to maintain the flux linking the ring. If the ring has a resistance R and an inductance L, this circulating current decays with a time constant $\tau = L/R$, and the magnetic field due to that current falls correspondingly. Kamerlingh Onnes found that over a period of many hours he could observe no discernible change in the field due to the circulating current. The experiment has since been repeated many times.[3] In one instance, such a current was maintained for more than two years, and would still be flowing were it not that a trucker's strike interrupted the supply of helium.

While persistent current is the best known, it is by no means the only, perhaps not even the most dramatic, property of a superconductor. To illustrate, consider the following two experiments.

We take a solid lead sphere, submerge it in liquid helium in the absence of a magnetic field, and then apply a small external field. We find that surface currents are induced in the sphere that establish within the sample a field that precisely cancels the external field so that the field inside the sample is exactly zero. There is nothing startling in this. It is just what we expect from Faraday's law of induction.

3. See, for example, J. File and R. G. Mills, *Phys. Rev. Lett.* **10,** 93 (1963); they find that the decay time of the circulating current is greater than 100,000 years!

TABLE 13.5 Transition Temperatures and Critical Fields of Some Type I Superconductors

Material	Transition Temperature (K)	Critical Field at 0 K (T)
Ti	0.39	0.0100
V	5.38	0.1420
Zn	0.85	0.0054
Ga	1.083	0.0058
Zr	0.546	0.0047
Nb	9.46	0.1980
Mo	0.92	0.0095
Tc	7.77	0.1410
Cd	0.56	0.0030
In	3.408	0.0281
Sn	3.722	0.0305
La	6.00	0.1100
Ta	4.47	0.0829
W	0.015	0.0001
Re	1.4	0.0198
Hg	4.153	0.0411
Tl	2.39	0.0171
Pb	7.193	0.0803

TABLE 13.6 Transition Temperatures and Critical Fields of Some Type II Superconductors

Material	Transition Temperature (K)	Upper Critical Field at 0 K (T)*
Nb_3Sn	18.0	24.5
Nb_3Ge	23.2	38
Nb_3Al	18.7	32.4
$Nb_3(AlGe)$	20.7	44
Nb_3Ga	20	35
NbTi	10	15
NbN	15.7	1.5
V_3Ga	14.8	2.1
V_3Si	16.9	2.35
PbMoS	14.4	6.0

*These are approximate values, based on extrapolations of data at temperatures generally above 6 K.

ing rare earth impurities remain superconducting to temperatures well above 90 K produced excitement among physicists and engineers rarely witnessed before.[11] The meeting of the American Physical Society that closely followed the announcement of these discoveries was characterized at the time as the "Woodstock of physics." In the following months and years the initial euphoria subsided. The compounds deteriorate slowly under atmospheric conditions; the materials are brittle and cannot

11. A. Mueller and G. Bednorz shared the 1987 Nobel Prize for their discovery of high-temperature superconductors.

be drawn into wires. Some of these problems also initially plagued the type II superconductors and were overcome in time. Assuredly, the new high-T_c materials will ultimately be employed in modern technology, though current opinion suggests that this may require another decade of diligent research.

Many countries are today supporting intensive efforts aimed toward the design and construction of superconducting electric generators, superconducting transmission lines, and high-speed transportation systems based on levitation of superconducting magnets. Modern particle accelerators, such as the ones at CERN and the Fermi Laboratory, as well as modern cyclotrons use superconducting magnets to bend particle beams. Smaller but still powerful superconducting magnets are standard items in most physics research laboratories. Superconducting magnets are also used in magnetic resonance imaging (see Section 14.6), fusion, and magnetohydrodynamic generators. The problems encountered in these applications are of a technical, engineering sort, and do not, for the most part, involve new physics concepts. A superconducting solenoid is, basically, no different from a solenoid made of copper coils; a superconducting generator relies on the same law of induction as a more conventional one. In this chapter we shall not consider these large-scale applications of superconductivity, though their potential practical importance in technology must not be underrated.

In the following pages we describe a few of the new superconducting devices that have revolutionized precision instrumentation, such as the SQUID (Superconducting **Qu**antum **I**nterference **D**evice). The breakthrough that sparked the new technology was the work of a young theorist, Brian Josephson, then a graduate student at Cambridge University.[12]

Josephson Junctions

In Chapter 7 we saw that because in quantum mechanics a particle is described by a wave function, it has a finite probability of tunneling through a barrier that, classically, would prevent its transit. In the early 1960s, Giaever had demonstrated tunneling of electrons through a thin insulating barrier. Although it was recognized that tunneling of electrons across an insulating barrier separating two superconductors could also occur, the conventional wisdom at the time was that the Cooper pairs responsible for the superconducting state would not survive the tunneling process; the tunnel junction could carry a normal current but not a supercurrent. Josephson showed theoretically that tunneling of Cooper pairs was as likely as tunneling of unpaired electrons. That is, a thin insulating film would support a superconducting current, and would present no resistance in such a circuit, provided the junction current was sufficiently small.

Josephson was able to characterize the tunneling current by two very simple equations. The supercurrent through the tunnel junction is given by

$$I_s = I_c \sin \theta \qquad \textbf{(13.37)}$$

and θ is related to the voltage across the junction by

$$\frac{d\theta}{dt} = \left(\frac{2e}{\hbar}\right) V(t) \qquad \textbf{(13.38)}$$

---

12. Leo Esaki, Ivar Giaever, and Brian D. Josephson shared the 1973 Nobel Prize for their work on electron tunneling.

The angle θ represents the phase change of the wave function that describes the Cooper pair as it tunnels through the junction.

According to Equation (13.37) the supercurrent through the junction is zero if $\theta = 0$ and increases to a maximum value of I_c for $\theta = \pi/2$. The critical current I_c that the junction can support is determined by its physical characteristics, the thickness and width of the insulating layer, and is normally very small. Note that a finite current will flow across the junction even though the voltage across the junction is zero; the junction is a "weak superconducting link." The adjective *weak* refers to the fact that the total superconducting current that the junction can support is very much less than that which can flow in the bulk superconductors that bound the junction.

Note also that if the voltage across the junction is finite, the phase angle increases linearly with time and, consequently, the supercurrent through the junction varies sinusoidally. The frequency of this sinusoidal current is determined by the ratio $2e/h$ and is 483.6 MHz/μV. This "ac Josephson effect"—the generation of an rf current whose frequency is proportional to the voltage across the junction—was used in a precision measurement of the ratio e/h. Alternatively, the effect can serve as a means for maintaining a constant-voltage source by monitoring the frequency of the ac Josephson current. By virtue of the ac Josephson effect, the volt has replaced the ampere as the fundamental unit of electromagnetism: one volt is now defined as that potential difference which, when impressed across a Josephson junction, will result in the generation of electromagnetic radiation of 483,597.9 gigahertz.

SQUID (Superconducting Quantum Interference Device)

The SQUID is basically a very sensitive magnetometer that employs a superconducting loop with a "weak link" to measure the flux changes within the loop. The functioning of the device rests on the fact that the weak link will become normal whenever the loop current exceeds a critical value I_c which is very much smaller than the critical current of the bulk superconductor.

Suppose the loop is initially in a field-free region so that the current in the loop is zero. If we now apply an external field, a current is induced in the loop such that the flux within the loop remains zero. As the external flux ϕ_x increases, so does the loop current I until the critical current I_c is reached. At that point, the weak link goes normal and one flux quantum ϕ_0 enters the loop. As a result, the loop current falls below I_c, since now the difference between the external flux and the flux within the loop has diminished by a single flux quantum. A further small increase in external flux will result in a small increase in loop current, but the loop will remain superconducting until the flux has been increased by one more flux quantum. Then the link goes normal again, admitting another flux quantum into the loop, and the loop current drops below I_c again.

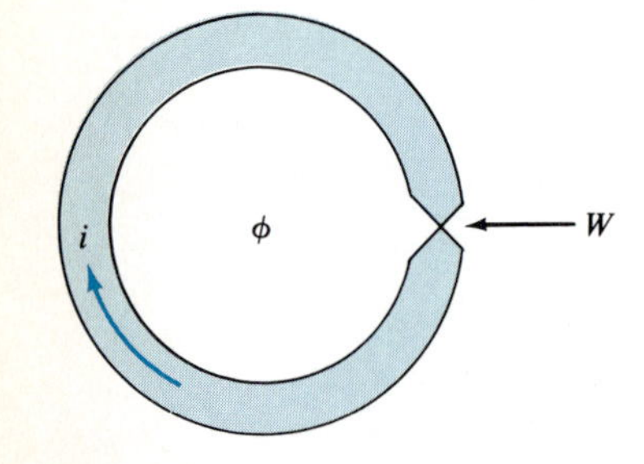

Figure 13.33 A superconducting loop with a "weak link" (W). The flux ϕ within the loop is the result of flux quanta that have entered the loop when the loop current exceeded the critical current of the weak link.

The pattern of flux ϕ within the loop as a function of external flux and of loop current as a function of external flux is shown in Figure 13.34. Note that the value of $m = \phi_x/\phi_0$ at which the first flux jump occurs may be quite large; thereafter, however, flux jumps and current discontinuities appear at regular intervals $\Delta m = 1$.

One could determine changes in external flux by simply counting the number of times the weak link goes normal. In practice the SQUID is operated in an ac mode, with the SQUID loop coupled inductively to an rf rank circuit. With the aid of sophisticated circuitry, it is possible to measure flux changes that are only a fraction of one fluxoid.

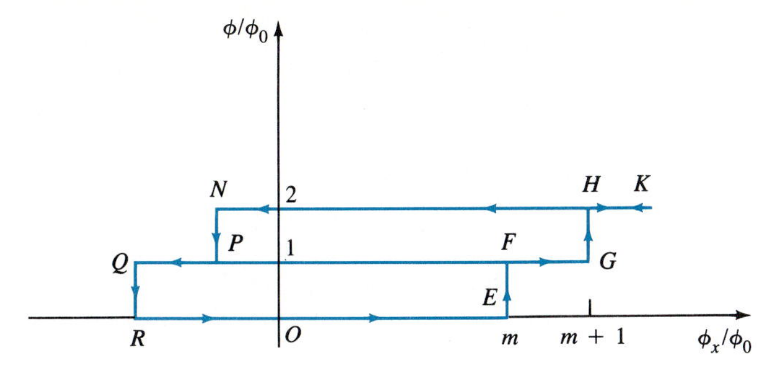

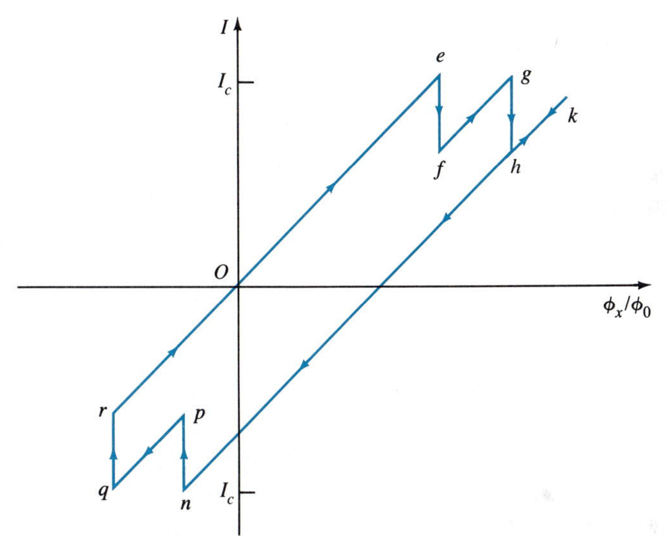

Figure 13.34 The flux ϕ within the SQUID loop and the current in the loop as functions of the external flux ϕ_x. As ϕ_x increases from 0 to $m\phi_0$, the induced loop current increases to I_c. A further increase of ϕ_x, i.e., an attempt to increase the loop current beyond I_c, renders the weak link momentarily normal, allowing one flux quantum ϕ_0 to enter the loop with a consequent decrease in loop current. At point K the external flux is reduced again. The loop current now decreases and reverses direction, and at N, when $I = -I_c$, the weak link again becomes normal. Note the hysteresis of the flux and current patterns.

SQUIDs have been used in a variety of measurement applications. In combination with high-precision current comparators it is now possible to make low-temperature resistance measurements to a relative accuracy of one part in 10^9. In low-impedance circuits such as thermocouples, voltages as small as 5×10^{-14} V are readily measured. And, of course, SQUIDs are the most sensitive magnetometers available.

One of the more exciting applications of SQUIDs outside the field of physics is in biomagnetism. That muscle and nerve activity is associated with local currents has been known for centuries, and electrocardiography and electroencephalography (measurement of electric potentials associated with cardiac and brain activity) have been standard clinical procedures for decades. Currents generate magnetic fields, and these can also be of value in diagnostics. Until the advent of SQUID detectors, however, the sensitivity of magnetometers was generally far too small to permit biomagnetic studies. Field strengths due to various biomagnetic activity range between about 10^{-13} and 10^{-11} T (10^{-9} to 10^{-7} G) and, typically, cover a frequency spectrum between 1 and 500 Hz. These field variations have to be detected in an environment of ambient noise due to sources such as motors, elevators, and automobiles, whose fields are generally orders of magnitude greater than that of the source to be studied.

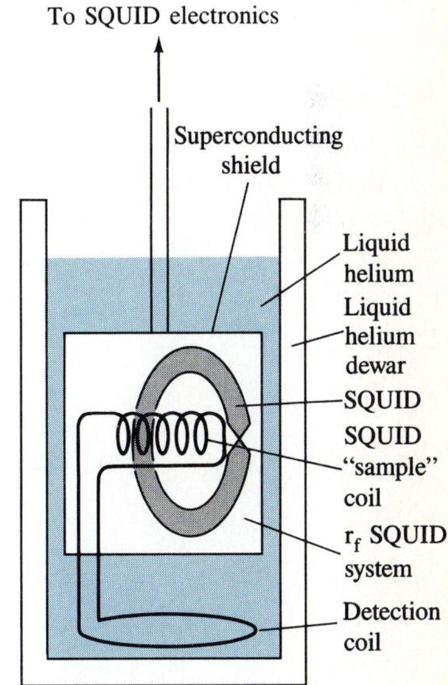

Figure 13.35 Schematic drawing of s SQUID system for biomagnetic measurements. The detection coil is mounted at the bottom of the dewar, as close as possible to the subject (see Figure 13.37). Current induced in the detection coil flows in the SQUID "sample" coil. Both sample and detection coils are made of superconducting wire.

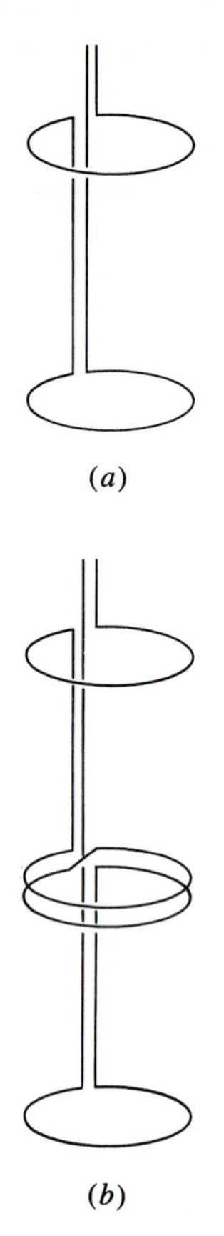

Figure 13.36 Detection coils for biomagnetic experiments. (a) Gradiometer detection coils; (b) second-order gradiometer detection coils.

The technique that allows discrimination between the desired signal and extraneous noise due to relatively distant sources depends on the fact that at the site of interest the field due to the noise source, though it may vary in time, is spatially almost uniform. The detection coils that are used to measure biomagnetic signals are therefore designed to sense not the total field but, rather, field gradients. For example, a pair of coils such as shown in Figure 13.36*a* will produce no signal if the time-varying magnetic field is spatially homogeneous; the current induced in one coil exactly cancels that induced in the coil wound in the opposing sense. A further refinement can be achieved with a second-order gradiometer, shown in Figure 13.36*b*. A magnetic field whose spatial variation along the axis of the coils is linear will produce no signal; only if the second or higher derivative of the time-varying field does not vanish will this second-order gradiometer coil generate an induced current in the SQUID sensing coil.

With the use of these techniques it has now been possible to observe and localize the origin of biomagnetic activity associated with cardiac function and with responses of the auditory and visual cortex to various stimuli, and to identify normal and abnormal patterns.[13] In many respects, biomagnetic signals are considerably more useful than their electrical counterparts. Whereas bioelectric signals reveal potential differences between contact points on the skin of the subject, biomagnetic signals permit the localization of the activity that generates the signals, often to within a millimeter.

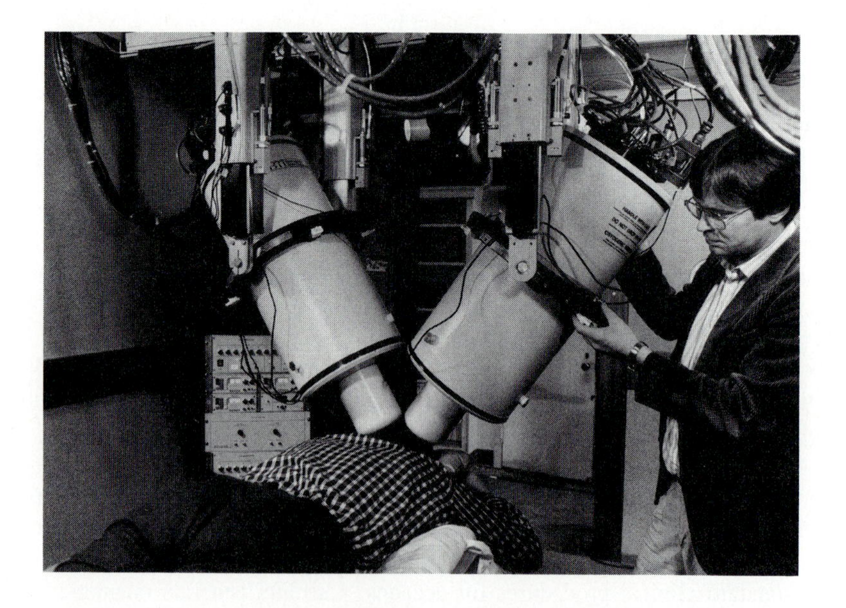

Figure 13.37 Equipment for biomagnetic measurements at the Center for Neuromagnetism at New York University. The two dewars, each containing a probe with 7 SQUID sensors are positioned just above the head of a subject *(S. J. Williamson.).*

13. D. S. Buchanan, D. Paulson, and S. J. Williamson, *Adv. in Cryogenic Engineering* **33,** 97 (1988).

Josephson Junction Logic Elements

Josephson junctions can also be designed for use as logic elements in computers. The simplest such logic element is a switch, shown in Figure 13.38. Two thin strips of superconducting metal are separated by a very thin film of insulating material, usually the oxide of the superconducting metal. One side of this tunnel junction is connected to ground, the other side to the signal source and the output device (see Figure 13.39). Another conductor, placed over the junction and separated from it by an insulating layer, carries the control current. If the control current is zero and the source current is less than the critical current of the junction, the junction provides a direct short circuit to ground, and none of the source current flows into the output device. If now a current is sent through the control line, the magnetic field generated by this current penetrates the junction and renders this weak superconducting link normal. In the normal state, the resistance of the junction may be several hundred ohms. Thus, the junction switches the source current from ground to the output device when a current is sent through the control line.

Multijunction devices can be constructed that act as OR gates or AND gates, the basic logic elements that perform the arithmetic functions in the central processing unit (CPU) of a digital computer.

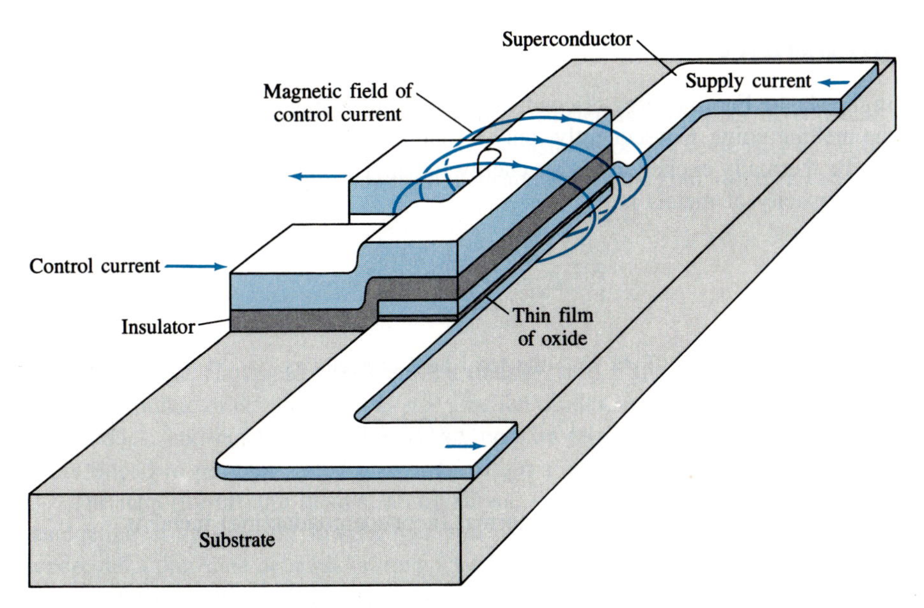

Figure 13.38 A Josephson junction switch.

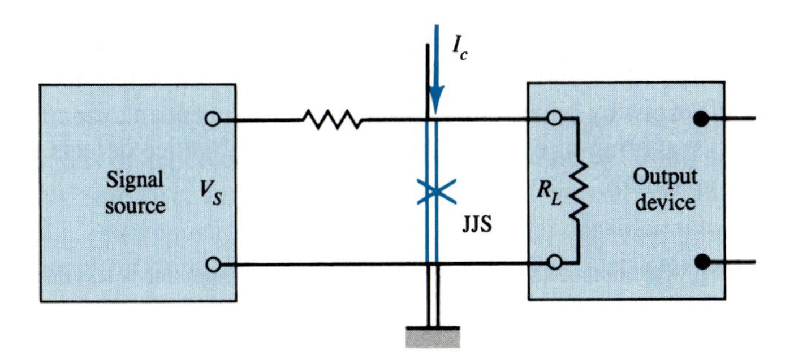

Figure 13.39 A circuit incorporating a Josephson junction switch (JJS). When the switch goes normal, the signal current flows mostly through the load resistor R_L of the output device. When the switch is superconducting, the voltage across R_L is zero.

Quite by accident, Becquerel had discovered radioactivity. A few months later, Becquerel reported, "All the uranium salts I have studied, be they phosphorescent or not, . . . have given me similar results." He also found that this radiation, like X rays, discharges electrified bodies, but that it is even more penetrating than "the rays emanating from the region of the anticathode of a Crookes tube."

Three years later, Rutherford studied this uranium radiation using an electrometer to measure the ionization of air surrounding the radiating substance. He concluded that

> *there are present at least two distinct types of radiation—one that is very readily absorbed, which will be termed for convenience the alpha radiation, and the other of a more penetrating character, which will be termed the beta radiation.*

The terminology has survived to this day, even though subsequent work demonstrated that α rays are helium nuclei and β rays are electrons.

While Rutherford was still at McGill University, he and Soddy, a young and remarkably ingenious chemist, electrified the scientific community with the announcement that the process of radioactivity involved the *transmutation of chemical elements:*

> *. . . the major part of the radioactivity of thorium—ordinarily about* 54 *percent—is due to a non-thorium type of matter, ThX, possessing distinct chemical properties, which is temporarily radioactive, its activity falling to half value in about four days. Since, therefore, radioactivity is at once an atomic phenomenon and accompanied by chemical changes in which new types of matter was produced, these changes must be occurring within the atom, and the radioactive elements must be undergoing spontaneous transformation.*

This was a truly revolutionary idea which contradicted the then firmly held conviction of the immutability of the elements.

At that time (1902) the constitution of atoms was, of course, not really understood. All that was known was that the atom consisted of electrons and a much more massive, positively charged component. The experiments of Geiger and Marsden, and their interpretation in terms of a nuclear atom by Rutherford, have already been described in Chapter 5. Nuclear physics, as a new and identifiable field of scientific study, thus commenced in 1913, the year that Geiger and Marsden gave clear experimental proof of the validity of Rutherford's scattering formula. This was also the year in which Soddy discovered that a given chemical element may have two or more isotopes, i.e., nuclei of different masses. Moreover, the appearance of isotopes was evidently not limited to radioactive elements; also in 1913, J. J. Thomson concluded on the basis of the deflection of ions in a magnetic field that "what has been called neon is not a simple gas but a mixture of two gases, one of which has an atomic weight about 20 and the other about 22."

Until about 1919 transmutation seemed to be a property of naturally occurring radioactive substances only. That year, however, marked yet another milestone in nuclear physics with the discovery by Rutherford that protons are released when nitrogen is bombarded with α particles. This was the first artificially induced nuclear reaction,

$$N + \alpha \rightarrow O + p$$

Rutherford correctly surmised that "if α particles—or similar projectiles—of still greater energy were available for experiment, we might expect to break down the nucleus structure of many of the lighter atoms."

Thus by 1920 the nuclear model that fitted the available data was that of a very small, extremely dense spherical object consisting of protons, in sufficient number to account for the total mass, and as many electrons as would be required to neutralize any excess positive charge and yield the correct charge-to-mass ratio. This proton-electron model seemed perfectly reasonable: Rutherford had demonstrated that "the hydrogen atom which is liberated [on bombarding nitrogen with α particles] formed a constituent part of the nitrogen nucleus," and Thomson and Becquerel had shown that the β rays emitted during radioactive disintegration are energetic electrons.

Yet this nuclear model led to serious problems. One surfaced only in 1927 with the publication of Heisenberg's uncertainty principle. If an electron were confined to a space as small as a nucleus, its momentum uncertainty and, hence, its kinetic energy would be enormous, several orders of magnitude greater than the observed energies of β rays. As Rutherford remarked, "Trying to put it [the electron] there is like trying to force an elephant into a dog house."

Another puzzle had to do with the electron's magnetic moment. If the number of electrons in the nucleus is odd, their magnetic moments cannot cancel and the nucleus should have a magnetic moment of the same magnitude as that of the electron. Yet spectroscopic data gave no indication of a large nuclear magnetic moment. Spectroscopists did, however, come up with results that further compounded the confusion.

Even before the turn of the century it had been observed that many atomic spectral lines possess a structure much finer than that due to electronic spin-orbit coupling (see Section 9.3), a so-called *hyperfine structure*. At first, following the discovery of isotopes, it was thought that the various components of the hyperfine structure were lines emitted by different isotopic constituents of the element. This explanation had to be abandoned when it became clear that the structure persisted also in isotopically pure samples. In 1924 Pauli suggested that a nucleus may have an *intrinsic nuclear spin* with spin quantum number I. Hyperfine structure would then result from the interaction between the associated nuclear magnetic moment and that of the orbiting electron. The number of hyperfine components would be $2I + 1$, the number of possible orientations of the nuclear spin, and thus be a measure of the spin quantum number I. The spin of the proton is $\frac{1}{2}$. It follows from the rules for the addition of angular momenta in quantum mechanics that a system for which the total number of electrons and protons is odd must have a nuclear spin which is an odd multiple of $\frac{1}{2}$. The spectrum of nitrogen, whose nucleus has a charge of $7e$ and a mass of about $14m_p$, showed that $I = 1$. But if this nucleus contained 14 protons and 7 electrons, i.e., 21 particles of spin $\frac{1}{2}$, its nuclear spin could not be an integer.

All of these difficulties were resolved with the discovery of the neutron, a particle of nearly the same mass as a proton but carrying no charge.[1] In 1920,

1. James Chadwick, *Proc. Roy. Soc. (London)* **A 136,** 692 (1932).

The nuclear density is

$$\rho = \frac{1\ \text{u}}{4\pi R_0^3/3} = \frac{1.66 \times 10^{-27}\ \text{kg}}{7.24 \times 10^{-45}\ \text{m}^3} = 2.3 \times 10^{17}\ \text{kg/m}^3$$

The mass of 1 mm^3 of nuclear matter is therefore

$$m = (2.3 \times 10^{17}\ \text{kg/m}^3)(10^{-9}\ \text{mm}^3/\text{m}^3) = 2.3 \times 10^{8}\ \text{kg}$$

14.3 NUCLEAR FORCE

Theoretical explanations of atomic, molecular, and solid-state phenomena may present formidable mathematical difficulties, but it is at least true that the interactions between the constituent particles are well understood. For these systems the forces the particles exert on each other are entirely of electromagnetic origin. We can write down the wave equation that must be solved to obtain eigenfunctions and eigenvalues, though the solution may require lengthy numerical calculation. The difficulty faced by nuclear physicists is not merely that the nucleus is a many-body system like an atom or molecule, but worse yet, the fundamental laws governing nuclear interactions are far more complex than those of electrodynamics.

We do not mean to suggest that the nuclear force is still shrouded in mystery; quite the contrary. Much is known about the details of this force from careful analysis of an enormous number of experiments. But this force cannot be couched in a simple formalism, nor can it be expressed in a closed analytic form like the electromagnetic force. Hence, in the description of nuclear properties one relies on various models, and no single model is completely adequate to reproduce all experimental data.

Probably the outstanding characteristics of the nuclear force are its enormous strength and very short range. We can appreciate the strength by calculating the electrostatic repulsion between two protons in a nucleus. The separation between their centers is about 2.4 fm, and a quick calculation shows that the Coulomb force is about 40 N, a force roughly equal to the weight of a 4-kg mass. This force acts not on a kilogram or microgram, but on a mass of only 1.67×10^{-27} kg. The fact that nuclei do not fly apart tells us that the nuclear force must be very strong, sufficient to more than balance this huge Coulomb repulsion.

Strong though it may be, the nuclear force must also be of short range, as demonstrated by scattering experiments designed to measure nuclear radii. The range of the nuclear force is in fact so small that a particular nucleon in a large nucleus interacts only with its immediate neighbors. As we shall see presently, this accounts for the observation that the average energy with which a nucleon is bound in a medium to heavy nucleus is nearly independent of the total number of nucleons.

More detailed knowledge of the nuclear force comes primarily from two sources. First, the properties of nuclei in their ground and excited states, such as energy levels, magnetic moments, spin states, and quadrupole moments, provide valuable information. Second, scattering experiments in which the projectile and target nuclei come within range of the nuclear force shed light on this force.

The simplest nontrivial nucleus is the deuteron, the nucleus of ''heavy'' hydrogen, which consists of a proton and a neutron in a bound-state configuration. We know the binding energy[7] in the ground state, 2.2246 MeV; we know that the spin of the deuteron is 1 and that its magnetic moment is $0.85742\mu_N$; we know that it has a small quadrupole moment. What conclusions can be drawn from these facts?

Even without detailed knowledge of the spatial dependence of the nuclear potential, one can obtain a number of important results using the simplest potential, a square well. For example, making a reasonable assumption concerning the well's radius, one can calculate the depth of the potential from the ground-state energy. Such a calculation suggests that the depth of the well is about 21 MeV.

From the spin of the deuteron and the fact that its magnetic moment is almost equal to the algebraic sum of μ_p and μ_n, we conclude that the proton and neutron spins are parallel and that the orbital angular momentum L of the deuteron in the ground state is zero. That is, in spectroscopic notation we would label the ground state $^3S_1$. However, the fact that the deuteron has a quadrupole moment tells us that the ground state cannot be a *pure* $L = 0$ state. If it were, the charge distribution would be spherically symmetric and the quadrupole moment would vanish. There must be some admixture of higher angular momentum states, and the most likely is the $L = 2$, the $^3D_1$ state. Detailed calculations indicate that the ground-state wave function is a linear superposition of $L = 0$ and $L = 2$ functions such that the probability of finding the deuteron in the $L = 2$ state is 0.07.

Why do we not consider the possibility of an $L = 1$ admixture? The answer is that to the best of our knowledge the strong nuclear interaction is parity-conserving so that it will admix only even states with the $L = 0$ state.

It would be nice if we could now proceed and confirm our result for the size and depth of the potential well by comparing the energies of excited states of the deuteron with calculated values. Alas, nature is not obliging; the deuteron does not have any bound excited states. To gather additional information on the two-body interaction between the proton and neutron, we must turn to scattering experiments.

Numerous such experiments extending over a wide range of energies have been performed over the years, and it is not our intention to review the results in detail here. Two crucial facts have emerged. First, the n-p force is *spin-dependent*. That is, if we try to express the results of such experiments in terms of a simple potential, like the square well, the depth of the potential depends on whether the spins of the two nucleons are parallel or antiparallel. We recall that the ground state of the deuteron is the triplet state, so that from deuteron data alone we can say nothing about the potential in the singlet state.

In addition to n-p scattering, one can, and indeed with greater ease, study p-p scattering. Once allowance is made for the Coulomb repulsion between the two protons, one finds that the same force parameters that account for n-p scattering also give excellent agreement for p-p scattering. In other words, the nuclear force is *charge-independent*. There is compelling evidence that the n-n force is no different from the n-p and p-p force. That evidence comes from a comparison of spins and energies of ground and excited states of so-called mirror nuclei, for example, $^{17}_{8}O$ $(Z = 8, N = 9)$ and $^{17}_{9}F$ $(Z = 9, N = 8)$. Once the difference in the Coulomb energy is taken into account, one finds that the energies of the ground and excited states are practically identical. Replacing a neutron by a proton does not affect the nuclear energy level structure significantly.

---

7. The binding energy of a nucleus is the energy required to separate it into its constituent protons and neutrons (see Section 14.4).

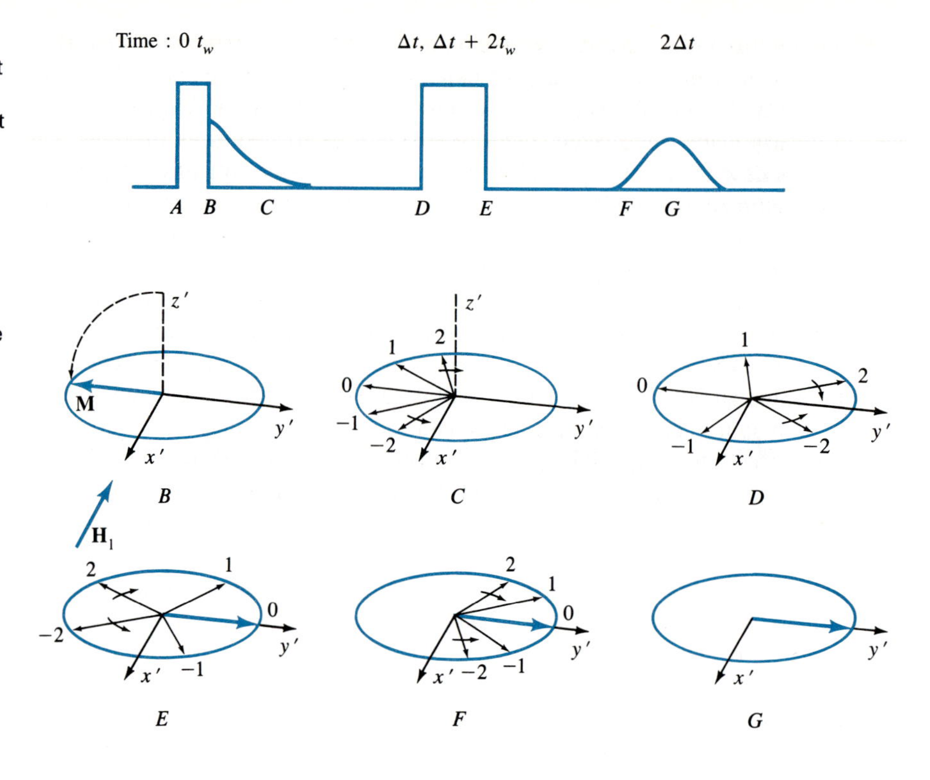

Figure 14.15 The spin-echo technique. A 90° pulse is applied at *A*, resulting in a free-induction signal that decays exponentially. At *D*, a time Δt following the initial 90° pulse, a 180° pulse is applied. At time $2\Delta t$ one then observes another free-induction signal, the spin echo. Here dephasing due to field inhomogeneities is assumed. The behavior of the nuclear magnetic moments during the time sequence is shown in parts *B* through *G*, which correspond to the points so marked in the upper figure.

If, as is often true, $T_1 \gg T_2$, T_2 can be obtained directly by simply observing the exponential decay of the free-induction signal. Let us assume, for the moment, that T_2 is short because of field inhomogeneities that dephase the individual nuclear magnetic moments as they precess about their own fields. Suppose that at a time Δt following the initial 90° pulse and after the free-induction signal has decayed to zero, but in a time short compared with T_1 or T_{2s}, we apply a "180° pulse," that is, an rf pulse that turns the magnetic moment of every nucleus through 180° in the rotating reference frame. Following application of this pulse, the nuclear moments will continue to precess about the field $\mathbf{H}_0$, but now in the opposite sense. The ones that between $t = 0$ and $t = \Delta t$ precessed, say, clockwise, too rapidly because their local magnetic field was greater than H_0 will again precess more rapidly than average, but now in a counterclockwise sense. As a result, all of the individual nuclear moments will again be in phase at the time $t = 2\Delta t$, and one will observe an exponential increase in the free-induction signal, followed by another exponential decay. In the limit as T_1 and $T_{2s} \rightarrow \infty$, the peak amplitude of this *spin echo* should be exactly the same as that of the original free-induction signal.

Suppose, however, that $T_2' > T_{2s} \simeq T_2 \ll T_1$. Since mutual spin flip is a random statistical process, dephasing of nuclear moments due to this process cannot be reversed by the spin-echo method. Thus, the signal pattern observed in a sequence consisting of an initial 90° pulse, followed at intervals $2\Delta t$ by successive 180° pulses, will appear as in Figure 14.17*b*. The exponential decay of each individual free-induction pulse is a measure of T_2, while the decay of the peak of the successive spin-echo signals is a measure of T_{2s}.

The pulse method can also be adapted to the measurement of T_1. We start with our sample at thermal equilibrium, apply a 90° pulse, and observe a free-induction signal. If we wait a long time $\Delta t \gg T_1$ before applying another 90° pulse, the

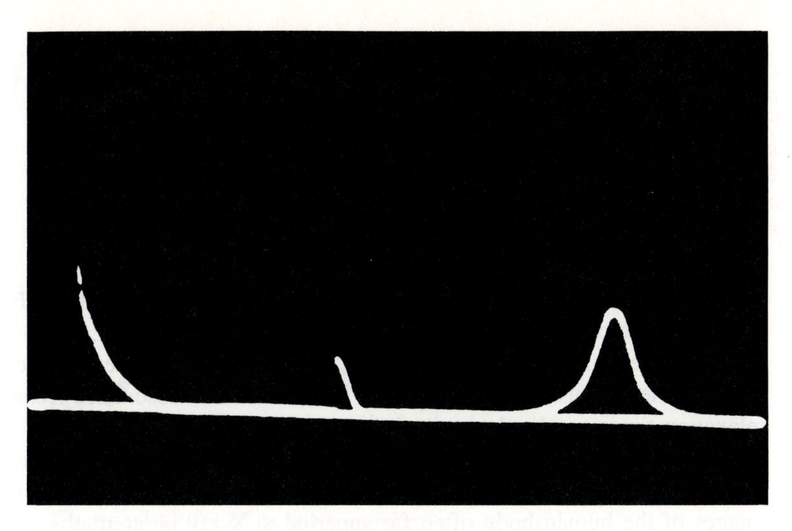

Figure 14.16 An experimental spin-echo signal from protons in water. The pulses are not shown but can be discerned from the free-induction decays. *(I. J. Lowe,* Solid States Physics, *vol. 2, Academic Press, New York, 1956, p. 58.)*

magnetization will have attained its equilibrium value again and the amplitude of the free-induction signal will be as great as initially. If, however, $\Delta t \simeq T_1$, the z component of the magnetization will not have recovered fully, and the ensuing free-induction signal will be of smaller amplitude than at first. In practice, one measures the amplitude of the free-induction signal following a large number of 90°

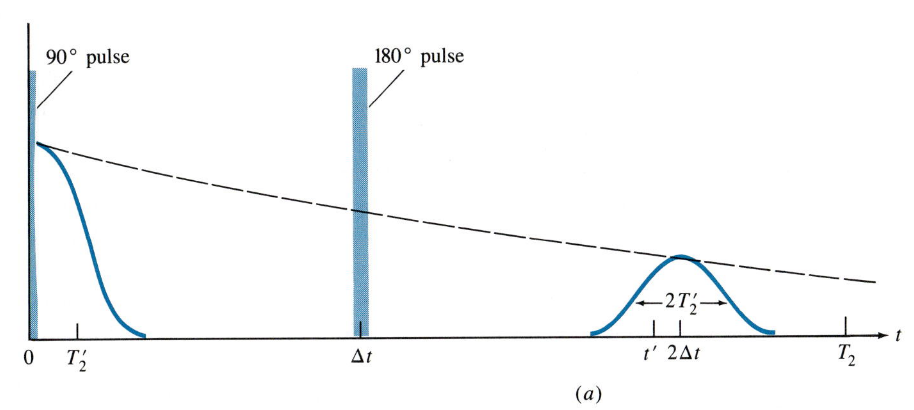

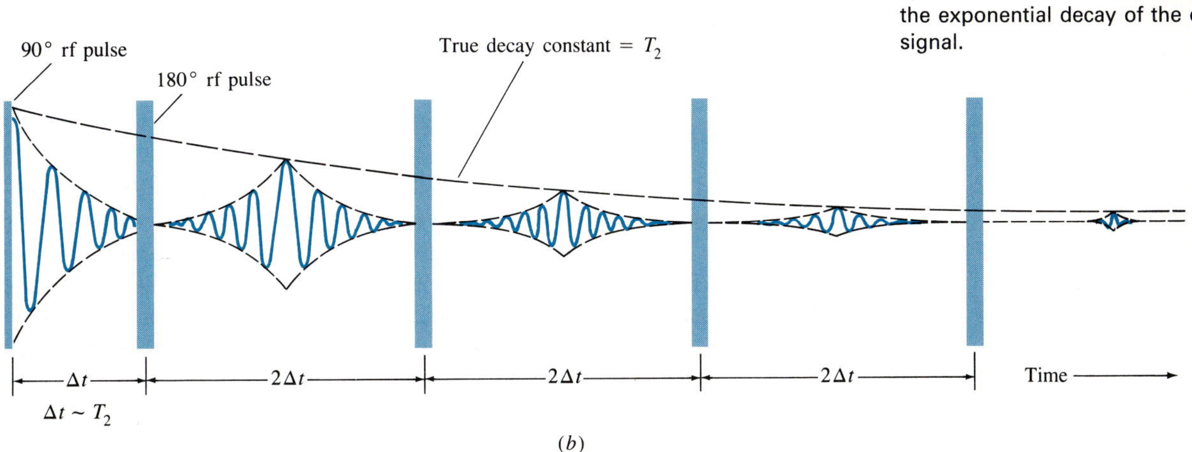

Figure 14.17 Measurement of the true spin-spin relaxation time T_{2s} and of T_2' using the spin-echo method, when $T_2' < T_{2s}$. (a) The free-induction decay and width of the spin echo give T_2'. The decay of the amplitude of the spin-echo signal is a measure of the true spin-spin relaxation time T_{2s}, provided $T_2 \ll T_1$. The exponential decay of the spin-echo amplitude, shown dashed in the figure, can be mapped out by varying the time interval between the 90° and 180° pulses. (b) An alternative procedure is to follow the initial 90° pulse with a series of equally spaced 180° pulses and observe the exponential decay of the echo signal.

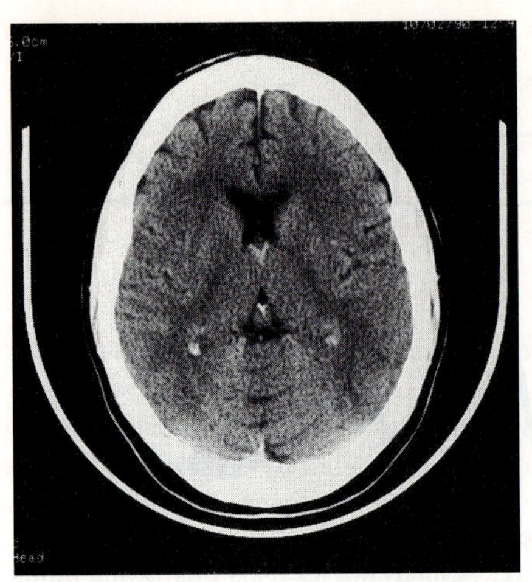

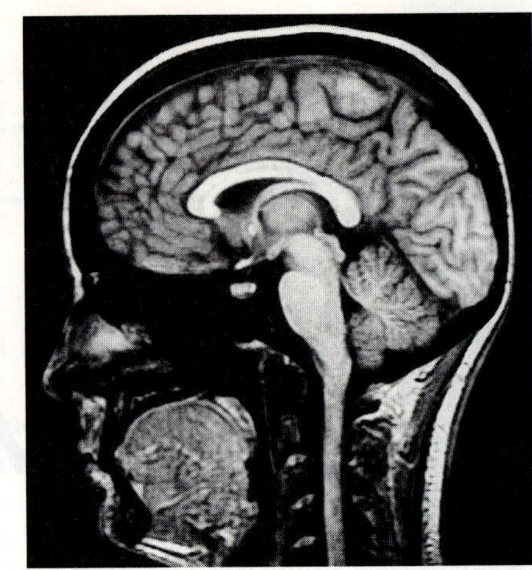

Figure 14.20 (*a*) An X-ray CAT scan image of the brain and (*b*) a MRI saggital image of the head. The excellent contrast between diverse soft tissue and fine anatomical detail demonstrate the superiority of MRI for medical applications. *(Courtesy GE Medical Systems.)*

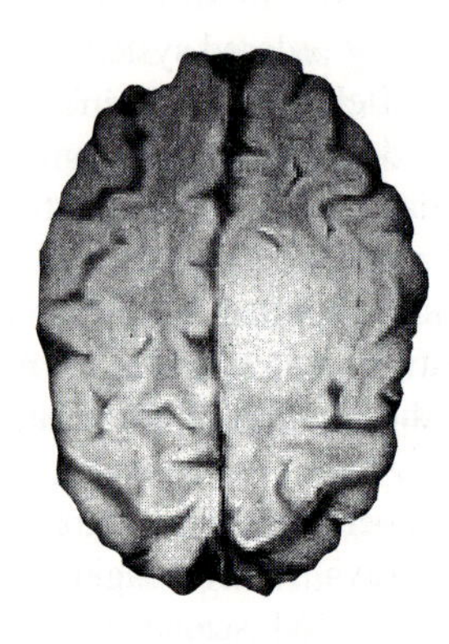

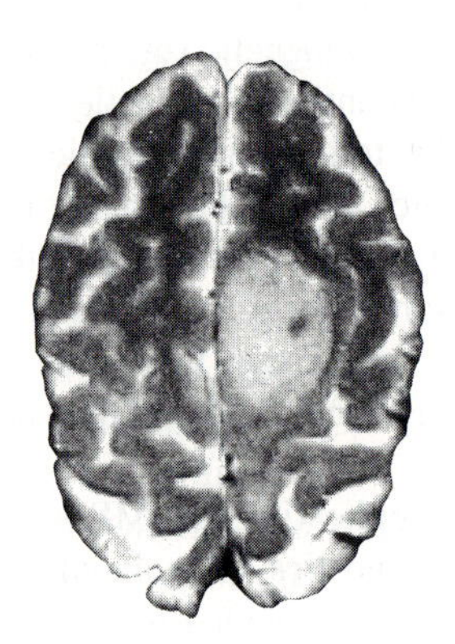

Figure 14.21 Two MRI images of the brain, using different imaging procedures. (*a*) A spin-density weighted image and (*b*) a T_2 weighted image. The tumor is much better revealed in the T_2 weighted image. *(Courtesy Drs. J. Creary and L. Partain, Vanderbilt University Medical Center.)*

between healthy tissue and a tumor.

Imaging is accomplished by placing the sample in a nonuniform field so that only that segment of the sample in the region where the local field equals the resonance field H_0 will give an NMR signal. In most modern systems the primary magnet is a large superconducting solenoid. A linear field gradient is then superimposed on this uniform field by means of additional coils. By controlling the currents in these coils the magnitude and direction of the field gradient can be adjusted, and scans in different directions can be obtained without physically rotating the patient in the field.

SUMMARY

Nuclei are composed of Z protons and N neutrons. The sum of Z and N is the *atomic mass number* A. A particular nuclide is represented by the symbol

$$^{A}_{Z}\text{X}$$

where X is the chemical symbol of the element. Nuclides with the same Z but different A are *isotopes* of the element of atomic number Z.

The proton has a mass of 1.007276 u ($1 \text{ u} = 1.6606 \times 10^{-27}$ kg $=$ 931.50 MeV/c^2), carries a charge of $e = 1.602 \times 10^{-19}$ C, has spin $\frac{1}{2}$ and a magnetic moment of $2.79\mu_N$, where $\mu_N = 5.051 \times 10^{-27}$ J/T is the *nuclear magneton*. The neutron has a mass of 1.008665 u, carries zero charge, has spin $\frac{1}{2}$ and a magnetic moment of $-1.91\mu_N$.

Nuclei are of roughly spherical shape with radius

$$R = R_0 A^{1/3} \qquad R_0 \simeq 1.2 \times 10^{-15} \text{ m} \tag{14.4}$$

The nuclear (strong) force is of very short range, and is charge-independent and spin-dependent. The force is mediated by pions.

The binding energy (BE) of a nucleus is given by

$$\text{BE} = (ZM_{\text{H}} + Nm_{\text{n}} - {}^{A}_{Z}M)c^2 \tag{14.8}$$

where M_{H}, m_{n}, and ${}^{A}_{Z}M$ are the masses of hydrogen, of the neutron, and of the atom of atomic mass number A and atomic number Z. The BE of stable nuclei with $A > 20$ is between about 8 and 9 MeV per nucleon. For $A > 56$, the BE of stable nuclei gradually decreases with increasing A. The BE of nuclei with $A > 30$ is represented quite well by the semiempirical mass formula, Equation (14.9), which is based on the liquid drop model.

The nuclear shell model explains the unusually large BE of "magic-number" nuclei and many features of nuclear energy-level patterns.

In nuclear magnetic resonance (NMR) transitions are induced between nuclear Zeeman levels by means of a radio-frequency magnetic field. NMR is widely employed in analytical chemistry, in solid-state physics, and in medical diagnostics.

Bibliography

Nuclear Physics

Roger C. Barrett and Daphne F. Jackson, *Nuclear Sizes and Structure,* Oxford University Press, New York, 1977.

John M. Blatt and Victor F. Weisskopf, *Theoretical Nuclear Physics,* Wiley, New York, 1952.

A. Bohr and B. R. Mottelson, *Nuclear Structure,* W. A. Benjamin, New York, 1975.

I. Kaplan, *Nuclear Physics,* Addison-Wesley, Reading, Mass., 1955.

Kenneth S. Krane, *Introductory Nuclear Physics,* Wiley, New York, 1987.

NMR and NMR Imaging

P. C. Lauterbur, "Image Formation by Induced Local Interactions: Examples Employing Nuclear Magnetic Resonance," *Nature* **242,** 190 (1973).

P. R. Moran, R. J. Nickles, and J. A. Zagzebski, "The Physics of Medical Imaging," *Physics Today,* July 1983, p. 36.

C. L. Partain, A. E. James, Jr., F. D. Rollo, and R. R. Price, *Nuclear Magnetic Resonance Imaging,* Saunders, Philadelphia, 1983.

I. L. Pykett, "NMR Imaging in Medicine," *Scientific American,* May 1982, p. 78.

R. T. Schumacher, *Introduction to Magnetic Resonance,* W. A. Benjamin, New York, 1970.

C. P. Slichter, *Principles of Magnetic Resonance,* Harper & Row, New York, 1963.

(a) Conservation of electric charge
(b) Conservation of mass-energy
(c) Conservation of linear momentum
(d) Conservation of angular momentum
(e) Conservation of nucleon number

For many years it was also believed that one more conservation law must be obeyed, known as conservation of parity.

It can be shown that conservation of energy, linear momentum, and angular momentum are direct consequences of fundamental symmetries of nature. Conservation of energy derives from the invariance of the laws of nature with respect to a time translation. Conservation of linear momentum follows from the fact that space is invariant under a simple translation, that is, to a shift of our coordinate origin. Conservation of angular momentum reflects the invariance of space under a rotation operation.

Conservation of parity derived from the apparent fact that space is invariant under an inversion operation, that is, replacing x by $-x$, y by $-y$, and z by $-z$. (It is conservation of parity that prohibits admixture of odd and even angular momentum states in the deuteron.) Since inversion is equivalent to a 180° rotation followed by a mirror reflection in the plane perpendicular to the rotation axis, parity conservation was tantamount to saying that nature does not distinguish between right-handed and left-handed systems. However, it is now well established that in some interactions, specifically, the interaction responsible for β decay, parity is not conserved. In β decay, nature does distinguish between right- and left-handedness!

Let us now examine the consequences of each conservation law.

(a) In all nuclear reactions, charge must be conserved. Thus in the α decay of the parent P to the daughter D

$$ {}^{A_P}_{Z_P}\mathrm{P} \rightarrow {}^{A_D}_{Z_D}\mathrm{D} + {}^{4}_{2}\mathrm{He} $$

we must have

$$ Z_P = Z_D + 2 \qquad A_P = A_D + 4 $$

whereas in β^- decay[1]

$$ {}^{A_P}_{Z_P}\mathrm{P} \rightarrow {}^{A_D}_{Z_D}\mathrm{D} + {}_{-1}\mathrm{e} + \bar{\nu}_e $$

$$ Z_P = Z_D - 1 \qquad A_P = A_D $$

(b) In radioactive decay, a parent nucleus P decays to a daughter nucleus D with the emission of a particle x. If P is initially at rest and x and D have some kinetic energy following the decay, conservation of mass-energy requires that

$$ M_P c^2 = (M_D + M_x)c^2 + E_D + E_x $$

---

1. There are some radioactive isotopes that decay by the emission of a positron, the antiparticle of the electron. Both processes are referred to as β decay, but are distinguished by means of a positive or negative superscript.

Beta decay is always accompanied by the emission of a neutrino (ν) or antineutrino ($\bar{\nu}$). For further details, see Section 15.3(b).

where E_D and E_x are the kinetic energies of D and x following the decay. (Both particles must carry off some kinetic energy to conserve linear momentum; see the discussion of conservation of linear momentum, which follows.) The total energy released in the decay process, conventionally denoted by the symbol Q, is

$$Q = (M_P - M_D - M_x)c^2 \tag{15.8}$$

If $Q > 0$, the process is *exothermic;* if $Q < 0$, it is *endothermic*. In the latter case, energy would have to be supplied from some external source to make the reaction proceed. Hence all spontaneous radioactive decays are exothermic; i.e., $Q > 0$. If the decay process involves the emission of a γ ray, M_x, the rest mass of the emitted particle, is zero, and E_x, the particle's kinetic energy, must be replaced by $h\nu$.

(c) To illustrate momentum conservation, we examine the decay of the radioactive isotope $^{227}_{90}\text{Th}$:

$$^{227}_{90}\text{Th} \rightarrow {}^{223}_{88}\text{Ra} + \alpha$$

The kinetic energy of the emitted α particle is 6.04 MeV. The atomic masses of the three participants are

^{227}Th . . . . . . 227.027703 u
^{223}Ra . . . . . . 223.018501 u
^{4}He . . . . . . 4.002603 u

If we now apply Equation (15.8), we find that for this reaction

$$Q = (0.006599 \text{ u})(931.5 \text{ MeV/u}) = 6.15 \text{ MeV}$$

which is slightly larger than the kinetic energy of the emitted α particle. The difference, 0.11 MeV, is the energy acquired by the daughter nucleus, which must recoil to conserve linear momentum. Since the parent nucleus was at rest before the disintegration, momentum conservation requires that

$$M_D v_D = M_\alpha v_\alpha \tag{15.9}$$

with v_D and v_α oppositely directed. We now square both sides of Equation (15.9) and divide by $2M_D$. Thus

$$E_D = \frac{M_D v_D^2}{2} = \frac{(M_\alpha/M_D)M_\alpha v_\alpha^2}{2} = \left(\frac{M_\alpha}{M_D}\right) E_\alpha$$

The total kinetic energy of the reaction products, $E_D + E_\alpha$, is

$$Q = E_\alpha\left(1 + \frac{M_\alpha}{M_D}\right) \tag{15.10}$$

For the disintegration of ^{227}Th,

$$Q = 6.04(1 + \tfrac{4}{223}) \text{ MeV} = 6.15 \text{ MeV}$$

in agreement with our calculation based on atomic masses.

Since the mass numbers of nuclei that are natural α emitters are much greater than 4, the difference between the Q of the reaction and the α particle's kinetic energy is only a few percent; small, but not negligible.

In β decay, the daughter nucleus also recoils. However, the mass of the electron (or positron) is only about 1/10,000 of the mass of the α particle, and in β decay the recoil energy carried off by the daughter nucleus is truly negligibly small. Even in the β decay of the neutron the recoil energy of the proton is no more than 0.05 percent of Q.

Following α or β decay, the daughter nucleus is often in an excited state, and then decays to the ground state with the emission of a γ ray. Again, momentum conservation must be obeyed. The momentum of the γ ray is

$$p_\gamma = h\nu \text{ cd}$$

and this must be the recoil momentum of the daughter. Hence its recoil energy is

$$E_\text{D} = \frac{p_\gamma^2}{2M_\text{D}} = \frac{E_\gamma^2}{2M_\text{D}c^2} \tag{15.11}$$

For typical γ-ray energies of 10^5 eV = 0.1 MeV and $M_\text{D} \simeq 50$ u,

$$E_\text{D} \simeq \frac{(10^{-1})^2}{2(10^4)(50)} = 10^{-8} \text{ MeV} = 0.01 \text{ eV}$$

Although very small compared to the γ ray energy, this recoil energy is often much greater than the natural line width of the γ ray. As we shall see (Section 15.4), this small but finite shift in γ ray energy due to recoil has important consequences.

(d) In the rest frame of the parent nucleus, the total angular momentum of the system is the spin angular momentum of the parent nucleus, $\hbar I_\text{P}$. The angular momentum after decay is the vector sum of the spin angular momenta of the decay products and their orbital angular momenta. Orbital angular momentum quantum numbers are always integers; the spins of nuclei may be integers or half integers, depending on whether A is even or odd. Since the spin of the α particle is 0, the spin of the daughter in α decay must be integral if the spin of the parent is integral, and the spin of the daughter must be half-integral if the spin of the parent is half-integral. The spins of daughter and parent may differ by an integer, however, since the decay products may come off with finite orbital angular momentum.

Conservation of angular momentum was a compelling argument for the emission of a neutrino in β decay. Since the neutron, the proton, and the electron are all spin-$\frac{1}{2}$ particles, it is impossible to conserve angular momentum in the decay of the neutron without the simultaneous emission of another spin-$\frac{1}{2}$ particle. Although the β decay of the neutron was not discovered until decades after the neutrino had been postulated by Pauli, it was known at that time that in every β decay process the difference between the spins of parent and daughter was an integer, never a half integer.

(e) In every radioactive decay and every low- to medium-energy nuclear reaction, the total number of protons plus neutrons, i.e., the total number of nucleons, is conserved. Apparently it is impossible to create nucleons by converting energy to rest mass. It is, of course, possible to create particle-antiparticle pairs; for example, electron-positron pairs can be created with the expenditure of $2m_ec^2$ of energy. Similarly, it is possible to create a proton-antiproton pair at a cost of $2m_pc^2$ of energy. In all such events charge is conserved, since the charge of the antiparticle is the negative of the charge of the particle of this pair. To allow for such events, the nucleon number that is assigned to an antiproton or an antineutron is -1. The creation of an antiproton-proton or antineutron-neutron pair leaves the nucleon number unchanged, and the same is true when a nucleon and its antinucleon annihilate.

Conservation of nucleon number is a special case of a more general conservation law, the conservation of baryon number. We shall examine this more general rule in Chapter 17 when we consider reactions at very high energies and the elementary particles that are created in such events.

15.3. RADIOACTIVE DECAY PROCESSES

15.3(a) Alpha Decay

Already in 1911 Geiger and Nuttall discovered a remarkable regularity in α decay. For some reason, the half-life of natural α emitters was a very sensitive function of the range of the α particles in air, that is, of their energy. In other words, there seemed to be an intimate relation between the probability for α decay and the Q value. Figure 15.4 is a plot of the logarithm of the half-life versus $E_\alpha^{-1/2}$, and it is evident from this plot that a functional relation of the form

$$\ln t_{1/2} = AE_\alpha^{-1/2} + B \tag{15.12}$$

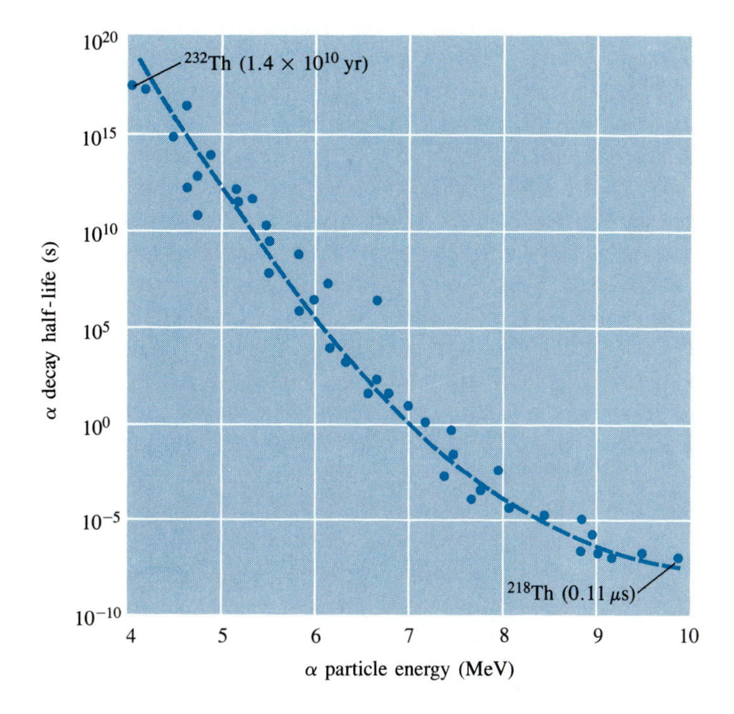

Figure 15.4 The Geiger-Nuttall relation for α emitters shown on a semilogarithmic plot.

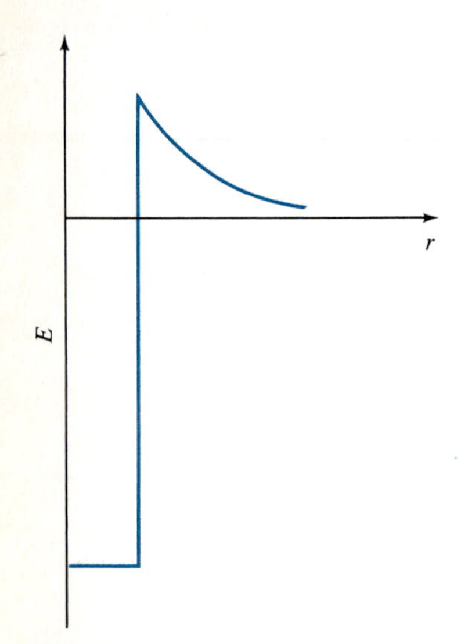

Figure 15.5 The potential energy seen by an α particle inside and near a nucleus (schematic diagram). Outside the nuclear radius the potential is the Coulomb potential $V(r) = 2(Z - 2)e^2/4\pi\epsilon_0 r$.

fits the data very well. It is worth noting here that although the range in kinetic energy of the α particles from the various decays is not very large, the variation in half-lives is enormous, from about 0.11 microsecond to 1.4×10^{10} years.

An explanation for this empirical rule was given some years later by Gamow,[2] and is based on a purely quantum mechanical phenomenon, tunneling through a potential barrier (see Section 7.8). Figure 15.5 is a crude approximation of the potential seen by an α particle in the vicinity of a heavy nucleus. When the α particle is outside the range of the nuclear forces it feels only the repulsive Coulomb force due to the Z protons of the nucleus. Once it is within the range of the nuclear force, however, the potential energy drops precipitously. From the binding energy per nucleon, roughly 7 MeV/nucleon, we can estimate that the depth of that potential well is in the neighborhood of 40 MeV.

Let us now consider a heavy nucleus, consisting of a large number of protons and neutrons. From our discussion of the Fermi gas model we know that the α particle is an especially stable configuration, not only as a free particle but also as a nuclear constituent; recall that the binding energies of "alpha particle nuclei" are exceptionally high. Hence, it is reasonable to hypothesize that within a heavy nucleus the motion of two protons and two neutrons may be highly correlated so that they simulate an α particle.

In Chapter 7 we saw that the solution of the wave equation for a particle confined to a potential well leads to a set of sharp energy levels. Although we performed detailed calculations only for the simplest case, a one-dimensional square well, we know from the study of the hydrogen spectrum that the same conclusion holds in three dimensions. The details of the energy-level structure for the bound states depend on the form of the potential, but that need not concern us here.

In Figure 15.6*a* we show three possible energy levels for an α particle inside the potential well of Figure 15.5. Our zero of energy is that of an α particle at rest at a great distance from the nucleus. Classically, if the energy of the α particle inside the nucleus were less than the height of the potential barrier it could never escape. However, we know from the analysis of the one-dimensional finite potential barrier that to satisfy the boundary conditions the wave function will have a finite though small amplitude outside the barrier. This implies that if its energy is greater than zero, i.e., if it is in the state of energy E_2 or E_3, there is a small but finite probability that the particle will appear outside the barrier. Since the kinetic energy of the emitted α particle is much smaller than the height of the Coulomb barrier (which is easily calculated), we know that the particle has not absorbed the requisite energy to surmount the barrier as a result of statistical fluctuations. Instead, it has tunneled through the barrier.

The rate of α particle emission due to tunneling is the product of two factors: the number of times per second that the α particle inside the nucleus strikes the potential barrier and the probability that it will tunnel through the barrier whenever it strikes. For an α particle of a few MeV kinetic energy, the time required to traverse the nuclear diameter is of the order of 10^{-22} s. Thus, to yield a lifetime of the order of seconds to years, the tunneling probability must be very, very small.

The tunneling probability depends on the height of the barrier, $V - E$, and on its width at the energy E. For a square barrier the tunneling probability is propor-

2. George Gamow, *Zeit. f. Phys.* **51,** 204 (1928).

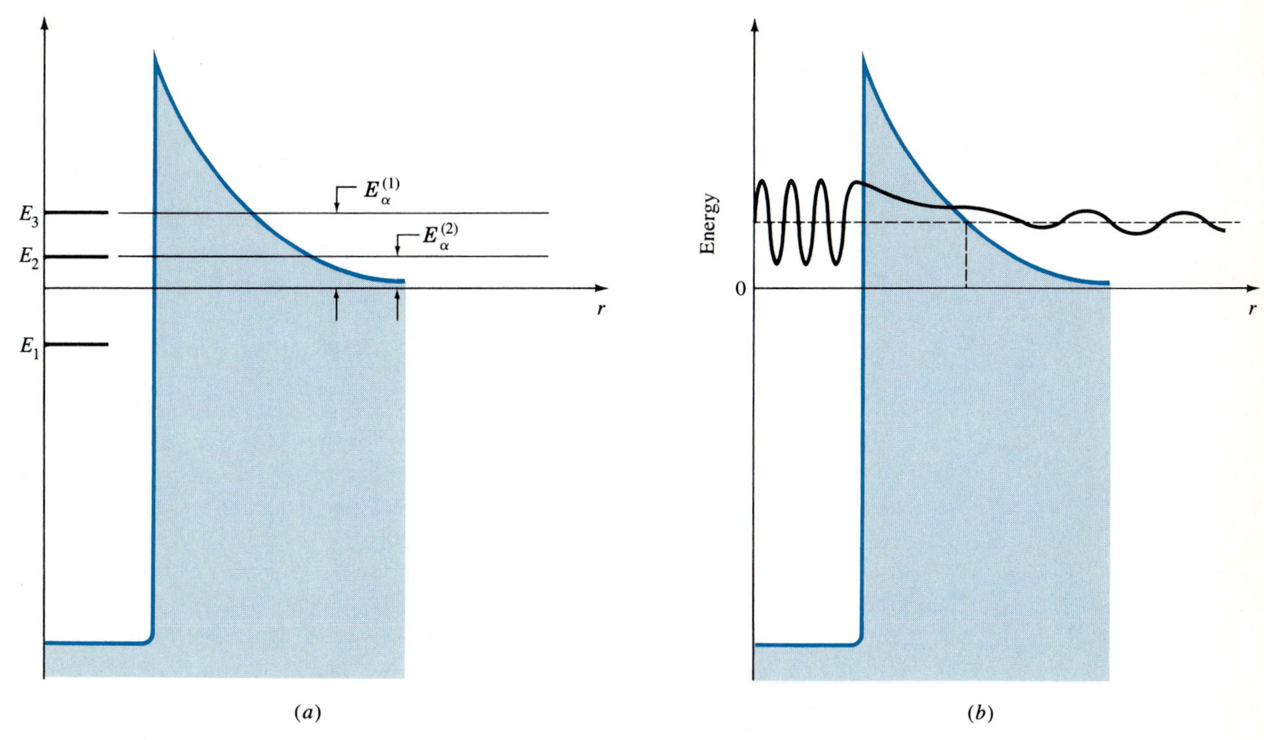

Figure 15.6 Tunneling of an α particle through the potential barrier. (*a*) Three possible energy states of an α particle in the potential well of Figure 15.5. If the α particle occupies the state E_1 it cannot escape. If it occupies the state E_2 or E_3 it can tunnel through the potential barrier and will have a kinetic energy as indicated. (*b*) The wave function of an α particle that can tunnel through the potential barrier.

tional to the exponential e^{-kD}, where D is the thickness of the barrier and $k = \sqrt{(2m/\hbar^2)(V - E)}$. Although the barrier that the α particle sees when inside the nucleus is not of this simple form, we may still expect that a relation of this sort will be valid. The empirical Geiger-Nuttall law follows directly from the exponential dependence of the tunneling probability on the parameter k. A detailed quantum mechanical solution of the problem yields values for the constants A and B of Equation (15.12) in good agreement with experiment.

15.3(b) Beta Decay

The term β decay refers to three different processes. In each, the mass number A of the nucleus remains constant but Z and N change by ± 1. In β^- decay a neutron changes into a proton in the parent nucleus and an electron is emitted. In β^+ decay a positron is emitted as a proton changes into a neutron.

The third process, which competes with positron emission, also results in the transformation of a proton into a neutron; an atomic electron is captured by the nucleus. This process can occur only if there is a finite probability that the atomic electron will be within the nucleus prior to capture. Consequently, it is principally the 1s and, with much lower probability, the 2s electrons that are involved in electron capture, and one often refers to *K-* or *L-capture,* depending on the orbital shell that the extranuclear electron occupied prior to capture.

The simplest example of β^- decay is that of the free neutron. The half-life for this decay is about 11 minutes. Since the decay products are a proton, a neutron, and an electron, the Q value of the reaction is

$$Q = (m_n - m_p - m_e)c^2 = (m_n - M_H)c^2 = 0.779 \text{ MeV}$$

where M_H is the mass of the hydrogen atom. In other β^- decay processes a nucleus of mass number A and atomic number Z_P changes to one of mass number A and atomic number $Z_D = Z_P + 1$ with the emission of an electron. As in the decay of the neutron, we can calculate the Q value from the difference between the *atomic masses* of parent and daughter. Using atomic masses rather than nuclear masses automatically takes account of the mass of the electron that is created in the decay.

If β decay were as simple as we have just described, the energy Q would be shared between the emitted electron and the recoiling daughter nucleus. From our earlier discussion of recoil energies we know that in this case the recoil energy of the daughter nucleus is only a very small fraction of Q. Hence, we should find that all electrons from a given β emitter have the same energy Q. However, the observed energy spectrum of the electrons encompasses the broad range from zero to Q (see Figure 15.7). The apparent violation of energy conservation in the vast majority of decays was a great conundrum until Pauli suggested that a third particle is also emitted, and that this third particle carries off some fraction of the decay energy. As was pointed out earlier, this particle was also needed to conserve angular momentum, and must have a spin of $\frac{1}{2}$. Since in some events the electron does have the expected kinetic energy Q, this third particle must have negligible or zero rest mass (no measurable energy is expended in its creation). Finally, it must have zero charge, for otherwise charge conservation would be violated. Fermi christened this hypothetical particle the *neutrino*.

It is now known that there are several kinds of neutrinos, the electron neutrino ν_e, which is involved in β decay; the muon neutrino ν_μ, which participates in the decay of the muon (see Section 17.6); and the ν_τ. All of these have their antiparti-

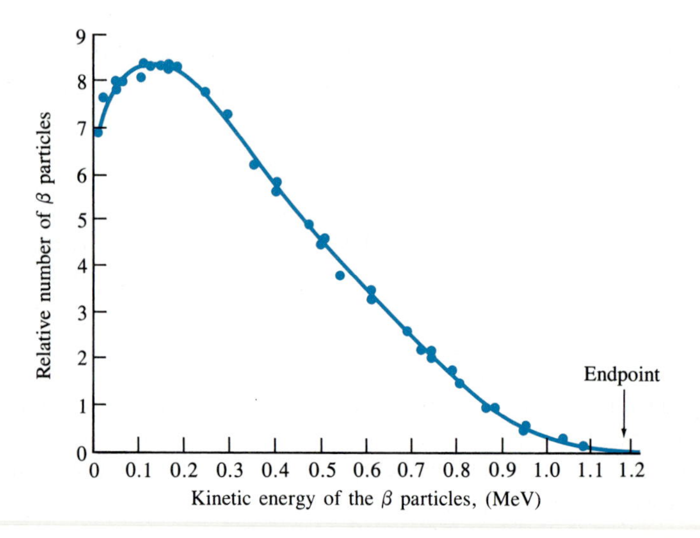

Figure 15.7 The energy spectrum of β particles emitted from ^{210}Bi.

cles, $\bar{\nu}_e$, $\bar{\nu}_\mu$, and $\bar{\nu}_\tau$; in β^- decay it is the electron antineutrino that is emitted. Thus the correct symbolic equation for the decay of the free neutron is

$$n \rightarrow p + e^- + \bar{\nu}_e \tag{15.13}$$

The spontaneous capture of an electron by a free proton is impossible; the exothermic reaction goes the other way. However, whenever

$$^{A}_{Z}M > {}^{A}_{Z-1}M$$

where, as before, the M's refer to atomic masses, electron capture is possible. Moreover, if

$$^{A}_{Z}M > {}^{A}_{Z-1}M + 2m_e$$

not only electron capture but also positron emission can occur. The latter, known as β^+ decay, has been observed in a number of artificially produced unstable isotopes. An example of β^+ decay is

$$^{22}_{11}\text{Na} \rightarrow {}^{22}_{10}\text{Ne} + e^+ + \nu_e$$

Only one natural positron emitter, ^{40}K, is known to exist; its half-life is more than a billion years, and only about one in 10^4 decays is via positron emission (see Appendix B).

EXAMPLE 15.3 As we have already stated, ^{231}Th is radioactive. It decays to ^{231}Pa—which is also unstable. What is the maximum energy of the electrons emitted by ^{231}Th?

SOLUTION The atomic masses of the parent and daughter are 231.036298 u and 231.035880 u. The Q of the decay is

$$Q = (231.036298 \text{ u} - 231.035880 \text{ u})(931.5 \text{ MeV/u}) = 0.389 \text{ MeV}$$

Since the mass of the emitted electron is negligibly small compared with that of the protactinium nucleus, we can neglect the recoil energy of the daughter. The maximum energy of the emitted electrons is 0.389 MeV.

EXAMPLE 15.4 What is the maximum energy of the positrons emitted in the decay of ^{22}Na?

SOLUTION The atomic masses of ^{22}Na and the decay product, ^{22}Ne, are 21.994434 u and 21.991383 u. The initial atomic mass includes that of eleven electrons; the atomic mass of ^{22}Ne includes only the mass of ten electrons. Hence, we must subtract from the right-hand side of the energy balance equation not only the mass of the positron that is created but also the mass of one more electron. The energy balance equation, Equation (15.8), therefore gives

$$Q = (M_P - M_D)c^2 - 2m_ec^2$$

and

$$Q = (21.994434 \text{ u} - 21.991383 \text{ u})(931.5 \text{ MeV/u}) - 2(0.51 \text{ MeV}) = 1.82 \text{ MeV}$$

As in the preceding example, we can again neglect the recoil energy of the daughter nucleus. The maximum energy of the emitted positron is therefore 1.82 MeV.

We must emphasize that neither the electron nor the positron emitted in β decay exists inside the nucleus prior to the decay. These particles are created in the process by the conversion of energy to rest mass. In that respect, β decay is fundamentally different from α decay, where the emitted particle is a constituent of the parent nucleus. In a manner of speaking, β decay is more nearly like the deexcitation of an excited atomic state in which a "particle," a photon, also not initially present, is emitted as the atom drops into a lower energy state.

There is, however, a basic difference between photon emission and the decay processes just described. The interaction between the nucleus and atomic electrons that results in bound atomic states and in the creation of photons during atomic transitions is the electromagnetic interaction. Electrons do not respond to the nuclear force, and since the neutron is uncharged, the electromagnetic interaction is not involved here. Indeed, if the electromagnetic interaction came into play, we should expect the inverse reaction, the combination of a proton and an electron to form a neutron, since the two opposite charges attract. To explain β decay one invokes a new interaction, and it must be a very weak one in view of the long half-lives of the decay on a nuclear time scale. It is therefore known as the *weak interaction,* or *weak force.*

Two aspects of this interaction deserve mention, although we can offer no more than that here. First, the weak interaction does not conserve parity. That is, in processes that involve this interaction, such as β decay, nature does distinguish between right- and left-handedness. Second, it has long been the dream of theoretical physicists to develop a scheme that could explain all the known forces of nature—the gravitational, weak, electromagnetic, and strong nuclear. Einstein devoted most of his lifetime to this task, but without success. In 1968 the first step in this direction was taken by Glashow, Salam, and Weinberg,[3] who devised a theory that encompasses the weak and electromagnetic interactions. This achievement has sparked renewed activity and new hope that a grand unification scheme will be found. In Chapter 17 we present a rather sketchy outline of these esoteric theories.

15.3(c) Gamma Rays and Nuclear Energy Levels

Almost without exception, the energies of α particles emitted by a radioactive sample fall into several sharply defined groups, as shown, for example, in

---

3. Sheldon L. Glashow, Abdus Salam, and Steven Weinberg shared the 1979 Nobel Prize in physics for "their unified model of the action of the weak and electromagnetic forces and for their prediction of the existence of neutral currents."

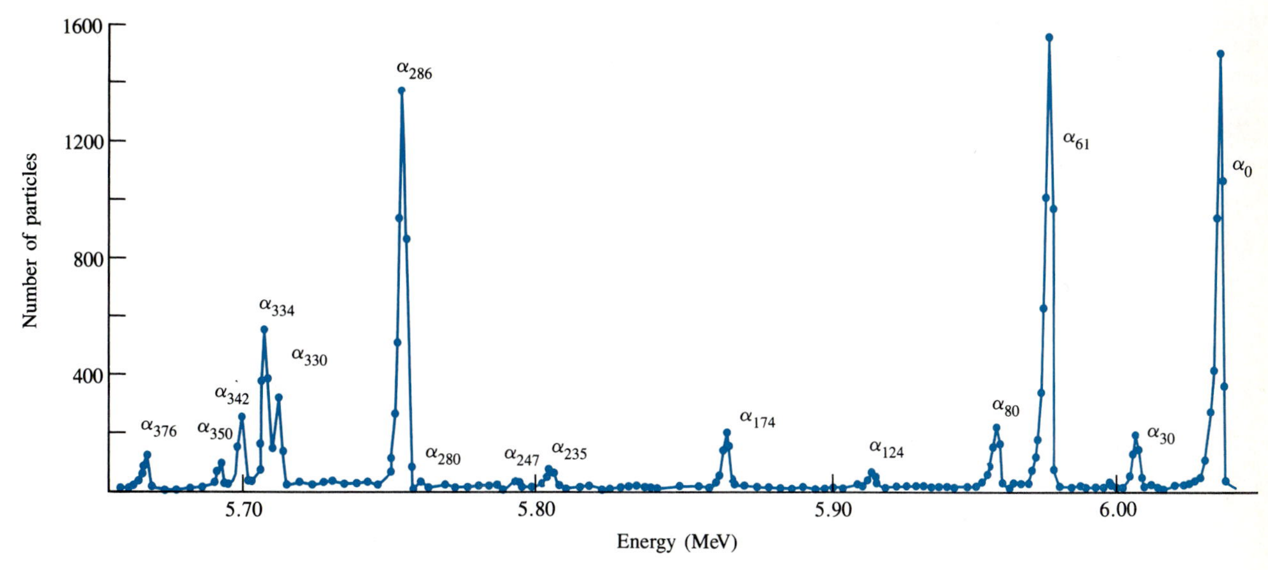

Figure 15.8 The spectrum of α particles emitted in the decay of ^{227}Th to ^{223}Ra. The various peaks in this spectrum are indicative of energy levels of the daughter nucleus.

Figure 15.8. Since presumably the parent nucleus was in its ground state prior to the decay, this spectrum of the emitted α particles suggests that the daughter nucleus can exist, at least for a brief time, in any of a large number of excited states. If that is so, we would expect the excited nucleus to decay to the ground state with the release of a photon, just as an atom in an excited state decays to the ground state with the emission of a quantum of radiation. The observation of such photons, that is, γ rays, whose energies correspond to the energy differences between the peaks of the α particle spectrum, supports that interpretation (see Figure 15.9). Emission of γ rays also generally accompanies β decay, showing that also in that case the daughter nucleus is often in an excited state.

15.4 MÖSSBAUER EFFECT

An atom in an excited state of energy E_1 decays to the ground state of energy E_0 with the emission of a photon, provided the transition is not forbidden by some selection rule. Similarly, an atom in the ground state can be raised to the excited state by absorbing a photon of the proper energy. In our earlier discussion of such processes we stated that the energy of the emitted photon is given by

$$h\nu = E_1 - E_0 \tag{15.14}$$

Equation (15.14) is, however, only approximate because in such a process not only energy but momentum must be conserved. The momentum of the photon that is emitted or absorbed is $h\nu/c$, and so the atom must recoil with that same momentum. The recoil energy is

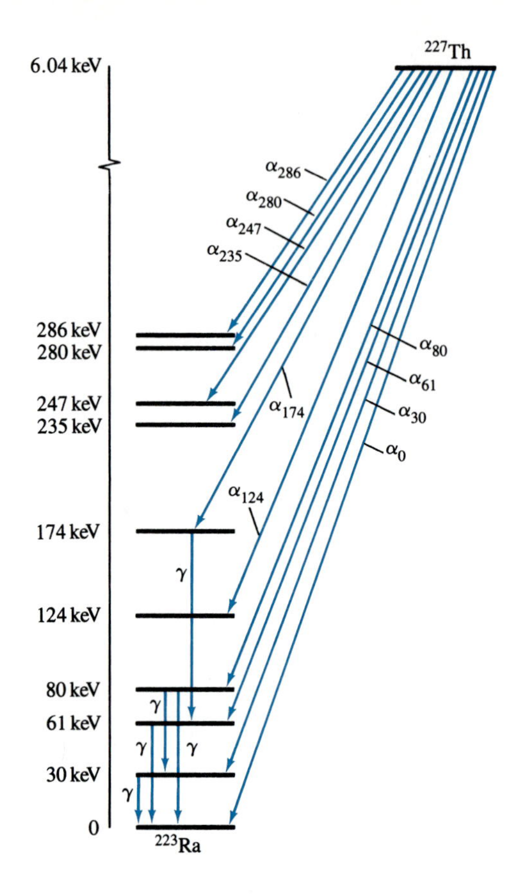

Figure 15.9 The energy levels of ^{223}Ra deduced from the results of Figure 15.8 and some of the γ ray transitions that follow the decay of ^{227}Th.

$$E_{\text{rec}} = \frac{p_{\text{rec}}^2}{2M} = \frac{(h\nu)^2}{2Mc^2} \tag{15.15}$$

For an atomic transition, $h\nu \simeq 1$ eV, and $Mc^2 \simeq A \times 10^9$ eV, where A is the atomic mass number. Thus the recoil energy is typically only about 10^{-10} eV.

Still, if the photon energy were precisely

$$h\nu = E_1 - E_0 - E_{\text{rec}}$$

it would be impossible to raise atoms from the ground state to an excited state by irradiating them with the light emitted by atoms that have made the inverse transition, a process known as resonance fluorescence. Yet resonance fluorescence does occur, and the phenomenon is widely used in the study of atomic energy states. This resonance absorption takes place because the energy of the excited state is broadened by the uncertainty relation

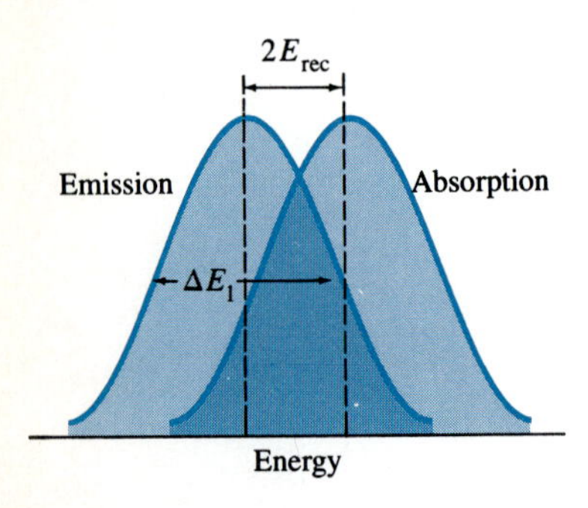

Figure 15.10 Atomic emission and absorption lines (schematic diagram). The energy shift due to recoil of the emitting or absorbing atom is much less than the line width.

$$\Delta E_1 \simeq \frac{\hbar}{\Delta t} \tag{15.16}$$

The lifetime of excited atomic states is about 10^{-8} s; hence the uncertainty broadening is about 10^{-7} eV, or roughly a thousand times greater than the energy shift due to recoil of the atom. Consequently, emission and absorption lines do overlap, and resonance fluorescence can be observed.

Consider now the analogous nuclear situation. Since the typical lifetime of an excited nuclear state is about 10^{-10} s, the uncertainty broadening is of the order of 10^{-5} eV, roughly a hundred times greater than for optical photons emitted by atoms. However, the recoil energy of a medium nucleus that emits a γ ray of about 100 keV is roughly 1 eV. Hence the energy shift due to recoil is now much greater than the natural line width of the γ ray, and resonance fluorescence cannot occur under ordinary circumstances. Prior to 1958 such resonance was studied by raising the temperature of the emitter and/or absorber so that the lines were broadened by the Doppler shift due to thermal motion of the nuclei. Another technique that proved successful was to place the source of γ rays on the perimeter of an ultracentrifuge, thus compensating for the recoil velocity.

In 1958 Mössbauer discovered that by placing the emitting and absorbing nuclei in a crystal, resonance fluorescence not only took place at ordinary tempera-

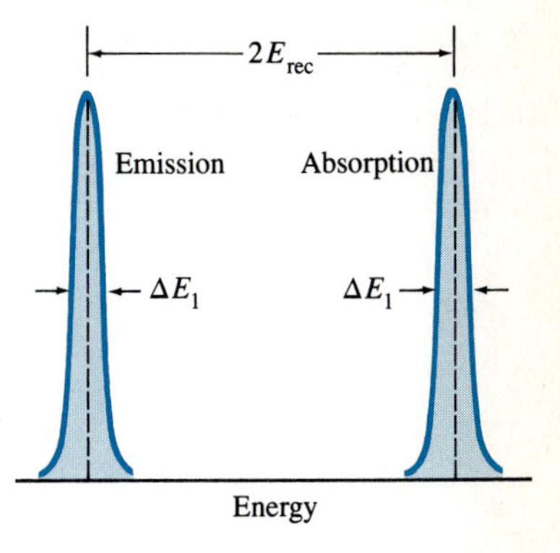

Figure 15.11 Emission and absorption lines for γ ray transitions of a nucleus (schematic diagram). The energy shift due to nuclear recoil is much greater than the natural line width.

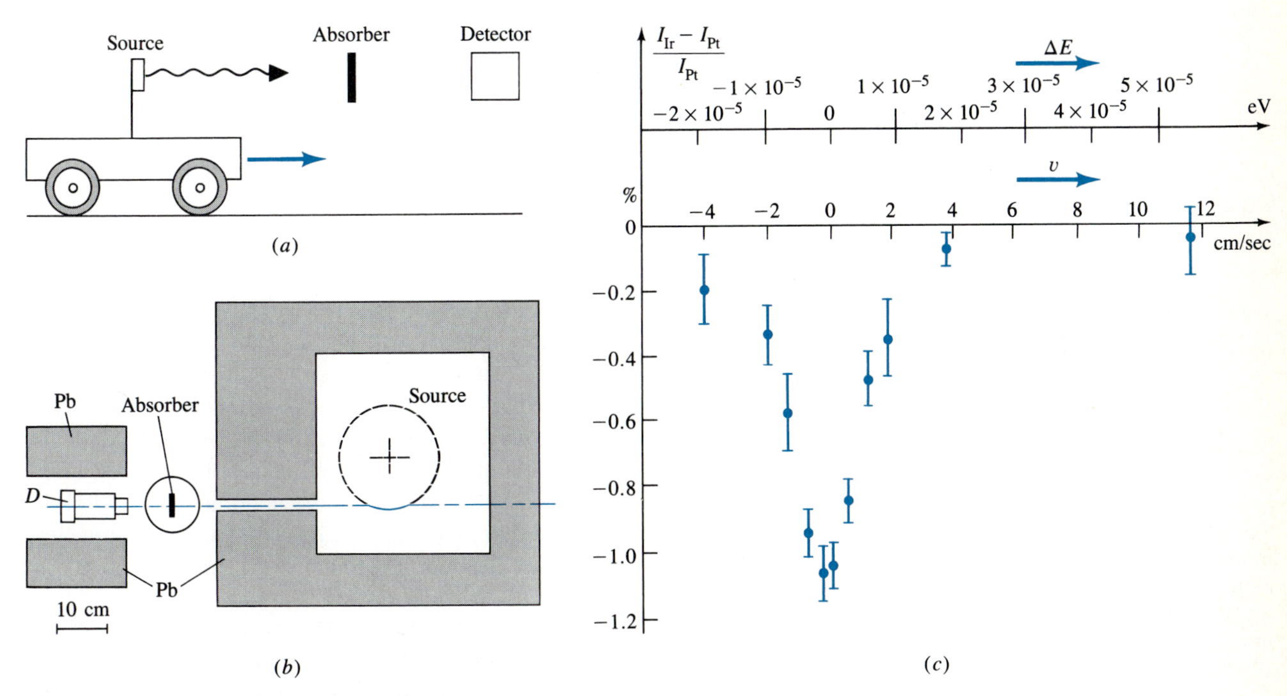

Figure 15.12 The Mössbauer effect. (*a*) Arrangement of source and absorber (schematic diagram). (*b*) The actual experimental arrangement used by Mössbauer. The source was deposited on the rim of a spinning wheel inside a lead shield. The detector *D* records only photons emitted when the source is in the solid portion of the path. (*c*) The shape of the emission line from the 129-keV γ transition in ^{191}Ir as measured using the apparatus shown in part *b*. Shown here is the relative difference in the intensity of the 129-keV γ ray when the detector is behind an iridium (resonance) absorber and when behind a platinum absorber for comparison, as a function of the relative velocity of source and absorber. Both source and absorber were at a temperature of 88 K. The half width of the line is approximately 0.65×10^{-5} eV.

tures but was, in fact, enhanced if the temperature of the crystals was lowered.[4] One can understand the Mössbauer effect by recognizing that the atoms in a crystal are so tightly bound to their lattice sites that the recoil energy is likely to be shared by all of the atoms of the crystal. It is as though the mass in the denominator of Equation (15.15) were now not the mass of a single nucleus but that of the entire crystal, about 10^{22} times as large. With the recoil energy thus diminished, emission and absorption lines again overlap. The probability of this so-called recoilless process is greater the lower the temperature of the crystal.

If the nuclear transition involves an isomeric state, the line is especially narrow. For example, ^{57}Fe has an isomeric state at 14.4 keV whose half-life is about 10^{-7} s. The line width is then only about 10^{-8} eV, or about one part in 10^{12}. With such a very narrow line it is possible to measure even the splitting of the γ ray transition due to the nuclear Zeeman effect, and one can thereby determine the magnetic field at the site of the emitting or absorbing nucleus.

In Mössbauer experiments it is common practice to ''sweep'' through the absorption line by moving either the emitter or absorber, thereby introducing a Doppler shift. The amount of Doppler shift required is quite small, so that the relative velocity of source and absorber is typically between 1 mm/s and 10 cm/s.

One of the more exciting applications of the Mössbauer effect involved the experimental verification of the gravitational red shift predicted by the theory of general relativity (see Chapter 3). In that experiment the source was placed about 20 m above the absorber and the change in frequency due to the minute change in gravitational potential was measured. The results were in perfect agreement with theory.

15.5 NUCLEAR REACTIONS

As Rutherford had foreseen, experimental nuclear physics really came into its own once machines capable of generating beams of energetic particles were available. Prior to 1932 information on nuclear structure was based on observations of radioactive decay and on scattering experiments using decay products, principally α particles, as projectiles. Newly constructed Van de Graaff electrostatic accelerators and cyclotrons gave physicists the freedom to study the interactions between nuclei and protons, deuterons, and α particles, and, even more important, to control the energies of these diverse projectiles over an ever increasing range. The first nuclear reaction produced by artificially accelerated particles was the reaction

$$\mathrm{p} + {}^{7}\mathrm{Li} \rightarrow {}^{8}\mathrm{Be} \rightarrow {}^{4}\mathrm{He} + {}^{4}\mathrm{He}$$

reported by Cockcroft and Walton in 1932.[5] Since then, literally thousands of nuclear reactions have been investigated in research laboratories in all parts of the world.

---

4. R. L. Mössbauer, *Z. Naturforsch.* **14,** 211 (1959). Rudolf L. Mössbauer shared the 1961 Nobel Prize in physics with Robert Hofstadter.

5. J. D. Cockcroft and E. T. S. Walton, *Proc. Roy. Soc.* **A137,** 229 (1932). Crockcroft and Walton shared the 1951 Nobel Prize for ''their pioneer work on the transmutation of atomic nuclei by artificially accelerated atomic particles.''

Nuclear reactions must meet the same conservation laws that apply to natural radioactive decay processes and which we have enumerated already. Energy conservation is at the heart of what may be called nuclear spectroscopy, the determination of nuclear energy levels by inelastic scattering. We shall consider this matter in some detail presently. In addition to energy conservation, a careful examination of the angular distribution of the reaction products yields information concerning angular momentum states and the parity of nuclear energy levels.

We can group the interaction of a projectile with a target nucleus into two categories. (1) The projectile may be elastically scattered by the target either through the Coulomb or the nuclear interaction, or both. (2) The incident projectile may interact with the target nucleus and either impart some energy to or gain some energy from the target; that is, the collision may be inelastic. In inelastic scattering the emerging particle may be of the same type as the incident, but have either more or less energy. The incident particle may also be absorbed by the target nucleus or may knock out one or more nucleons from the target. Last, if the energy of the incident projectile is great enough, it may create particles by mass-energy conversion. In the following pages we shall examine all of these processes except the last, which we shall consider in Chapter 17.

Schematically, we can represent the typical nuclear reaction by the symbolic equation

$$\mathrm{x} + \mathrm{X} \rightarrow \mathrm{Y} + \mathrm{y} + Q \qquad \textbf{(15.17)}$$

Here X represents the target nucleus, x the incident projectile, Y the nucleus of the reaction product, y an emitted particle (or possibly two or more particles), and Q the mass-energy of the reaction. The Q value is given by

$$Q = (m_\mathrm{x} + M_\mathrm{X} - M_\mathrm{Y} - m_\mathrm{y})c^2 \qquad \textbf{(15.18)}$$

and is positive if the reaction is exothermic, negative if it is endothermic.

Until recently, the incident projectile was almost always a nucleon, a deuteron, an α particle, or a γ ray, and the same was true for the emerging particles of the reaction. The shorthand notation

$$\mathrm{X(x,\ y)Y}$$

is a convenient way to express the symbolic Equation (15.17). For example, the first artificially induced nuclear transmutation was the $^{14}\mathrm{N}(\alpha, \mathrm{p})^{17}\mathrm{O}$ reaction; the reaction studied by Chadwick from which he deduced the existence of the neutron was $^{9}\mathrm{Be}(\alpha, \mathrm{n})^{12}\mathrm{C}$. In this notation, elastic proton scattering by a nucleus is represented by X(p, p)X and inelastic proton scattering by X(p, p′)X.

In an elastic interaction, x = y, X = Y, and $Q = 0$. We have seen already how a particular form of interaction, the Coulomb force between charged particles, leads to a very specific scattering law. If the nature of the interaction is not known, as was the case for the nuclear force, it is possible to invert the procedure and deduce some information on the details of this force by studying the distribution of elastically scattered particles with scattering angle and as a function of energy. This motivated the numerous elastic scattering experiments performed between about 1940 and 1950.

Inelastic events fall, broadly, into two groups. The first, known as *direct inter-*

actions, are those in which the incoming projectile interacts with a nucleon of the target nucleus, perhaps knocking it out of the target. Another form of direct interaction is the stripping process; if the incident particle is a deuteron, it may be stripped of either its proton or neutron, and the particle that emerges is then a neutron or proton. The inverse process is also possible, and is known as a pickup reaction; an incoming proton may pick up a neutron from the target, and a deuteron will then come off. Generally, direct interactions are more likely at higher energies. At low energies, the de Broglie wavelength of the projectile is so great that it is not sufficiently localized to interact with just one nucleon.

In the second type of nuclear reaction, the incident particle interacts with the entire target nucleus. That is, it shares its energy with all the nucleons of the target and one then speaks of the formation of a *compound nucleus*. This compound nucleus is in an excited state and decays in a time of the order of 10^{-18} s. Generally, there are several possible decay schemes, and which happens to take place is a matter of statistical probability. The crucial point about interactions involving compound nucleus formation is that the relative probability for a particular decay process does not depend on the manner in which the compound nucleus was formed. In other words, the compound nucleus has ''forgotten'' how it was formed. That this loss of memory could occur in just 10^{-18} s may seem surprising. However, even a slow neutron will have a kinetic energy of about 6 MeV once it enters the target nucleus since the average binding energy per nucleon is between 6 and 8 MeV. It will then move with a speed of about 10^7 m/s and traverse the nuclear diameter of 10^{-14} m in a time of about 10^{-21} s. In 10^{-18} s the incident particle will have bounced back and forth within the target some thousand times, and its energy will have been shared with all the other nucleons; it is then just one of the constituents of the compound nucleus.

15.5(a) Reaction Cross Section

It is common practice to express the probability for a particular nuclear reaction in terms of its *reaction cross section* σ, which is defined as follows:

The cross section for a reaction is the number of reactions per unit time per target particle divided by the incident flux (number of incident particles per unit time per unit area).

That is,

$$\sigma = \frac{R}{NI} \tag{15.19}$$

where R is the reaction rate, N is the number of target particles in the beam, and I is the incident flux. Note that since the dimension of R and I are $[T]^{-1}$ and $[L]^{-2}[T]^{-1}$ and N is dimensionless, σ has the dimension of an area, $[L]^2$.

The cross section may be thought of as the effective area presented by the target nucleus to the incident particles for a specific reaction. Since this effective area is often of the same order of magnitude as the geometric cross section of the nucleus, i.e., about 10^{-28} m^2, this is a convenient unit for measuring reaction cross sections, and has been given the name *barn;* 1 barn $= 10^{-28}$ m^2. It must be emphasized, however, that reaction cross sections may differ vastly from geometric cross sections. If the particular reaction has a low probability, its reaction cross section will be correspondingly small. On the other hand, some reaction cross sections are orders of magnitude greater than the geometric cross section. For instance, the cross section for capture of slow neutrons by ^{113}Cd is about 55,000 barns, roughly 10^4 times the geometric cross section of ^{113}Cd. This large reaction cross section makes cadmium a very effective material for controlling nuclear reactors. By inserting cadmium control rods into a reactor, the neutron flux and, thus, the number of fission reactions can be reduced.

EXAMPLE 15.5 The cross section for the ^{60}Ni(α, n)^{63}Zn reaction at an α particle energy of 18 MeV is 0.7 barn. This reaction is produced in a cyclotron using a nickel foil of 2.5 μm thickness. The density of nickel is 8.8 g/cm^3. The beam current is 8 μA. Find the rate at which this reaction proceeds.

SOLUTION We first find the incident flux of α particles. Since each α particle has a charge of $2 \times 1.6 \times 10^{-19}$ C and the amount of charge that passes through the foil is 8×10^{-6} C/s, the α particle flux is

$$I_0 = \frac{8 \times 10^{-6}}{(3.2 \times 10^{-19})A} = \frac{2.5 \times 10^{13}}{A} \ (\text{cm}^2\cdot\text{s})^{-1}$$

where A is the cross-sectional area of the beam.

Next we need to calculate the number of ^{60}Ni nuclei in the target. The active volume of the foil is $(2.5 \times 10^{-4})A$ cm^3. The number of nickel atoms in the active region is therefore

$$N_{\text{ni}} = \frac{(6.02 \times 10^{23} \text{ atoms/mol})(2.5 \times 10^{-4} \times A \text{ cm}^3)(8.8 \text{ g/cm}^3)}{58.7 \text{ g/mol}}$$

$$= A(2.26 \times 10^{19}) \text{ atoms}$$

However, since the natural abundance of ^{60}Ni is only 26.2 percent, the number of nuclei of that isotope in the active region of the nickel foil is

$$N = 0.262 N_{\text{Ni}} = A(5.92 \times 10^{18})$$

From the definition of reaction cross section, Equation (15.19), we now find that the number of reactions per second, R, is

$$R = (5.92 \times 10^{18})(0.7 \times 10^{-24} \text{ cm}^2)(2.5 \times 10^{13}/\text{cm}^2\cdot\text{s}) = 1.04 \times 10^8 \text{ s}^{-1}$$

15.5(b) Nuclear Spectroscopy

The techniques used in the determination of nuclear energy levels are fundamentally the same as those employed in atomic and molecular spectroscopy. The system under investigation is excited into one of its higher quantum states and the energy of that state relative to the ground state is measured by observing either the energy that the system absorbs in that transition or the energy released as the system returns to

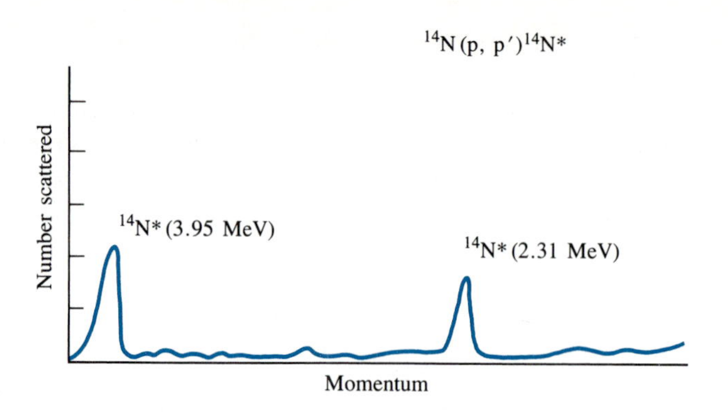

Figure 15.13 Momentum spectrum of protons scattered inelastically from nitrogen in the reaction $^{14}N(p, p')^{14}N^*$, showing two energy levels of the ^{14}N nucleus.

the ground state. The principal difference between atomic spectroscopy and nuclear spectroscopy is that in the former the primary method for excitation is via the electromagnetic interaction, whereas in the latter it is the strong nuclear interaction that is employed in the majority of investigations.

The most direct method for determining the energy levels in a nucleus is through inelastic scattering. Figure 15.13 shows the spectrum of protons scattered from the target nucleus in the reaction $^{14}N(p, p')^{14}N^*$ using protons of 6.92-MeV kinetic energy.[6] Just as in the Frank-Hertz experiment (see Section 5.7) excitation of the target atoms revealed itself by a resonant energy absorption, so here, too, the energies of the excited nuclear states are manifested by resonant scattering of the incident protons. Once excited to a higher energy state, the nucleus soon decays to the ground state with the emission of a γ ray. Measuring the energies of the emitted γ rays is another way by which nuclear energy levels can be established.

In all interactions, energy and momentum must be conserved. If the event is the decay of a radioactive nucleus that is initially at rest, the momentum of the center of mass is zero, and so is the kinetic energy associated with the center-of-mass motion. The energy Q is then the sum of the kinetic energies of the decay products as measured in the laboratory. However, in the typical nuclear reaction

$$x + X \rightarrow Y + y + Q \qquad \textbf{(15.20)}$$

the projectile x is moving at high speed toward the target nucleus X, which is at rest in the laboratory. The products of the reaction, Y + y, must then have velocities such that total momentum is conserved. In Equation (15.20), Q is the *net* energy that is released in the reaction, and does not include the kinetic energy of the center of mass.

If Q is negative, that is, if the reaction is endothermic, energy must be supplied to make the reaction proceed. Hence, there is a threshold, a minimum kinetic energy that the incident particle must have. However, not all of the energy of the incident projectile is available for the reaction, because the momentum of the reaction products must equal that of the incident particle. The way to deal with the problem is to transform to the center-of-mass reference frame, i.e., to a reference frame in which the center of mass is initially at rest and, therefore, also at rest after the reaction. In that reference frame, the kinetic energy of the center of mass is zero,

6. The asterisk (*) denotes an excited nuclear state.

and Q is then just equal to the difference in the rest energies of the initial and final reaction components.

Here we consider only the nonrelativistic case. In the center-of-mass (CM) frame, the projectile of mass m_x is moving with a velocity v toward the target nucleus of mass M_X, which is moving with velocity V in the opposite direction. Since in this reference frame the total momentum must vanish,

$$m_x v = M_X V = p$$

The kinetic energy in the CM frame is

$$E_{CM} = \frac{p^2}{2m_x} + \frac{p^2}{2M_X} = p^2 \frac{m_x + M_X}{2m_x M_X} \tag{15.21}$$

In the laboratory, however, X is at rest, and hence the velocity of the incident projectile is

$$v_x = v + V \tag{15.22}$$

while that of the target is zero.

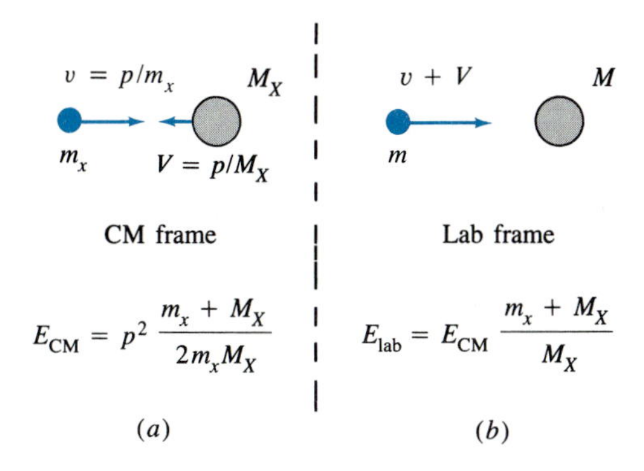

Figure 15.14 Momentum and energy relations in a head-on collision (*a*) in the center-of-mass reference frame and (*b*) in the laboratory reference frame.

The kinetic energy in the laboratory frame is

$$E_{lab} = \frac{1}{2} m_x v_x^2 = \frac{p^2}{2m_x} \left(\frac{m_x + M_X}{M_X}\right)^2 = E_{CM} \left(\frac{m_x + M_X}{M_X}\right) \tag{15.23}$$

Following the reaction, the particles Y and y will have velocities u and U in the CM frame, and these must meet the conservation condition

$$m_y u = M_Y U$$

At the reaction threshold, $u = U = 0$, and the total kinetic energy in the CM system following the reaction is zero. By definition, Q is positive if the reaction is exothermic, negative if it is endothermic. In the latter instance, the energy loss in the reaction must be supplied by the kinetic energy of the reactants. Hence the energy E_{CM} at threshold is just equal to $-Q$. The threshold kinetic energy of the incident projectile in the laboratory frame is thus given by

$$E_{th} = -Q\left(1 + \frac{m_x}{M_X}\right) \tag{15.24}$$

EXAMPLE 15.6 What is the threshold energy for the reaction $^{13}C(n, \alpha)^{10}Be$?

SOLUTION The atomic masses of ^{13}C and ^{10}Be are 13.003355 and 10.013534. Hence in the CM system the energy released in the reaction is

$$[(13.003355 + 1.008665 - 10.013534 - 4.002603)\ u](931.5\ MeV/u)$$
$$= -3.835\ MeV$$

The negative sign tells us that the reaction is endothermic, and that this amount of energy must be supplied in the CM system before the reaction can proceed. The threshold energy in the laboratory is given by Equation (15.24). We find

$$E_{th} = 3.835(1 + \tfrac{1}{13}) = 4.13\ MeV$$

15.5(c) Compound Nucleus

The characteristic feature of a nuclear reaction that involves the formation of a compound nucleus is that the decay of the compound nucleus does not depend on the manner in which it was formed. For example, the compound nucleus $^{64}Zn^*$ that is the intermediate state in the reaction

$$^{60}Ni + {}^{4}He \rightarrow {}^{64}Zn^* \rightarrow {}^{63}Zn + n + Q$$

can also be formed by bombarding ^{63}Cu with protons. Depending on the state of excitation of the compound nucleus, there are other modes of decay, for example,

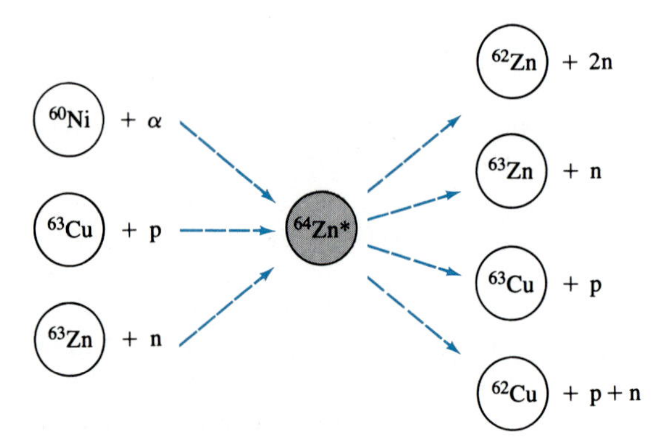

Figure 15.15 Several possible "channels" for the formation and decay of the compound nucleus $^{64}Zn^*$.

$$^{64}\text{Zn}^* \rightarrow {}^{62}\text{Cu} + \text{p} + \text{n} + Q$$

$$^{64}\text{Zn}^* \rightarrow {}^{62}\text{Zn} + 2\text{n} + Q$$

According to Bohr's postulate of the compound nucleus, the probabilities for these several decay modes of ^{64}Zn* should be independent of the method by which the compound nucleus was produced, provided the nucleus is in the same excited state. The energy difference between the ground states of ^{60}Ni and ^{63}Cu is such that an α particle of energy $E_\alpha = E_\text{p} + 7$ MeV that is absorbed by the nickel nucleus will produce the same excited state of ^{64}Zn as a proton of energy E_p that is absorbed by the copper nucleus. Hence the ratio of the cross sections for the reactions ^{60}Ni(α, n)^{63}Zn, ^{60}Ni(α, pn)^{62}Cu, and ^{60}Ni(α, 2n)^{62}Zn should be the same as for the reactions ^{63}Cu(p, n)^{63}Zn, ^{63}Cu(p, pn)^{62}Cu, and ^{63}Cu(p, 2n)^{62}Zn at proton energies $E_\text{p} = E_\alpha - 7$ MeV. The experimental results shown in Figure 15.16 provide convincing confirmation of this prediction.

It is apparent from Figure 15.16 that the reaction cross section is a function of energy. A peak in the cross section is indicative of an excited state of the compound nucleus, and the study of cross sections as functions of energy is one method for finding the excited states of compound nuclei. The lifetime of the excited state is obtained from the width of the "resonance."

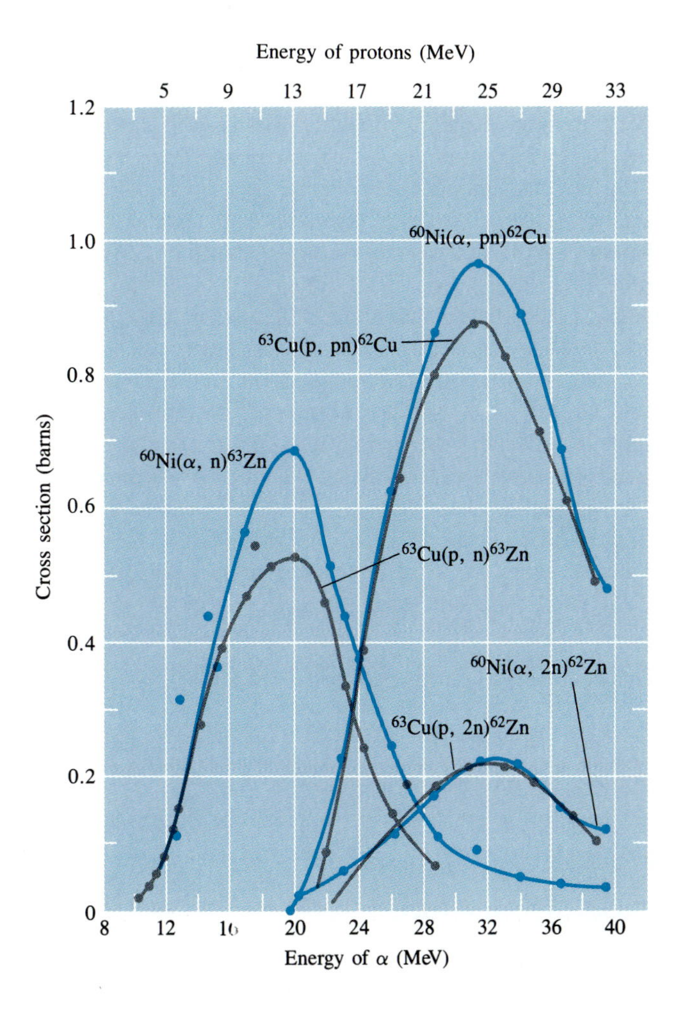

Figure 15.16 Cross sections for reactions involving the compound nucleus ^{64}Zn* as a function of the incident α particle energy. The incident proton energy for the Cu(p, y)Y reactions has been adjusted by 7 MeV. The results confirm the predictions of the theory of compound nucleus formation in nuclear reactions.

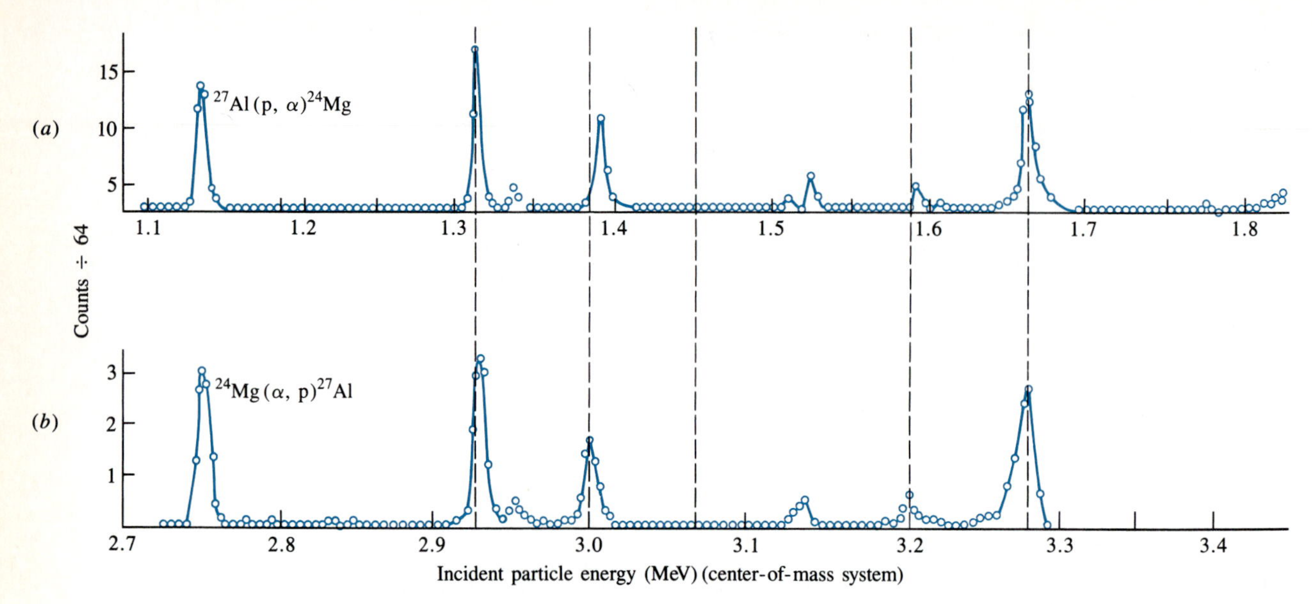

Figure 15.17 Resonances in the ^{27}Al(p, α)^{24}Mg reaction and in the inverse reaction, ^{24}Mg(α, p)^{27}Al. The resonances correspond to excited levels of the compound nucleus ^{28}Si.

It follows from microscopic reversibility that the resonance pattern should be the same for the forward and backward reactions. For example, the reaction ^{27}Al(p, α)^{24}Mg exhibits a number of resonances, as shown in Figure 15.17*a*. These resonances are due to excited states of the compound nucleus ^{28}Si, and an identical resonance pattern is displayed when the inverse reaction ^{24}Mg(α, p)^{27}Al is examined (see Figure 15.17*b*). The energy difference of 1.6 MeV by which one curve is displaced relative to the other is due to the difference in rest energies; the total rest energy of ^{24}Mg + ^{4}He is 1.6 MeV less than the rest energy of ^{27}Al + ^{1}H, so that an additional 1.6 MeV of energy must be supplied to form the same excited state of the compound nucleus ^{28}Si if we use the ^{24}Mg(α, p)^{27}Al reaction. Resonances of the compound nucleus yield data only on higher-lying excited states, because the binding energy of the incident particle contributes about 7 MeV per nucleon. Consider, for instance, the excited level of ^{28}Si that corresponds to the resonance at 1.31 MeV of Figure 15.17*a*. Its energy relative to the ground state of ^{28}Si is obtained by subtracting the rest energy of ^{28}Si from the sum of 1.31 MeV and the rest energies of ^{27}Al and ^{1}H. Thus, the excited state is at an energy of

$$1.31 \text{ MeV} + [(26.981539 + 1.007825 - 27.976927) \text{ u}](931.5 \text{ MeV/u}) = 12.86 \text{ MeV}$$

above the ground state of ^{28}Si.

15.6 FISSION AND FUSION

15.6(a) Fission; Reactors

In June 1939, a few months before the start of World War II, the journal *Naturwissenschaften* printed an article by O. Hahn and F. Strassmann in which the authors reported the results of meticulous chemical analyses on the decay products that result when uranium is subjected to bombardment by neutrons.[7] They presented convincing evidence that the element barium is one of the decay products, and that some rare earth elements also appear as a consequence of this reaction. Still, they found it difficult to state categorically that these elements are indeed produced by the bombardment of uranium with neutrons.[8]

> *As chemists we should, on the basis of the data presented here, write the symbols Ba, La and Ce in place of Ra, Ac, and Th. As nuclear chemists we find it difficult to take this step which contradicts all previous experiences in nuclear physics.*

Lise Meitner, a former collaborator of Hahn, and her nephew Otto Frisch, who had fled to Sweden from Nazi Germany, recognized almost immediately the far-reaching significance of those results—the uranium nucleus had split into two large fragments.[9] Hahn and Strassmann had observed neutron-induced nuclear fission!

A glance at Figure 14.4 shows that fission is energetically favorable because the binding energy per nucleon is approximately 1 MeV greater for nuclei with $A \simeq 100$ than for nuclei with $A \simeq 200$. Thus, by splitting into two more tightly bound systems, the potential energy is reduced by about 200 MeV, and this energy is then released as kinetic energy. Further experiments showed that fission does not always yield the exact same reaction products; it is a statistical event and the atomic mass numbers of the reaction products range between about 80 and 160 (see Figure 15.18).

A semiquantitative understanding of the fission process can be gained from the liquid drop model. If a ^{235}U nucleus captures a slow neutron (a neutron whose kinetic energy is negligibly small), the ^{236}U so formed will not be in its ground state. The mass of the ^{236}U nucleus created by the capture of the slow neutron is

$$^{235}M + m_{\text{n}} = 235.043924\ \text{u} + 1.008665\ \text{u} = 236.052589\ \text{u}$$

The mass of ^{236}U in its ground state is 236.045562 u (see Appendix B). The ^{236}U nucleus formed when ^{235}U captures a slow neutron is therefore in an excited state $(236.052589 - 236.045562)(931.5) = 6.5465$ MeV above the ground state. This excitation energy frequently appears as a collective mode of vibration wherein the nucleus oscillates between an elongated (prolate) and flattened (oblate) spheroid. Departure from sphericity increases the surface tension energy of the liquid drop; on the other hand, by increasing the average separation between the positively charged protons, elongation reduces the Coulomb energy. For small oscillations the surface tension term dominates, and the energy of the nucleus at first increases with increasing distortion from sphericity; however, for larger amplitudes of oscillation the decrease in Coulomb energy more than balances the increase in surface tension energy (see Figure 15.19). The prolate spheroid then develops a narrow neck and

---

7. O. Hahn and F. Strassmann, *Naturwiss.* **27,** 11; 89; 163 (1939)

8. Four years earlier, Ida Noddack had speculated that in Fermi's experiments in which he bombarded uranium with slow neutrons, the uranium nucleus had split into two fragments. At that time, Hahn expressed his skepticism to her in the strongest terms and advised Noddack to put aside these wild ideas. No wonder that Hahn hesitated to draw the correct conclusion from his data.

9. L. Meitner and O. R. Frisch, *Nature* **143,** 239 (1939).

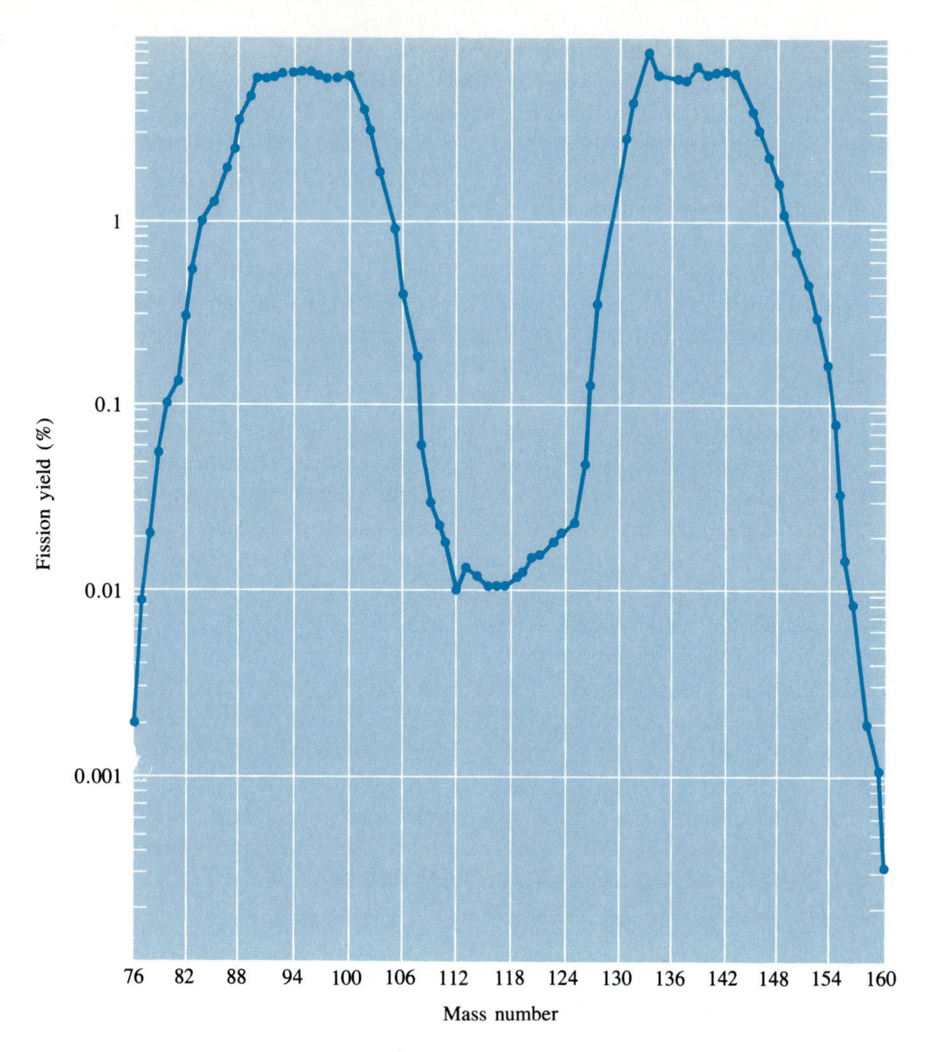

Figure 15.18 Relative abundance of fission products from slow-neutron–induced fission of ^{235}U.

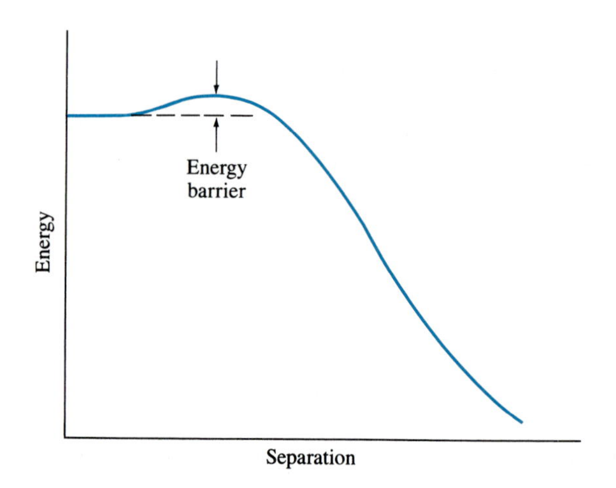

Figure 15.19 The potential energy for spontaneous fission as a function of the average separation of the two fission fragments (schematic diagram). The potential energy barrier that must be overcome to result in fission of nuclei with $A \simeq 240$ is about 6 MeV.

the "liquid drop" separates into two smaller droplets that repel one another and fly apart (see Figure 15.20). Calculations indicate that for a typical nucleus of $A \simeq 240$ the fission barrier is about 6 MeV. Thus the excitation energy available from the capture of a slow neutron by ^{235}U is sufficient to cause fission of the ^{236}U nucleus. The time interval between neutron capture and fission is about 10^{-15} s.

Uranium has two long-lived isotopes, ^{235}U and ^{238}U, whose natural abundances are 0.7 percent and 99.3 percent. Only ^{235}U fissions on capture of a slow neutron. A calculation like the foregoing shows that when ^{238}U captures a slow neutron the excitation energy of ^{239}U is only 4.807 MeV, well below the fission threshold. The difference in excitation energy between ^{236}U and ^{239}U formed by slow-neutron capture is principally the pairing energy, the last term of Equation (14.9). Capture of a neutron by ^{235}U changes an odd-even into an even-even nucleus, whereas the reverse happens on capture of a neutron by ^{238}U. The difference in pairing energy in the two cases amounts to about 1.5 MeV. Although ^{238}U does fission on absorption of energetic neutrons (KE > 1 MeV), the cross section for that process is about three orders of magnitude smaller than the slow-neutron–induced fission cross section of ^{235}U.

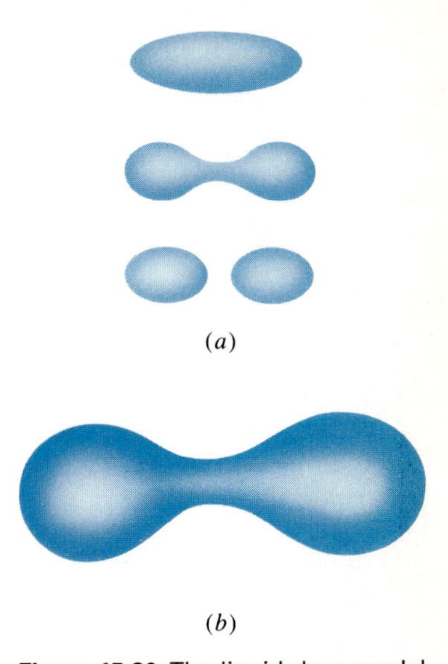

Figure 15.20 The liquid drop model of fission. *(a)* Presumed sequence of events leading to fission of the excited nucleus. *(b)* A drop of water suspended in oil just before fission; fission was induced by application of a potential difference across the oil, thereby producing the initial deformation of the drop, i.e., giving it the excitation energy. *(Lawrence Berkeley Laboratory.)*

It is apparent from a glance at the Segrè plot of Figure 14.6 that when the ^{236}U nucleus breaks into two fragments, both fragments will be well above the line of stability, i.e., will be extremely neutron-rich. The fission reaction, therefore, results in simultaneous release of several neutrons, and these can induce further fission in other ^{235}U nuclei. However, fission neutrons are not slow but have energies of several MeV and must be slowed to thermal energies before they have a high probability of inducing fission in other ^{235}U nuclei.

Two processes are active in slowing, or *moderating,* fast neutrons. First, energetic neutrons lose energy by inelastic collisions with ^{238}U, raising that nucleus to an excited state from which it decays by emission of γ rays. Once the neutron energy is below about 1 MeV, the dominant energy loss mechanism is through elastic collisions in which the incident neutron imparts some of its energy to the nucleus with which it collides. However, just as a marble that bounces off a billiard ball does not impact much energy to the billiard ball and rebounds with almost as much energy as it had before the collision, so a neutron that scatters elastically from a U nucleus loses very little of its kinetic energy. Sharing of energy in elastic collisions is most efficient if the target nucleus has the same mass as the neutron. Hydrogen would, therefore, be the best moderator were it not for the $^{1}H(n, \gamma)^{2}H$ reaction which has a fairly large cross section at low neutron energies; hydrogen not only slows down the neutrons but also competes with ^{235}U in capturing slow neutrons. Ordinary water can be used as the moderator in a nuclear reactor only if the concentration of ^{235}U is increased, i.e., if "enriched" uranium, containing 3 to 4 percent ^{235}U, is used as fuel. Since the two uranium isotopes are chemically indistinguishable, the enrichment process must rely on the small mass difference. Diffusion of a gas is a physical process that depends on the mass of the molecules, and it is by this means that enrichment is accomplished.

One alternative to using enriched uranium is to employ "heavy" water, D_2O, as moderator. The first nuclear reactor, the "pile" constructed by Fermi and coworkers under the stands of the football field at the University of Chicago in 1942, used another alternative, employing graphite blocks as moderators. Carbon is a fairly light, stable, and abundant element and has a very small cross section for neutron capture.

On the average, 2.5 neutrons are released when ^{235}U fissions. This does not mean, however, that two ^{235}U fissions will, on the average, induce fission in five other ^{235}U nuclei. Neutrons may leave the reactor region by escaping through the walls of the reactor or by being captured by other nuclei. For example, the reaction

$$^{238}U + n \rightarrow {}^{239}U^* \rightarrow {}^{239}Np + e^- + \bar{\nu}$$

$$^{239}Np \rightarrow {}^{239}Pu + e^- + \bar{\nu}$$

which produces the fissionable plutonium, is one such competing mechanism.

The reproduction constant k is defined as the number of thermal neutrons per fission that induce further fission of ^{235}U nuclei. If $k < 1$, the reactor is subcritical; if $k = 1$, the reactor is said to be critical, and the fission process will proceed at constant rate; if $k > 1$, each fission will initiate more than one subsequent fission and an avalanche of fission reactions will result. The time required for thermalization of fast neutrons is only about 1 ms. If $k = 1.01$, the fission rate would increase by a factor of $(1.01)^{1000} \simeq 2 \times 10^4$ in just one second. Thus, in a very short time an enormous amount of energy would be released. Such an uncontrolled chain reaction at a far more rapid rate is, of course, what happens when a fission bomb is detonated, but this is exactly what must be prevented in the operation of a nuclear reactor.

Control of the reproduction constant in a reactor is achieved by means of control rods made of a material such as cadmium which has a very large cross section for neutron absorption. When the reactor is first started, the control rods are fully inserted and $k < 1$. As the rods are gradually withdrawn, k increases and the reaction becomes self-sustaining when $k = 1$. As we have just seen, it is crucial that k not be allowed to exceed 1 by even a percent or two for more than a fraction of a second. If that did happen, the rate of energy release would be more than the cooling system could handle and the reactor core would melt.

If all the neutrons released in fission were "prompt" neutrons, i.e., created at the instant of the fission, control of the reactor would be virtually impossible. The response time of the mechanical device that raises and lowers the rods could not be

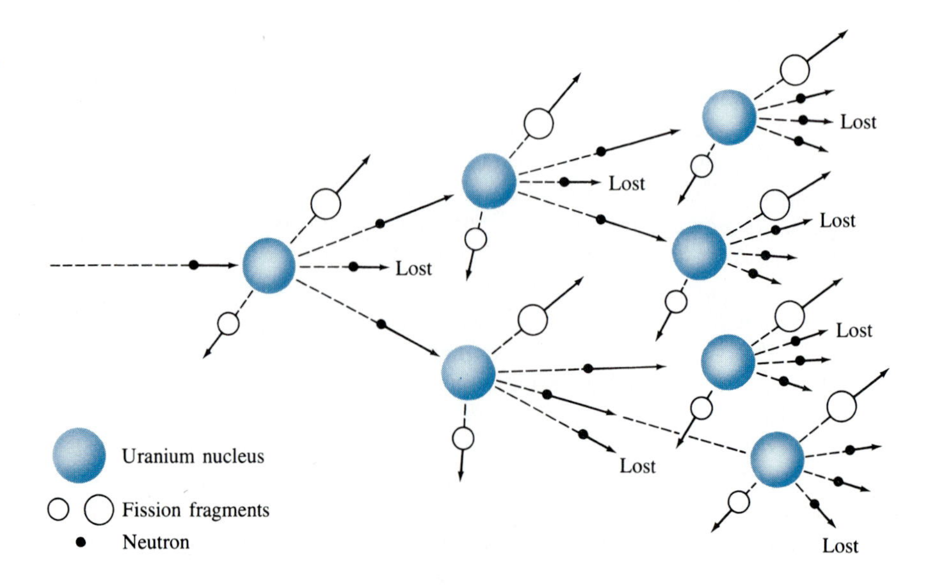

Figure 15.21 A fission chain reaction (schematic diagram). The reproduction constant $k = 2$; the moderator is not shown.

sufficiently short to respond to small statistical fluctuations in the fission rate. It is a happy accident of nature that a small fraction of the neutrons produced in fission are "delayed" neutrons, emitted several seconds after fission by some of the fission fragments. For example, one of the fission reactions is

$$^{235}\text{U} + \text{n} \rightarrow {}^{93}\text{Rb} + {}^{141}\text{Cs} + 2\text{n}$$

Here the two neutrons are prompt neutrons, emitted at the instant of fission. Both ^{93}Rb and ^{141}Cs are unstable, with half-lives of about 6 s and 25 s, respectively. The principal decay mode of each is β decay; however, 1.3 percent of the time ^{93}Rb decays to ^{92}Rb with the emission of a neutron, and 0.03 percent of the time ^{141}Cs emits a neutron. Indeed, many neutron-rich fission fragments contribute to the delayed-neutron flux (see Appendix B). A small fraction of the neutron flux in a reactor comes from these delayed neutrons, and that contribution is sufficiently sluggish that it can be controlled by mechanical motion of the control rods.

The final question relating to reactor design concerns the conversion of nuclear energy into electrical energy. Practically all of the energy released in fission goes into kinetic energy of the fission products. These fragments quickly lose energy by collisons within the reactor core, raising the temperature of the core. This energy is extracted by a heat exchanger and used to drive an electric generator in much the same way that the heat of chemical combustion is used in a conventional generating plant.

Most of today's reactors use one of two schemes.

1. Boiling-water reactors. Water is circulated through the reactor core and the heat of the reactor boils the water. The steam, generally at fairly high pressures, is then used to drive a generator. The major problem inherent in this scheme is that the water becomes highly radioactive, and if the pipes were to rupture at the turbines the resulting spread of radioactive material would constitute a major hazard.
2. Pressurized-water reactors. Here the heat exchange involves a two-step process. Water, under very high pressure, is circulated through the reactor core where it is raised to high temperature. This water is then passed through a

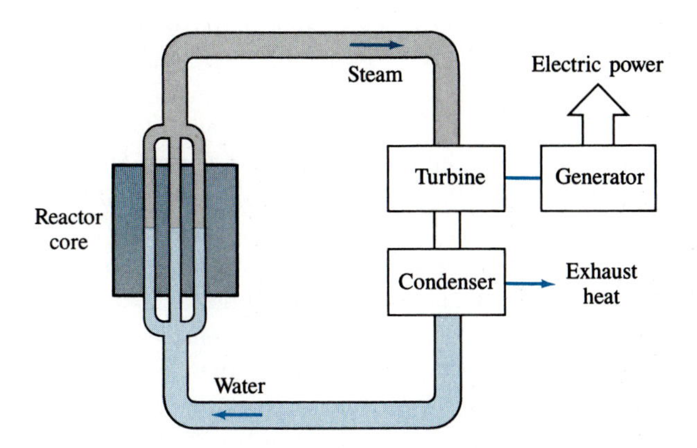

Figure 15.22 Schematic diagram of a boiling-water reactor.

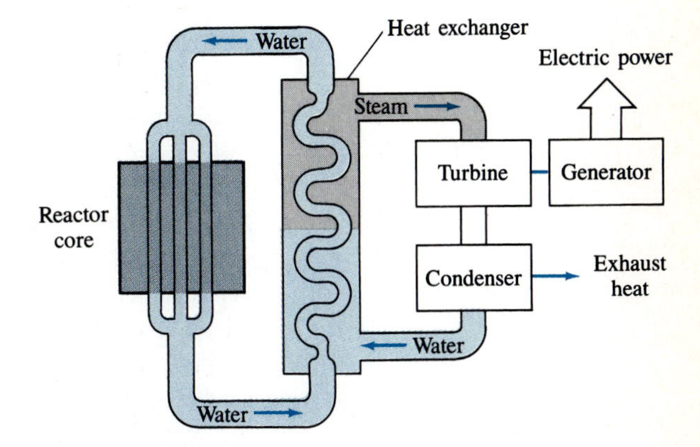

Figure 15.23 Schematic diagram of a pressurized-water reactor.

heat exchanger where it boils the water of another system and the steam thus produced drives the turbines of the generating plant. The advantage of this scheme is that the water that circulates through the reactor core remains in a closed system within the containment building.

In both approaches, the water that extracts the heat from the reactor also serves as the moderator.

One of the major and, as yet unsolved problems is the safe disposal of the radioactive waste products of nuclear reactors. Fission spawns a gamut of radioactive isotopes whose half-lives range from fractions of a second to thousands of years. For example, one fission reaction is

$$^{235}\mathrm{U} + \mathrm{n} \rightarrow {}^{144}\mathrm{Ba} + {}^{89}\mathrm{Kr} + 3\mathrm{n}$$

$$^{89}\mathrm{Kr} \rightarrow {}^{89}\mathrm{Rb} \rightarrow {}^{89}\mathrm{Sr} \rightarrow {}^{89}\mathrm{Y}$$

$$^{144}\mathrm{Ba} \rightarrow {}^{144}\mathrm{La} \rightarrow {}^{144}\mathrm{Ce} \rightarrow {}^{144}\mathrm{Pr} \rightarrow {}^{144}\mathrm{Nd}$$

^{89}Y is stable; however, ^{144}Nd is an α emitter with a half-life of 2×10^{15} years. Other long-lived fission fragments are ^{99}Tc ($t_{1/2} = 2.2 \times 10^5$ yr) and ^{98}Tc ($t_{1/2} = 4.2 \times 10^6$ yr). Some of the fission products are extremely toxic; others present a major health problem by virtue of their radioactivity. For example, one dangerous fission product is ^{90}Sr whose half-life is 29 years. This β emitter is chemically similar to calcium and, if ingested, concentrates in normally calcium-rich regions of the body, such as bone. There the β particles can damage cells of bone marrow, causing leukemia.

One of the toxic and long-lived by-products of a fission reactor is ^{239}Pu. This transuranic element is produced in a nuclear reaction involving the abundant uranium isotope ^{238}U:

$$^{238}\mathrm{U} + \mathrm{n} \rightarrow {}^{239}\mathrm{U}^* \rightarrow {}^{239}\mathrm{Np} + \mathrm{e}^- + \bar{\nu}$$

$$^{239}\mathrm{Np} \rightarrow {}^{239}\mathrm{Pu} + \mathrm{e}^- + \bar{\nu}$$

Plutonium is an α emitter with a half-life of 2.4×10^4 yr. It is also susceptible to fission by slow neutrons and can serve as the fuel in a nuclear reactor or as the explosive charge in a fission bomb. In a *breeder reactor,* some of the neutrons from the fission of ^{235}U are used to transmute ^{238}U to ^{239}Pu by the above reaction. The ^{239}Pu can then be extracted chemically and used as fuel in a plutonium reactor. Of course, the breeder reactor does not produce an inexhaustible supply of fission fuel; it does convert the abundant isotope of uranium into fissionable plutonium.

Safe operation of a breeder reactor still poses many problems. Some of these are strictly technical and will surely be solved in due time, but others are not so susceptible to satisfactory answers by the scientific community. Plutonium is the active material of the common fission bomb, and the presence of breeder reactors in all parts of the world providing the raw material for nuclear bombs is a grave concern. Another worrisome aspect is that plutonium is extremely toxic and, once produced, will remain in the environment for thousands of years.

Reactors are not only important sources of energy but also valuable research tools. A host of radioactive isotopes that are used in biological research and medical therapy are produced by irradiating stable isotopes with neutrons. The reaction

$^{32}S(n, p)^{32}P$ produces the radioactive isotope of phosphorus that is a common tracer in biological studies; ^{99}Tc, a fission product, is used in radiation therapy; the isotope ^{131}I, employed in radiation therapy of thyroid cancer, is obtained from the $^{130}Te(n, \beta)^{131}I$ reaction. The neutrons themselves are also highly prized by physicists as probes to explore the structure of solids, especially those with magnetic ions. Slow neutrons have de Broglie wavelengths comparable to the lattice spacing in crystals, and can be used like X rays to investigate crystal structure. However, since a neutron carries no charge, it interacts with matter (other than in nuclear reactions) only because its magnetic moment senses the local magnetic field. Neutron diffraction reveals the disposition of magnetic ions and the orientation of their magnetic moments in a crystal.

Another use of reactors is in *neutron activation analysis* (NAA), an extremely sensitive and nondestructive analytical technique. In NAA the sample is subjected to a flux of thermal neutrons from a reactor. Nearly all stable isotopes capture thermal neutrons in an (n, γ) reaction. The resulting neutron-rich isotope then generally decays by β^- emission, and the daughter nuclide decays to its ground state with the emission of one or more γ rays. These γ rays are as characteristic of that nuclide as are the atomic spectral lines of a particular element. Using NAA, it is possible to determine not only which trace elements are present in a sample but also their concentration. Generally nanograms and in some cases as little as a few picograms of trace material can be detected.

15.6(b) Fusion

Scientists have known for centuries that the ancients' picture of the sun as a ball of fire, fueled by wood or some other combustible material, was incorrect. The major difficulty is the apparently inexhaustible supply of fuel. The clue to the solution of this age-old mystery was the mass-energy equivalence of the special theory of relativity. Rutherford, Curie, and others quickly realized that the energy produced in radioactive decay derives from this relativistic mass-energy conversion. By 1930 it had been established that the helium nucleus is about 1 percent less massive than four protons, and Eddington suggested that fusion of protons to form helium might be the source of stellar energy.

A glance at Figure 14.4 shows that substantial energy must be released when four nucleons combine to form a helium nucleus. There are a number of ways in which four protons can combine to form a helium nucleus. One possible sequence is

1. $$^1H + {}^1H \rightarrow {}^2H + e^+ + \nu + 0.42 \text{ MeV}$$
2. $$^1H + {}^2H \rightarrow {}^3He + \gamma + 5.49 \text{ MeV}$$
3. $$^3He + {}^3He \rightarrow {}^4He + 2\,{}^1H + 12.86 \text{ MeV}$$

Reactions 1 and 2 must occur twice before the third reaction can take place. The net result can be expressed by the single reaction equation

$$4\,{}^1H \rightarrow {}^4He + 2e^+ + 2\nu + 24.68 \text{ MeV}$$

The energy released in the fusion of four protons to form a helium nucleus must include the 2.04 MeV liberated by the annihilation of the two positrons with two of the four hydrogenic electrons, yielding a total of 26.72 MeV. Fusion is clearly a more "mass-energy efficient" process than fission, yielding about six times as much energy per gram of active material as fission. Fusion is, however, much more difficult to achieve than is fission. The nuclei that amalgamate in the first step of this *proton-proton cycle* must overcome or tunnel through a Coulomb barrier of about 1 MeV. At 1.5×10^7 K, the temperature of the sun's core, kT is only about 1 keV, and the probability for fusion of two ^{1}H nuclei is extremely small even at that temperature. Moreover, the process involves the weak interaction and, consequently, proceeds at a very slow rate. The first step of the proton-proton cycle is sometimes referred to as the "deuteron bottleneck." Large fusion energy is generated in the sun only because the mass and density of hydrogen in the sun's core is extremely large.

At temperatures above about 10^8 K another fusion sequence, the *carbon cycle* first proposed in a brief but justly celebrated paper by Hans Bethe,[10] is dominant. The following sequence of six steps uses ^{12}C, which is formed by the fusion of three ^{4}He nuclei, as a catalyst and proceeds much more rapidly than the proton-proton cycle:

1. $^1\text{H} + {}^{12}\text{C} \rightarrow {}^{13}\text{N} + Q_1$
2. ${}^{13}\text{N} \rightarrow {}^{13}\text{C} + e^+ + \nu + Q_2$
3. $^1\text{H} + {}^{13}\text{C} \rightarrow {}^{14}\text{N} + Q_3$
4. $^1\text{H} + {}^{14}\text{N} \rightarrow {}^{15}\text{O} + Q_4$
5. ${}^{15}\text{O} \rightarrow {}^{15}\text{N} + e^+ + \nu + Q_5$
6. $^1\text{H} + {}^{15}\text{N} \rightarrow {}^{12}\text{C} + {}^4\text{He} + Q_6$

The net result of the six steps of the carbon cycle is again the fusion of four protons to form a ^{4}He nucleus and two positrons, with a yield of 26.72 MeV per helium nucleus.

That is what happens in a star, in the very hot, dense, and enormous plasma of the central core. The same sequence of reactions could not be maintained on earth. Nevertheless, fusion reactions can be initiated on our planet in a small space. To achieve fusion requires a temperature of about 10^8 K and extremely high pressure. These conditions must be maintained for a time that may be brief by normal reckoning, but long enough to allow the fusion reaction to proceed.

Of the various possible fusion reactions, the one that is most promising and also yields substantial energy is

$$^2\text{H} + {}^3\text{H} \rightarrow {}^4\text{He} + \text{n} + 17.6 \text{ MeV}$$

With Edward Teller as its principal advocate, scientists began work on a fusion bomb, a thermonuclear "device," as early as 1942. Spurred on by the cold war atmosphere of the early 1950s, this activity culminated in the explosion of the first H-bomb on November 1, 1952, in which a small island of the Eniwetok atoll was pulverized and vanished from the face of the earth. That bomb, though it demonstrated that fusion could be made to "work," was not a practical weapon, since the

10. H. A. Bethe, *Phys. Rev.* **55,** 103 (1938).

Figure 15.24 Changing the surface of our planet: vaporization of a small island. *(UPI/Bettmann Newsphotos.)*

heavy hydrogen and tritium had to be liquefied before the explosion, i.e., the bomb had to be kept at very low temperature until the fatal instant. Soon, however, a better technique was found, based on an idea proposed already in 1946 by the Viennese physicist Thirring. A high concentration of deuterium could be maintained at ordinary temperature by incorporating the deuterium in a solid, specifically, LiD. Moreover, once the fission bomb that triggers the thermonuclear event by providing the enormous temperature and pressure needed for fusion is exploded, the huge flux of neutrons will produce the tritium needed for deuterium-tritium fusion via the reaction

$$^{6}\mathrm{Li} + \mathrm{n} \rightarrow {}^{4}\mathrm{He} + {}^{3}\mathrm{H} + Q$$

In early 1954 the United States tested such a bomb in a 15-megaton burst of destruction.

Still, fusion could be the salvation as well as annihilation of the human species. The raw material is cheap and abundant. Each 50 liters of water contains about one gram of deuterium, whose fusion would liberate more energy than is stored chemically in 500 gallons of gasoline. Because deuterium is twice as massive as ordinary hydrogen, the extraction of this isotope is a simple matter, not in any way comparable to the difficulty in enriching uranium. And most important, fusion does not produce the hazardous and long-lived radioactive wastes that accompany fission. The problem, which now appears to be near solution, is how to maintain the necessary high temperature and high density in an environment other than a fusion bomb.

Two approaches have been pursued. One is based on a doughnut-shaped device invented by the Soviets and named *tokamak*. The toroid is surrounded by powerful magnets which confine the deuterons and tritium nuclei within the doughnut and prevent them from reaching the walls. The gas within the tokamak is heated to about 10^8 K by sending brief pulses of enormous current through this plasma.

The other method is known as *inertial confinement*. The fusion fuel is introduced into the reaction chamber in the form of a tiny pellet of solid deuterium-tritium mixture. This pellet is then hit from all sides with a burst of energy by the simultaneous firing of high-power pulsed lasers. As the surface layer of the pellet

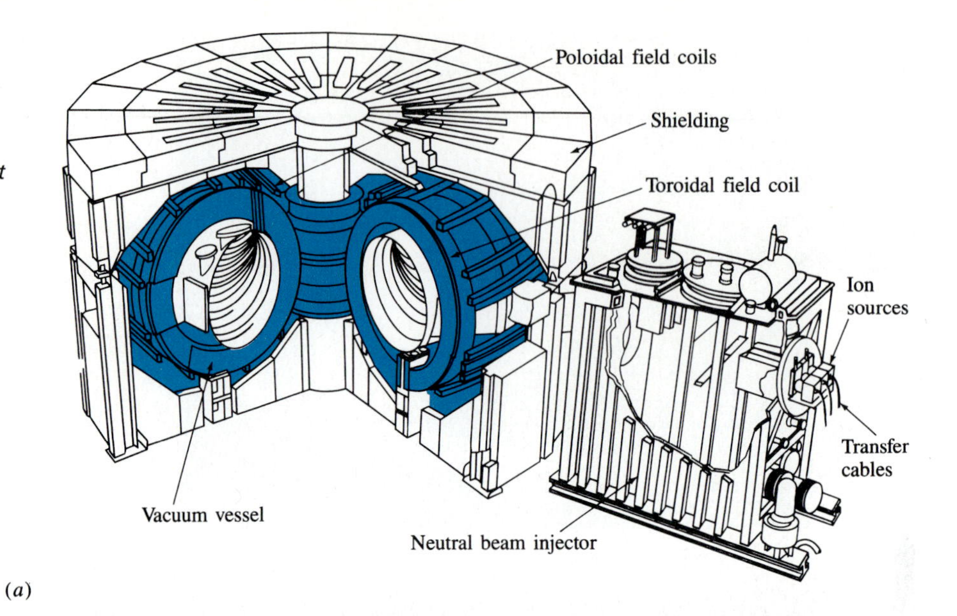

(*a*)

(*b*)

Figure 15.25 (*a*) Diagram of the tokamak device for controlled fusion. (*b*) The tokamak at Princeton University. *(Princeton University, Plasma Physics Laboratory, Tokomak Fusion Test Reactor, Princeton, NJ.)*

vaporizes, the outward motion of the surface atoms compresses the remaining material by the reaction force. The effective pressure is about 10^{11} atmospheres, and at that pressure the density of the pellet approaches more than a thousand times normal density. At the same time, the implosion and laser beams raise the temperature to more than 10^8 K, and at that instant fusion is initiated thermally, and a burst of about 10^6 joules is released. All this takes place in 10^{-11} s or less.

Figure 15.26 The inertial-confinement fusion facility at Lawrence Livermore Laboratory, showing five of the ten laser tubes surrounding the target chamber. *(Lawrence Livermore National Laboratory.)*

During the past several decades, research teams at a number of laboratories have come ever closer to the break-even point where the energy that is expended to make the fusion reaction go is balanced by the energy released in that reaction. It now appears increasingly probable that controlled fusion will indeed be a practical source of energy for future generations.

*15.7 APPLICATIONS OF NUCLEAR PHYSICS

15.7(a) Radioisotopes in Medicine

The use of nuclear radiation in medicine falls into two main categories, diagnosis and therapy. The principal purpose of diagnosis is to ascertain possible malfunction of an organ or to locate lesions or abnormalities, such as tumors. Radiotherapy relies on radiation damage to destroy malignancies.

Diagnosis

As a general rule, in diagnosis a radiopharmaceutical is administered to the patient either orally or intravenously. The compound selected is determined by the organ to be examined; for example, for investigating thyroid function, radioactive iodine is administered because the body concentrates this element in the thyroid. Following the administration of the radioisotope, the activity is measured by suitable detectors, and, in many instances, the distribution of the radioisotope within the organ is determined by scanning or other techniques. Evidently, isotopes used for diagnosis must meet three primary requirements. (1) The radiation emitted by the isotope must be able to penetrate at least 10 to 30 cm of body tissue so that it can be detected by instruments placed near but outside the body of the patient. (2) The radiation emitted by the isotope should not itself cause significant damage to healthy body tissues. (3) The effective half-life should be of modest duration, but long enough to permit the necessary measurements. The effective half-life is given by

$$\frac{1}{t_{1/2(\mathrm{eff})}} = \frac{1}{t_{1/2}} + \frac{1}{t_{1/2(b)}}$$

where $t_{1/2}$ is the natural half-life of the radioisotope and $t_{1/2(b)}$ is its "biological" half-life. The biological half-life is a measure of the time required for the normal body functions to eliminate the radioisotope (urine, feces, diffusion into the lung cavity).

Of the three types of radiation emitted by radioisotopes, only γ rays have the requisite penetrating power; even several MeV α and β particles are absorbed within a few millimeters of tissue. Most γ ray activity follows directly the emission of charged particles; the radioactive parent decays to an excited state of the daughter, which promptly drops to the ground state with the emission of one or more γ rays of sharply defined energy. Since α and β rays do damage cells by disrupting molecular bonds, the use of isotopes in which γ rays are accompanied by α or β emission involves a potential hazard to the patient. Until the end of the Second World War, however, these were the only γ-active isotopes available.

The radioisotope now favored for diagnostic purposes is ^{99m}Tc. As the superscript m indicates, this is a metastable or *isomeric* state of ^{99}Tc. ^{99m}Tc decays to the ground state of ^{99}Tc by emitting a 140-keV γ ray, with a half-life of 6 hours. No charged particle emission takes place; of course, some energetic charged particles will be produced by Compton scattering, a process that, in itself, is biologically damaging. Still, the radiation damage resulting from Compton scattering is far less than that due to β or α emission. Moreover, the half-life of this isotope is particularly well suited for diagnostic studies. It is long enough that the isotope is carried to all parts of the body by the bloodstream in a time short compared with its half-life. Yet, within a day (4 half-lives) the activity decays to a few percent of the initial value.

While the use of a short-lived radioisotope is advantageous in that it reduces the radiation dose to the patient, this does create some technical problems. Clearly, the time available for shipment of the isotope, the preparation of a suitable pharmaceutical incorporating this element, and its administration to the patient is severely limited. Fortunately, ^{99m}Tc, as well as a number of other short-lived radioisotopes, can be prepared in the hospital pharmacy using a "generator." The process works as follows.

^{99m}Tc is the daughter of another radioactive isotope, ^{99}Mo. The decay scheme of ^{99}Mo, whose half-life is 67 hours, is shown in Figure 15.27. Except for the excited state at 142 keV, all other excited states decay promptly to the ground state of ^{99}Tc.[11] The parent, ^{99}Mo, is produced by the reaction

$$^{98}\mathrm{Mo} + \mathrm{n} \rightarrow {}^{99}\mathrm{Mo} + \gamma$$

^{99}Mo is also one of the many fission products of ^{235}U.

---

11. ^{99}Tc is itself radioactive. Its half-life is extremely long, 2.1×10^5 years; consequently, the activity from this isotope, in amounts typically administered during radiodiagnosis, is negligibly small.

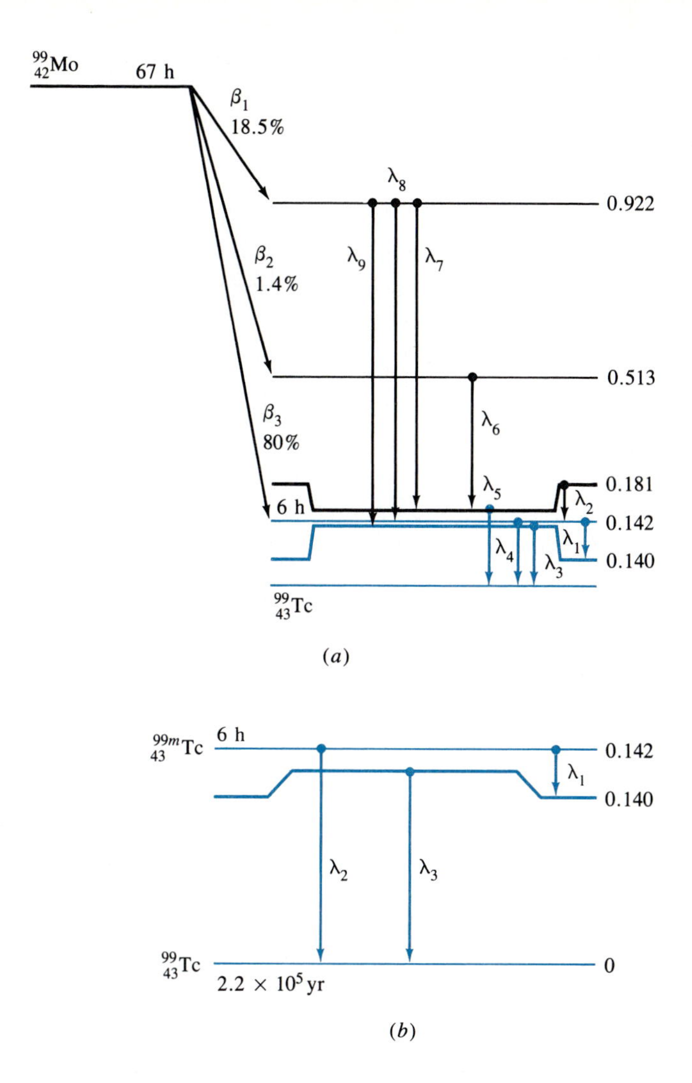

Figure 15.27 (*a*) The decay scheme of ^{99}Mo. (*b*) The decay of the isomeric state ^{99m}Tc.

The ^{99m}Tc generator consists of a column of alumina on which ^{99}Mo in the form of sodium molybdate has been adsorbed. Since the half-life of ^{99}Mo is relatively long, the generator can be prepared at the reactor site and shipped to the hospital, where it maintains reasonable activity for several days.

As ^{99}Mo decays, ^{99m}Tc is continuously produced, and, of course, also decays with a much shorter half-life. Within a few half-lives of the daughter a transient secular or dynamic equilibrium is established, such that the rate of production of ^{99m}Tc equals the rate of decay of this isotope. At secular equilibrium the activities of daughter and parent are related by

$$\frac{\mathscr{A}_{\mathrm{D}}}{\mathscr{A}_{\mathrm{P}}} = \frac{t_{1/2(\mathrm{P})}}{t_{1/2(\mathrm{P})} + t_{1/2(\mathrm{D})}}$$

Since the half-life of the parent is much longer than that of the daughter, the activity of the daughter at dynamic equilibrium is nearly the same as that of the parent [see Equation (15.7)].

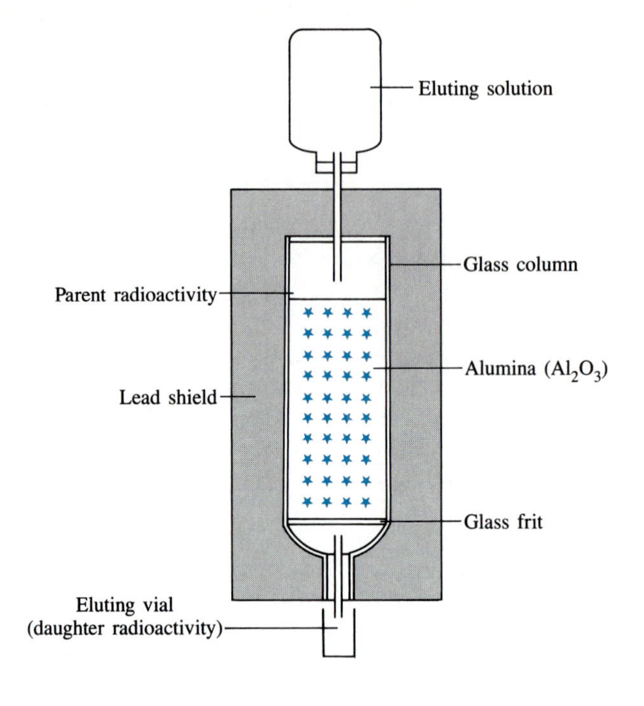

Figure 15.28 Schematic diagram of a radionuclide generator.

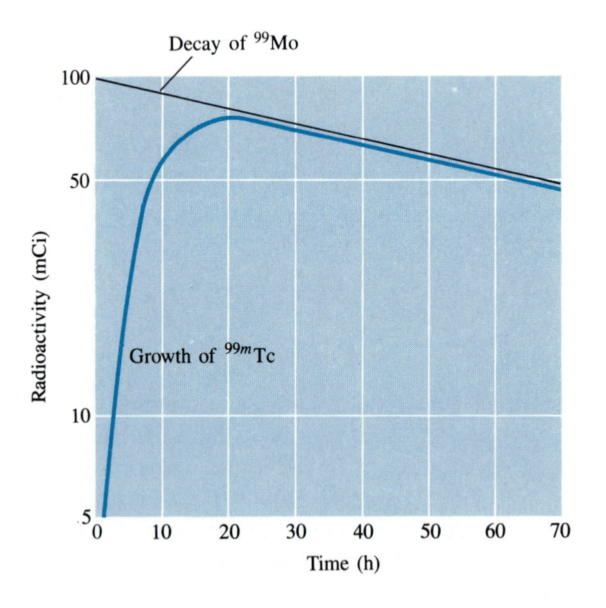

Figure 15.29 Activity of ^{99m}Tc as a function of time, starting with a sample of pure ^{99}Mo. Dynamic equilibrium is attained within about 24 hours, or four half-lives of ^{99m}Tc.

The daughter isotope is extracted from the generator by elution with a saline solution passed through the column. Following this process, the activity of ^{99m}Tc again builds up in the generator until after about 24 hours it is roughly equal to that of the parent, which is, of course, less than a day earlier. One need not wait the full 24 hours before another elution; however, if the generator is eluted in a shorter time, the activity of ^{99m}Tc will not have attained its optimum value.

Elution with saline solution yields a solution of sodium pertechnate, which may be administered directly or used to prepare ^{99m}Tc–labeled pharmaceuticals with desired properties. For example, for brain imaging, sodium pertechnate is generally used. Diffusion of the pertechnate through damaged cell membranes, signaling a breakdown of the blood-brain barrier, results in localized accumulation of the radioactive tracer, which then appears as a "hot spot" on a brain scan. On the other hand, small colloidal particles are filtered out of the bloodstream by reticuloendothelial cells, notably in the spleen and liver. The radiopharmaceutical of preference for diagnosis of liver malfunction is a sulfur colloid containing ^{99m}Tc. When this is administered intravenously, 70 percent to 80 percent of the dose is localized in the liver within 10 to 20 minutes. The remainder concentrates in the spleen and in bone marrow.

For some purposes, total uptake of a radiopharmaceutical may provide sufficient data. For instance, the thyroid uses iodine in producing hormones that control the metabolic rate. In an individual with a hypothyroid (underactive) condition, the thyroid will accumulate less iodine than in a normal person, and the converse is true for one with a hyperthyroid condition. To test for overall thyroid function, the subject is given a small amount of ^{123}I orally (iodine is readily absorbed through the stomach), and the concentration of ^{123}I in the thyroid is then determined 24 hours later using standard scintillation detectors.

EXAMPLE 15.7 A patient is given 30 μCi of ^{123}I. Twenty-four hours after ingestion of the radiopharmaceutical, the activity emanating from the thyroid region is monitored and found to be 4 μCi. What fraction of the ingested ^{123}I was concentrated in the thyroid during the 24-hour period?

SOLUTION The half-life of ^{123}I is 13 hours. Twenty-four hours after ingestion, the activity of the ^{123}I will have been reduced to

$$(30\ \mu\text{Ci})(\tfrac{1}{2})^{24/13} = 8.344\ \mu\text{Ci}$$

Since the activity from the thyroid region is 4 μCi, 4/8.344 = 48 percent of the iodine has been concentrated in the thyroid. In the normal person, between 15 and 40 percent of the iodine intake is concentrated in the thyroid, and so this result could be indicative of a hyperactive thyroid condition.

Often the physician needs to know not only the amount but also the distribution of the tracer in the organ. To this end, several techniques are available. The oldest, and now rarely used, is the rectilinear scan. A NaI scintillation detector, fitted with a collimator (see Figure 15.30), is moved in a raster pattern over the region of interest and the counting rate recorded. The output of the detector is either displayed directly on a monitor or stored in a computer memory for analysis. Imaging using rectilinear scanning is a slow process, often taking 30 minutes or more, during which the patient should remain immobile. Since ordinary breathing shifts the position of body organs, spatial resolution from rectilinear scans is generally poor.

The device that has displaced the rectilinear scan is the Anger camera, also referred to as a γ ray camera. This unit consists of a large single NaI crystal placed between a lead collimator plate and an array of photomultiplier (PM) tubes. The PM

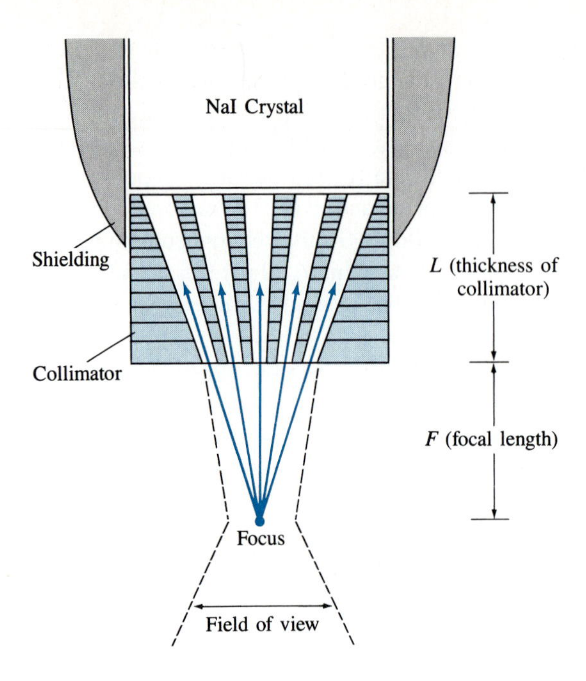

Figure 15.30 A lead collimator fitted to a NaI scintillation detector for use in rectilinear scanning.

tubes not only measure the total amount of radiation but also sense the origin of the individual scintillations and thus form a picture of the distribution of the radioisotope in the area examined. Generally, the output of the PM tubes is digitized and subjected to computer analysis.

In recent years the development and advances in nuclear magnetic resonance imaging (see Section 14.6) have made this the preferred imaging method. NMR imaging is often more sensitive and gives better spatial resolution than γ ray imaging; moreover, it does not subject the patient to potentially harmful radiation.

Radioisotopes for Therapy

The biological hazards of nuclear radiation were recognized already in the early years of radioactivity research. In her doctoral dissertation, Marie Curie reported that a capsule of radium strapped to her husband's arm for a few hours resulted in a lesion that did not heal for several weeks. It was not long before the potential of radium and other radioactive isotopes for treating cancer was investigated. In contrast to radiodiagnosis, the aim of radiotherapy is the destruction of cells by radiation damage. A major problem now is to confine the damage to a limited region of the body, to maximize the damage to tumor cells while minimizing damage to surrounding healthy tissue. In some treatment, γ rays, typically from an intense ^{60}Co source, are used. In that case the collimated beam of radiation is centered on the tumor by rotating the source around the patient, thereby reducing damage to the rest of the body.

A preferred method of radiation therapy is to use energetic charged particles from a small accelerator, a betatron for energetic electrons or a cyclotron for protons, deuterons, or α particles. Fairly high energies are required, except for treatment of skin cancer, so that these particles will penetrate to the location of the malignancy. Particle beams have a number of advantages over γ rays. First, the beam is well defined and can be aimed precisely at the malignancy. Second, the

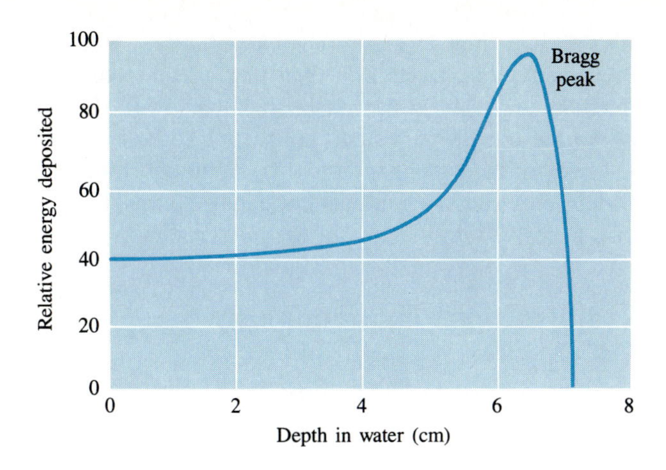

Figure 15.31 Energy loss of a high-energy α particle as a function of penetration. Note the pronounced peak (Bragg peak) at the end of the range.

energy loss of these particles, and hence their efficacy in producing biological damage, reaches a fairly sharp maximum near the end of their range, the so-called Bragg peak. How far the particles penetrate is determined by their charge, mass, and energy. Thus, by controlling the energy of the incident beam, damage can be maximized at the desired location. As with γ ray therapy, it is common practice to either rotate the patient in the beam or, preferably, rotate the beam about the patient to minimize damage to healthy tissue.

A particularly attractive particle for cancer therapy is the π^- (negative pion). This particle, given sufficient energy, also penetrates considerable distance and also exhibits the characteristic Bragg peak at the end of its range. Moreover, there is a high probability that at the end of its travel the π^- will interact with a nucleus, producing additional reaction fragments that also deposit their energy in the immediate vicinity of this "star." Thus the effective dose is greatly enhanced, as indicated in Figure 15.32.

Unfortunately, cyclotrons are expensive and cumbersome machines, and relatively few medical centers have acquired such units. Machines capable of producing

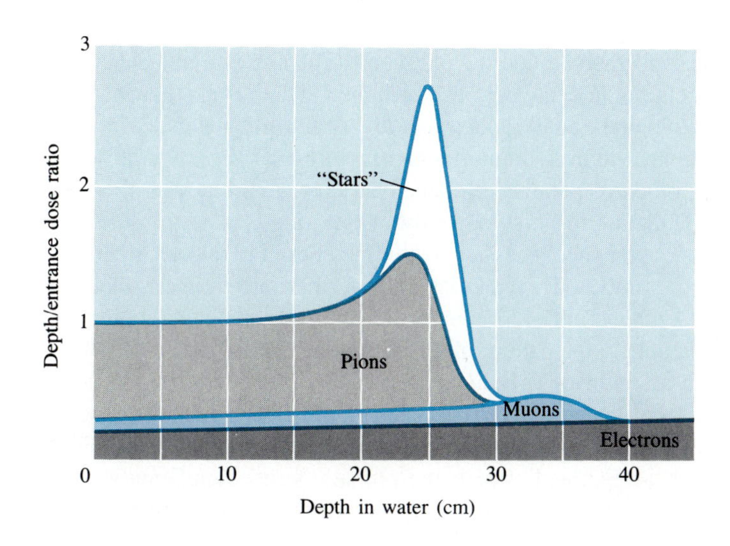

Figure 15.32 Energy deposited by 95-MeV π^- particles as a function of penetration. The normal Bragg peak is further enhanced by the energy deposited due to capture of the π^- by a nucleus and the subsequent nuclear reaction (star).

useful π^- beams are even rarer; there is one at Los Alamos, N.M., and another in Vancouver, B.C., and only a fraction of their time is allocated to medical work.

Another method employed in cancer treatment, known as brachytherapy, involves the insertion of one or more small radioactive sources, usually α emitters, directly into the body of the patient at the site of the tumor. Since the range of the ionizing radiation is only about one millimeter, damage is well localized and can be confined almost exclusively to the area of the malignancy.

15.7(b) Radioisotopes in Archeology

Archeology is a relatively young science. It was as recent as 1850 that amateur "antiquarians" and geologists found incontrovertible evidence that humans lived at times when the earth was populated by animals long extinct. These discoveries, in England and France, of human remains and primitive tools alongside bones and teeth of mammoths, gave support to Darwinians, who maintained that the human race as well as other life forms had evolved over many millennia. However, for about a century, archeologists were unable to establish the age of remains found at ancient sites with any accuracy. Relative dating was based on the location of bones, potsherds, tools, and other artifacts in layers of soil in caves or ancient cities. Objects lying below others were presumed to be of earlier origin, but assignment of absolute dates was next to impossible. In some regions, for example, Egypt, historical records could be used to trace human activities to about 3000 B.C., but in most others only sketchy, indirect evidence was available. Heated arguments and controversy concerning dates assigned to various cultures abound in archeological writing prior to 1950.

What was needed was an accurate archeological "clock." One of the first such clocks was discovered by a young American astronomer, Andrew Douglass, who had been asked by Percival Lowell, Director of the Harvard College Observatory, to find a good site for the Lowell Observatory under the clear skies of Arizona. Douglass was interested in sunspots; these appear at all times, but are most prominent at regular intervals of about 11.4 years. It was believed, then as now, that sunspot activity influences the climate on earth, and Douglass looked for evidence in support of that hypothesis. He found it by investigating the width of tree rings. During the spring and summer growth period, a new layer of cambium forms below the bark. In winter, growth stops, the cells die, and the following spring a new cambium layer is formed around the old wood. The age of a tree can be readily determined by counting the number of these light and dark rings. However, closer examination reveals that not all rings are of the same width. A wide ring indicates more than normal growth in a summer, a narrow ring relatively little growth; these variations suggest corresponding climatic variations, and it was this which at first prompted Douglass to pursue a study of tree rings.

It is, of course, not necessary to confine oneself to the examination of tree rings of freshly cut timber. To carry the study further into the past, Douglass collected samples of timber from old Indian settlements. By matching patterns of tree rings from one sample to the pattern of a somewhat older sample, he was ultimately able to establish a continuous chronology as far back as A.D. 700. An accurate clock, whose technical name is *dendrochronology,* had been found, a clock that did not rely on human historical records. Still, it was not a universal clock. For example, in regions such as England, where the climate varies significantly over small distances, the method cannot be applied. The tree ring patterns from two locations only a few miles apart are often quite different. In other regions, for example, large portions of the Middle East, there are practically no trees; stones and mud bricks were used for building.

The first, and still widely used universal archeological clock is carbon-14, discovered and exploited by Willard F. Libby, who was awarded the 1961 Nobel Prize in chemistry for this work which revolutionized archeology. Carbon-14 dating has an interesting and instructive history.

It was known since the early part of the twentieth century that there is an electrically charged layer in the upper atmosphere, the ionosphere. It is that layer which makes long-distance radio communication possible by reflecting waves back down to earth. By the mid-1930s it was also known that this layer is created by a continuous stream of energetic particles that strike the upper atmosphere, some of them originating from the sun, while others, the most energetic, come from greater distances. These are the cosmic rays. The vast majority of these energetic particles never reach the ground, but interact with nuclei of oxygen and nitrogen, resulting in other particles, such as muons and electrons. Shortly before the Second World War, Serge Korff placed neutron detectors in a high-altitude balloon and found that 15 miles above the earth's surface the neutron flux was surprisingly large, about 2 neutrons per square centimeter per second. What happens to these high-altitude neutrons could be deduced from laboratory studies; they interact with nitrogen nuclei via the (n, p) reaction

$$^{14}\mathrm{N} + \mathrm{n} \rightarrow {}^{14}\mathrm{C} + {}^{1}\mathrm{H}$$

forming the radioactive isotope ^{14}C, which subsequently decays to ^{14}N by β emission with a half-life of 5730 ± 30 years:

$$^{14}\mathrm{C} \rightarrow {}^{14}\mathrm{N} + \beta^{-} + \bar{\nu}$$

As a result of the constant creation and simultaneous decay of ^{14}C, a steady-state concentration of this isotope will be established in a time of several ^{14}C half-lives. Knowing the neutron flux in the upper atmosphere and the half-life of ^{14}C, one could calculate the relative atmospheric concentration of ^{14}C, that is, the ratio of the number of ^{14}C to ^{12}C atoms; it proved to be very small. The currently accepted value of this ratio is $1/7.6 \times 10^{11}$; only about one atmospheric carbon atom in 10^{12} is ^{14}C. Although ^{14}C had been produced in the laboratory, no one had found a trace of this isotope in nature. Nevertheless, Libby developed the theory of radiocarbon dating during the war years, though he had to wait until 1945 before he could pursue these ideas. It then took another five years of intensive effort to show that ^{14}C does occur naturally in the atmosphere in the form of $^{14}CO_2$ and, consequently, also in all organisms that rely directly or indirectly on the uptake of atmospheric carbon dioxide.

Carbon dioxide, used by plants in photosynthesis, enters all organisms that live off plants, be they insects, reptiles, fish, or mammals. Thus the carbon in a living plant or animal contains one ^{14}C atom for every 7.6×10^{11} ^{12}C atoms. However, once the organism dies, there is no further uptake of atmospheric carbon. Hence, over the years the ^{14}C nuclei in the organism decay to ^{14}N; 5730 years following death, the concentration of ^{14}C will be only $1/3.8 \times 10^{11}$, and another 5730 years later the number of ^{14}C nuclei will have diminished by another factor of 2. Consequently, a measurement of the ^{14}C activity of a sample of wood, bone, leather, or

Bibliography

For general references on nuclear physics, see the bibliography to Chapter 14.

Radioactivity and Nuclear Reactions

G. M. Lewis, *Neutrinos,* Wykeham, London, 1970.

G. R. Satchler, *Introduction to Nuclear Reactions,* Wiley, New York, 1980.

K. Siegbahn, (ed.), *Alpha-, Beta- and Gamma-Ray Spectroscopy,* North-Holland, Amsterdam, 1965.

C. Strachan, *The Theory of Beta Decay,* Pergamon, Oxford, 1969.

G. K. Wertheim, *Mössbauer Effect: Principles and Applications,* Academic Press, New York, 1964.

Fission and Fusion

I. R. Cameron, *Nuclear Fission Reactors,* Plenum, New York, 1982.

Robert W. Conn, ''The Engineering of Magnetic Fusion Reactors,'' *Scientific American,* October 1983, p. 60.

R. S. Craxton, R. L. McCrory, and J. M. Soures, ''Progress in Laser Fusion,'' *Scientific American,* August 1986, p. 68.

J. J. Duderstadt and G. A. Moses, *Inertial Confinement Fusion,* Wiley, New York, 1982.

Harold P. Furth, ''Progress toward a Tokamak Fusion Reactor,'' *Scientific American,* August 1979, p. 50.

Harold W. Lewis, ''The Safety of Fission Reactors,'' *Scientific American,* March 1980, p. 53.

W. M. Stacey, Jr., *Fusion: An Introduction to the Physics and Technology of Magnetic Confinement Fusion,* Wiley, New York, 1984.

Applications of Nuclear Physics

E. G. A. Aird, *An Introduction to Medical Physics,* Heinemann, London, 1975.

J. T. Andrews and M. J. Milne, *Nuclear Medicine,* Wiley, New York, 1977.

N. A. Dyson, *An Introduction to Nuclear Physics with Applications in Medicine and Biology,* Ellis Horwood, Chichester, England, 1981.

P. Kruger, *Principles of Activation Analysis,* Wiley, New York, 1971.

Willard F. Libby, *Radiocarbon Dating,* University of Chicago Press, Chicago, 1952.

Richard A. Muller, ''Radioisotope Dating with Accelerators,'' *Physics Today,* February 1979, p. 23.

David Wilson, *The New Archeology,* New American Library, New York, 1974.

PROBLEMS

15.1 The relative abundance of the natural radioactive isotope of potassium, ^{40}K, is 1.2×10^{-4}. The half-life of this isotope is 1.3×10^9 yr. Potassium is an essential element in a living cell, and constitutes approximately 0.37 percent of the human body, by weight. What is the activity of this radioactive source in the average person?

15.2 The activity of a radioactive source at $t = 0$ is 4×10^6 Bq. Half an hour later the activity has diminished to 2.5×10^5 Bq. What is the half-life of this radioactive material?

15.3 Suppose nucleus A has a half-life that is twice that of nucleus B. If at some moment there are twice as many B nuclei in a sample as there are A nuclei, and if the half-life of nuclide A is 4 h, when, if ever, will the activity of the sample be due in equal measure to the decays of A and B nuclei? When will the number of A and B nuclei be equal?

15.4 Nuclide A decays to B, and nuclide B decays to C. Prove that after secular equilibrium has been reached, the activities of nuclides A and B are the same.

15.5 Nuclide A decays to nuclide B with a half-life $t_{1/2}(\mathrm{A})$, and nuclide B then decays with a half-life $t_{1/2}(\mathrm{B})$. Suppose that the half-lives are related by $t_{1/2}(\mathrm{B}) = 2t_{1/2}(\mathrm{A})$. Plot $N_\mathrm{A}(t)$ and $N_\mathrm{B}(t)$ between $t = 0$ and $t = 4t_{1/2}(\mathrm{B})$, assuming that at $t = 0$, $N_\mathrm{A} = N_0$ and $N_\mathrm{B} = 0$.

•15.6 The parent nuclide P decays to the daughter D with a half-life $t_{1/2}(\mathrm{P})$; nuclide D, in turn, decays to C with a half-life $t_{1/2}(\mathrm{D})$. A sample contains only nuclide P at $t = 0$. Derive an expression for the time, T_{max}, at which the number of daughter nuclides is a maximum.

15.7 ^{214}Pb has a half-life of 26.8 min, decaying to ^{214}Bi, which then decays to ^{214}Po or ^{210}Tl with a half-life of 19.7 min. If at $t = 0$ a sample of ^{214}Pb contains no ^{214}Bi, at what time will the activity due to ^{214}Bi reach a maximum?

15.8 The activity of a radioactive sample is measured with a counter and the data shown below are obtained. Plot the activity as a function of time on semilogarithmic paper and determine the disintegration constant and half-life of the radioisotope.

Time (min)	Activity	Time (min)	Activity
0	2156	20	448
5	1420	30	209
10	988	40	95
15	670	60	20

15.9 Calculate the mass of ^{226}Ra and of ^{231}Pa in 10 kg of uranium ore, U_3O_8.

15.10 Consider the reactions $^{13}C(p, \alpha)^{10}B$, $^{13}C(p, d)^{12}C$, and $^{13}C(p, \gamma)^{14}N$. Determine the Q value for each of the reactions. Which are exothermic, which are endothermic?

•15.11 Calculate the threshold energy for the $^{13}C(p, \alpha)^{10}B$ reaction. Remember that the proton must have enough energy to come within the nuclear radius of the ^{13}C target nucleus.

•15.12 If the reaction of Problem 15.11 proceeds at the calculated threshold energy of the proton, what is then the energy of the emitted α particle in the laboratory? Assume the α particle is emitted in the direction of the incident proton.

15.13 What is the threshold energy for the $^{66}Zn(n, p)^{66}Cu$ reaction? What is the energy of the proton in the laboratory frame?

15.14 ^{64}Cu is unstable against β^- and β^+ decay. What are the maximum energies of the electrons and positrons emitted in the two decay schemes?

15.15 Calculate the energy released in the β decay

$$^{14}C \rightarrow {}^{14}N + e^- + \bar{\nu}_e$$

15.16 The nuclide pairs $^{11}C-{}^{11}B$, $^{15}O-{}^{15}N$, and $^{17}F-{}^{17}O$ are "mirror" nuclei. The first of each pair is a positron emitter. Calculate the energy released in the β^+ decay using only Equation (14.4) and the result, derived in Example 14.5, that the electrostatic energy of a uniformly charge sphere is

$$E_{el} = \frac{1}{4\pi\epsilon_0}\frac{3Q^2}{5R}$$

Observed values of the maximum β^+ energies are given in Appendix B.

15.17 Complete the equation for the following reactions:

$$^{32}Sn + n \rightarrow {}^{32}P + \quad ; {}^{2}H + p \rightarrow {}^{3}He + \quad ; {}^{223}Fr \rightarrow {}^{223}Ra +$$

15.18 For each of the following reactions give the compound nucleus and indicate three possible decay schemes:

$$^{32}Ge + {}^{4}He \rightarrow \qquad ; {}^{50}Cr + {}^{1}H \rightarrow \qquad ; {}^{128}Sn + n \rightarrow$$

15.19 In the reaction $^{14}N(n, p)^{14}C$ a resonance is observed when the energy of the incident neutron, in the laboratory frame, is 3.1 MeV. What is the compound nucleus formed in this reaction, and what is the energy of the excited state of that nucleus?

•15.20 The same compound nucleus that is formed in the reaction of Problem 15.19 can also be produced by bombarding ^{11}B with α particles. What should be the incident α particle energy in the laboratory frame so that the same excited compound nucleus state is formed by this reaction?

15.21 The cross section for the reaction $^{107}Ag(p, n)^{107}Cd$ is 0.08 barn. How many neutrons are released per second if a 15-μA beam of protons is incident on a target whose thickness is 15 μm? What fraction of the incident protons are removed from the beam in this process?

•15.22 The cross section for the reaction $^{59}Co(n, \gamma)^{60}Co$ at low neutron energies is 120 barns. ^{60}Co is unstable to β decay with a half-life of 5.27 yr. A cube of cobalt, 2 cm on edge, is placed into a beam of neutrons whose flux is 2×10^{12} neutrons/cm^2·s and left for 6 hours. What is the activity of the sample when it is removed from the reactor?

15.23 What are the Q values of the six reactions of the carbon cycle given on page 386?

15.24 A saline solution containing ^{24}Na whose activity is 300 kBq is injected into the vein of a man. Ten hours later the activity of one cubic centimeter of blood is 30 Bq. Calculate the volume of blood in the man's body.

15.25 Calculate the relative abundance of ^{235}U two billion years ago.

15.26 Cadmium has eight stable isotopes, some with large absorption cross sections for slow neutrons. If the average neutron absorption cross section is 4000 barns, what thickness of cadmium would be needed to absorb 95 percent of the incident neutron flux? The atomic weight of cadmium is 112.4, and its density is 8.64 g/cm^3.

15.27 ^{239}Pu decays by α emission to ^{235}U. The half-life of ^{239}Pu is 24,360 yr. What is the Q value for the decay of ^{239}Pu? At what rate is energy released from one gram of pure ^{239}Pu?

15.28 What is the most likely decay of the compound nucleus formed when ^{11}B absorbs a slow neutron? What is the Q value of this reaction?

15.29 What fraction of the energy of a fast neutron is lost in a head-on elastic collision with a deuteron that is initially at rest?

15.30 How many head-on elastic collisions are required to reduce the energy of a 2-MeV neutron to 0.1 eV if the moderating nuclei are (a) deuterons? (b) ^{12}C? Is it a reasonable approximation to neglect the thermal motion of these nuclei?

15.31 The reaction $^{9}Be(\alpha, n)^{12}C$ has a cross section of 0.1 barn for 4-MeV α particles. If a beryllium target of 2 μm thickness is placed in a 10-μA beam of 4-MeV α particles, at what rate are neutrons emitted from the target?

•15.32 ^{60}Co is frequently used in radiation therapy. This isotope decays by β^- emission with a half-life of 5.24 yr, and the decay is generally accompanied by the emission of γ rays from the daughter nucleus. ^{60}Co can be produced by irradiating natural cobalt with thermal neutrons from a reactor. The cross section for the reaction $^{59}Co(n, \gamma)^{60}Co$ is 120 barns. A slab of natural cobalt 4 cm × 5 cm × 0.1 cm is subjected to a neutron flux of 1.5×10^{12} neutrons/cm^2·s. How long must the cobalt sample be kept in the neutron beam if the activity due to the ^{60}Co produced is to be 5 Ci?

15.33 A piece of wood of unknown age is burned in an atmosphere of pure oxygen and a total of 0.3 mol of CO_2 collected.

(*a*)

Figure 16.20 (*a*) The Big European Bubble Chamber (BEBC) at CERN. The bubble chamber itself is inside the large magnet and not visible in the photograph. (*b*) Photograph of events initiated by an inelastic neutrino-proton collision. The lifetime of the "charmed" meson D (see Chapter 17) is so short that its presence cannot be seen; all other particles originating from the decay of the D*, with the exception of neutrinos, leave visible tracks in the bubble chamber. (*c*) Drawing of the events shown in *b*. Below and to the left of the principal vertex, the K^- produced in the decay of the $D°$ interacts with a proton in the hydrogen bubble chamber; the neutron from the decay of the Σ^- leaves no visible track. *(CERN.)*

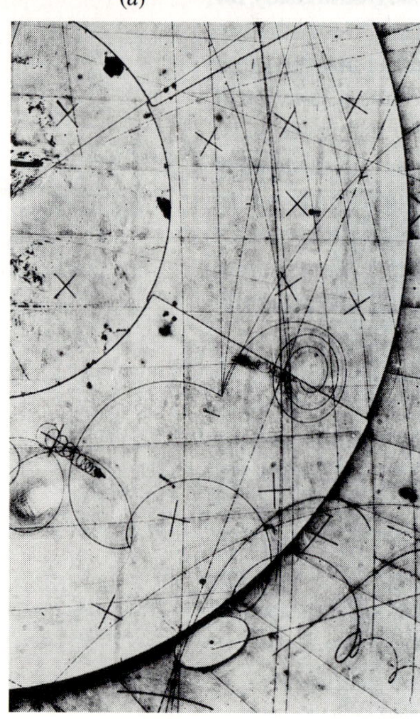

(*b*)

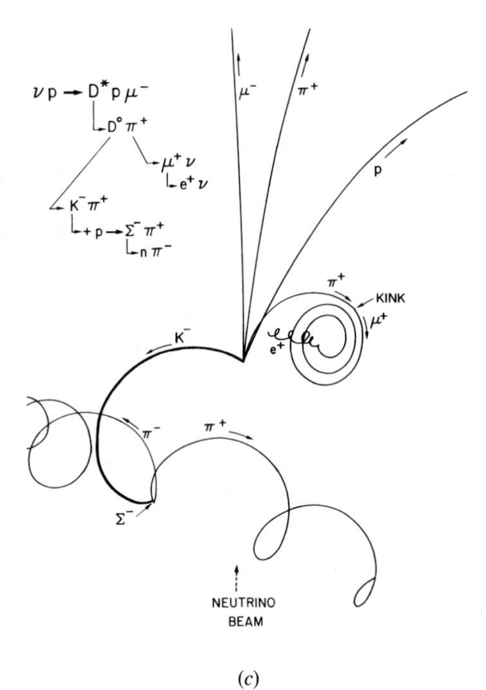

(*c*)

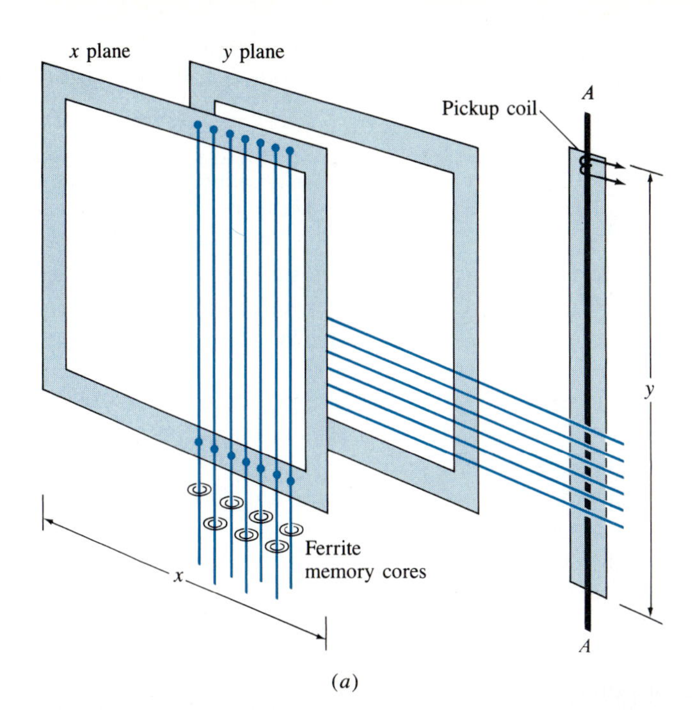

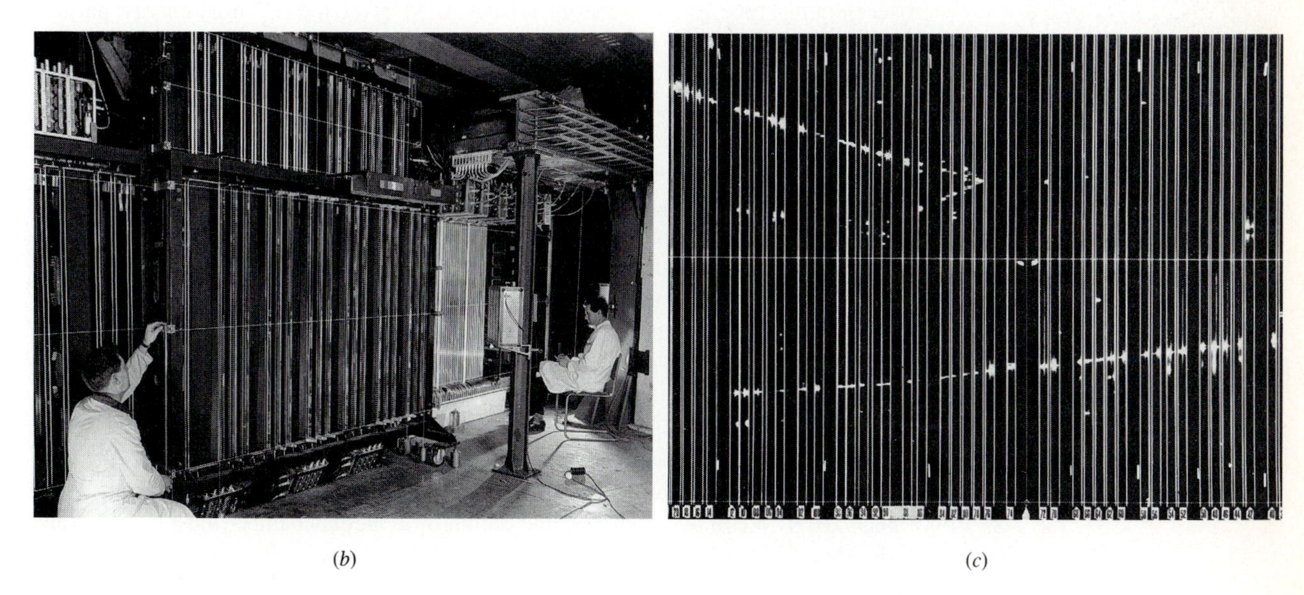

Figure 16.21 (*a*) Schematic diagram of a digitized wire-mesh spark chamber. (*b*) Assembly of a large spark chamber. *(CERN.)* (*c*) Tracks of particles in a spark chamber. *(CERN.)*

nate system, the location of each spark discharge is readily determined, and the data can be stored and analyzed by computer. If desired, visual records of interesting events can be obtained later by computer simulation.

Finally, we should mention one detector which has the virtue of providing direct information on the particle's velocity. Suppose a charged particle enters a transparent dielectric whose index of refraction n is greater than $c/v = 1/\beta$. In that case, light emitted by excited atoms or molecules along the particle's track propagates at a speed less than that of the particle itself. The situation is completely analogous to what happens when a source of sound travels faster than the speed of sound in air: a shock wave forms. The cone angle of this coherent wavefront is a direct measure of βn. The device that makes use of this effect is known as a Cerenkov counter, named after the Russian physicist who first observed and studied the phenomenon.

SUMMARY

The earliest particle accelerators were electrostatic machines. Although the maximum proton energy that can be achieved is about 25 MeV, Van de Graaff accelerators are still in use today because they provide beams of high intensity and excellent energy resolution.

High-energy accelerators use time-varying electric fields. These fall into two categories, *linear accelerators (linacs),* and *circular accelerators*. In a linac the charged particles progress through successively longer drift tubes that are connected to radio-frequency oscillators so that adjoining drift tubes are of opposite polarity. As the charged particles traverse the region between adjoining drift tubes, they are accelerated by the electric field between the drift tubes.

In circular accelerators the particles are confined to circular or near-circular paths by static or time-varying magnetic fields. *Cyclotrons* use static magnetic fields. *Synchrotrons* use a series of magnets spaced along the circumference of the fixed circular path whose fields increase so that the particles continue to travel in the accelerating tube as their energy escalates. Most of today's high-energy particle accelerators are of the colliding-beam type.

Particle detectors fall into two broad categories, those that sense the presence of an energetic particle and those that provide a visual record of the passage of the particle. Among the first group are *ionization counters* and *scintillation counters*. Ionization counters produce a pulse of current as a result of ionization of gas within the counter by the charged particle. In scintillation counters a pulse of light is generated by the passage of a charged particle; the light is sensed by a photomultiplier tube whose output is a current pulse.

Bubble chambers and *spark chambers* provide a visual record of the passage of charged particles. Both chambers are almost always placed in a strong magnetic field to permit the measurement of the particles' momenta. In a bubble chamber the passage of an ionizing particle produces a track of bubbles in a superheated liquid. A spark chamber consists of a multitude of fine wires with a high potential between the wires. Passage of an ionizing particle is revealed by a series of sparks between wires. Because the output of a spark chamber can be fed into, stored by, and analyzed by a computer whereas the output of a bubble chamber must be examined visually, the spark chamber is the preferred detector at colliding-beam machines.

Bibliography

Accelerators

S. L. Glashow and L. M. Lederman, "The SSC: A Machine for the Nineties," *Physics Today,* March 1985, p. 29.

M. Goldsmith and E. Shaw, *Europe's Giant Accelerator,* Taylor and Francis, London, 1977.

R. Gourian, *Particles and Accelerators,* McGraw-Hill, New York, 1967.

J. David Jackson, Maury Tigner, and Stanley Wojcicki, "The Superconducting Supercollider," *Scientific American,* March 1986, p. 66.

E. Persico, E. Ferrari, and S. E. Segrè, *Principles of Particle Accelerators,* W. A. Benjamin, New York, 1968.

John R. Rees, "The Stanford Linear Collider," *Scientific American,* October 1989, p. 58.

A. M. Sessler, "New Particle Acceleration Techniques," *Physics Today,* January 1988, p. 26.

Robert R. Wilson, "The Next Generation of Particle Accelerators," *Scientific American,* January 1980, p. 42.

Particle Detectors

R. C. Fernow, *Introduction to Experimental Particle Physics,* Cambridge University Press, Cambridge, 1986.

Glenn F. Knoll, *Radiation Detection and Measurement,* Wiley, New York, 1979.

R. E. Lapp and H. L. Andrews, *Nuclear Radiation Physics,* 4th ed., Prentice-Hall, Englewood Cliffs, N.J., 1972.

W. Price, *Nuclear Radiation Detection,* 2d ed., McGraw-Hill, New York, 1964.

PROBLEMS

•**16.1** Electrons are injected into a drift-tube linear accelerator with an initial energy of 60,000 eV. The rf oscillators connected to the drift tubes operate at 1.5 GHz with an amplitude of 20,000 V.

(a) Determine the correct length of the first, fifth, tenth, and fiftieth drift tube.
(b) Estimate the energy of the emerging beam if the total length of the accelerator is 1 km.

16.2 The frequency of the oscillator that drives a cyclotron is fixed at 12 MHz. The accelerated particles are extracted at a radius of 60 cm. What is the energy of the extracted particles if they are (a) protons and (b) deuterons?

16.3 A conventional cyclotron of 1.5 m diameter is used to accelerate protons to an energy of 15 MeV. What should be the strength of the magnetic field and the frequency of the rf oscillator?

16.4 If the machine of Problem 16.3 is used to accelerate deuterons, what will be their energy if the frequency of the rf oscillator is not changed?

16.5 Repeat Problem 16.4 if α particles are accelerated.

16.6 If dephasing precludes acceleration of particles in a conventional cyclotron beyond an energy such that $m/m_0 = 1.1$, what are the maximum beam energies that can be achieved with such a machine that accelerates (a) protons, (b) deuterons, (c) α particles?

16.7 Protons are injected into a proton synchrotron at an energy of 20 MeV. If the initial field of the guide magnets is 0.015 T, what must be the diameter of the ring? If the magnetic field at the end of the cycle is 1.2 T, what is the energy of the beam at that time?

16.8 A sector-focused cyclotron produces protons of 400 MeV. The diameter of the pole faces is 2.4 m. What must be the average magnetic field at the periphery and near the center of the machine, and what should be the frequency of the rf oscillator?

16.9 Repeat Problem 16.8 for a machine that accelerates α particles to the same energy.

16.10 Suppose that electrons are injected into a synchrotron at an energy of 150 MeV. The initial field of the guide magnets is 0.4 T and the maximum field is 1.2 T.

(a) Show that a fixed-frequency oscillator can be used to accelerate the particles, and determine its frequency.
(b) What is the calculated energy of the electrons at the end of a cycle?
(c) Why is the energy of the electrons actually less than that calculated in part (b)?

16.11 Typical photomultiplier tubes have an amplification factor of 10^5; that is, for every electron emitted by the primary cathode, 10^5 electrons will be collected at the final anode. Suppose that the output of the photomultiplier is connected to a high-impedance amplifier whose input capacitance is 2 pF and whose gain is 10^4. What will be the magnitude of the output voltage pulse if 10 photons strike the photocathode whose conversion efficiency is 8%?

Chapter 17 Elementary Particles

All science is the search for unity in hidden likeness.

J. Bronowski

17.1 INTRODUCTION

In this chapter we embark on a brief excursion to the edge of the known, a world of bizarre phenomena, populated by exotic inhabitants whose relationships are governed by strange rules of behavior. These rules derive from a framework that brings order to an apparently chaotic proliferation of physical entities, the *elementary particles*. The foundation of that framework is far removed not only from everyday experience but even from previously accepted postulates of quantum mechanics. Bizarre it may be, yet this model has been so successful in explaining and **predicting** phenomena that it is now known as the *standard model*. One's first reaction is understandably skepticism, no different from that of physicists at the start of the twentieth century who faced the onslaught of relativity and quantum theory on their comfortable world of Newtonian mechanics.

The world around us has almost infinite diversity. There is an enormous multitude of life forms; within any one species each individual shows a special character that sets this one apart from other members of the class. We take pleasure in this variety, but we also look for similarities that can help us understand the broad pattern of nature. We firmly believe that despite this diversity the world is constructed from just a few basic elements. The ancients thought that these were air, earth, fire, and water. For Dalton and the chemists of the nineteenth century, atoms were the basic elements, the fundamental particles from which all matter could be constructed by proper combination. Early in the twentieth century atoms gave way to electrons, protons, and photons as fundamental particles, and neutrons and neutrinos had to be included two decades later.

The underlying idea is that of economy of effort, of an aesthetically pleasing simplicity that we find intellectually attractive. There is no guarantee that nature really is the way we like it to be; we trust that it is so. We know or sense that we shall understand it better if its diversity is of form, not of substance. And so we look for the ultimate building blocks, the elementary particles, and trust that they are few in number.

Atoms consist of electrons and nuclei; nuclei consist of nucleons, protons and neutrons. If nucleons have a structure, are combinations of more elementary particles, these must be smaller than the nucleon. If we want to "see" something very small, we must use light of very short wavelength. Short wavelength means high frequency or high energy. This is why elementary particle physics and high-energy physics are, at least for the experimentalist, practically synonymous. However, not only are high energies a requisite for probing the submicroscopic structure of particles, but high energies are also required if we wish to verify that elementary particles that have been postulated actually exist in nature. For instance, the meson, proposed by Yukawa as the mediator of the strong force, whose mass was estimated to be $230m_e$, can be created in a high-energy collision, but only if the excess energy available is greater than $230m_ec^2$. Today mesons are produced in great number in high-energy accelerators.

A theory of elementary particles must involve a marriage of quantum mechanics (the particles are small) and relativity (they have high energy, travel at relativistic speeds), a union that has produced remarkable offspring.

17.2 RELATIVISTIC QUANTUM MECHANICS

The Schrödinger equation is the quantum analog of Newton's second law. It correctly describes the behavior of particles at nonrelativistic speeds, but is not invariant under a Lorentz transformation. The reason for this is not difficult to find. Relativity demands a symmetry of space and time coordinates; events are described in a four-dimensional space whose coordinates are x, y, z, and ct. Since the Schrödinger equation involves the first derivative with respect to time and the second derivative with respect to space coordinates, it cannot be invariant under a Lorentz transformation. In 1928, P. A. M. Dirac devised an equation for a quantum system that is. A proper discussion of that equation is well beyond the scope of this book. Here we focus on the consequences of that equation.

Figure 17.1 P. A. M. Dirac (1902–1984). *(The Bettmann Archive/ Bettmann Newsphotos.)*

We recall that the Schrödinger equation is consistent with the Planck and de Broglie postulates,

$$E = h\nu = \frac{hc}{\lambda} \tag{17.1}$$

$$p = \frac{h}{\lambda} \tag{17.2}$$

We can eliminate λ and obtain $E = pc$, which is correct for photons, i.e., for massless particles. For particles with nonzero rest mass m_0 the total relativistic energy is not pc but is given by

$$E^2 = p^2c^2 + m_0^2c^4 \tag{17.3}$$

Evidently, Equation (17.3) is not consistent with Equations (17.1) and (17.2).

Dirac's relativistically invariant equation yields the result

$$E = \pm(p^2c^2 + m_0^2c^4)^{1/2} \tag{17.4}$$

Normally, we would discard the negative-energy solutions of Equation (17.4) as being "unphysical," but we cannot do this quite so cavalierly and still retain the relativistic wave equation. Of course, one might well question the wisdom of preserving an equation that gives such peculiar results. However, another consequence of that equation was not peculiar at all: it **predicted** an intrinsic spin of $\frac{1}{2}$, placing the ad hoc postulate of Goudsmit and Uhlenbeck on firm theoretical footing. Still, the negative-energy states posed a real dilemma. How could an atom exist in its ground state if states of much lower energy are available to an electron? Since the negative-energy states extend to infinity (or, to be precise, to minus infinity), all matter should vanish as electrons drop into this infinite energy pit.

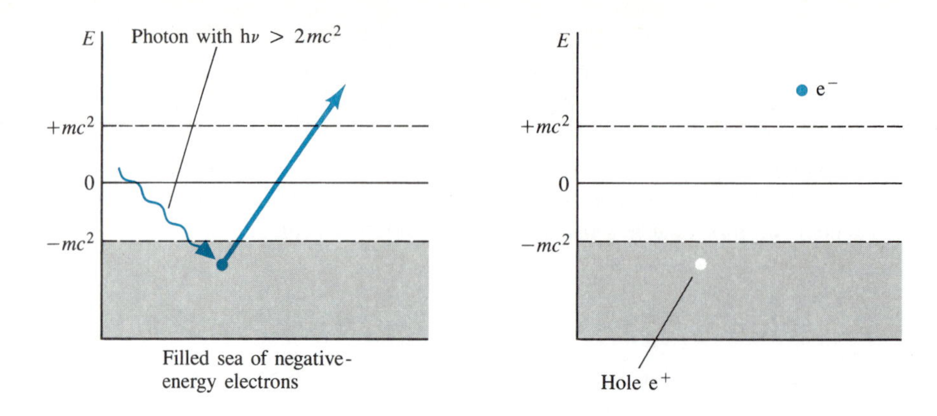

Figure 17.2 Schematic diagram showing the absorption of a photon by an electron in the negative-energy sea. This electron is excited to a positive-energy state, and a hole is left in the negative-energy sea. The result is an electron-positron pair.

Dirac extricated himself from this morass by postulating that all negative-energy states are already filled. According to the Pauli exclusion principle, electrons in the real world cannot then fall into any of these states. However, if there are electrons in negative-energy states it should be possible to excite them to unoccupied positive-energy states, as indicated in Figure 17.2, provided we conserve energy in the process. Like the hole in the valence band of a semiconductor, the absence of an electron in the previously filled infinite sea of negative-energy states should manifest itself as a positively charged particle of mass m_0, the *positron* or *antielectron*. In 1932, positrons were discovered by C. D. Anderson and S. Neddermeyer in cosmic ray tracks.[1] "Pair creation" is a common event whenever

Figure 17.3 (*a*) Bubble chamber photograph *(Lawrence Berkeley Laboratory)* and (*b*) tracing showing the tracks of electron-positron pairs created by a beam of energetic protons incident on a thin metal plate.

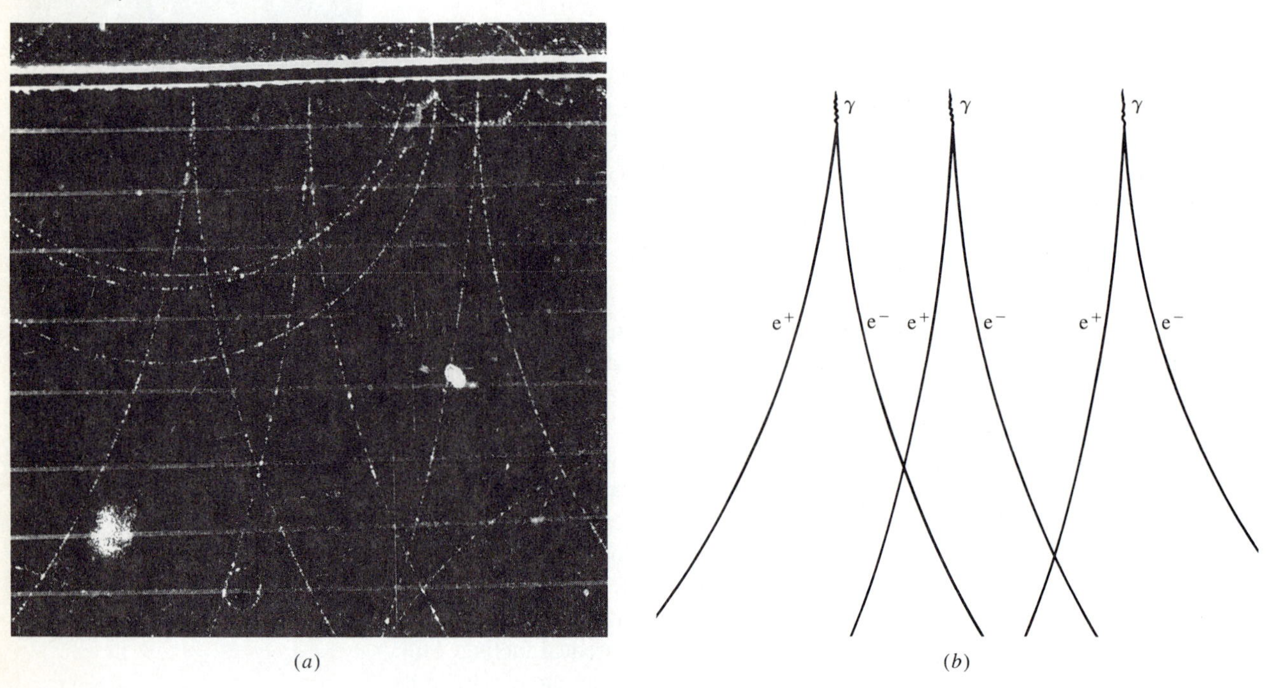

1. Carl D. Anderson, *Science* **76,** 238 (1932). C. D. Anderson and S. H. Neddermeyer, *Phys. Rev.* **43,** 1034 (1933).

high-energy γ rays pass through matter. The inverse, "pair annihilation," also must conserve energy and momentum. Except in very rare cases, whenever an electron and positron annihilate, two photons appear which, in the center-of-mass frame of the electron-positron pair, propagate in diametrically opposite directions and have each an energy $h\nu = m_0c^2$.

Dirac's ideas have more far-reaching consequences. His equation should apply with equal validity to other particles of spin $\frac{1}{2}$, for example, protons and neutrons. Antiprotons and antineutrons have been observed and studied. In fact this bifurcation of nature is universal. Every particle, whatever its charge state or spin, has its antiparticle (though some neutral particles are their own antiparticles).[2] As we shall see, these antiparticles play an essential role in elementary particle physics.

17.3 CLASSIFICATION OF ELEMENTARY PARTICLES

Once pions had been produced by high-energy collisions in the laboratory, physicists set about the task of constructing ever more powerful accelerators in the expectation that other new particles could be created. Events amply justified these expectations. Already in the early 1960s the "elementary" particles numbered in the hundreds. With such a proliferation of particles to contend with, high-energy physics threatened to descend into a veritable chaos. The goal of elegant simplicity seemed to recede ever farther from reality.

One way out of this sort of confusion is to devise classification schemes in the hope that certain regularities may point the way toward the fundamental principles that govern the interactions among these particles. In other words, what was sought was something like the periodic table of the elements, which not only served to classify the elements according to their valence in chemical reactions but, more important, pointed the way to a successful theory of atomic structure.

Elementary particle taxonomy is based on the type of interaction in which a particle may be involved. There are four basic forces, or interactions, in nature: the strong, or nuclear force; the electromagnetic force; the weak force that operates in β decay; and the gravitational force. Let us look at these more carefully.

The *strong interaction* is characterized by an extremely short interaction time, about 10^{-23} s. Particles that interact via the strong force are nucleons (protons and neutrons), many elementary particles such as the Δ (delta), the Σ (sigma), and the ξ (xi), and, of course, the π (pion), which is the *mediator* of the strong force. Particles that participate in the strong interaction are called *hadrons*.[3]

Particles that interact via the *electromagnetic force* are, first of all, any and all particles with charge. Moreover, even neutral particles may interact via the electromagnetic force. The neutron, for example, "exists" for part of the time as a proton through the exchange of a virtual pion in the interaction

2. The construct of a filled negative-energy sea fails for bosons. To allow for antibosons, modern theory rests on the fact that only *changes* in energy of a system are observable. The creation of an antiparticle of positive energy is equivalent to the destruction of a particle of negative energy, and vice versa.

3. Note, however, that in the presently accepted, highly successful model the fundamental constituents of hadrons are quarks, and the force that holds quarks together is mediated by so-called gluons.

$$n \rightarrow p + \pi^-$$

Another neutral "particle" that is involved in the electromagnetic interaction is the photon. According to the quantum theory of electrodynamics, known as *quantum electrodynamics* (QED), the photon is the mediator of the electromagnetic interaction. The characteristic interaction time for the electromagnetic force is about 10^{-18} s.

The *weak interaction* has a characteristic time of about 10^{-16} to 10^{-10} s. As far as we know, all particles can participate in the weak interaction. In fact, the only particle that is unique in that it responds *only* to the weak force is the neutrino (and antineutrino). Particles of half-integral spin that do not interact via the strong force are called *leptons*.

All particles participate in the *gravitational interaction*. This force is, however, so very weak that we need not consider it here. The quantum of this interaction is the *graviton,*[4] which must be massless and have spin 2.

A pattern that already emerges from this analysis concerns the character of the mediators. The photon, the pion, and the graviton are all bosons. One of the fundamental postulates of elementary particle theory is that every interaction is mediated by a boson.

Because the pion has a mass intermediate between that of the electron and nucleon, it was given the name *meson*. Today, all elementary particles of integral spin (bosons) that interact via the strong force are referred to as mesons.[5] The photon, though a boson, is not a meson but belongs to a separate class that also includes the mediators of the weak force, the Z^0 and $W^{\pm}$.

In approximate order of mass, *leptons* are the least massive, and have spin $\frac{1}{2}$: the neutrinos, electron, and muon. The next group are the *mesons,* which have zero or integral spin. Here we find the pions (π^+, π^0, π^-) and a number of other particles that were discovered more recently (K^+, K^0, K^-, η, ϕ, ρ, ψ, etc.). The last group are the *baryons,* the proton, the neutron, and all particles whose rest mass is greater than the nucleon's and that have half-integral spin (Λ, Σ, Δ, Ξ, Ω^-).

Classification by mass has a certain appeal but also some flaws. Not all leptons are light, nor are all mesons of intermediate mass. For example, the tau (τ), discovered in 1975, has all the earmarks of a lepton but a rest mass of 1784 MeV/c^2, nearly twice that of the nucleon. There are other deviants, mostly among the mesons. Numerous particles have been discovered whose rest masses are greater than that of a nucleon, but whose properties clearly identify them as mesons, i.e., bosons that participate in the strong interaction.

Table 17.1 is a partial list of the known baryons, mesons, and leptons.

Only a few of the great multitude of elementary particles are stable, namely, the proton,[6] the electron, and the neutrino. All the others decay into less massive parti-

---

4. See Chapter 3 on evidence for the existence of gravitational waves (the wave manifestation of gravitational quanta).

5. As mentioned earlier (Section 14.3), the first intermediate-mass particle discovered was the μ and was referred to as the mu meson. According to currently accepted classification, the muon is a lepton, identical to the electron except for its mass, and use of the term *mu meson* is discouraged.

6. Extensions of the standard model that attempt to include the strong interaction predict a finite lifetime of about 10^{32} years for the proton. So far, all attempts to observe proton decay have met with failure.

TABLE 17.1 A (Much Abbreviated) List of Elementary Particles

Leptons

Particle	Mass (MeV/c^2)	Spin	Charge (e)	Lifetime (s)
e	0.511003	$\frac{1}{2}$	-1	stable
μ	105.659	$\frac{1}{2}$	-1	2.19703×10^{-6}
τ	1784	$\frac{1}{2}$	-1	3.4×10^{-23}
ν_e	0	$\frac{1}{2}$	0	stable
ν_μ	0	$\frac{1}{2}$	0	stable
ψ_τ	0	$\frac{1}{2}$	0	stable

Baryons

Particle	Mass (MeV/c^2)	Spin, parity	Charge (e)	Isospin	Lifetime (s) or width (MeV)
		Nucleon Resonances (N); $S = 0$, $T = \frac{1}{2}$			
N	939	$\frac{1}{2}+$	1, 0	$\frac{1}{2}$	p: $>10^{32}$ yr n: 898 s
N	1440	$\frac{1}{2}+$	1, 0	$\frac{1}{2}$	200 MeV
N	1520	$\frac{3}{2}-$	1, 0	$\frac{1}{2}$	125 MeV
N	1535	$\frac{1}{2}-$	1, 0	$\frac{1}{2}$	150 MeV
		Delta Resonances (Δ); $S = 0$, $T = \frac{3}{2}$			
Δ	1232	$\frac{3}{2}+$	2, 1, 0, -1	$\frac{3}{2}$	115 MeV
Δ	1620	$\frac{1}{2}-$	2, 1, 0, -1	$\frac{3}{2}$	140 MeV
Δ	1700	$\frac{3}{2}-$	2, 1, 0, -1	$\frac{3}{2}$	250 MeV
Δ	1900	$\frac{1}{2}-$	2, 1, 0, -1	$\frac{3}{2}$	150 MeV
		Lambda Resonances (Λ); $S = -1$, $T = 0$			
Λ	1116	$\frac{1}{2}+$	0	0	2.63×10^{-10} s
Λ	1405	$\frac{1}{2}-$	0	0	40 MeV
Λ	1520	$\frac{3}{2}-$	0	0	15.6 MeV
Λ	1600	$\frac{1}{2}+$	0	0	150 MeV
		Sigma Resonances (Σ); $S = -1$; $T = 1$			
Σ	1193	$\frac{1}{2}+$	1, 0, -1	1	Σ^+: 0.80×10^{-10} s Σ: 1.48×10^{-10} s Σ: 6×10^{-20} s
Σ	1385	$\frac{3}{2}+$	1, 0, -1	1	35 MeV
Σ	1670	$\frac{3}{2}-$	1, 0, -1	1	60 MeV
Σ	1775	$\frac{5}{2}-$	1, 0, -1	1	120 MeV
		Cascade Resonances (Ξ); $S = -2$, $T = \frac{1}{2}$			
Ξ	1318	$\frac{1}{2}+$	0, -1	$\frac{1}{2}$	Ξ^0: 2.9×10^{-10} s Ξ^-: 1.64×10^{-10} s
Ξ	1530	$\frac{3}{2}+$	0, -1	$\frac{1}{2}$	9.1 MeV
		$S = -3$ Resonance Particle			
Ω^-	1672	$\frac{3}{2}+$	-1	0	0.82×10^{-10} s

Mesons

Particle	Mass (MeV/c^2)	Spin, parity	Charge (e)	Isospin	Lifetime (s) or width (MeV)
		$S = 0$ Mesons			
π	138	$0-$	1, 0, -1	1	$\pi^\pm$: 2.603×10^{-8} s π^0: 0.87×10^{-16} s
η	549	$0-$	0	0	6×10^{-19} s
ρ	770	$1-$	1, 0, -1	1	154 MeV

(continued)

Mesons					
Particle	**Mass (MeV/c²)**	**Spin, parity**	**Charge (e)**	**Isospin**	**Lifetime (s) or width (MeV)**
ω	783	1 −	0	0	9.9 MeV
δ	980	0 +	1, 0, −1	1	54 MeV
ρ	1600	1 −	1, 0, −1	1	260 meV
J/ψ	3100	1 −	0	0	63 keV
ψ	3685	1 −	0	0	215 keV
			$S = \pm 1$ Mesons		
K	496	0 −	1, 0	$\frac{1}{2}$	$K^{\pm}$: 1.237×10^{-8} s K^0: 5.18×10^{-6} s K^0: 89.2×10^{-12} s
K*	892	1 −	1, 0	$\frac{1}{2}$	51 MeV
Q	1280	1 +	1, 0	$\frac{1}{2}$	90 MeV
L	1770	2 −	1, 0	$\frac{1}{2}$	200 MeV
			$S = 0$ Charmed Mesons		
D	1867	0 −	1, 0	$\frac{1}{2}$	$D^{\pm}$: 0.92×10^{-12} s D^0: 0.44×10^{-12} s
			$S = \pm 1$ Charmed Mesons		
F^+	1971	0 −	1	0	0.28×10^{-12} s
			$S = 0$ Bottom Mesons		
B	5272	0 −	1,0	$\frac{1}{2}$	B^0: 0.14×10^{-12} s

cles, often by a variety of decay modes. An examination of the decay schemes provides further clues to the nature of these inhabitants of the elementary particle zoo.

17.4 CONSERVATION LAWS, SYMMETRY, AND SELECTION RULES

We are already familiar with the conservation laws of classical physics:

1. Conservation of energy
2. Conservation of linear momentum
3. Conservation of angular momentum
4. Conservation of charge (Q)

In addition, we postulated another conservation law in Chapter 15, conservation of nucleon number, which is a special case of the more general

5. Conservation of baryon number (B)

As far as we know, these laws are always obeyed. Conservation laws, however, cannot tell us how a system will evolve in time. In fact, the equations of classical as well as quantum physics are invariant with respect to time reversal,[7] so

7. There is evidence that time reversal symmetry may be violated in weak interactions.

that we cannot tell a priori how things will proceed. Yet there is a law that does define "the arrow of time," the second law of thermodynamics. According to the second law, any change must be in such a direction that the universe becomes more disordered. But, you may argue, the second law is only statistical. Does it make sense to invoke this law when we look at microscopic events such as the decay of unstable particles? Well, if these elementary particles are in fact not really elementary at all but are constituted of still smaller entities, perhaps it does.

Consider, for example, an enclosure of volume V with a total energy content E. If this energy is in the form of electromagnetic radiation, it can be distributed among an infinite number of modes. If, on the other hand, all of the energy is concentrated in a blob of mass $m = E/c^2$, it is highly localized. In other words, the entropy of the system is much greater in the former than the latter case. According to the second law, we should expect that whenever a particle of finite rest mass can decay into several particles, such an event should occur. This idea has been elevated to a universal principle, the so-called *totalitarian principle:*

Every process that is not forbidden must occur.

Now one can conceive a multitude of ways by which a massive particle, for example, a Σ^+ might decay. For instance,

$$\Sigma^+ \rightarrow \mathrm{p} + K^0 \quad \textbf{(17.5a)}$$
$$\Sigma^+ \rightarrow \mathrm{n} + \pi^+ \quad \textbf{(17.5b)}$$
$$\Sigma^+ \rightarrow \pi^0 + \pi^+ \quad \textbf{(17.5c)}$$
$$\Sigma^+ \rightarrow \Lambda^0 + \pi^+ \quad \textbf{(17.5d)}$$
$$\Sigma^+ \rightarrow K^0 + \pi^+ \quad \textbf{(17.5e)}$$

Decay schemes (b) and (d) occur, as do some others. But (a), (c), and (e) have never been observed.

Failure to observe a decay that does not violate a known conservation law implies that some other fundamental conservation law forbids that event. A careful survey of creation and decay processes suggested the following conservation rules in addition to the ones already enumerated:

6. Conservation of lepton number (L)
7. Conservation of parity (P)
8. Conservation of isospin (T)
9. Conservation of strangeness or hypercharge (S, Y)

We have here introduced several new terms. The first, *lepton number,* is the number of leptons in the system. Every lepton is assigned lepton number $L = +1$, every antilepton $L = -1$. For example, in the β decay of the neutron

$$\mathrm{n} \rightarrow \mathrm{p} + \mathrm{e}^- + \bar{\nu}_\mathrm{e} \quad \textbf{(17.6)}$$

it must be an antineutrino that is created (as indicated) in order that the total lepton number on the right-hand side be zero.[8]

---

8. As we shall see presently, conservation of lepton number is more restrictive than we have indicated.

We encountered *parity* already in Chapter 7 when we discussed the stationary states of a particle in a one-dimensional well. Parity refers to the behavior of a system under mirror reflection. Since mirror reflection transforms a right-handed rotation into a left-handed rotation, conservation of parity means that the process does not distinguish between right- and left-handedness. Parity is always conserved in strong and electromagnetic interactions,[9] but not in weak interactions.

The concept of *isospin* (or isotopic spin) derives from the charge independence of the strong interaction (see Chapter 14) and the empirical fact that the known hadrons (baryons and mesons) tend to form groups of nearly the same mass. For example, the π^0, π^-, and π^+ have almost identical mass; the neutron and proton have nearly the same mass; and the masses of Σ^+, Σ^-, and Σ^0 are 1189, 1197, and 1193 MeV/c^2, respectively. In analogy with angular momentum, the three charge states of the π and Σ and the two charge states of the nucleon (proton and neutron) are viewed as the multiplicity, $2T + 1$, of a single hadronic state of isospin $\mathbf{T}$. Thus the nucleon is assigned $\mathbf{T} = \frac{1}{2}$, and the Σ and π an isospin $\mathbf{T} = 1$. The "z-component" of $\mathbf{T}$, generally denoted by T_3, is then related to the charge Q, measured in units of e. For the nucleon, $T_3 = -\frac{1}{2}$ corresponds to $Q = 0$ (neutron), $T_3 = +\frac{1}{2}$ to $Q = 1$ (proton); for the π and Σ triplets, $T_3 = 1, 0, -1$ correspond to charges $+1$, 0, and -1. The splitting of the mass multiplets arises because hadrons also interact with the electromagnetic field.

While this formalism may seem to complicate the familiar concept of charge, conservation of isospin in hadronic interactions leads to selection rules above and beyond that of charge conservation. In such interactions, $\mathbf{T}$ and T_3 are conserved. For example, the process

$$\Sigma^+ \rightarrow \mathrm{p} + \eta^0 \tag{17.7}$$

thought it does conserve charge as well as angular momentum and baryon number, is forbidden ($\mathbf{T}_\Sigma = 1$, $\mathbf{T}_\mathrm{p} = \frac{1}{2}$, $\mathbf{T}_\eta = 0$) and has never been observed, whereas

$$\Sigma^+ \rightarrow \mathrm{p} + \pi^0 \tag{17.8}$$

is allowed and is, in fact, the dominant decay process.

Strangeness came about as follows. Not long after the discovery of the pion physicists observed other unusual events in cloud chamber photographs and photographic emulsions exposed to cosmic rays. These events appeared as tracks in the shape of V's, similar to those shown in Figure 17.4. The presence of a pair of tracks that originate at a point and are of opposite curvature in a magnetic field signifies that a neutral particle, which leaves no ionizing track, has decayed at the vertex of the V into two particles of opposite charge. In a number of instances it was possible to identify the charged particles as π^+ and π^-, and from their momenta determine the energy and momentum of the original neutral particle. This particle is now known as the K^0, whose rest mass is about 500 MeV/c^2.

Another heavy neutral particle discovered not long after the K^0 is the lambda (Λ^0), which decays into a proton and a π^-. Using energetic particles from large accelerators these and other baryons could be created in the laboratory and their

9. Selection rules for atomic transitions in which a photon is created or absorbed include conservation of parity (see Chapter 9).

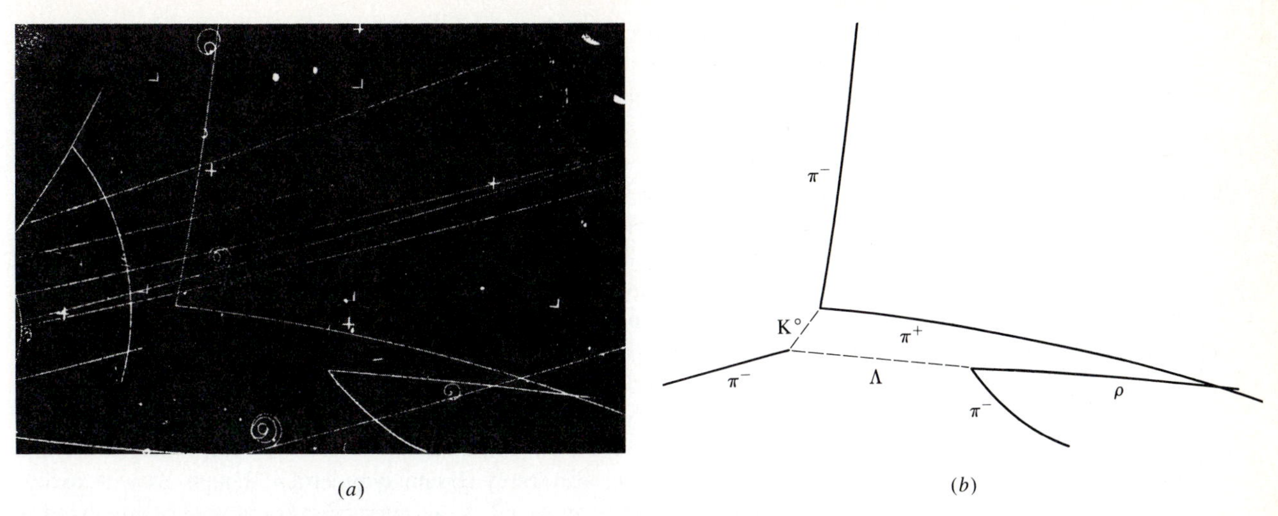

Figure 17.4 Bubble chamber photograph (*a*) and tracing (*b*) showing the production of strange particles and their decay. A 1.2-GeV π^- enters from the lower left and interacts with a proton in the chamber, creating two neutral strange particles, the K^0 and Λ^0. These particles are uncharged and, therefore, leave no visible track in the chamber. Their decay products ($K^0 \rightarrow \pi^+ + \pi^-$, $\Lambda^0 \rightarrow \text{p} + \pi^-$) can be seen. *(Lawrence Berkeley Laboratory.)*

properties studied. Some of them decay ultimately into leptons and are, therefore, grouped with the pions as mesons; others decay into products of which one is a proton, and are classified as baryons. Both types are hadrons because they respond to the strong interaction.

The "strange" behavior of these particles is their very long lifetime. Since they respond to the strong force, the decay times of the K^0 or Λ^0 should be about 10^{-23} s. In that time, even traveling at nearly the speed of light, they would cover a distance less than 10^{-14} m, not several centimeters. Their half-life of about 10^{-10} s suggests that they decay via the weak interaction, not the strong. Why should particles that are created in strong interactions decay only by the weak interaction? Here was a situation that violated the "totalitarian principle" and implied the operation of a new selection rule.

The new property, called *strangeness* and characterized by the new quantum number S, was proposed by M. Gell-Mann and K. Nishijima following the suggestion of A. Pais that strange particles are always created in pairs. For example, in the reaction shown in Figure 17.4, a high-energy π^- interacts with a proton in the hydrogen bubble chamber creating, via the strong interaction, a K^0 and a Λ^0. If we arbitrarily assign $S = 0$ to the nucleon and $S = +1$ to the K^0, strangeness is conserved if the Λ^0 has $S = -1$. Strangeness appears to be conserved in strong and electromagnetic interactions, but not in weak interactions. Strangeness quantum numbers are deduced from a study of the strong interactions in which the strange particle participates. For example, the reaction $\text{p} + K^- \rightarrow \text{n} + \Lambda^0$ implies that the strangeness of the K^- is $S = -1$; from the reaction $\text{p} + \text{p} \rightarrow \text{p} + \text{p} + K^- + K^+$ it then follows that $S = +1$ for the K^+.

Strangeness is in fact related to the charge, T_3, and the baryon number, B. The relation is

$$S = 2(Q - T_3) - B \qquad \textbf{(17.9)}$$

Conservation of strangeness is seen to be not a new conservation law but a combination of conservation of charge, isospin, and baryon number. In recent years, the use of S has been largely replaced by another quantity, the *hypercharge,* defined by

$$Y = S + B = 2(Q - T_3) \tag{17.10}$$

for reasons that will become apparent shortly.

A famous theorem, known as Noether's theorem, states that every conservation principle is intimately related to a symmetry property of nature. For example, consider conservation of angular momentum, $\mathbf{L} = I\omega = I(d\phi/dt)$. If the potential energy U of the system is independent of angle, the torque $\tau = -\partial U/\partial\phi$ must vanish and $d\mathbf{L}/dt = 0$. Similarly, translational symmetry implies conservation of linear momentum, temporal symmetry conservation of energy, and reflection symmetry conservation of parity. The new conservation laws of isospin, strangeness, and lepton and baryon number also imply certain symmetries, though, as with parity, they may not be universally applicable to all interactions or as readily visualized as spatial symmetries. Consideration of these implied symmetries was a critical ingredient in the construction of the quark model, described in Section 17.7.

17.5 RESONANCE PARTICLES

We mentioned earlier that the characteristic time for the strong interaction is about 10^{-23} s. Indeed, the concept of strangeness came about because some particles known to respond to the strong force had a lifetime many orders of magnitude longer. An elementary particle that decays hadronically, even if moving at nearly the speed of light, will traverse only about 10^{-14} m, a distance far too small to leave a detectable track in a bubble or spark chamber. Its presence can be inferred only indirectly from resonances in scattering experiments that produce that particle.

An example of the experimental evidence for such a short-lived particle is Figure 17.5, which shows the cross sections of various reactions involving a K^- and a proton as functions of energy in the center-of-mass reference frame. The remarkable fact that all of the reactions proceed with maximum probability when the center-of-mass energy is 1.52 GeV suggests that what we are seeing is the formation of an intermediate state, somewhat analogous to the formation of the compound nucleus in nuclear reactions. That intermediate state then decays via several "channels" into the observed reaction products. The short-lived intermediate state is known as a *resonance particle,* or simply a *resonance*. Note that the resonance shown in Figure 17.5 is not sharp; its width is about 16 MeV, and is not the result of instrument deficiency. This width is real, and is a direct consequence of the uncertainty principle. From the width we can deduce that the lifetime of this resonance particle is

$$\Delta t \simeq \frac{\hbar}{\Delta E} \simeq 4 \times 10^{-23} \text{ s}$$

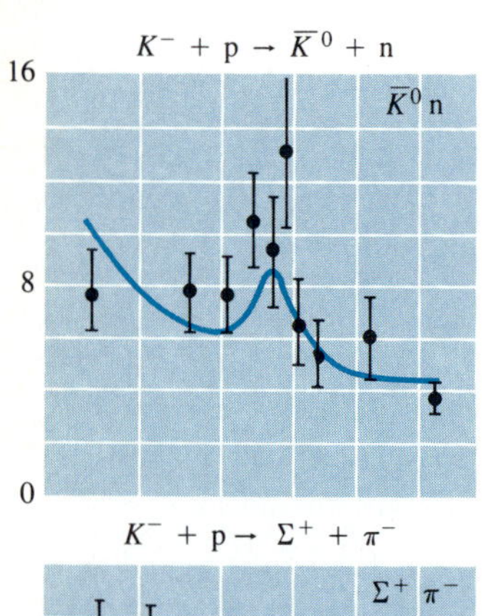

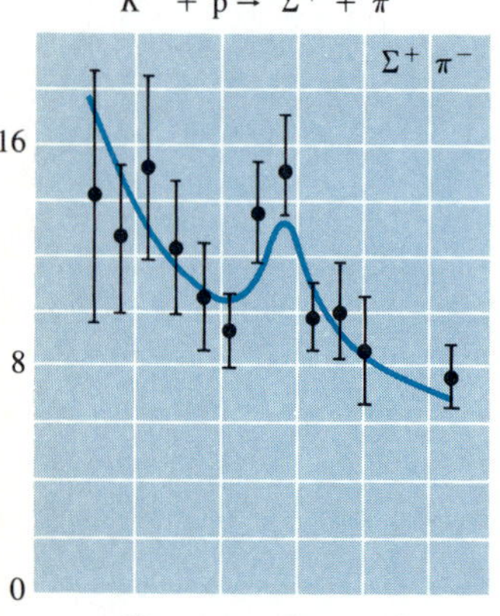

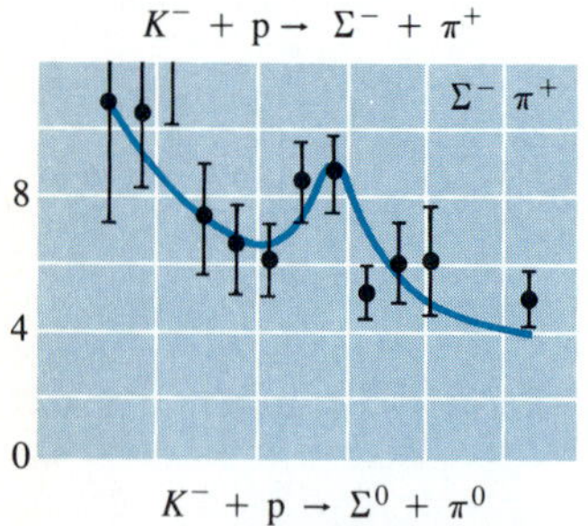

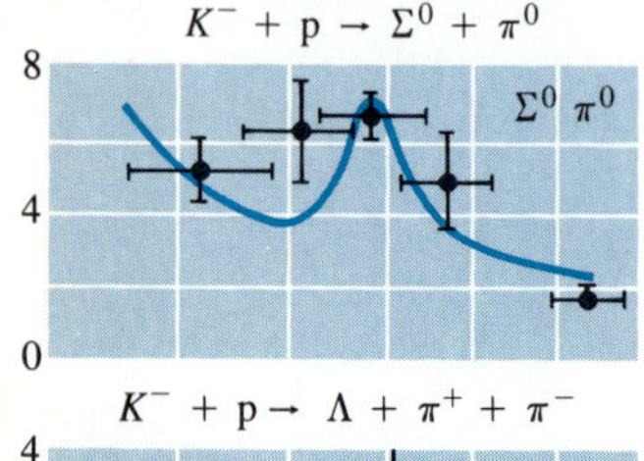

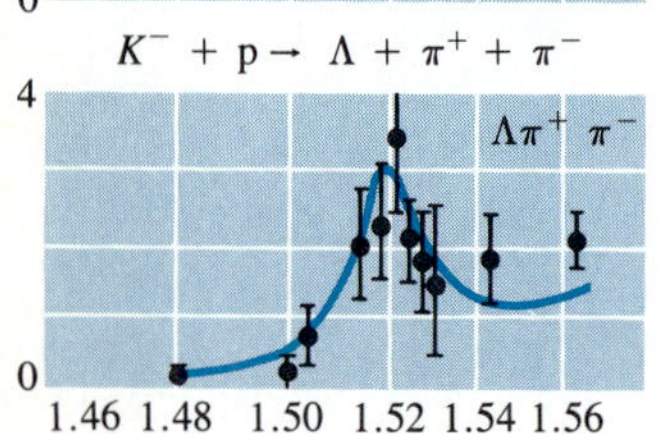

Figure 17.5 Cross sections for inelastic K^--p scattering as functions of energy in the center-of-mass frame for various reactions. The reactions are indicated next to each plot. Note that all reaction cross sections peak at 1.52 GeV, indicative of the formation of a "resonance particle" at that energy.

EXAMPLE 17.1 From the width of the resonance in the $\pi^- p$ scattering data of Figure 17.6, estimate the lifetime of the $\rho(770)$ resonance particle.

SOLUTION The width of the resonance at half maximum indicated in Figure 17.6 is about 150 MeV. The lifetime of the particle is given by

$$\tau \simeq \frac{\hbar}{\Delta E} = \frac{6.58 \times 10^{-22} \text{ MeV·s}}{150 \text{ MeV}} = 4.4 \times 10^{-24} \text{ s}$$

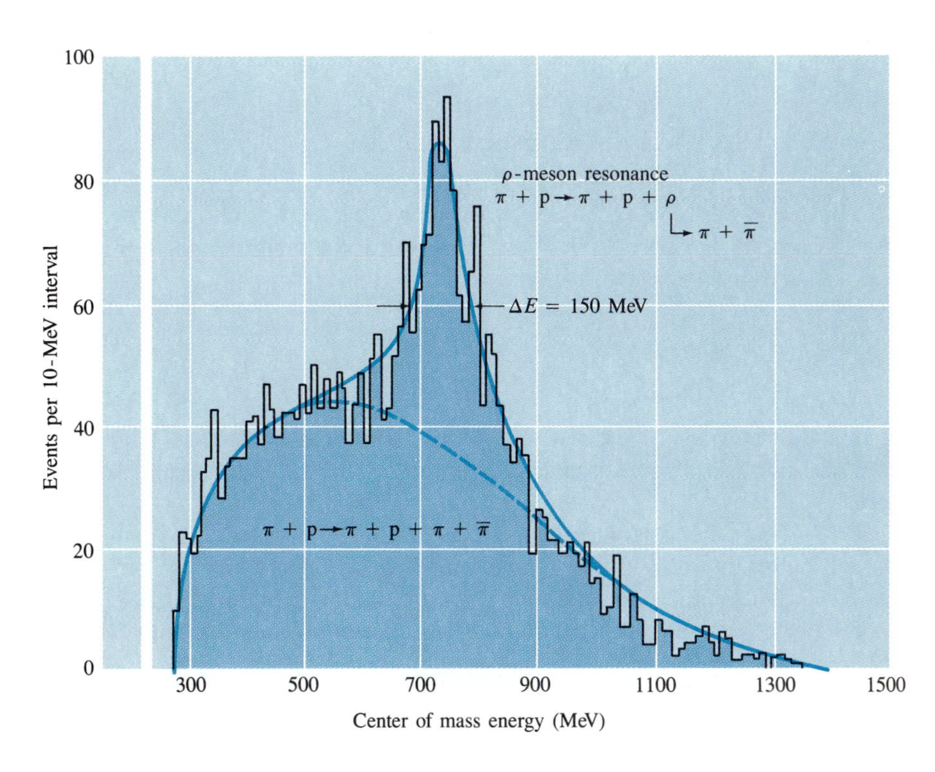

Figure 17.6 Pion-proton scattering resulting in the creation of two additional pions ($\pi + p \rightarrow \pi + p + \pi + \bar{\pi}$). The resonance at 770 MeV is due to the formation of the ρ-meson, which decays hadronically into two pions.

17.6 MORE ABOUT LEPTONS

The pion is unstable, decaying to a muon, which, in turn, decays to an electron. In these decay processes, the μ and e are not the only particles created; they are, however, the only charged objects that leave a track in a bubble chamber.

A possible decay scheme for the muon might be

$$\mu^- \rightarrow e^- + \gamma \tag{17.11}$$

which conserves energy and charge as well as angular momentum, since the spins of the muon and electron are $\frac{1}{2}$. Yet, this does not happen. If it did, the energy of the electron in the rest frame of the muon would always have the same value. Experimentally, the electron energy spectrum is broad, as in β decay, characteristic of a decay into three or more rather than two particles. Since the only particle that is detected is the electron, the other particles must be two neutrinos; the decay is

$$\mu^- \rightarrow e^- + \bar{\nu}_e + \nu_\mu \tag{17.12a}$$

$$\mu^+ \rightarrow e^+ + \nu_e + \bar{\nu}_\mu \tag{17.12b}$$

Note the subscripts e and μ on the neutrinos in Equations (17.12). The neutrino (and antineutrino) associated with muon decay is different from that associated with beta decay. Both types have spin $\frac{1}{2}$, zero rest mass, and no charge. How do we know that they are not identical?

Consider the following reaction:

$$n + \nu \rightarrow p + e^- \tag{17.13}$$

which is the inverse of the β decay of the neutron. If the neutrinos from muon decay and β decay were identical, this reaction should proceed using neutrinos generated by muon decay. Sensitive experiments have shown that reaction (17.13) proceeds if the neutrinos are produced from β decay but not if the neutrinos come from the decay of muons. Conversely, the reaction

$$n + \nu \rightarrow p + \mu^- \tag{17.14}$$

is observed only if the neutrinos come from muon decay. Evidently, muon and electron neutrinos are different; consequently, conservation of lepton number is more specific than previously indicated: lepton numbers L_e and L_μ must be conserved separately.

In 1975, another lepton, called the tau (τ) was discovered. This particle is unusual in that its mass is 1784 MeV/c^2, nearly twice that of a proton, yet it behaves like a lepton. Its decay scheme is like that of the muon, except that since its mass is greater than that of the muon as well as electron it can decay into either of these leptons according to the reactions

$$\tau^+ \rightarrow e^+ + \nu_e + \bar{\nu}_\tau \tag{17.15a}$$

$$\tau^+ \rightarrow \mu^+ + \nu_\mu + \bar{\nu}_\tau \tag{17.15b}$$

$$\tau^- \rightarrow e^- + \bar{\nu}_e + \nu_\tau \tag{17.15c}$$

$$\tau^- \rightarrow \mu^- + \bar{\nu}_\mu + \nu_\tau \tag{17.15d}$$

or into a charged pion according to

$$\tau^+ \rightarrow \pi^+ + \bar{\nu}_\tau \tag{17.16a}$$

$$\tau^- \rightarrow \pi^- + \nu_\tau \tag{17.16b}$$

Note that since the lepton number of the pion is zero, only one neutrino is needed in the decays of reactions (17.16).

17.7 QUARKS

By 1960, the study of elementary particles, which was expected to clarify the nature of the strong interaction, had instead spawned a bewildering array of particles whose production and decay modes defied understanding within the framework of existing models. In the late 1950s and early 1960s various schemes were proposed and then discarded with disconcerting regularity. The grouping of these particles into isospin multiplets, the concepts of strangeness and hypercharge, and conservation of lepton and baryon numbers suggested some fundamental pattern, but it took a few false starts to hit upon what now appears to be the correct model.

The quark model was proposed independently by M. Gell-Mann and George Zweig in 1964. We can gain some insight into the regularities that led to this model by plotting hypercharge Y against the isospin component T_3 for particles of the same spin and parity. Such plots are shown in Figures 17.8 through 17.10; their distinc-

Figure 17.7 M. Gell-Mann (b. 1929). *(Serge Collection/Courtesy AIP, Niels Bohr Library.)*

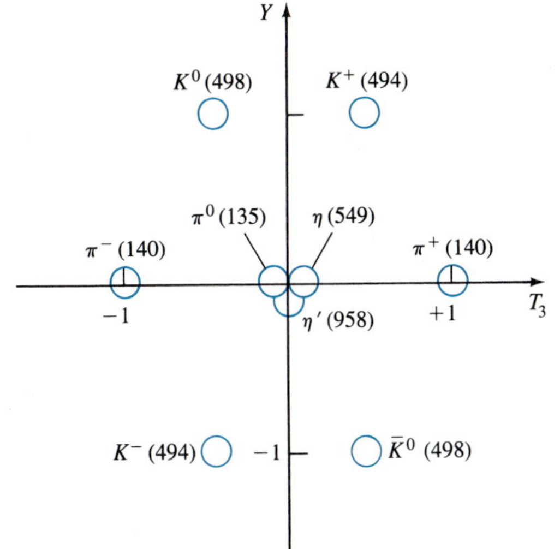

Figure 17.8 A plot of hypercharge versus electric charge (T_3) for some spin-0 mesons.

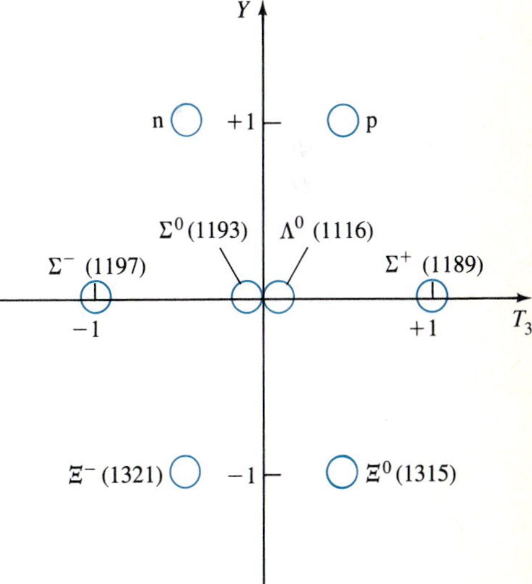

Figure 17.9 A plot of hypercharge versus T_3 for some spin-$\frac{1}{2}$ baryons.

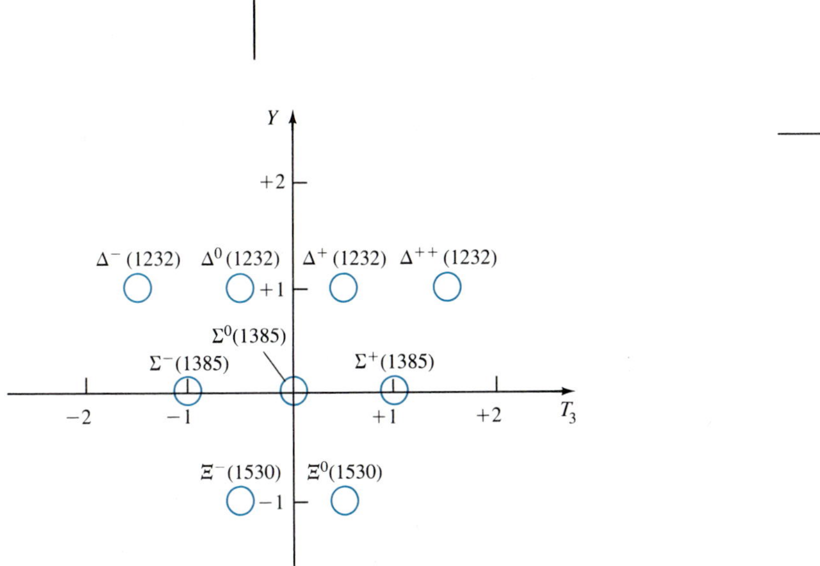

Figure 17.10 A plot of hypercharge versus T_3 for some spin-$\frac{3}{2}$ baryons.

TABLE 17.2 Quantum Numbers of the Three Quarks u, d, and s

Name	Symbol	Charge	Spin	Baryon number	Strangeness	Antiquark
Up	u	$+\frac{2}{3}$	$\frac{1}{2}$	$\frac{1}{3}$	0	$\bar{u}$
Down	d	$-\frac{1}{3}$	$\frac{1}{2}$	$\frac{1}{3}$	0	$\bar{d}$
Strange	s	$-\frac{1}{3}$	$\frac{1}{2}$	$\frac{1}{3}$	-1	$\bar{s}$

tive patterns suggest a threefold rotational symmetry in Y-T_3 space. Zweig and Gell-Mann recognized that these patterns could be generated if it was assumed that the elementary particles were, in fact, composed of three fundamental particles and their antiparticles. Gell-Mann chose the name *quark* for these new constituents of matter. Quarks are hadrons, having spin $\frac{1}{2}$; they are designated by the letters u, d, and s, for "up," "down," and "strange." The u quark has a charge of $+\frac{2}{3}e$, the d and s quarks have charges of $-\frac{1}{3}e$. The u and d quarks have $S = 0$, the s (strange) quark has $S = -1$. All quarks have baryon number $B = \frac{1}{3}$. (See Table 17.2.)

Since mesons have zero baryon number, they must be combinations of quarks and antiquarks. Of course, a quark-antiquark pair need not have spin 0 but could have spin 1 as well,[10] as indicated in Table 17.3, which lists the charge, spin, and strangeness of the nine quark-antiquark combinations. If we now make a plot of Y versus T_3 for these pairs,[11] we obtain a pattern strikingly similar to Figure 17.8.

Three quarks combine to give a particle with baryon number $B = 1$. The total spin of three spin-$\frac{1}{2}$ quarks may be either $\frac{1}{2}$ or $\frac{3}{2}$. Table 17.4 lists possible three-quark combinations and gives their charge, spin, and hypercharge. If we again make a two-dimensional plot of hypercharge versus T_3 for the spin-$\frac{1}{2}$ and for the spin-$\frac{3}{2}$ combinations, the resulting patterns are identical to Figures 17.9 and 17.10.

The quark model thus provides a rational basis for the observed multiplet structure of mesons and baryons. However, that is only part of the story. The proof of the pudding is in the eating, and the merit of a new theory rests on its ability not only to

TABLE 17.3 Quantum Numbers of the Nine Quark-Antiquark Combinations

Combination	Charge	Spin	Baryon number	Strangeness	Hypercharge Y	T_3
$u\bar{u}$	0	0, 1	0	0	0	0
$u\bar{d}$	+1	0, 1	0	0	0	+1
$u\bar{s}$	+1	0, 1	0	+1	+1	$+\frac{1}{2}$
$d\bar{u}$	−1	0, 1	0	0	0	−1
$d\bar{d}$	0	0, 1	0	0	0	0
$d\bar{s}$	0	0, 1	0	+1	+1	$-\frac{1}{2}$
$s\bar{u}$	−1	0, 1	0	−1	−1	$-\frac{1}{2}$
$s\bar{d}$	0	0, 1	0	−1	−1	$+\frac{1}{2}$
$s\bar{s}$	0	0, 1	0	0	0	0

10. Higher integral intrinsic angular momenta of mesons are possible if we allow for excited states of quark combinations with "orbital" angular momenta.

11. For mesons, $B = 0$. Consequently, $Y = S$ and the charge $Q = T_3$.

TABLE 17.4 Quantum Numbers of Three-Quark Combinations

Combination	Charge	Spin	Baryon number	Strangeness	Hyper-charge Y	T_3
uuu	+2	$\frac{3}{2}$	1	0	+1	$+\frac{3}{2}$
uud	+1	$\frac{1}{2}, \frac{3}{2}$	1	0	+1	$+\frac{1}{2}$
udd	0	$\frac{1}{2}, \frac{3}{2}$	1	0	+1	$-\frac{1}{2}$
uus	+1	$\frac{1}{2}, \frac{3}{2}$	1	−1	0	+1
uss	0	$\frac{1}{2}, \frac{3}{2}$	1	−2	−1	$+\frac{1}{2}$
uds	0	$\frac{1}{2}, \frac{3}{2}$	1	−1	0	0
ddd	−1	$\frac{3}{2}$	1	0	+1	$-\frac{3}{2}$
dds	−1	$\frac{1}{2}, \frac{3}{2}$	1	−1	0	−1
dss	−1	$\frac{1}{2}, \frac{3}{2}$	1	−2	−1	$-\frac{1}{2}$
sss	−1	$\frac{3}{2}$	1	−3	−2	0

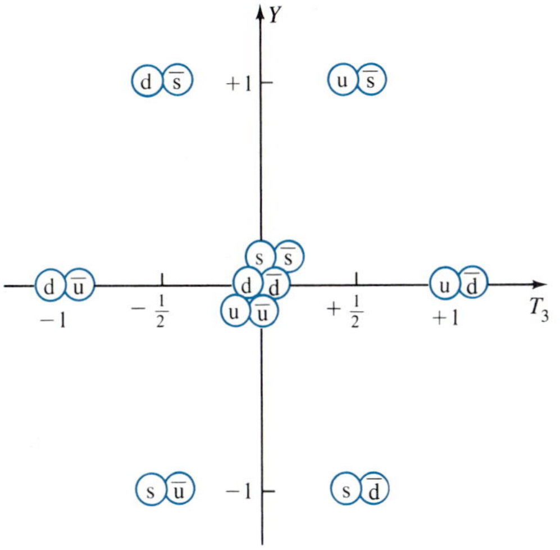

Figure 17.11 A plot of hypercharge versus T_3 for the quark-antiquark combinations of Table 17.3.

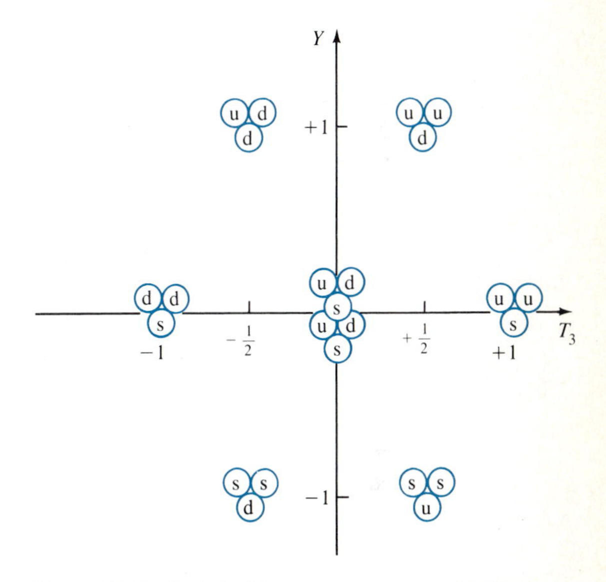

Figure 17.12 A plot of hypercharge versus T_3 for three-quark combinations of spin $\frac{1}{2}$.

explain old data but to predict new observations. In this regard, too, the quark theory has been eminently successful. When the quark model was introduced, only seven spin-0 mesons were known to exist. On the basis of the model, Gell-Mann wrote, "The most clear-cut prediction for the [spin-0] mesons is the existence of [one more] which should decay into two photons like the π^0." A few months later, the last of the spin-0 octet was discovered and named the η meson; it decays into two photons as predicted.

Consider the spin-$\frac{3}{2}$ decuplet of Figure 17.10. You will note that the masses increase in a regular way as the hypercharge changes from 1 to −2. At the time the quark model was proposed, the Ω^- had not yet been seen. Gell-Mann predicted its existence and, moreover, its mass, isospin, and strangeness. Note that unlike the other members of this decuplet, which could decay hadronically, i.e., without vio-

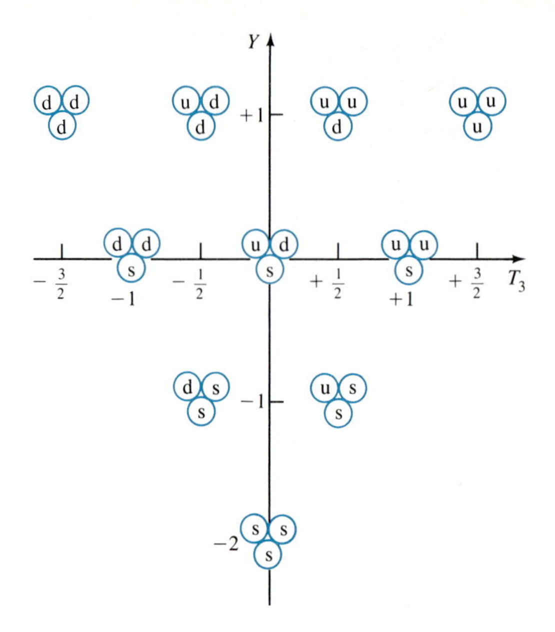

Figure 17.13 A plot of hypercharge versus T_3 for three-quark combinations of spin $\frac{3}{2}$.

lating strangeness conservation, the Ω^- must be a "strange" or long-lived particle. The track of the Ω^- and the complex cascade of decays to stable particles in which S changes successively by one unit are shown in Figure 17.14. The measured mass of the Ω^- is in excellent agreement with Gell-Mann's prediction.

Reactions among elementary particles can also be understood in terms of the quark model. The strangeness conservation rule for hadronic and electromagnetic interactions follows from the rule that in these interactions quark-antiquark pairs can be created or annihilated but the transformation of one quark into another is forbidden. Only the weak interaction can change one type of quark into another, subject to charge conservation.

EXAMPLE 17.2 Write the reaction $\pi^- + \mathrm{p} \rightarrow \Sigma^0 + K^0$ and the decay modes $\Sigma^0 \rightarrow \Lambda + \gamma$ and $\Lambda \rightarrow \mathrm{p} + \pi^-$ in terms of the quark constituents.

SOLUTION The reaction in which the two strange particles are created can be written as

$$\mathrm{d\bar{u}} + \mathrm{uud} \rightarrow \mathrm{uds} + \mathrm{d\bar{s}}$$

The reaction is one in which a $\mathrm{u\bar{u}}$ pair is annihilated and an $\mathrm{s\bar{s}}$ pair is created; it is an allowed hadronic reaction.

For the decay of the Σ^0 we can write

$$\mathrm{uds} \rightarrow \mathrm{uds} + \gamma$$

where the three-quark combination on the right-hand side represents the Λ. Again, the reaction fulfills the requirement of hadronic or electromagnetic interactions; in this case, since the other particle that is released is a photon, which can only

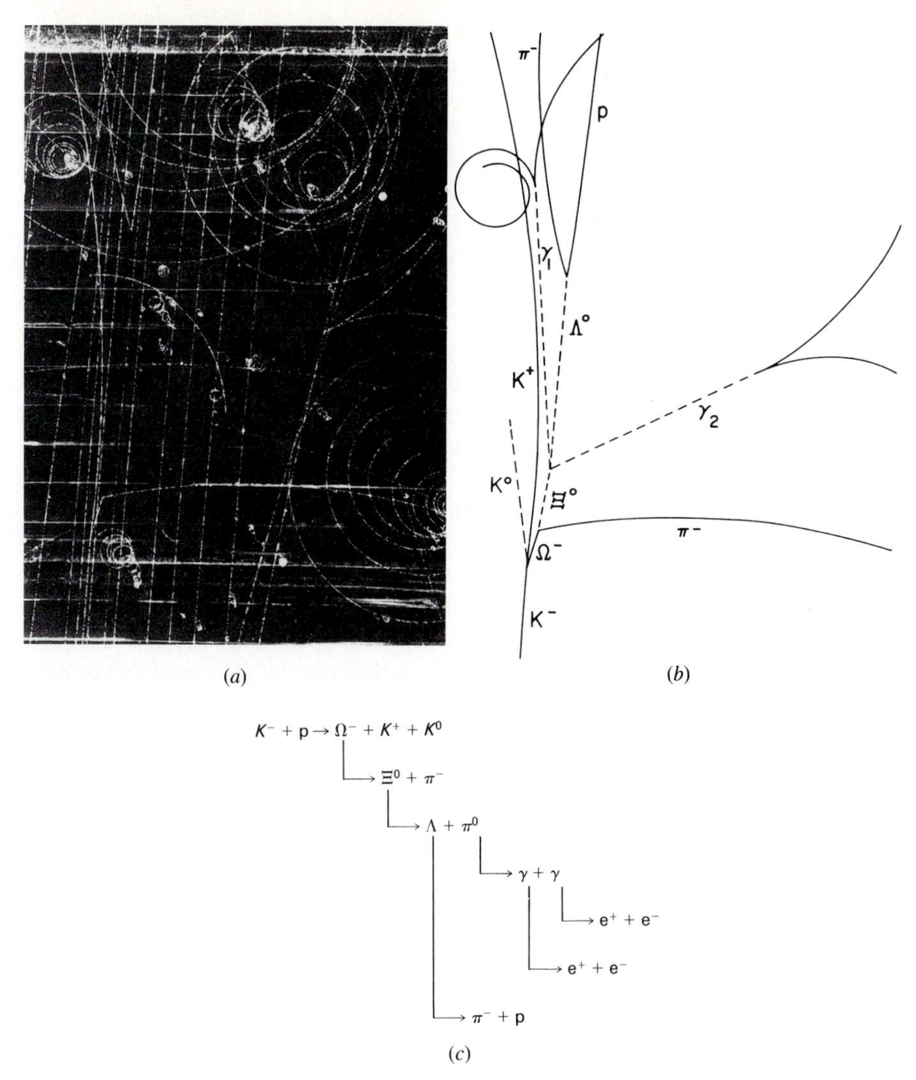

Figure 17.14 (*a*) Bubble chamber photograph showing the creation and decay of the Ω^- baryon. *(Brookhaven National Laboratory.)* (*b*) Line tracing of the significant tracks of charged particles and paths of neutral particles of the bubble chamber photograph. (*c*) The reaction in which the Ω^- is created and the subsequent cascade of decays. Note that at the point where the Ξ^0 decays to the $\Lambda + \pi^0$ three paths are shown, corresponding to the Λ and two gamma rays, γ_1 and γ_2. The neutral pion decays via the electromagnetic interaction and has a correspondingly short lifetime of about 10^{-16} s. The two γ rays therefore appear to come from the same spot as the Λ. The electron-positron pairs created by these γ rays are visible in the photograph.

participate in the latter, the decay involves the electromagnetic interaction. This is confirmed by the lifetime of 5.8×10^{-20} s, which is characteristic of this interaction.

The decay of the Λ can be written as

$$\text{uds} \rightarrow \text{uud} + \text{d}\bar{\text{u}}$$

Here a $\text{u}\bar{\text{u}}$ pair has been created and, more importantly, an s has been changed into a d quark. This transformation requires that the decay proceed via the weak interaction; the lifetime is correspondingly long, 2.6×10^{-10} s.

The discovery in 1974 by experimenters at Brookhaven National Laboratory and at the Stanford Linear Accelerator Center of a new heavy meson, the J/ψ, confirmed the presence of a fourth quark, the so-called *charmed* quark, c, postulated by theorists several years before. Three years later, the discovery of another heavy meson, the Y, with a mass of about 10^4 MeV/c^2, called for yet another pair of quarks, named *top* and *bottom*. So we now have six quarks and their antiquarks with which to construct the various elementary particles.

Moreover, to explain certain features of the quark model it is necessary to assign an intrinsic property, called *color,* to each quark. Accordingly, each quark may come in one of three ''colors'' labeled red, green, and blue. Hadrons are ''colorless'' quark combinations; that is, the three quarks that constitute a baryon must be of different color, one red, one green, and one blue. Similarly, the mesons are ''colorless'' pairs of quarks and antiquarks of a particular color. This property of color may be viewed as somewhat analogous to the two charges of electromagnetism, where stability demands a balance between positive and negative charges, i.e., a ''chargeless'' system.[12] Quarks are said to be bound by the ''color force'' that is mediated by ''gluons.''

There is a crucial difference between the character of the electromagnetic and color forces. As the separation between two quarks increases, the force of attraction due to their color charge remains constant or increases. Since this attractive force does not diminish with distance, an infinite amount of energy would be required to separate a quark from its colorless group. According to this hypothesis, it is impossible to separate a quark from its companions, which may be the reason that despite every effort no isolated quark has ever been observed.

A new pattern appears to emerge from what may at first seem an unwelcome proliferation of quarks. At the level of the basic building blocks of matter nature presents us with triplets as well as the more familiar pairs—positive and negative charge, particle and antiparticle. There are **three** families of fundamental particles—leptons, mesons, and baryons. There are **three** sets of leptons—e and ν_e, μ and ν_μ, τ and ν_τ. There are **three** groups of quarks—u and d, c and s, t and b. Each of these six quarks appears in one of **three** colors—red, green, or blue. All but one of the six quarks have been observed, and the hunt for the top quark is being pursued with confidence and determination.

Are there really quarks, or are these objects merely convenient artifacts of a sophisticated mathematical model? In a sense, the same question could be asked with regard to electrons, neutrons, neutrinos, or other elementary particles. None of these have been ''seen'' in the usual meaning of the term, that is, visually, under a microscope. These particles were postulated to explain certain phenomena, and their tracks in photographic emulsions and bubble chambers are evidence of their passing. It is as though we lived on some strange island, populated by all sorts of fauna, birds, insects, reptiles, and mammals, all interacting with one another, but invisible to our eyes. All we find are their tracks in the sand, each clearly identifiable as the track of a particular species. Some of these animals are long-lived; others barely make their presence known before they vanish.

Quarks are a peculiar genus on that island. They seem to travel about only in pairs or triplets, never alone. Despite numerous ingenious efforts to find isolated quarks, none have been observed. There is, however, good evidence that quarks exist within baryons. Researchers at the Stanford Linear Accelerator Center have

---

12. These ideas have led to a quantum field theory analogous to quantum electrodynamics called *quantum chromodynamics* (QCD).

studied the differential scattering cross section of protons bombarded with electrons of extremely high energy, i.e., very short wavelength. The results of these deep inelastic scattering experiments show that the proton is not a simple glob of uniform positive charge but contains a charge distribution indicative of three separate charged objects.[13]

17.8 UNIFIED FIELD THEORIES

Prior to 1820, electricity and magnetism were regarded as two separate, unrelated phenomena. That year, Oersted demonstrated that an electric current generates a magnetic field. A decade later, Faraday discovered electromagnetic induction, the appearance of an electric field due to a time-varying magnetic field. These observations led Maxwell to his famous equations which unified electric and magnetic phenomena into a single comprehensive theory. At the beginning of the twentieth century, Einstein achieved a further unification by demonstrating that the electric and magnetic fields are components of a single six-dimensional field that transforms according to the Lorentz transformation of the special theory of relativity. In his later years Einstein sought to unify the electromagnetic and gravitational interactions, but without success. The goal of unification remains today the holy grail of theorists, and there is now some hope that it may yet be attained.

An important step in this direction is the *electroweak* theory developed in the late 1970s by Glashow, Salam, and Weinberg.[14] A discussion of this theory is well beyond the scope of this text; suffice to say that it has been strikingly successful not only in explaining within a single framework both the electromagnetic and weak interactions, but also in predicting new particles.

We mentioned earlier that in quantum field theory the electromagnetic and strong force are mediated by photons and pions, two bosons. According to the electroweak theory, the weak force is also mediated by bosons, the *intermediate vector bosons* Z^0 and $W^{\pm}$, whose masses were predicted to be 93 GeV/c^2 and 83 GeV/c^2. In 1983 these particles were discovered at CERN in proton-antiproton collisions at very high energy,[15] and the measured masses of 92.6 GeV/c^2 and 81.8 GeV/c^2 are in excellent agreement with the theory.

Spurred by these successes, physicists have proposed several variants of a theory that would unify the electroweak and strong interactions. A common feature of the *grand unified theories (GUTs)* is that the photon, the intermediate vector bosons, and the gluons—the mediators of the strong interaction in the quark model—appear as components of a single multicomponent field. It then follows that the proton is also unstable. Different GUTs suggest somewhat different lifetimes, generally in the neighborhood of 10^{32} years. Since 1980 several experiments have sought evidence of such decay, but as of this date no decays have been observed.

A GUT does not, however, unify all the forces of nature since it fails to include gravity. Already in 1921, before Schrödinger and Heisenberg published their theories of quantum mechanics, Theodor Kaluza, a German mathematician, proposed a

---

13. J. I. Friedman, H. W. Kendall, and R. E. Taylor were awarded the 1990 Nobel prize for the "breakthrough in our understanding of matter" by experiments that confirmed the reality of quarks.

14. Sheldon L. Glashow, Abdus Salam, and Steven Weinberg shared the 1979 Nobel Prize for this work.

15. Carlo Rubbia and Simon van der Meer shared the 1984 Nobel Prize for their work in designing the experimental system for high-energy proton-antiproton collision.

scheme whereby Einstein's general theory of relativity—the theory of gravitation—and Maxwell's equations of the electromagnetic field emerge from a single theoretical formulation. The scheme required appending a fifth dimension to the four dimensions of space-time. Since we perceive only the four space-time dimensions, the added dimension of the theory presented a conceptual conundrum. Five years later Oscar Klein suggested that this fifth dimension is "curled up" upon itself in a space so minute that it cannot be observed. From the values of the fundamental electric charge e and the universal gravitational constant G, Klein deduced that the curled up loops of this fifth dimension would have a diameter of only about 10^{-32} m, about 10^{-17} times the size of a nucleus.

Following the work of Dirac, it was recognized that unification of all four fundamental forces of nature required that gravity be reformulated within the context of relativistic quantum theory. The marriage of quantum theory with gravity has proved a particularly difficult task, and until about 1980 all attempts to construct a quantum theory of gravity failed dismally.

Then, in the latter part of the 1970s, several imaginative and mathematically versatile theorists resurrected the idea of a multidimensional manifold that had lain dormant for half a century. In *superstring theory,* quantum gravity, electromagnetism, and the weak and strong forces appear as components of a ten-dimemsional field. As in the Klein-Kaluza theory, six of these dimensions are "curled upon themselves" in a space of about 10^{-32} m. Superstring theory has been hailed by some of its advocates as a *Theory of Everything*. However, though mathematically consistent and devoid of the infinities that plagued previous attempts to construct a quantum theory of gravity, superstring theory has yet to make a single experimentally accessible prediction. Not surprisingly, the theory is viewed in some quarters with considerable skepticism.

SUMMARY

There are four basic forces in nature, the strong (nuclear) force, the electromagnetic force, the weak force, and the gravitational force. All but the last play important roles at the level of elementary particles.

Hadrons are particles that respond to the strong, electromagnetic, and weak forces. *Leptons* respond only to the electromagnetic, and weak forces. There are two categories of hadrons: baryons and mesons. *Baryons* are fermions—particles with odd half-integral spin ($\frac{1}{2}$, $\frac{3}{2}$, . . .); their rest mass is at least as great as that of the proton. *Mesons* are bosons—particles with zero or integral spin; the lightest meson is the pion.

Interactions among elementary particles obey certain conservation laws. All known interactions obey conservation of energy, charge, linear and angular momentum, and baryon and lepton number. In addition to these, *parity, isospin,* and *strangeness, (hypercharge)* are conserved in strong interactions. Isospin conservation may be violated in electromagnetic and weak interactions. Parity and strangeness conservation may be violated in weak interactions.

All hadrons are composed of *quarks* (and antiquarks). Quarks are fermions with charges $\frac{2}{3}$e and $-\frac{1}{3}$e, and come in three "colors"—red, green, and blue. There are six quarks, the *up* and *down* (u, d), *charmed* and *strange* (c, s), and *top* and *bottom* (t, b) quarks. Baryons are "colorless" (red + green + blue) combinations of three quarks. Mesons are colorless (e.g., red-antired) quark-antiquark combinations.

In hadronic interactions quark-antiquark pairs may be created or annihilated, but the transformation of one type of quark to another is forbidden. The weak interaction may transform one type of quark to another, subject to conservation of charge (e.g., u to c, or d to s, but not u to s).

Forces between fundamental particles are mediated by bosons: photons are the mediators of the electromagnetic force, pions mediate the force between nucleons, the *intermediate vector bosons,* Z^0 and $W^{\pm}$, are the mediators of the weak force, and *gluons* have been postulated as the mediators of the color force between quarks. The *electroweak* theory unifies the weak and electromagnetic interactions. The intermediate vector bosons predicted by that theory have been observed. Physicists are currently exploring *grand unified theories* (GUTs) that encompass the electroweak and strong interactions.

Superstring theory is a mathematically consistent unified quantum formulation of the four fundamental forces of nature.

Bibliography

Jeremy Bernstein, *The Tenth Dimension,* McGraw-Hill, New York, 1989.

David B. Cline, "A Fourth Family of Particles beyond Truth and Beauty," *Scientific American,* August 1988, p. 60.

David B. Cline, Carlo Rubbia, and Simon van der Meer, "The Search for Intermediate Vector Bosons," *Scientific American,* March 1982, p. 48.

R. P. Crease and C. C. Mann, *The Second Creation,* Macmillan, New York, 1986.

P. C. W. Davies and Julian Brown (eds.), *Superstrings; A Theory of Everything?,* Cambridge University Press, Cambridge, 1988.

P. C. W. Davies, *The Forces of Nature,* Cambridge University Press, Cambridge, 1979.

J. E. Dodd, *The Ideas of Particle Physics,* Cambridge University Press, Cambridge, 1984.

Sheldon L. Glashow, *Interactions,* Warner Books, New York, 1988.

Haim Harari, "The Structure of Quarks and Leptons," *Scientific American,* April 1983, p. 56.

P. Langacker and A. K. Mann, "The Unification of Electromagnetism with the Weak Force," *Physics Today,* December 1989, p. 22.

J. M. LoSecco, F. Reines, and D. Sinclair, "The Search for Proton Decay," *Scientific American,* June 1985, p. 54.

Hans C. Ohanian, *Modern Physics,* Prentice-Hall, Englewood Cliffs, N.J., 1987.

J. C. Polkinghorne, *The Particle Play,* W. H. Freeman, San Francisco, 1979.

Chris Quigg, "Elementary Particles and Forces," *Scientific American,* April 1985, p. 84.

James S. Trefil, *From Atoms to Quarks,* Scribner, New York, 1980.

Steven Weinberg, "The Decay of the Proton," *Scientific American,* June 1981, p. 64.

PROBLEMS

17.1 Consider the following decay schemes:

(a) $\Lambda \rightarrow \pi^- + \text{p}$
(b) $\Lambda \rightarrow K^- + \text{p}$
(c) $\pi^0 \rightarrow \beta^+ + \beta^- + \nu_e + \bar{\nu}_e$
(d) $\text{n} \rightarrow \text{p} + \beta^- + \bar{\nu}_e$
(e) $\eta \rightarrow \gamma + \gamma$
(f) $\Delta^+ \rightarrow \text{n} + \pi^+$

Which of these decays are allowed? For those that are not possible, state which selection rule(s) would be violated. For the allowed decays, identify the interaction responsible for the decay.

17.2 Each of the following decays is forbidden. State the conservation law(s) that would be violated in each case.

(a) $\Lambda \rightarrow \text{n} + K^0$
(b) $\Lambda \rightarrow \pi^+ + \pi^-$
(c) $\Lambda \rightarrow \text{n} + \gamma$
(d) $\Omega^- \rightarrow \Sigma^0 + \pi^-$
(e) $\Sigma^- \rightarrow K^0 + \mu^-$
(f) $\pi^- \rightarrow \beta^- + \gamma$

17.3 Consider the following reactions. Which of these cannot involve the strong interaction, and which are strictly forbidden?

(a) $p + n \rightarrow \Sigma^+ + K^+ + K^-$
(b) $p + n \rightarrow \Delta^+ + \eta$
(c) $p + p \rightarrow K^+ + \Sigma^+ + \Sigma^0$
(d) $p + p \rightarrow p + \Delta^+ + K^0$
(e) $p + p \rightarrow \Sigma^+ + K^0 + p$
(f) $\pi^- + p \rightarrow \Delta^+ + K^- + K^0$

17.4 The following reactions are impossible because one or more conservation laws would be violated. Identify the conservation law(s) that preclude each of these interactions.

(a) $p + p \rightarrow p + p + K^0$
(b) $\overline{\nu}_e + n \rightarrow p + \beta^-$
(c) $p + p \rightarrow p + n + \Lambda + K^+$
(d) $\pi^- + p \rightarrow \Lambda + \overline{K}^0$
(e) $p + \overline{p} \rightarrow n + \Lambda + K^0$

17.5 In each of the following decays identify the missing particle:

(a) $\eta \rightarrow \pi^+ + \pi^- +$
(b) $K^+ \rightarrow \pi^0 + \beta^+ +$
(c) $\Sigma^+ \rightarrow \pi^+ +$
(d) $\pi^- \rightarrow \mu^- +$
(e) $\Xi^- \rightarrow \pi^- +$

17.6 For each of the following particles give at least two possible decay modes that satisfy the conservation laws:

(a) Σ^0
(b) Σ^-
(c) η
(d) Δ^+

17.7 Supply the missing particle(s) for each of the following reactions:

(a) $p + p \rightarrow \Lambda +$
(b) $p + p \rightarrow n + \Lambda +$
(c) $\pi^- + p \rightarrow \Xi^0 + K^0 +$
(d) $p + \overline{p} \rightarrow \Lambda + \pi^0 +$

17.8 Give at least one possible decay scheme for the antiparticles of (a) Ω^-, (b) n, (c) Λ, and (d) Σ^+.

17.9 What are the uncertainty widths in the masses of the following particles?

(a) Λ (1116)
(b) η (549)
(c) Σ^+ (1193)
(d) Ω^-

17.10 What are the approximate lifetimes of the following particles?

(a) Σ (1670)
(b) ρ (770)
(c) ρ (1600)
(d) J/ψ

17.11 Express the following decays in terms of the quark model:

(a) $\pi^- \rightarrow \mu^- + \overline{\nu}_\mu$
(b) $n \rightarrow p + \beta^- + \overline{\nu}_e$
(c) $\Sigma^{*+} \rightarrow \Lambda + \pi^+$

17.12 Express the following reactions in terms of the quark model:

(a) $K^- + p \rightarrow \Omega^- + K^+ + K^0$
(b) $\pi^- + p \rightarrow \Lambda + K^0$
(c) $\pi^- + p \rightarrow \Sigma^0 + K^0$
(d) $\pi^+ + p \rightarrow \Sigma^+ + K^+$
(e) $p + \overline{p} \rightarrow \overline{n} + \pi^0 + \Lambda + K^0$

17.13 The rest masses of the π^+ and π^- are the same, but the rest masses of the Σ^+ and Σ^- differ slightly. Give a reason for this difference in behavior.

17.14 The lifetime of the Σ^+ is 0.8×10^{-10} s; that of the Σ^- is nearly twice as great (1.48×10^{-10} s). By contrast, the lifetime of the Σ^0 is ten orders of magnitude shorter, 6×10^{-20} s. How can one account for these differences in the lifetimes of members of the same isospin multiplet?

Appendix A Fundamental Physical Constants

Quantity	Symbol	Value	Units	Relative Uncertainty (ppm)
Universal constants				
Speed of light in vacuum	c	299 792 458	m $\sec^{-1}$	Exact
Permeability of vacuum	μ_0	$4\pi \times 10^{-7}$	N A^{-2}	
		= 12.566 370 614 . . .	10^{-7} N A^{-2}	Exact
Permittivity of vacuum	ϵ_0	$1/\mu_0 c^2$		
		= 8.854 187 817 . . .	10^{-12} F m^{-1}	Exact
Newtonian constant of gravitation	G	6.672 59(85)	10^{-11} m^3 kg^{-1} $\sec^{-2}$	128
Planck constant	h	6.626 075 5(40)	10^{-34} J sec	0.60
in electronvolts, h/e		4.135 669 2(12)	10^{-15} eV sec	0.30
$h/2\pi$	$\hbar$	1.054 572 66(63)	10^{-34} J sec	0.60
in electronvolts, $\hbar/e$		6.582 122 0(20)	10^{-16} eV sec	0.30
Electromagnetic constants				
Elementary charge	e	1.602 177 33(49)	10^{-19} C	0.30
	e/h	2.417 988 36(72)	10^{14} A J^{-1}	0.30
Magnetic flux quantum, $\hbar/2e$	Φ_0	2.067 834 61(61)	10^{-15} Wb	0.30
Josephson frequency-voltage quotient	$2e/h$	4.835 976 7(14)	10^{14} Hz V^{-1}	0.30
Quantized Hall resistance, $h/e^2 = \mu_0 c/2\alpha$	R_H	25 812.805 6(12)	Ω	0.045
Bohr magneton, $e\hbar/2m_e$	μ_B	9.274 015 4(31)	10^{-24} J T^{-1}	0.34
in electronvolts, μ_B/e		5.788 382 63(52)	10^{-5} eV T^{-1}	0.089
in hertz, μ_B/h		1.399 624 18(42)	10^{10} Hz T^{-1}	0.30
Nuclear magneton, $e\hbar/2m_p$	μ_N	5.050 786 6(17)	10^{-27} J T^{-1}	0.34
in electronvolts, μ_N/e		3.152 451 66(28)	10^{-8} eV T^{-1}	0.089
in hertz, μ_N/h		7.622 591 4(23)	MHz T^{-1}	0.30
Atomic constants				
Fine-structure constant, $\mu_0 ce^2/2h$	α	7.297 353 08(33)	10^{-3}	0.045
Inverse fine-structure constant	α^{-1}	137.035 989 5(61)		0.045
Rydberg constant, $m_e ce^2/2h$	R_∞	10 973 731.534(13)	m^{-1}	0.0012
in eV, $R_\infty hc/e$		13.605 698 1(40)	eV	0.30
Bohr radius, $\alpha/4\pi R_\infty$	a_0	0.529 177 249(24)	10^{-10} m	0.045
Electron				
Mass	m_e	9.109 389 7(54)	10^{-31} kg	0.59
		5.485 799 03(13)	10^{-4} u	0.023
in electronvolts, $m_e c^2/e$		0.510 999 06(15)	MeV	0.30
Electron-proton mass ratio	m_e/m_p	5.446 170 13(11)	10^{-4}	0.020
Specific charge	$-e/m_e$	$-$ 1.758 819 62(53)	10^{11} C kg^{-1}	0.30
Compton wavelength, $h/m_e c$	λ_C	2.426 310 58(22)	10^{-12} m	0.089
Magnetic moment	μ_e	928.477 01(31)	10^{-26} J T^{-1}	0.34
in Bohr magnetons	μ_e/μ_B	1.001 159 652 193(10)		1×10^{-5}
g-factor	g_e	2.002 319 304 386(20)		1×10^{-5}
Muon				
Mass	m_μ	1.883 532 7(11)	10^{-28} kg	0.61
		0.113 428 913(17)	u	0.15
in electronvolts, $m_\mu c^2/e$		105.658 389(34)	MeV	0.32

(continued)

Quantity	Symbol	Value	Units	Relative Uncertainty (ppm)
Magnetic moment	μ_μ	4.490 451 4(15)	10^{-26} J T^{-1}	0.33
in Bohr magnetons	μ_μ/μ_B	4.841 970 97(71)	10^{-3}	0.15
in nuclear magnetons	μ_μ/μ_N	8.890 598 1(13)		0.15
g-factor	g_μ	2.002 331 846(17)		0.0084
Proton				
Mass	m_p	1.672 623 1(10)	10^{-27} kg	0.59
		1.007 276 470(12)	u	0.012
in electronvolts, m_pc^2/e		938.272 31(28)	MeV	0.30
Proton-electron mass ratio	m_p/m_e	1836.152 701(37)		0.020
Specific charge	e/m_p	9.578 830 9(29)	10^7 C kg^{-1}	0.30
Molar mass	$M(p)$	1.007 276 470(12)	10^{-3} kg mol^{-1}	0.012
Compton wavelength, h/m_pc	$\lambda_{C,p}$	1.321 410 02(12)	10^{-15} m	0.089
Magnetic moment	μ_p	1.410 607 61(47)	10^{-26} J T^{-1}	0.34
in Bohr magnetons	μ_p/μ_B	1.521 032 202(15)	10^{-3}	0.010
in nuclear magnetons	μ_p/μ_N	2.792 847 386(63)		0.023
Neutron				
Mass	m_n	1.674 928 6(10)	10^{-27} kg	0.59
		1.008 664 904(14)	u	0.014
in electronvolts, m_nc^2/e		939.565 63(28)	MeV	0.30
Neutron-electron mass ratio	m_n/m_e	1838.683 662(40)		0.022
Neutron-proton mass ratio	m_n/m_p	1.001 378 404(9)		0.009
Compton wavelength, h/m_nc	$\lambda_{C,n}$	1.319 591 10(12)	10^{-15} m	0.089
Magnetic moment*	μ_n	0.966 237 07(40)	10^{-26} J T^{-1}	0.41
in Bohr magnetons	μ_n/μ_B	1.041 875 63(25)	10^{-3}	0.24
in nuclear magnetons	μ_n/μ_N	1.913 042 75(45)		0.24
Deuteron				
Mass	m_d	3.343 586 0(20)	10^{-27} kg	0.59
		2.013 553 214(24)	u	0.012
in electronvolts, m_dc^2/e		1875.613 39(57)	MeV	0.30
Magnetic moment*	μ_d	0.433 073 75(15)	10^{-26} J T^{-1}	0.34
in Bohr magnetons	μ_d/μ_B	0.466 975 447 9(91)	10^{-3}	0.019
in nuclear magnetons	μ_d/μ_N	0.857 438 230(24)		0.028
Physicochemical constants				
Avogadro constant	N_A, L	6.022 136 7(36)	10^{23} mol^{-1}	0.59
Atomic mass constant, $m(^{12}C)/12$	m_u	1.660 540 2(10)	10^{-27} kg	0.59
in electronvolts, m_uc^2/e		931.494 32(28)	MeV	0.30
Faraday constant	F	96 485.309(29)	C mol^{-1}	0.30
Molar gas constant	R	8.314 510(70)	J mol^{-1} K^{-1}	8.4
Boltzmann constant, R/N_A	k	1.380 658(12)	10^{-23} J K^{-1}	8.5
in electronvolts, k/e		8.617 385(73)	10^{-5} eV K^{-1}	8.4
in hertz, k/h		2.083 674(18)	10^{10} Hz K^{-1}	8.4
Stefan-Boltzmann constant, $(\pi^2/60)k^4/\hbar^3c^2$	σ	5.670 51(19)	10^{-8} W m^{-2} K^{-4}	34
Conversion factors and units				
Electronvolt, $(e/C)J = eJ$	eV	1.602 177 33(49)	10^{-19} J	0.30
Atomic mass unit (unified), $m_u = m(^{12}C)/12$	u	1.660 540 2(10)	10^{-27} kg	0.59
Standard atmosphere	atm	101 325	Pa	Exact
Standard acceleration of gravity	g_n	9.806 65	m sec^{-2}	Exact

Digits in parentheses indicate the standard deviation uncertainty in the last digits of the given value.

*The scalar magnitude of the neutron moment is listed here. The neutron magnetic dipole is directed oppositely to that of the proton, and corresponds to the dipole associated with a spinning negative charge distribution. The vector sum $\mu_d = \mu_p + \mu_n$ is approximately satisfied.

Appendix B Table of Isotopes

Chemical Symbol	Z	A	Atomic Mass (u)	Spin and Parity	Decay Modes*	Half-Life $t_{1/2}$	Natural Abundance (%)
n	0	1	1.008665	$\frac{1}{2}+$	β^-	12 min	
H	1	1	1.007825	$\frac{1}{2}+$			99.985
H (D)	1	2	2.014102	$1+$			0.015
H (T)	1	3	3.016049	$\frac{1}{2}+$	β^-	12.26 yr	
He	2	3	3.016029	$\frac{1}{2}+$			0.00014
He	2	4	4.002603	$0+$			99.99986
He	2	6	6.018886	$0+$	β^-	0.808 s	
He	2	8	8.033920	$0+$	β^-	0.122 s	
Li	3	5	5.012539	$\frac{3}{2}-$	p, α		
Li	3	6	6.015121	$1+$			7.5
Li	3	7	7.016003	$\frac{3}{2}-$			92.5
Li	3	8	8.022485	$2+$	β^-, α	0.844 s	
Li	3	9	9.026789	$\frac{3}{2}-$	β^-	0.178 s	
Be	4	7	7.016928	$\frac{3}{2}-$	E.C.	53.29 d	
Be	4	8	8.005305	$0+$	2α	0.067 fs	
Be	4	9	9.012182	$\frac{3}{2}-$			100
Be	4	10	10.013534	$0+$	β^-	1.6×10^6 yr	
Be	4	11	11.021658	$\frac{1}{2}+$	β^-	13.8 s	
B	5	8	8.024605	$2+$	β^+, 2α	0.772 s	
B	5	9	9.013328		p, 2α	0.85 as	
B	5	10	10.012937	$3+$			19.8
B	5	11	11.009305	$\frac{3}{2}-$			80.2
B	5	12	12.014352	$1+$	β^-	0.0202 s	
B	5	13	13.017780	$\frac{3}{2}-$	β^-	0.0173 s	
C	6	9	9.031039		β^+, p, 2α	127 ms	
C	6	10	10.016858	$0+$	β^+	19.3 s	
C	6	11	11.011433	$\frac{3}{2}-$	β^+, E.C.	20.3 min	
C	6	12	12.000000	$0+$			98.90
C	6	13	13.003355	$\frac{1}{2}-$			1.10
C	6	14	14.003241	$0+$	β^-	5730 yr	
C	6	15	15.010599	$\frac{1}{2}+$	β^-	2.45 s	
C	6	16	16.014701		β^-	0.75 s	
N	7	12	12.018613	$1+$	β^+	11.0 ms	
N	7	13	13.005738	$\frac{1}{2}-$	β^+	9.97 min	
N	7	14	14.003074	$1+$			99.63
N	7	15	15.000109	$\frac{1}{2}-$			0.37
N	7	16	16.006099	$2-$	β^-, α	7.13 s	
N	7	17	17.008450	$\frac{1}{2}-$	β^-	4.17 s	
N	7	18	18.014081		β^-	0.63 s	
N	7	19	19.017040		β^-	0.42 s	
O	8	13	13.028100		β^+, p	8.9 ms	

*The following symbols are used: α = alpha particle emission; β^- = electron emission; β^+ = positron emission; E.C. = electron capture; n = neutron emission; p = proton emission; S.F. = spontaneous fission.

(continued)

Chemical Symbol	Z	A	Atomic Mass (u)	Spin and Parity	Decay Modes*	Half-Life $t_{1/2}$	Natural Abundance (%)
O	8	14	14.008595	$0+$	β^+	70.6 s	
O	8	15	15.003065	$\frac{1}{2}-$	β^+	122 s	
O	8	16	15.994915	$0+$			99.762
O	8	17	16.999131	$\frac{5}{2}+$			0.038
O	8	18	17.999160	$0+$			0.200
O	8	19	19.003577	$\frac{5}{2}+$	β^-	26.9 s	
O	8	20	20.004075	$0+$	β^-	13.5 s	
O	8	21	21.008730		β^-	3.14 s	
F	9	17	17.002095	$\frac{5}{2}+$	β^+	64.7 s	
F	9	18	18.000937	$1+$	β^+, E.C.	109.8 min	
F	9	19	18.998403	$\frac{1}{2}+$			100
F	9	20	19.999981	$2+$	β^-	11.0 s	
F	9	21	20.999948	$\frac{5}{2}+$	β^-	4.33 s	
F	9	22	22.003030	$4+$	β^-	4.23 s	
F	9	23	23.003600	$\frac{5}{2}+$	β^-	2.2 s	
Ne	10	17	17.017690	$\frac{1}{2}-$	β^+, p	109 ms	
Ne	10	18	18.005710	$0+$	β^+	1.67 s	
Ne	10	19	19.001879	$\frac{1}{2}+$	β^+	17.22 s	
Ne	10	20	19.992435	$0+$			90.51
Ne	10	21	20.993843	$\frac{3}{2}+$			0.21
Ne	10	22	21.991383	$0+$			9.22
Ne	10	23	22.994465	$\frac{5}{2}+$	β^-	37.2 s	
Ne	10	24	23.993613	$0+$	β^-	3.38 min	
Ne	10	25	24.997690	$\frac{1}{2}+$	β^-	0.61 s	
Na	11	19	19.013879		β^+, p	0.03 s	
Na	11	20	20.007344	$2+$	β^+, α	0.446 s	
Na	11	21	20.997650	$\frac{3}{2}+$	β^+	22.5 s	
Na	11	22	21.994434	$3+$	$\beta+$	2.605 yr	
Na	11	23	22.989767	$\frac{3}{2}+$			100
Na	11	24	23.990961	$4+$	β^-	14.97 h	
Na	11	25	24.989953	$\frac{5}{2}+$	β^-	59.3 s	
Na	11	26	25.992586		β^-	1.07 s	
Na	11	27	26.993940		β^-	0.29 s	
Na	11	28	27.978780	$1+$	β^-	30 ms	
Na	11	29	29.002830		β^-, n	43 ms	
Na	11	30	30.008800		β^-, n	53 ms	
Na	11	31	31.012680		β^-, n	17 ms	
Mg	12	20	20.018864		β^+, p	0.1 s	
Mg	12	21	21.011716	$\frac{5}{2}+$	β^+, p	122 ms	
Mg	12	22	21.999574	$0+$	β^+	3.86 s	
Mg	12	23	22.994124	$\frac{3}{2}+$	β^+	11.32 s	
Mg	12	24	23.985042	$0+$			78.99
Mg	12	25	24.985837	$\frac{5}{2}+$			10.00
Mg	12	26	25.982593	$0+$			11.01
Mg	12	27	26.984341	$\frac{1}{2}+$	β^-	9.45 min	
Mg	12	28	27.983876	$0+$	β^-	21.0 h	
Mg	12	29	28.988480	$\frac{3}{2}+$	β^-	1.3 s	
Mg	12	30	29.990230	$0+$	β^-	0.33 s	
Mg	12	31	30.995930		β^-, n	0.25 s	
Al	13	22	22.079370		β^+	70 ms	

Chemical Symbol	Z	A	Atomic Mass (u)	Spin and Parity	Decay Modes*	Half-Life $t_{1/2}$	Natural Abundance (%)
Al	13	23	23.007265		β^+	0.47 s	
Al	13	24	23.999941	4 +	β^+	2.07 s	
Al	13	25	24.990429	$\frac{5}{2}$ +	β^+	7.17 s	
Al	13	26	25.986892	5 +	β^+	7.2×10^5 yr	
Al	13	27	26.981539	$\frac{5}{2}$ +			100
Al	13	28	27.981910	3 +	β^-	2.25 min	
Al	13	29	28.980446	$\frac{5}{2}$ +	β^-	6.5 min	
Al	13	30	29.982940		β^-	3.69 s	
Al	13	31	30.983800		β^-	0.64 s	
Si	14	24	24.011546		β^+, p	0.1 ms	
Si	14	25	25.004109		β^+, p	220 ms	
Si	14	26	25.992330	0 +	β^+	2.20 s	
Si	14	27	26.986704	$\frac{5}{2}$ +	β^+	4.14 s	
Si	14	28	27.976927	0 +			92.23
Si	14	29	28.976495	$\frac{1}{2}$ +			4.67
Si	14	30	29.973770	0 +			3.10
Si	14	31	30.975362	$\frac{3}{2}$ +	β^-	2.62 h	
Si	14	32	31.974148	0 +	β^-	650 yr	
Si	14	33	32.977920		β^-	6.2 s	
Si	14	34	33.976360	0 +	β^-	2.8 s	
P	15	26	26.012080		β^+, p	20 ms	
P	15	28	27.992313	3 +	β^+	270 ms	
P	15	29	28.981308	$\frac{1}{2}$ +	β^+	4.14 s	
P	15	30	29.978307	1 +	β^+	2.5 min	
P	15	31	30.973762	$\frac{1}{2}$ +			100
P	15	32	31.973907	1 +	β^-	14.28 d	
P	15	33	32.971725	$\frac{1}{2}$ +	β^-	25.3 d	
P	15	34	33.973636	1 +	β^-	12.4 s	
P	15	35	34.973232		β^-	47.0 s	
P	15	36	35.977570		β^-	5.9 s	
S	16	29	28.996610	$\frac{5}{2}$ +	β^+	0.19 s	
S	16	30	29.984903	0 +	β^+	1.18 s	
S	16	31	30.979554	$\frac{1}{2}$ +	β^+	2.55 s	
S	16	32	31.972070	0 +			95.02
S	16	33	32.971456	$\frac{3}{2}$ +			0.75
S	16	34	33.967866	0 +			4.21
S	16	35	34.969031	$\frac{3}{2}$ +	β^-	87.2 d	
S	16	36	35.967080	0 +			0.02
S	16	37	36.971125		β^-	5.05 min	
S	16	38	37.971162	0 +	β^-	2.84 h	
S	16	39	38.975310		β^-	11.5 s	
Cl	17	31	30.992410		β^+	0.15 s	
Cl	17	32	31.985690	1 +	β^+	297 ms	
Cl	17	33	32.977451	$\frac{3}{2}$ +	β^+	2.51 s	
Cl	17	34	33.973763	0 +	β^+	1.53 s	
Cl	17	35	34.968852	$\frac{3}{2}$ +			75.77
Cl	17	36	35.968306	2 +	β^-	3.0×10^5 yr	
Cl	17	37	36.965903	$\frac{3}{2}$ +			24.23
Cl	17	38	37.968010	2 −	β^-	37.2 min	
Cl	17	39	38.968005	$\frac{3}{2}$ +	β^-	55.7 min	

(continued)

Chemical Symbol	Z	A	Atomic Mass (u)	Spin and Parity	Decay Modes*	Half-Life $t_{1/2}$	Natural Abundance (%)
Cl	17	40	39.970440	2 −	β^-	1.35 min	
Cl	17	41	40.970590		β^-	34 s	
Ar	18	32	31.997660		β^+, p	0.1 s	
Ar	18	33	32.989930	$\frac{1}{2}$ +	β^+	17 ms	
Ar	18	34	33.980269	0 +	β^+	0.844 s	
Ar	18	35	34.975256	$\frac{3}{2}$ +	β^+	1.77 s	
Ar	18	36	35.967545	0 +			0.337
Ar	18	37	36.966776	$\frac{3}{2}$+	E.C.	34.8 d	
Ar	18	38	37.962732	0 +			0.063
Ar	18	39	38.964314	$\frac{7}{2}$ −	β^-	269 yr	
Ar	18	40	39.962384	0 +			99.60
Ar	18	41	40.964501	$\frac{7}{2}$ −	β^-	1.83 h	
Ar	18	42	41.963050	0 +	β^-	33 y	
Ar	18	43	42.965670		β^-	5.4 min	
Ar	18	44	43.963650	0 +	β^-	11.9 min	
Ar	18	45	44.968090	$\frac{7}{2}$ −	β^-	21 s	
K	19	35	34.988011	$\frac{3}{2}$ +	β^+	0.19 s	
K	19	36	35.981293	2 +	β^+	0.342 s	
K	19	37	36.973377	$\frac{3}{2}$ +	β^+	1.23 s	
K	19	38	37.969080	3 +	β^+	7.63 min	
K	19	39	38.963707	$\frac{3}{2}$ +			93.2581
K	19	40	39.963999	4 −	β^-, β^+, E.C.	1.25×10^9 yr	0.0117
K	19	41	40.961825	$\frac{3}{2}$ +			6.7302
K	19	42	41.962402	2 −	β^-	12.36 h	
K	19	43	42.960717	$\frac{3}{2}$ +	β^-	22.3 h	
K	19	44	43.960156	2 −	β^-	22.1 min	
K	19	45	44.960696	$\frac{3}{2}$ +	β^-	17.3 min	
K	19	46	45.961976	2 −	β^-	107 s	
K	19	47	46.961677	$\frac{1}{2}$ +	β^-	17.5 s	
K	19	48	47.965514	2 −	β^-	69 s	
K	19	49	48.966940		β^-, n	1.3 s	
Ca	20	36	35.993090		β^+	0.1 s	
Ca	20	37	36.985873	$\frac{3}{2}$ +	β^+	175 ms	
Ca	20	38	37.976318	0 +	β^+	0.45 s	
Ca	20	39	38.970718	$\frac{3}{2}$ +	β^+	0.86 s	
Ca	20	40	39.962591	0 +			96.941
Ca	20	41	40.962278	$\frac{7}{2}$ −	E.C.	1.0×10^5 yr	
Ca	20	42	41.958618	0 +			0.647
Ca	20	43	42.958766	$\frac{7}{2}$ −			0.135
Ca	20	44	43.955480	0 +			2.086
Ca	20	45	44.956185	$\frac{7}{2}$ −	β^-	163.8 d	
Ca	20	46	45.953689	0 +			0.004
Ca	20	47	46.954543	$\frac{7}{2}$ −	β^-	4.536 d	
Ca	20	48	47.952533	0 +			0.187
Ca	20	49	48.955672	$\frac{3}{2}$ −	β^-	8.72 min	
Ca	20	50	49.957519	0 +	β^-	14 s	
Ca	20	51	50.961420		β^-	10 s	
Sc	21	40	39.977963	4 −	β^+	0.182 s	
Sc	21	41	40.969250	$\frac{7}{2}$ −	β^+	0.596 s	
Sc	21	42	41.965514	0 +	β^+	0.68 s	

Chemical Symbol	Z	A	Atomic Mass (u)	Spin and Parity	Decay Modes*	Half-Life $t_{1/2}$	Natural Abundance (%)
Sc	21	43	42.961150	$\frac{7}{2}$ −	β^+, E.C.	3.89 h	
Sc	21	44	43.959404	2 +	β^+, E.C.	3.93 h	
Sc	21	45	44.955910	$\frac{7}{2}$ −			100
Sc	21	46	45.955170	4 +	β^-	83.8 d	
Sc	21	47	46.952409	$\frac{7}{2}$ −	β^-	3.42 d	
Sc	21	48	47.952235	6 +	β^-	43.7 h	
Sc	21	49	48.950022	$\frac{7}{2}$ −	β^-	57.3 min	
Sc	21	50	49.952186	5 +	β^-	1.71 min	
Sc	21	51	50.953602	$\frac{7}{2}$ −	β^-	12.4 s	
Ti	22	41	40.983150	$\frac{3}{2}$ +	β^+, p	80 ms	
Ti	22	42	41.973031		β^+	0.20 s	
Ti	22	43	42.968523	$\frac{7}{2}$ −	β^+	0.49 s	
Ti	22	44	43.959690	0 +	E.C.	47 yr	
Ti	22	45	44.958124	$\frac{7}{2}$ −	β^+	3.078 h	
Ti	22	46	45.952429	0 +			8.2
Ti	22	47	46.951764	$\frac{5}{2}$ −			7.4
Ti	22	48	47.947947	0 +			73.7
Ti	22	49	48.947871	$\frac{7}{2}$ −			5.4
Ti	22	50	49.944792	0 +			5.2
Ti	22	51	50.946616	$\frac{3}{2}$ −	β^-	5.76 min	
Ti	22	52	51.946898	0 +	β^-	1.7 min	
Ti	22	53	52.949730	$\frac{3}{2}$ −	β^-	32.7 s	
V	23	44	43.974450		β^+, α	0.09 s	
V	23	46	45.960198	0 +	β^+	0.22 s	
V	23	47	46.954906	$\frac{3}{2}$ −	β^+, E.C.	31.3 min	
V	23	48	47.952257	4 +	β^+	15.98 d	
V	23	49	48.948517	$\frac{7}{2}$ −	E.C.	331 d	
V	23	50	49.947161	6 +	E.C.	$>4 \times 10^{17}$ yr	0.25
V	23	51	50.943962	$\frac{7}{2}$ −			99.75
V	23	52	51.944778	3 +	β^-	3.76 min	
V	23	53	52.944340	$\frac{7}{2}$ −	β^-	1.61 min	
V	23	54	53.946442	5 +	β^-	49.8 s	
V	23	55	54.947240		β^-	6.5 s	
Cr	24	45	44.979110	$\frac{7}{2}$ −	β^+, p	0.05 s	
Cr	24	46	45.968360	0 +	β^+	0.26 s	
Cr	24	47	46.962905	$\frac{3}{2}$ −	β^+	0.46 s	
Cr	24	48	47.954033	0 +	E.C.	21.6 h	
Cr	24	49	48.951338	$\frac{5}{2}$ −	β^+, E.C.	42.1 min	
Cr	24	50	49.946046	0 +			4.35
Cr	24	51	50.944768	$\frac{7}{2}$ −	E.C.	27.70 d	
Cr	24	52	51.940510	0 +			83.79
Cr	24	53	52.940651	$\frac{3}{2}$ −			9.50
Cr	24	54	53.938882	0 +			2.36
Cr	24	55	54.940842	$\frac{3}{2}$ −	β^-	3.50 min	
Cr	24	56	55.940643	0 +	β^-	5.9 min	
Cr	24	57	56.943440	$\frac{3}{2}$ −	β^-	21 s	
Mn	25	49	48.951338	$\frac{5}{2}$ −	β^+	0.38 s	
Mn	25	50	49.954239	0 +	β^+	0.238 s	
Mn	25	51	50.948213	$\frac{5}{2}$ −	β^+, E.C.	46.2 min	
Mn	25	52	51.945568	6 +	β^+, E.C.	5.59 d	

(continued)

Chemical Symbol	Z	A	Atomic Mass (u)	Spin and Parity	Decay Modes*	Half-Life $t_{1/2}$	Natural Abundance (%)
Mn	25	53	52.941291	$\frac{7}{2}-$	E.C.	3.7×10^6 yr	
Mn	25	54	53.940361	3 +	E.C.	312 d	
Mn	25	55	54.938047	$\frac{5}{2}-$			100
Mn	25	56	55.938906	3 +	β^-	2.579 h	
Mn	25	57	56.938285	$\frac{5}{2}-$	β^-	1.45 min	
Mn	25	58	57.940060	3 +	β^-	65 s	
Mn	25	59	58.940440		β^-	4.6 s	
Mn	25	60	59.943210	3 +	β^-	1.8 s	
Fe	26	51	50.956825	$\frac{5}{2}-$	β^+	0.25 s	
Fe	26	52	51.948114	0 +	β^+, E.C.	8.28 h	
Fe	26	53	52.945310	$\frac{7}{2}-$	β^+	8.51 min	
Fe	26	54	53.939612	0 +			5.8
Fe	26	55	54.938296	$\frac{3}{2}-$	E.C.	2.7 yr	
Fe	26	56	55.934939	0 +			91.72
Fe	26	57	56.935396	$\frac{1}{2}-$			2.2
Fe	26	58	57.933277	0 +			0.28
Fe	26	59	58.934877	$\frac{3}{2}-$	β^-	44.51 d	
Fe	26	60	59.934078	0 +	β^-	1.5×10^6 yr	
Fe	26	61	60.936748	$\frac{3}{2}-$	β^-	6.0 min	
Fe	26	62	61.936773	0 +	β^-	68 s	
Fe	26	63	62.940750		β^-	4.9 s	
Co	27	53	52.954225	$\frac{7}{2}-$	β^+	0.26 s	
Co	27	54	53.948460	0 +	β^+	0.19 s	
Co	27	55	54.942001	$\frac{7}{2}-$	β^+, E.C.	17.5 h	
Co	27	56	55.939841	0 +	β^+, E.C.	77.7 d	
Co	27	57	56.936294	$\frac{7}{2}-$	E.C.	271 d	
Co	27	58	57.935755	2 +	β^+, E.C.	70.91 d	
Co	27	59	58.933198	$\frac{7}{2}-$			100
Co	27	60	59.933819	5 +	β^-	5.272 yr	
Co	27	61	60.932478	$\frac{7}{2}-$	β^-	1.65 h	
Co	27	62	61.934060	2 +	β^-	1.5 min	
Co	27	63	62.933614	$\frac{7}{2}-$	β^-	27.5 s	
Co	27	64	63.935812	1 +	β^-	0.30 s	
Ni	28	53	52.968430	$\frac{7}{2}-$	β^+, p	0.05 s	
Ni	28	55	54.951336	$\frac{7}{2}-$	β^+	0.19 s	
Ni	28	56	55.943124	0 +	E.C.	6.10 d	
Ni	28	57	56.939799	$\frac{3}{2}-$	β^+, E.C.	36.1 h	
Ni	28	58	57.935346	0 +			68.27
Ni	28	59	58.934349	$\frac{3}{2}-$	E.C.	7.6×10^4 yr	
Ni	28	60	59.930788	0 +			26.10
Ni	28	61	60.931058	$\frac{3}{2}-$			1.13
Ni	28	62	61.928346	0 +			3.59
Ni	28	63	62.929669	$\frac{1}{2}-$	β^-	100 yr	
Ni	28	64	63.927968	0 +			0.91
Ni	28	65	64.930086	$\frac{5}{2}-$	β^-	2.52 h	
Ni	28	66	65.929116	0 +	β^-	54.8 h	
Ni	28	67	66.931570		β^-	20 s	
Cu	29	58	57.944538	1 +	β^+, E.C.	3.21 s	
Cu	29	59	58.939503	$\frac{3}{2}-$	β^+	82 s	
Cu	29	60	59.937366	2 +	β^+, E.C.	23.2 min	

Chemical Symbol	Z	A	Atomic Mass (u)	Spin and Parity	Decay Modes*	Half-Life $t_{1/2}$	Natural Abundance (%)
Cu	29	61	60.933461	$\frac{3}{2}-$	β^+	3.41 h	
Cu	29	62	61.932586	$1+$	β^+	9.74 min	
Cu	29	63	62.929598	$\frac{3}{2}-$			69.17
Cu	29	64	63.929765	$1+$	β^-, β^+, E.C.	12.701 h	
Cu	29	65	64.927793	$\frac{3}{2}-$			30.83
Cu	29	66	65.928872	$1+$	β^-	5.10 min	
Cu	29	67	66.927747	$\frac{3}{2}-$	β^-	61.9 h	
Cu	29	68	67.929620	$1+$	β^-	31 s	
Cu	29	69	68.929425	$\frac{3}{2}-$	β^-	3.0 min	
Cu	29	70	69.932386	$1+$	β^-	5 s	
Cu	29	71	70.932560	$\frac{3}{2}-$	β^-	20 s	
Zn	30	59	58.949270	$\frac{3}{2}-$	β^+, p	184 ms	
Zn	30	60	59.941830	$0+$	β^+	2.4 min	
Zn	30	61	60.939514	$\frac{3}{2}-$	β^+	89.1 s	
Zn	30	62	61.934332	$0+$	E.C.	9.26 h	
Zn	30	63	62.933214	$\frac{3}{2}-$	β^+	38.1 min	
Zn	30	64	63.929145	$0+$			48.6
Zn	30	65	64.929243	$\frac{5}{2}-$	β^+	243.8 d	
Zn	30	66	65.926034	$0+$			27.9
Zn	30	67	66.927129	$\frac{5}{2}-$			4.1
Zn	30	68	67.924846	$0+$			18.8
Zn	30	69	68.926552	$\frac{1}{2}-$	β^-	57 min	
Zn	30	70	69.925325	$0+$			0.6
Zn	30	71	70.927727	$\frac{1}{2}-$	β^-	2.4 min	
Zn	30	72	71.926856	$0+$	β^-	46.5 h	
Zn	30	73	72.929780	$\frac{3}{2}-$	β^-	24 s	
Zn	30	74	73.929461		β^-	96 s	
Zn	30	75	74.932690		β^-	10.2 s	
Zn	30	76	75.932940		β^-	5.7 s	
Zn	30	77	76.936750	$\frac{7}{2}+$	β^-	1.4 s	
Zn	30	78	77.937780		β^-	1.5 s	
Ga	31	62	61.944178	$0+$	β^+, E.C.	0.116 s	
Ga	31	63	62.939140		β^+, E.C.	32 s	
Ga	31	64	63.936836	$0+$	β^+	2.63 min	
Ga	31	65	64.932738	$\frac{3}{2}-$	β^+	15.2 min	
Ga	31	66	65.931590	$0+$	β^+, E.C.	9.4 h	
Ga	31	67	66.928204	$\frac{3}{2}-$	E.C.	78.25 h	
Ga	31	68	67.927981	$1+$	β^+	68.1 min	
Ga	31	69	68.925580	$\frac{3}{2}-$			60.1
Ga	31	70	69.926028	$1+$	β^-	21.1 min	
Ga	31	71	70.924700	$\frac{3}{2}-$			39.9
Ga	31	72	71.926365	$3-$	β^-	13.95 h	
Ga	31	73	72.925169	$\frac{3}{2}-$	β^-	4.87 h	
Ga	31	74	73.926940	$3-$	β^-	8.1 min	
Ga	31	75	74.926499	$\frac{3}{2}-$	β^-	2.1 min	
Ga	31	76	75.928670	$3-$	β^-	29.1 s	
Ga	31	77	76.928700		β^-	13 s	
Ga	31	78	77.931760	$3+$	β^-	5.09 s	
Ga	31	79	78.932530		β^-, n	3.0 s	
Ga	31	80	79.936250		β^-, n	1.66 s	
Ga	31	81	80.937750		β^-, n	1.23 s	

(continued)

Chemical Symbol	Z	A	Atomic Mass (u)	Spin and Parity	Decay Modes*	Half-Life $t_{1/2}$	Natural Abundance (%)
Ge	32	64	63.941570	0 +	β^+, E.C.	63 s	
Ge	32	65	64.939440		β^+, E.C.	31 s	
Ge	32	66	65.933847	0 +	β^+, E.C.	2.27 h	
Ge	32	67	66.932737	$\frac{1}{2}$ −	β^+	19.0 min	
Ge	32	68	67.928096	0 +	E.C.	288 d	
Ge	32	69	68.927969	$\frac{5}{2}$ −	β^+, E.C.	39.1 h	
Ge	32	70	69.924250	0 +			20.5
Ge	32	71	70.924953	$\frac{1}{2}$ −	E.C.	11.2 d	
Ge	32	72	71.922079	0 +			27.4
Ge	32	73	72.923463	$\frac{9}{2}$ +			7.8
Ge	32	74	73.921177	0 +			36.5
Ge	32	75	74.922858	$\frac{1}{2}$ −	β^-	82.8 min	
Ge	32	76	75.921401	0 +			7.8
Ge	32	77	76.923548	$\frac{7}{2}$ +	β^-	11.30 h	
Ge	32	78	77.922853	0 +	β^-	1.45 h	
Ge	32	79	78.925360	$\frac{1}{2}$ −	β^-	42 s	
Ge	32	80	79.925520	0 +	β^-	29 s	
Ge	32	81	80.928820	$\frac{1}{2}$ +	β^-	7.6 s	
Ge	32	82	81.929810	0 +	β^-	4.6 s	
Ge	32	83	82.934250		β^-	1.9 s	
As	33	67	66.939190	$\frac{5}{2}$ −	β^+, E.C.	43 s	
As	33	68	67.936790	3 +	β^+	2.53 min	
As	33	69	68.932280	$\frac{5}{2}$ −	β^+	15.1 min	
As	33	70	69.930929	4 +	β^+, E.C.	52.6 min	
As	33	71	70.927114	$\frac{5}{2}$ −	β^+, E.C.	62 h	
As	33	72	71.926755	2 −	β^+, E.C.	26.0 h	
As	33	73	72.923827	$\frac{3}{2}$ −	E.C.	80.3 d	
As	33	74	73.923827	2 −	β^+, β^-, E.C.	17.78 d	
As	33	75	74.921594	$\frac{3}{2}$ −			100
As	33	76	75.922393	2 −	β^-	26.3 h	
As	33	77	76.920646	$\frac{3}{2}$ −	β^-	38.8 h	
As	33	78	77.921830	2 −	β^-	1.515 h	
As	33	79	78.920946	$\frac{3}{2}$ −	β^-	9.0 min	
As	33	80	79.922528	1 +	β^-	16 s	
As	33	81	80.922131	$\frac{3}{2}$ −	β^-	33 s	
As	33	82	81.924769	1 +	β^-	19 s	
As	33	83	82.924980		β^-, n	13 s	
As	33	84	83.929060	1 −	β^-, n	5.5 s	
As	33	85	84.931820	$\frac{3}{2}$ −	β^-, n	2.03 s	
Se	34	69	68.939570		β^+, E.C.	27.4 s	
Se	34	70	69.933880	0 +	β^+	41.1 min	
Se	34	71	70.932270	$\frac{5}{2}$ −	β^+, E.C.	4.7 min	
Se	34	72	71.927110	0 +	E.C.	8.4 d	
Se	34	73	72.926768	$\frac{9}{2}$ +	β^+, E.C.	7.1 h	
Se	34	74	73.922475	0 +			0.9
Se	34	75	74.922522	$\frac{5}{2}$ +	E.C.	118.5 d	
Se	34	76	75.919212	0 +			9.0
Se	34	77	76.919913	$\frac{1}{2}$ −			7.6
Se	34	78	77.917308	0 +			23.5
Se	34	79	78.918498	$\frac{7}{2}$ +	β^-	6×10^4 yr	
Se	34	80	79.916520	0 +			49.6

Chemical Symbol	Z	A	Atomic Mass (u)	Spin and Parity	Decay Modes*	Half-Life $t_{1/2}$	Natural Abundance (%)
Se	34	81	80.917990	$\frac{1}{2}-$	β^-	18.5 min	
Se	34	82	81.916698	0 +			9.4
Se	34	83	82.919117	$\frac{9}{2}+$	β^-	22.3 min	
Se	34	84	83.918463	0 +	β^-	3.3 min	
Se	34	85	84.922260	$\frac{5}{2}+$	β^-	32 s	
Se	34	86	85.924270		β^-	15 s	
Se	34	87	86.928390		β^-, n	5.6 s	
Br	35	72	71.936630	3 +	β^+	1.31 min	
Br	35	73	72.931680	$\frac{3}{2}-$	β^+	3.4 min	
Br	35	74	73.929898		β^+	25.3 min	
Br	35	75	74.925753	$\frac{3}{2}-$	β^+, E.C.	98 min	
Br	35	76	75.924528	1 −	β^+, E.C.	16.1 h	
Br	35	77	76.921378	$\frac{3}{2}-$	E.C.	57.0 h	
Br	35	78	77.921144	1 +	β^+	6.46 min	
Br	35	79	78.918336	$\frac{3}{2}-$			50.69
Br	35	80	79.918528	1 +	β^-, β^+, E.C.	17.6 min	
Br	35	81	80.916289	$\frac{3}{2}-$			49.31
Br	35	82	81.916802	5 −	β^-	35.3 h	
Br	35	83	82.915179	$\frac{3}{2}-$	β^-	2.39 h	
Br	35	84	83.916503	2 −	β^-	31.8 min	
Br	35	85	84.915612	$\frac{3}{2}-$	β^-	2.87 min	
Br	35	86	85.918800	2 −	β^-	1.76 min	
Br	35	87	86.920690	$\frac{3}{2}-$	β^-, n	56.1 s	
Br	35	88	87.924080	1 −	β^-, n	16.4 s	
Br	35	89	88.926550	$\frac{3}{2}-$	β^-, n	4.4 s	
Br	35	90	89.931010	2 −	β^-, n	1.9 s	
Kr	36	72	71.942060	0 +	β^+, E.C.	17 s	
Kr	36	73	72.938920		β^+, E.C.	27 s	
Kr	36	74	73.933290	0 +	β^+, E.C.	11.5 min	
Kr	36	75	74.931029		β^+, E.C.	4.5 min	
Kr	36	76	75.925595	0 +	E.C.	14.8 h	
Kr	36	77	76.924610	$\frac{5}{2}+$	β^+, E.C.	1.24 h	
Kr	36	78	77.920396	0 +			0.35
Kr	36	79	78.920084	$\frac{1}{2}-$	β^+, E.C.	35.0 h	
Kr	36	80	79.916380	0 +			2.25
Kr	36	81	80.916590	$\frac{7}{2}+$	E.C.	2.1×10^5 yr	
Kr	36	82	81.913482	0 +			11.6
Kr	36	83	82.914135	$\frac{9}{2}+$			11.5
Kr	36	84	83.911507	0 +			57.0
Kr	36	85	84.912531	$\frac{9}{2}+$	β^-	10.72 yr	
Kr	36	86	85.910616	0 +			17.3
Kr	36	87	86.913360	$\frac{5}{2}+$	β^-	76.3 min	
Kr	36	88	87.914553	0 +	β^-	2.84 h	
Kr	36	89	88.917640	$\frac{5}{2}+$	β^-	3.16 min	
Kr	36	90	89.919520	0 +	β^-	32.3 s	
Kr	36	91	90.923380	$\frac{5}{2}+$	β^-	8.6 s	
Kr	36	92	91.926270		β^-, n	1.84 s	
Rb	37	75	74.938510		β^+	17 s	
Rb	37	76	75.934960		β^+	17 s	
Rb	37	77	76.930280	$\frac{3}{2}-$	β^+	3.8 min	

(continued)

Chemical Symbol	Z	A	Atomic Mass (u)	Spin and Parity	Decay Modes*	Half-Life $t_{1/2}$	Natural Abundance (%)
Rb	37	78	77.928090	0 +	β^+, E.C.	17.66 min	
Rb	37	79	78.923954	$\frac{5}{2}$ −	β^+, E.C.	23 min	
Rb	37	80	79.922519	1 +	β^+	34 s	
Rb	37	81	80.918990	$\frac{3}{2}$ −	β^+, E.C.	4.58 h	
Rb	37	82	81.918195	1 +	β^+	1.273 min	
Rb	37	83	82.915144	$\frac{5}{2}$ −	E.C.	86.2 d	
Rb	37	84	83.914390	2 −	β^+, E.C.	32.9 d	
Rb	37	85	84.911794	$\frac{5}{2}$ −			72.17
Rb	37	86	85.911172	2 −	β^-	18.63 d	
Rb	37	87	86.909187	$\frac{3}{2}$ −			27.83
Rb	37	88	87.911326	2 −	β^-	17.7 min	
Rb	37	89	88.912278	$\frac{3}{2}$ −	β^-	15.2 min	
Rb	37	90	89.914811	1 −	β^-	2.6 min	
Rb	37	91	90.916485	$\frac{3}{2}$ −	β^-	58.4 s	
Rb	37	92	91.919661	1 −	β^-	8.12 s	
Rb	37	93	92.921977	$\frac{5}{2}$ −	β^-, n	5.85 s	
Rb	37	94	93.926432		β^-, n	2.73 s	
Rb	37	95	94.929352		β^-, n	0.38 s	
Rb	37	96	95.934370		β^-, n	0.20 s	
Rb	37	97	96.937440		β^-, n	0.17 s	
Rb	37	98	97.941960		β^-, n	0.13 s	
Sr	38	79	78.929860		β^+	2.1 min	
Sr	38	80	79.924650	0 +	β^+	106 min	
Sr	38	81	80.923270	$\frac{1}{2}$ +	β^+, E.C.	22.2 min	
Sr	38	82	81.918414	0 +	E.C.	25.6 d	
Sr	38	83	82.917566	$\frac{7}{2}$ +	β^+, E.C.	32.4 h	
Sr	38	84	83.913430	0 +			0.56
Sr	38	85	84.912937	$\frac{9}{2}$ +	E.C.	64.8 d	
Sr	38	86	85.909267	0 +			9.86
Sr	38	87	86.908884	$\frac{9}{2}$ +			7.00
Sr	38	88	87.905619	0 +			82.58
Sr	38	89	88.907450	$\frac{5}{2}$ +	β^-	50.52 d	
Sr	38	90	89.907738	0 +	β^-	29 yr	
Sr	38	91	90.910187	$\frac{5}{2}$ +	β^-	9.5 h	
Sr	38	92	91.910944	0 +	β^-	2.71 h	
Sr	38	93	92.913987	$\frac{5}{2}$ +	β^-	7.5 min	
Sr	38	94	93.915367	0 +	β^-	75 s	
Sr	38	95	94.919380		β^-	25 s	
Sr	38	96	95.921170	0 +	β^-	1.06 s	
Sr	38	97	96.926140		β^-, n	0.4 s	
Sr	38	98	97.928620		β^-, n	0.65 s	
Y	39	80	79.936250		β^+	36 s	
Y	39	81	80.929200		β^+	72 s	
Y	39	82	81.926810	1 +	β^+	9.5 s	
Y	39	83	82.922300	$\frac{1}{2}$ −	β^+, E.C.	7.1 min	
Y	39	84	83.920310	5 −	β^+, E.C.	40 min	
Y	39	85	84.916437	$\frac{1}{2}$ −	β^+, E.C.	2.6 h	
Y	39	86	85.914893	4 −	β^+, E.C.	14.74 h	
Y	39	87	86.910882	$\frac{1}{2}$ −	E.C.	80.3 h	
Y	39	88	87.909508	4 −	E.C.	106.61 d	
Y	39	89	88.905849	$\frac{1}{2}$ −			100

Chemical Symbol	Z	A	Atomic Mass (u)	Spin and Parity	Decay Modes*	Half-Life $t_{1/2}$	Natural Abundance (%)
Y	39	90	89.907152	2 −	β^-	64.0 h	
Y	39	91	90.907303	$\frac{1}{2}$ −	β^-	58.5 d	
Y	39	92	91.908917	2 −	β^-	3.54 h	
Y	39	93	92.909571	$\frac{1}{2}$ −	β^-	10.2 h	
Y	39	94	93.911597	2 −	β^-	18.7 min	
Y	39	95	94.912814	$\frac{1}{2}$ −	β^-	10.3 min	
Y	39	96	95.915940	0 −	β^-	6.2 s	
Y	39	97	96.918120	$\frac{1}{2}$ −	β^-, n	3.7 s	
Y	39	98	97.922300	1 +	β^-, n	0.65 s	
Y	39	99	98.924720	$\frac{1}{2}$ −	β^-, n	1.5 s	
Zr	40	82	81.931100		β^+	2.5 min	
Zr	40	83	82.928760		β^+	44 s	
Zr	40	84	83.923320	0 +	β^+, E.C.	28 min	
Zr	40	85	84.921470	$\frac{7}{2}$ +	β^+, E.C.	7.9 min	
Zr	40	86	85.916290	0 +	E.C.	16.5 h	
Zr	40	87	86.914817	$\frac{9}{2}$ +	β^+, E.C.	1.73 h	
Zr	40	88	87.910225	0 +	E.C.	83.4 d	
Zr	40	89	88.908890	$\frac{9}{2}$ +	β^+, E.C.	78.4 h	
Zr	40	90	89.904703	0 +			51.45
Zr	40	91	90.905644	$\frac{5}{2}$ +			11.27
Zr	40	92	91.905039	0 +			17.17
Zr	40	93	92.906474	$\frac{5}{2}$ +	β^-	1.5×10^6 yr	
Zr	40	94	93.906314	0 +			17.33
Zr	40	95	94.908042	$\frac{5}{2}$ +	β^-	64.03 d	
Zr	40	96	95.908275	0 +			2.78
Zr	40	97	96.910950	$\frac{1}{2}$ −	β^-	16.8 h	
Zr	40	98	97.912735	0 +	β^-	30.7 s	
Zr	40	99	98.916540	$\frac{1}{2}$ +	β^-	2.1 s	
Zr	40	100	99.917750	0 +	β^-	7.1 s	
Zr	40	101	100.921520		β^-	2.0 s	
Nb	41	86	85.925310		β^+	1.45 min	
Nb	41	87	86.920370		β^+, E.C.	2.6 min	
Nb	41	88	87.917950	8 +	β^+, E.C.	14.3 min	
Nb	41	89	88.913449	$\frac{1}{2}$ −	β^+, E.C.	66 min	
Nb	41	90	89.911263	8 +	β^+, E.C.	14.6 h	
Nb	41	91	90.906991	$\frac{5}{2}$ +	E.C.	700 yr	
Nb	41	92	91.907192	7 +	E.C.	3×10^7 yr	
Nb	41	93	92.906377	$\frac{9}{2}$ +			100
Nb	41	94	93.907280	6 +	β^-	2.4×10^4 yr	
Nb	41	95	94.906835	$\frac{9}{2}$ +	β^-	34.98 d	
Nb	41	96	95.908100	6 +	β^-	23.4 h	
Nb	41	97	96.908096	$\frac{9}{2}$ +	β^-	73.6 min	
Nb	41	98	97.910330	1 +	β^-	2.8 s	
Nb	41	99	98.911619	$\frac{9}{2}$ +	β^-	15.0 s	
Nb	41	100	99.914180		β^-	3.1 s	
Nb	41	101	100.915320		β^-	7.1 s	
Nb	41	102	101.918040		β^-	4.3 s	
Nb	41	103	102.919370		β^-	1.5 s	
Mo	42	88	87.921820	0 +	β^+, E.C.	8.0 min	
Mo	42	89	88.919480	$\frac{9}{2}$ +	β^+, E.C.	2.2 min	

(continued)

Chemical Symbol	Z	A	Atomic Mass (u)	Spin and Parity	Decay Modes*	Half-Life $t_{1/2}$	Natural Abundance (%)
Mo	42	90	89.913933	0 +	β^+, E.C.	5.67 h	
Mo	42	91	90.911755	$\frac{9}{2}$ −	β^+, E.C.	15.5 min	
Mo	42	92	91.906808	0 +			14.84
Mo	42	93	92.906813	$\frac{5}{2}$ +	E.C.	3.5×10^6 yr	
Mo	42	94	93.905085	0 +			9.25
Mo	42	95	94.905840	$\frac{5}{2}$ +			15.92
Mo	42	96	95.904678	0 +			16.68
Mo	42	97	96.906020	$\frac{5}{2}$ +			9.55
Mo	42	98	97.905406	0 +			24.13
Mo	42	99	98.907711	$\frac{1}{2}$ +	β^-	65.94 h	
Mo	42	100	99.907477	0 +			9.63
Mo	42	101	100.910345	$\frac{1}{2}$ +	β^-	14.6 min	
Mo	42	102	101.910297	0 +	β^-	11.2 min	
Mo	42	103	102.913470		β^-	68 s	
Mo	42	104	103.913600	0 +	β^-	60 s	
Mo	42	105	104.917190		β^-	50 s	
Mo	42	106	105.917950	0 +	β^-	8.4 s	
Tc	43	90	89.923810	1 +	β^+	8.3 s	
Tc	43	92	91.915257	8 +	β^+, E.C.	4.4 min	
Tc	43	93	92.910246	$\frac{9}{2}$ +	β^+, E.C.	2.83 h	
Tc	43	94	93.909654	7 +	β^+, E.C.	4.88 h	
Tc	43	95	94.907657	$\frac{9}{2}$ +	E.C.	20.0 h	
Tc	43	96	95.907870	7 +	E.C.	4.3 d	
Tc	43	97	96.906364	$\frac{9}{2}$ +	E.C.	2.6×10^6 yr	
Tc	43	98	97.907215	6 +	β^-	4.2×10^6 yr	
Tc	43	99	98.906254	$\frac{9}{2}$ +	β^-	2.13×10^5 yr	
Tc	43	100	99.907657	1 +	β^-	15.8 s	
Tc	43	101	100.907327	$\frac{9}{2}$ +	β^-	14.2 min	
Tc	43	102	101.909208	1 +	β^-	5.3 s	
Tc	43	103	102.909172		β^-	54 s	
Tc	43	104	103.911460		β^-	18.3 min	
Tc	43	105	104.911820	$\frac{5}{2}$ +	β^-	7.6 min	
Tc	43	106	105.914510		β^-	36 s	
Tc	43	107	106.915230		β^-	21.2 s	
Tc	43	108	107.918420		β^-	5.0 s	
Ru	44	92	91.920120	0 +	β^+, E.C.	3.7 min	
Ru	44	93	92.917050	$\frac{9}{2}$ +	β^+, E.C.	60 s	
Ru	44	94	93.911361	0 +	E.C.	52 min	
Ru	44	95	94.910414	$\frac{5}{2}$ +	β^+, E.C.	1.64 h	
Ru	44	96	95.907599	0 +			5.52
Ru	44	97	96.907556	$\frac{5}{2}$ +	E.C.	2.89 d	
Ru	44	98	97.905287	0 +			1.88
Ru	44	99	98.905939	$\frac{5}{2}$ +			12.7
Ru	44	100	99.904219	0 +			12.6
Ru	44	101	100.905582	$\frac{5}{2}$ +			17.0
Ru	44	102	101.904348	0 +			31.6
Ru	44	103	102.906323	$\frac{5}{2}$ +	β^-	39.24 d	
Ru	44	104	103.905424	0 +			18.7
Ru	44	105	104.907744	$\frac{3}{2}$ +	β^-	4.44 h	
Ru	44	106	105.907321	0 +	β^-	372.6 d	
Ru	44	107	106.910130		β^-	3.8 min	

Chemical Symbol	Z	A	Atomic Mass (u)	Spin and Parity	Decay Modes*	Half-Life $t_{1/2}$	Natural Abundance (%)
Ru	44	108	107.910140		β^-	4.6 min	
Ru	44	109	108.913240		β^-	35 s	
Ru	44	110	109.913760		β^-	15 s	
Rh	45	94	93.921670	8 +	β^+	25.8 s	
Rh	45	95	94.915900	$\frac{9}{2}$ +	β^+	5.0 min	
Rh	45	96	95.914515	5 +	β^+, E.C.	9.9 min	
Rh	45	97	96.911320	$\frac{9}{2}$ +	β^+	31.0 min	
Rh	45	98	97.910716	5 +	β^+	8.6 min	
Rh	45	99	98.908192	$\frac{1}{2}$ −	β^+, E.C.	16.1 d	
Rh	45	100	99.908116	1 −	β^+, E.C.	20.8 h	
Rh	45	101	100.906159	$\frac{1}{2}$ −	E.C.	3.3 yr	
Rh	45	102	101.906814	6 +	E.C.	2.9 yr	
Rh	45	103	102.905500	$\frac{1}{2}$ −			100
Rh	45	104	103.906651	1 +	β^-	41.8 s	
Rh	45	105	104.905686	$\frac{7}{2}$ +	β^-	35.4 h	
Rh	45	106	105.907279	1 +	β^-	29.8 s	
Rh	45	107	106.906751	$\frac{5}{2}$ +	β^-	21.7 min	
Rh	45	108	107.908650	1 +	β^-	6.0 min	
Rh	45	109	108.908734	$\frac{5}{2}$ +	β^-	81 s	
Rh	45	110	109.910960		β^-	29 s	
Rh	45	111	110.911630		β^-	11 s	
Rh	45	112	111.914410		β^-	0.8 s	
Pd	46	96	95.918010		E.C.	2.0 min	
Pd	46	97	96.916480	$\frac{5}{2}$ +	β^+, E.C.	3.1 min	
Pd	46	98	97.912722	0 +	β^+, E.C.	18 min	
Pd	46	99	98.911763	$\frac{5}{2}$ +	β^+, E.C.	21.4 min	
Pd	46	100	99.908527	0 +	E.C.	3.6 d	
Pd	46	101	100.908287	$\frac{5}{2}$ +	β^+, E.C.	8.4 h	
Pd	46	102	101.905634	0 +			1.02
Pd	46	103	102.906114	$\frac{5}{2}$ +	E.C.	16.97 d	
Pd	46	104	103.904029	0 +			11.14
Pd	46	105	104.905079	$\frac{5}{2}$ +			22.33
Pd	46	106	105.903478	0 +			27.33
Pd	46	107	106.905127	$\frac{5}{2}$ +	β^-	6.5×10^6 yr	
Pd	46	108	107.903895	0 +			26.46
Pd	46	109	108.905954	$\frac{5}{2}$ +	β^-	13.4 h	
Pd	46	110	109.905167	0 +			11.72
Pd	46	111	110.907660	$\frac{5}{2}$ +	β^-	22 min	
Pd	46	112	111.907323	0 +	β^-	21.03 h	
Pd	46	113	112.910110	$\frac{5}{2}$ +	β^-	98 s	
Pd	46	114	113.910310	0 +	β^-	2.48 min	
Pd	46	115	114.913590	$\frac{5}{2}$ +	β^-	47 s	
Pd	46	116	115.914000	0 +	β^-	12.7 s	
Ag	47	97	96.923890		β^+, E.C.	19 s	
Ag	47	98	97.921560	5 +	β^+, E.C.	46.7 s	
Ag	47	99	98.917590	$\frac{9}{2}$ +	β^+, E.C.	2.07 min	
Ag	47	100	99.916140	5 +	β^+, E.C.	2.0 min	
Ag	47	101	100.912810	$\frac{9}{2}$ +	β^+, E.C.	11.1 min	
Ag	47	102	101.911950	5 +	β^+, E.C.	13 min	
Ag	47	103	102.908980	$\frac{7}{2}$ +	β^+, E.C.	66 min	
Ag	47	104	103.908628	5 +	β^+, E.C.	69 min	

(continued)

Chemical Symbol	Z	A	Atomic Mass (u)	Spin and Parity	Decay Modes*	Half-Life $t_{1/2}$	Natural Abundance (%)
Ag	47	105	104.906520	$\frac{1}{2}-$	E.C.	41.3 d	
Ag	47	106	105.906662	$1+$	β^+, E.C.	24.0 min	
Ag	47	107	106.905092	$\frac{1}{2}-$			51.84
Ag	47	108	107.905952	$1+$	β^-, β^+, E.C.	2.42 min	
Ag	47	109	108.904757	$\frac{1}{2}-$			48.16
Ag	47	110	109.906111	$1+$	β^-	24.6 s	
Ag	47	111	110.905295	$\frac{1}{2}-$	β^-	7.47 d	
Ag	47	112	111.907010	$2-$	β^-	3.14 h	
Ag	47	113	112.906558	$\frac{1}{2}-$	β^-	5.3 h	
Ag	47	114	113.908760	$1+$	β^-	4.5 s	
Ag	47	115	114.908800	$\frac{1}{2}-$	β^-	20.0 min	
Ag	47	116	115.911200	$2-$	β^-	2.68 min	
Ag	47	117	116.911700	$\frac{1}{2}-$	β^-	73 s	
Ag	47	118	117.914570		β^-	4.0 s	
Ag	47	119	118.915630	$\frac{1}{2}-$	β^-	2.1 s	
Ag	47	120	119.918650		β^-	1.2 s	
Ag	47	121	120.919970		β^-	0.8 s	
Cd	48	99	98.924860		β^+, E.C.	16 s	
Cd	48	100	99.920230		β^+, E.C.	1.1 min	
Cd	48	101	100.918740	$\frac{5}{2}+$	β^+, E.C.	1.2 min	
Cd	48	102	101.914440	$0+$	β^+, E.C.	5.5 min	
Cd	48	103	102.913451	$\frac{5}{2}+$	β^+, E.C.	7.7 min	
Cd	48	104	103.909851	$0+$	E.C.	58 min	
Cd	48	105	104.909459	$\frac{5}{2}+$	β^+, E.C.	55.3 min	
Cd	48	106	105.906461	$0+$			1.25
Cd	48	107	106.906613	$\frac{5}{2}+$	β^+, E.C.	6.5 h	
Cd	48	108	107.904176	$0+$			0.89
Cd	48	109	108.904953	$\frac{5}{2}+$	E.C.	462.3 d	
Cd	48	110	109.903005	$0+$			12.49
Cd	48	111	110.904182	$\frac{1}{2}+$			12.80
Cd	48	112	111.902758	$0+$			24.13
Cd	48	113	112.904400	$\frac{1}{2}+$			12.22
Cd	48	114	113.903357	$0+$			28.73
Cd	48	115	114.905430	$\frac{1}{2}+$	β^-	53.5 h	
Cd	48	116	115.904754	$0+$			7.49
Cd	48	117	116.907228	$\frac{1}{2}+$	β^-	2.49 h	
Cd	48	118	117.906914	$0+$	β^-	50.3 min	
Cd	48	119	118.909890	$\frac{1}{2}+$	β^-	2.69 min	
Cd	48	120	119.909852	$0+$	β^-	50.8 s	
Cd	48	121	120.913100	$\frac{3}{2}+$	β^-	13.5 s	
Cd	48	122	121.913500	$0+$	β^-	5.8 s	
In	49	102	101.924400		E.C.	24 s	
In	49	103	102.920110	$\frac{9}{2}+$	β^+, E.C.	1.1 min	
In	49	104	103.918440	$6+$	β^+, E.C.	1.82 min	
In	49	105	104.914558	$\frac{9}{2}+$	β^+, E.C.	4.9 min	
In	49	106	105.913490	$6+$	β^+, E.C.	5.3 min	
In	49	107	106.910284	$\frac{9}{2}+$	β^+, E.C.	32.5 min	
In	49	108	107.909678	$6+$	β^+, E.C.	57 min	
In	49	109	108.907133	$\frac{9}{2}+$	β^+, E.C.	4.2 h	
In	49	110	109.907230	$2+$	β^+, E.C.	69 min	
In	49	111	110.905109	$\frac{9}{2}+$	E.C.	2.806 d	

Chemical Symbol	Z	A	Atomic Mass (u)	Spin and Parity	Decay Modes*	Half-Life $t_{1/2}$	Natural Abundance (%)
In	49	112	111.905536	1 +	β^+, E.C.	14.4 min	
In	49	113	112.904061	$\frac{9}{2}$ +			4.3
In	49	114	113.904916	1 +	β^-, E.C.	71.9 s	
In	49	115	114.903880	$\frac{9}{2}$ +	β^-	4.4×10^{14} yr	95.7
In	49	116	115.905264	1 +	β^-	14.1 s	
In	49	117	116.904517	$\frac{9}{2}$ +	β^-	43.1 min	
In	49	118	117.906120	1 +	β^-	5.0 s	
In	49	119	118.905819	$\frac{9}{2}$ +	β^-	2.4 min	
In	49	120	119.907890	5 +	β^-	44 s	
In	49	121	120.907847	$\frac{9}{2}$ +	β^-	23 s	
In	49	122	121.910280		β^-	10.1 s	
In	49	123	122.910450	$\frac{9}{2}$ +	β^-	6.0 s	
In	49	124	123.912980	3 +	β^-	3.2 s	
In	49	125	124.913670	$\frac{9}{2}$ +	β^-	2.33 s	
In	49	126	125.916470	3 +	β^-	1.53 s	
In	49	127	126.917320	$\frac{9}{2}$ +	β^-	1.12 s	
In	49	128	127.920560	3 +	β^-	0.9 s	
In	49	129	128.921600		β^-, n	0.59 s	
In	49	130	129.924870		β^-, n	0.51 s	
In	49	131	130.926410	$\frac{9}{2}$ +	β^-, n	0.28 s	
Sn	50	106	105.917030		β^+, E.C.	2.1 min	
Sn	50	107	106.915870		β^+, E.C.	2.9 min	
Sn	50	108	107.911880	0 +	β^+, E.C.	10.3 min	
Sn	50	109	108.911294	$\frac{7}{2}$ +	β^+, E.C.	18.0 min	
Sn	50	110	109.907858	0 +	E.C.	4.0 h	
Sn	50	111	110.907741	$\frac{7}{2}$ +	β^+, E.C.	35.3 min	
Sn	50	112	111.904826	0 +			1.01
Sn	50	113	112.905176	$\frac{1}{2}$ +	E.C.	115.1 d	
Sn	50	114	113.902784	0 +			0.67
Sn	50	115	114.903348	$\frac{1}{2}$ +			0.38
Sn	50	116	115.901747	0 +			14.6
Sn	50	117	116.902956	$\frac{1}{2}$ +			7.75
Sn	50	118	117.901609	0 +			24.3
Sn	50	119	118.903310	$\frac{1}{2}$ +			8.6
Sn	50	120	119.902200	0 +			32.4
Sn	50	121	120.904238	$\frac{3}{2}$ +	β^-	27.0 h	
Sn	50	122	121.903440	0 +			4.56
Sn	50	123	122.905722	$\frac{11}{2}$ −	β^-	129.2 d	
Sn	50	124	123.905274	0 +			5.64
Sn	50	125	124.907785	$\frac{11}{2}$ −	β^-	9.62 d	
Sn	50	126	125.907654	0 +	β^-	1×10^5 yr	
Sn	50	127	126.910355	$\frac{11}{2}$ −	β^-	2.1 h	
Sn	50	128	127.910560	0 +	β^-	59.1 min	
Sn	50	129	128.913440	$\frac{3}{2}$ +	β^-	2.5 min	
Sn	50	130	129.913920	0 +	β^-	3.7 min	
Sn	50	131	130.916940		β^-	61 s	
Sn	50	132	131.917760		β^-	40 s	
Sb	51	109	108.918143		β^+, E.C.	18.3 s	
Sb	51	110	109.916770	3 +	β^+, E.C.	23.5 s	
Sb	51	111	110.913220	$\frac{5}{2}$ +	β^+, E.C.	75 s	
Sb	51	112	111.912411	3 +	β^+, E.C.	51 s	

(continued)

Chemical Symbol	Z	A	Atomic Mass (u)	Spin and Parity	Decay Modes*	Half-Life $t_{1/2}$	Natural Abundance (%)
Sb	51	113	112.909372	$\frac{5}{2}+$	β^+, E.C.	6.7 min	
Sb	51	114	113.909110	$3+$	β^+, E.C.	3.5 min	
Sb	51	115	114.906601	$\frac{5}{2}+$	β^+, E.C.	32.1 min	
Sb	51	116	115.906800	$3+$	β^+, E.C.	16 min	
Sb	51	117	116.904841	$\frac{5}{2}+$	β^+, E.C.	2.80 h	
Sb	51	118	117.905534	$1+$	β^+, E.C.	3.6 min	
Sb	51	119	118.903948	$\frac{5}{2}+$	E.C.	36.1 h	
Sb	51	120	119.905077	$1+$	β^+, E.C.	15.9 min	
Sb	51	121	120.903821	$\frac{5}{2}+$			57.3
Sb	51	122	121.905179	$2-$	β^-, β^+	2.71 d	
Sb	51	123	122.904216	$\frac{7}{2}+$			42.7
Sb	51	124	123.905038	$3-$	β^-	60.20 d	
Sb	51	125	124.905252	$\frac{1}{2}+$	β^-	2.76 yr	
Sb	51	126	125.907250	$8-$	β^-	12.4 d	
Sb	51	127	126.906919	$\frac{7}{2}+$	β^-	3.84 d	
Sb	51	128	127.909180	$8-$	β^-	9.1 h	
Sb	51	129	128.909146		β^-	4.4 h	
Sb	51	130	129.911590	$8-$	β^-	38.4 min	
Sb	51	131	130.911950	$\frac{7}{2}+$	β^-	23.0 min	
Sb	51	132	131.914410	$8-$	β^-	4.12 min	
Sb	51	133	132.915150	$\frac{7}{2}+$	β^-	2.5 min	
Sb	51	134	133.920550		β^-, n	10.4 s	
Sb	51	135	134.924520	$\frac{7}{2}+$	β^-, n	1.71 s	
Te	52	108	107.929550	$0+$	α, β^+, E.C.	2.1 s	
Te	52	109	108.927370		α, β^+, E.C.	4.2 s	
Te	52	110	109.922560	$0+$	β^+, E.C.	18.5 s	
Te	52	111	110.921130	$\frac{7}{2}+$	β^+, E.C.	19.3 s	
Te	52	112	111.917020	$0+$	β^+, E.C.	2.0 min	
Te	52	113	112.915920	$\frac{7}{2}+$	β^+, E.C.	1.7 s	
Te	52	114	113.912230	$0+$	β^+, E.C.	15 min	
Te	52	115	114.911700	$\frac{7}{2}+$	β^+, E.C.	5.8 min	
Te	52	116	115.908450	$0+$	E.C.	2.5 h	
Te	52	117	116.908630	$\frac{1}{2}+$	β^+, E.C.	62 min	
Te	52	118	117.905908	$0+$	E.C.	6.00 d	
Te	52	119	118.906411	$\frac{1}{2}+$	β^+, E.C.	16.05 h	
Te	52	120	119.904048	$0+$			0.096
Te	52	121	120.904947	$\frac{1}{2}+$	E.C.	16.8 d	
Te	52	122	121.903054	$0+$			2.60
Te	52	123	122.904271	$\frac{1}{2}+$			0.903
Te	52	124	123.902823	$0+$			4.816
Te	52	125	124.904433	$\frac{1}{2}+$			7.14
Te	52	126	125.903314	$0+$			18.95
Te	52	127	126.905227	$\frac{3}{2}+$	β^-	9.5 h	
Te	52	128	127.904463	$0+$			31.69
Te	52	129	128.906594	$\frac{3}{2}-$	β^-	69.5 min	
Te	52	130	129.906229	$0+$			33.80
Te	52	131	130.908528	$\frac{3}{2}+$	β^-	25.0 min	
Te	52	132	131.908517	$0+$	β^-	78.2 h	
Te	52	133	132.910910	$\frac{3}{2}+$	β^-	12.5 min	
Te	52	134	133.911520	$0+$	β^-	42 min	
Te	52	135	134.916420		β^-	19.2 s	

Chemical Symbol	Z	A	Atomic Mass (u)	Spin and Parity	Decay Modes*	Half-Life $t_{1/2}$	Natural Abundance (%)
Te	52	136	135.920120	0 +	β^-, n	18 s	
Te	52	137	136.925410		β^-, n	4 s	
I	53	110	109.935060		α, β^+, E.C.	0.65 s	
I	53	111	110.930250		β^+, E.C.	7.5 s	
I	53	112	111.927940		β^+, E.C.	3.4 s	
I	53	113	112.923650		β^+, E.C.	5.9 s	
I	53	114	113.921790		β^+, E.C.	2.1 s	
I	53	115	114.918090		β^+, E.C.	28 s	
I	53	116	115.916780	1 +	β^+, E.C.	2.9 s	
I	53	117	116.913460	$\frac{5}{2}$ +	β^+, E.C.	2.3 min	
I	53	118	117.912780		β^+, E.C.	14.3 min	
I	53	119	118.910030	$\frac{5}{2}$ +	β^+, E.C.	19.2 min	
I	53	120	119.909840	2 −	β^+, E.C.	1.35 h	
I	53	121	120.907394	$\frac{5}{2}$ +	β^+, E.C.	2.12 h	
I	53	122	121.907595	1 +	β^+, E.C.	3.6 min	
I	53	123	122.905594	$\frac{5}{2}$ +	E.C.	13.1 h	
I	53	124	123.906207	2 −	β^+, E.C.	4.17 d	
I	53	125	124.904620	$\frac{5}{2}$ +	E.C.	59.9 d	
I	53	126	125.905624	2 −	β^+, β^-, E.C.	13.0 d	
I	53	127	126.904473	$\frac{5}{2}$ +			100
I	53	128	127.905810	1 +	β^-, E.C.	25.00 min	
I	53	129	128.904986	$\frac{7}{2}$ +	β^-	1.6×10^7 yr	
I	53	130	129.906713	5 +	β^-	12.36 h	
I	53	131	130.906114	$\frac{7}{2}$ +	β^-	8.040 d	
I	53	132	131.907987	4 +	β^-	2.30 h	
I	53	133	132.907780	$\frac{7}{2}$ +	β^-	20.8 h	
I	53	134	133.909850	4 +	β^-	52.5 min	
I	53	135	134.910023	$\frac{7}{2}$ +	β^-	6.585 h	
I	53	136	135.914650	2 −	β^-	83.6 s	
I	53	137	136.917870	$\frac{7}{2}$ +	β^-, n	24.5 s	
I	53	138	137.922370		β^-, n	6.4 s	
I	53	139	138.926050		β^-, n	2.3 s	
Xe	54	114	113.928170	0 +	β^+, E.C.	10.3 s	
Xe	54	115	114.926280		β^+, E.C.	18 s	
Xe	54	116	115.921610	0 +	β^+, E.C.	57 s	
Xe	54	117	116.920250		β^+, E.C.	61 s	
Xe	54	118	117.916210	0 +	β^+, E.C.	4 min	
Xe	54	119	118.915390		β^+, E.C.	5.8 min	
Xe	54	120	119.911940	0 +	β^+, E.C.	40 min	
Xe	54	121	120.911450	$\frac{5}{2}$ +	β^+, E.C.	39 min	
Xe	54	122	121.908170	0 +	E.C.	20.0 h	
Xe	54	123	122.908469	$\frac{1}{2}$ +	β^+, E.C.	2.0 h	
Xe	54	124	123.905894	0 +			0.10
Xe	54	125	124.906397	$\frac{1}{2}$ +	E.C.	17.1 h	
Xe	54	126	125.904281	0 +			0.09
Xe	54	127	126.905182	$\frac{1}{2}$ +	E.C.	36.341 d	
Xe	54	128	127.903531	0 +			1.91
Xe	54	129	128.904780	$\frac{1}{2}$ +			26.4
Xe	54	130	129.903509	0 +			4.1
Xe	54	131	130.905072	$\frac{3}{2}$ +			21.2
Xe	54	132	131.904144	0 +			26.9

(continued)

Chemical Symbol	Z	A	Atomic Mass (u)	Spin and Parity	Decay Modes*	Half-Life $t_{1/2}$	Natural Abundance (%)
Xe	54	133	132.905888	$\frac{3}{2}+$	β^-	5.25 d	
Xe	54	134	133.905395	$0+$			10.4
Xe	54	135	134.907130	$\frac{3}{2}+$	β^-	9.10 h	
Xe	54	136	135.907214	$0+$			8.9
Xe	54	137	136.911557	$\frac{7}{2}-$	β^-	3.84 min	
Xe	54	138	137.913980	$0+$	β^-	14.1 min	
Xe	54	139	138.918740	$\frac{7}{2}-$	β^-	40.4 s	
Xe	54	140	139.921620	$0+$	β^-	13.6 s	
Xe	54	141	140.926610		β^-, n	1.72 s	
Xe	54	142	141.929630	$0+$	β^-, n	1.2 s	
Cs	55	114	113.941270		β^+, E.C.	0.57 s	
Cs	55	115	114.936070		β^+, E.C.	1.4 s	
Cs	55	116	115.933110	$5+$	β^+, E.C.	3.8 s	
Cs	55	117	116.928900		β^+, E.C.	6.7 s	
Cs	55	118	117.926740		β^+, E.C.	15 s	
Cs	55	119	118.922490	$\frac{9}{2}+$	β^+, E.C.	38 s	
Cs	55	120	119.920800		β^+, E.C.	64 s	
Cs	55	121	120.917250	$\frac{3}{2}+$	β^+, E.C.	136 s	
Cs	55	122	121.916090	$1+$	β^+, E.C.	21 s	
Cs	55	123	122.912990	$\frac{1}{2}+$	β^+, E.C.	5.9 min	
Cs	55	124	123.912270	$1+$	β^+, E.C.	30.8 s	
Cs	55	125	124.909725	$\frac{1}{2}+$	β^+, E.C.	45 min	
Cs	55	126	125.909465	$1+$	β^+, E.C.	64 min	
Cs	55	127	126.907428	$\frac{1}{2}+$	β^+, E.C.	6.2 h	
Cs	55	128	127.907755	$1+$	β^+, E.C.	3.62 min	
Cs	55	129	128.906027	$\frac{1}{2}+$	E.C.	32.3 h	
Cs	55	130	129.906753	$1+$	β^+, β^-, E.C.	29.2 min	
Cs	55	131	130.905444	$\frac{5}{2}+$	E.C.	9.69 d	
Cs	55	132	131.906431	$3+$	β^+, β^-, E.C.	6.47 d	
Cs	55	133	132.905429	$\frac{7}{2}+$			100
Cs	55	134	133.906696	$4+$	β^-	2.065 yr	
Cs	55	135	134.905885	$\frac{7}{2}+$	β^-	3×10^6 yr	
Cs	55	136	135.907289	$5+$	β^-	13.1 d	
Cs	55	137	136.907073	$\frac{7}{2}+$	β^-	30.17 yr	
Cs	55	138	137.911004	$3-$	β^-	32.2 min	
Cs	55	139	138.913349	$\frac{7}{2}+$	β^-	4.2 min	
Cs	55	140	139.917256	$1-$	β^-	63.7 s	
Cs	55	141	140.920006	$\frac{7}{2}+$	β^-, n	24.9 s	
Cs	55	142	141.924220		β^-, n	1.8 s	
Cs	55	143	142.927220	$\frac{3}{2}+$	β^-, n	1.78 s	
Cs	55	144	143.931930	$1-$	β^-, n	1.01 s	
Cs	55	145	144.935320	$\frac{3}{2}+$	β^-, n	0.58 s	
Ba	56	120	119.926490	$0+$	β^+, E.C.	32 s	
Ba	56	121	120.924700		β^+, E.C.	30 s	
Ba	56	122	121.920170	$0+$	β^+, E.C.	2.0 min	
Ba	56	123	122.919210		β^+, E.C.	2.7 min	
Ba	56	124	123.915380		β^+, E.C.	11.4 min	
Ba	56	125	124.914640		β^+, E.C.	3.5 min	
Ba	56	126	125.911260	$0+$	β^+, E.C.	99 min	
Ba	56	127	126.911130	$\frac{1}{2}+$	β^+, E.C.	12 min	
Ba	56	128	127.908237	$0+$	E.C.	2.43 d	

Chemical Symbol	Z	A	Atomic Mass (u)	Spin and Parity	Decay Modes*	Half-Life $t_{1/2}$	Natural Abundance (%)
Ba	56	129	128.908642	$\frac{1}{2}+$	β^+, E.C.	2.5 h	
Ba	56	130	129.906282	$0+$			0.106
Ba	56	131	130.906902	$\frac{1}{2}+$	E.C.	11.8 d	
Ba	56	132	131.905042	$0+$			0.101
Ba	56	133	132.905988	$\frac{1}{2}+$	E.C.	10.53 yr	
Ba	56	134	133.904486	$0+$			2.417
Ba	56	135	134.905665	$\frac{3}{2}+$			6.592
Ba	56	136	135.904553	$0+$			7.854
Ba	56	137	136.905812	$\frac{3}{2}+$			11.23
Ba	56	138	137.905232	$0+$			71.70
Ba	56	139	138.908826	$\frac{7}{2}-$	β^-	82.9 min	
Ba	56	140	139.910581	$0+$	β^-	12.76 d	
Ba	56	141	140.914363	$\frac{3}{2}-$	β^-	18.3 min	
Ba	56	142	141.916360	$0+$	β^-	10.7 min	
Ba	56	143	142.920480	$\frac{3}{2}-$	β^-	15 s	
Ba	56	144	143.922840	$0+$	β^-	11.5 s	
Ba	56	145	144.926960	$\frac{5}{2}-$	β^-	4.0 s	
Ba	56	146	145.930120	$0+$	β^-	2.2 s	
Ba	56	147	146.934230		β^-, n	0.70 s	
Ba	56	148	147.937190		β^-, n	0.47 s	
La	57	127	126.916280		β^+, E.C.	3.8 min	
La	57	128	127.915320	$5-$	β^+, E.C.	4.6 min	
La	57	129	128.912640	$\frac{3}{2}+$	β^+, E.C.	11.60 min	
La	57	130	129.912400	$3-$	β^+, E.C.	8.7 min	
La	57	131	130.910080	$\frac{3}{2}+$	β^+, E.C.	59 min	
La	57	132	131.910100	$2-$	β^+, E.C.	4.8 h	
La	57	133	132.908140	$\frac{5}{2}+$	β^+, E.C.	3.91 h	
La	57	134	133.908460	$1+$	β^+, E.C.	6.5 min	
La	57	135	134.906953	$\frac{5}{2}+$	E.C.	19.5 h	
La	57	136	135.907630	$1+$	β^+, E.C.	9.87 min	
La	57	137	136.906460	$\frac{7}{2}+$	E.C.	6×10^4 yr	
La	57	138	137.907105	$5+$			0.09
La	57	139	138.906346	$\frac{7}{2}+$			99.91
La	57	140	139.909471	$3-$	β^-	40.28 h	
La	57	141	140.910896	$\frac{7}{2}+$	β^-	3.93 h	
La	57	142	141.914090	$2-$	β^-	92 min	
La	57	143	142.915920		β^-	14.1 min	
La	57	144	143.919650		β^-	40 s	
La	57	145	144.921650		β^-	25 s	
La	57	146	145.925530		β^-	10 s	
La	57	147	146.928100		β^-	4.1 s	
La	57	148	147.931390		β^-	2.6 s	
Ce	58	131	130.914270		β^+, E.C.	9.5 min	
Ce	58	132	131.911490	$0+$	E.C.	3.5 h	
Ce	58	133	132.911360	$\frac{9}{2}-$	β^+, E.C.	5.4 h	
Ce	58	134	133.908890	$0+$	E.C.	76 h	
Ce	58	135	134.909117	$\frac{1}{2}+$	β^+, E.C.	17.8 h	
Ce	58	136	135.907140	$0+$			0.19
Ce	58	137	136.907780	$\frac{3}{2}+$	β^+, E.C.	9.0 h	
Ce	58	138	137.905985	$0+$			0.25
Ce	58	139	138.906631	$\frac{3}{2}+$	E.C.	137.2 d	

(continued)

Chemical Symbol	Z	A	Atomic Mass (u)	Spin and Parity	Decay Modes*	Half-Life $t_{1/2}$	Natural Abundance (%)
Ce	58	140	139.905433	0 +			88.48
Ce	58	141	140.908271	$\frac{7}{2}$ −	β^-	32.5 d	
Ce	58	142	141.909241	0 +			11.08
Ce	58	143	142.912383	$\frac{3}{2}$ −	β^-	33.0 h	
Ce	58	144	143.913643	0 +	β^-	284.4 d	
Ce	58	145	144.917230	$\frac{5}{2}$ −	β^-	2.9 min	
Ce	58	146	145.918670	0 +	β^-	13.6 min	
Ce	58	147	146.922530		β^-	56 s	
Ce	58	148	147.924410	0 +	β^-	48 s	
Ce	58	149	148.927760		β^-	5.2 s	
Ce	58	150	149.929670		β^-	4.4 s	
Ce	58	151	150.933170		β^-	1.0 s	
Pr	59	132	131.919120		β^+, E.C.	1.6 min	
Pr	59	133	132.916190	$\frac{5}{2}$ +	β^+, E.C.	6.7 min	
Pr	59	134	133.915440	2 +	β^+, E.C.	17 min	
Pr	59	135	134.913140	$\frac{3}{2}$ +	β^+, E.C.	25 min	
Pr	59	136	135.912640	2 +	β^+, E.C.	13.1 min	
Pr	59	137	136.910680	$\frac{5}{2}$ +	β^+, E.C.	77 min	
Pr	59	138	137.910748	1 +	β^+, E.C.	1.5 min	
Pr	59	139	138.908917	$\frac{5}{2}$ +	β^+, E.C.	4.41 h	
Pr	59	140	139.909071	1 +	β^+, E.C.	3.39 min	
Pr	59	141	140.907647	$\frac{5}{2}$ +			100
Pr	59	142	141.910039	2 −	β^-	19.13 h	
Pr	59	143	142.910814	$\frac{7}{2}$ +	β^-	13.58 d	
Pr	59	144	143.913301	0 −	β^-	17.3 min	
Pr	59	145	144.914501	$\frac{7}{2}$ +	β^-	5.98 h	
Pr	59	146	145.917570		β^-	24.1 min	
Pr	59	147	146.918980	$\frac{5}{2}$ +	β^-	13.4 min	
Pr	59	148	147.922210		β^-	2.28 min	
Pr	59	149	148.923792	$\frac{5}{2}$ +	β^-	2.3 min	
Pr	59	150	149.926360	1 +	β^-	6.2 s	
Pr	59	151	150.927910		β^-	4 s	
Pr	59	152	151.930700		β^-	3.2 s	
Nd	60	134	133.918870	0 +	β^+, E.C.	8.5 min	
Nd	60	135	134.918190	$\frac{9}{2}$ −	β^+, E.C.	12 min	
Nd	60	136	135.915010	0 +	β^+, E.C.	50.7 min	
Nd	60	137	136.914760	$\frac{1}{2}$ +	β^+, E.C.	38 min	
Nd	60	138	137.911820	0 +	E.C.	5.1 h	
Nd	60	139	138.911920	$\frac{3}{2}$ +	β^+, E.C.	30 min	
Nd	60	140	139.909306	0 +	E.C.	3.37 d	
Nd	60	141	140.909594	$\frac{3}{2}$ +	β^+, E.C.	2.5 h	
Nd	60	142	141.907719	0 +			27.13
Nd	60	143	142.909810	$\frac{7}{2}$ −			12.18
Nd	60	144	143.910083	0 +	β^-	2.1×10^{15} yr	23.80
Nd	60	145	144.912570	$\frac{7}{2}$ +			8.30
Nd	60	146	145.913113	0 +			17.19
Nd	60	147	146.916097	$\frac{5}{2}$ −	β^-	10.99 d	
Nd	60	148	147.916889	0 +			5.76
Nd	60	149	148.920145	$\frac{5}{2}$ −	β^-	1.73 h	
Nd	60	150	149.920887	0 +			5.64
Nd	60	151	150.923825	$\frac{3}{2}$ +	β^-	12.4 min	

Chemical Symbol	Z	A	Atomic Mass (u)	Spin and Parity	Decay Modes*	Half-Life $t_{1/2}$	Natural Abundance (%)
Nd	60	152	151.924680	0 +	β^-	11.4 min	
Nd	60	153	152.929400		β^-	40 s	
Pm	61	136	135.923980		β^+, E.C.	1.8 min	
Pm	61	137	136.920450	$\frac{11}{2}$ −	β^+, E.C.	2.4 min	
Pm	61	138	137.919340	3 +	β^+, E.C.	3.2 min	
Pm	61	139	138.916780	$\frac{5}{2}$ +	β^+, E.C.	4.1 min	
Pm	61	140	139.915820	1 +	β^+, E.C.	9.2 s	
Pm	61	141	140.913600	$\frac{5}{2}$ +	β^+, E.C.	20.9 min	
Pm	61	142	141.912970	1 +	β^+, E.C.	40.5 s	
Pm	61	143	142.910930	$\frac{5}{2}$ +	E.C.	265 d	
Pm	61	144	143.912588	5 −	E.C.	363 d	
Pm	61	145	144.912743	$\frac{5}{2}$ +	E.C.	17.7 yr	
Pm	61	146	145.914708	3 −	E.C., β^-	5.53 yr	
Pm	61	147	146.915135	$\frac{7}{2}$ +	β^-	2.6234 yr	
Pm	61	148	147.917473	1 −	β^-	5.37 d	
Pm	61	149	148.918332	$\frac{7}{2}$ +	β^-	53.1 h	
Pm	61	150	149.920981	1 −	β^-	2.69 h	
Pm	61	151	150.921203	$\frac{5}{2}$ +	β^-	28.4 h	
Pm	61	152	151.923490	1 +	β^-	4.1 min	
Pm	61	153	152.924134	$\frac{5}{2}$ −	β^-	5.4 min	
Pm	61	154	153.926500		β^-	1.7 min	
Pm	61	155	154.927960		β^-	48 s	
Sm	62	138	137.923420	0 +	β^+, E.C.	3.0 min	
Sm	62	139	138.922600	$\frac{1}{2}$ +	β^+, E.C.	2.6 min	
Sm	62	140	139.919040	0 +	β^+, E.C.	14.8 min	
Sm	62	141	140.918473	$\frac{1}{2}$ +	β^+, E.C.	10.2 min	
Sm	62	142	141.915206	0 +	β^+, E.C.	72.5 min	
Sm	62	143	142.914626	$\frac{3}{2}$ +	β^+, E.C.	8.83 min	
Sm	62	144	143.911998	0 +			3.1
Sm	62	145	144.913409	$\frac{7}{2}$ −	E.C.	340 d	
Sm	62	146	145.913053	0 +	α	1.03×10^8 yr	
Sm	62	147	146.914895	$\frac{7}{2}$ −	α	1.08×10^{11} yr	15.0
Sm	62	148	147.914820	0 +	α	7×10^{15} yr	11.3
Sm	62	149	148.917181	$\frac{7}{2}$ −	α	1×10^{16} yr	13.8
Sm	62	150	149.917273	0 +			7.4
Sm	62	151	150.919929	$\frac{5}{2}$ −	β^-	90 yr	
Sm	62	152	151.919729	0 +			26.7
Sm	62	153	152.922094	$\frac{5}{2}$ +	β^-	46.7 h	
Sm	62	154	153.922206	0 +			22.7
Sm	62	155	154.924636	$\frac{3}{2}$ −	β^-	22.2 min	
Sm	62	156	155.925518	0 +	β^-	9.4 h	
Sm	62	157	156.928210	$\frac{3}{2}$ −	β^-	8.1 min	
Sm	62	158	157.930000	0 +	β^-	5.5 min	
Eu	63	141	140.924870	$\frac{5}{2}$ +	β^+, E.C.	40 s	
Eu	63	142	141.923150	1 +	β^+, E.C.	1.22 min	
Eu	63	143	142.920150	$\frac{5}{2}$ +	β^+, E.C.	2.62 min	
Eu	63	144	143.918792	1 +	β^+, E.C.	10.2 s	
Eu	63	145	144.916267	$\frac{5}{2}$ +	β^+, E.C.	5.93 d	
Eu	63	146	145.917215	4 −	β^+, E.C.	4.58 d	
Eu	63	147	146.916742	$\frac{5}{2}$ +	β^+, E.C.	24.3 d	
Eu	63	148	147.918125	5 −	E.C.	54.5 d	

(continued)

Chemical Symbol	Z	A	Atomic Mass (u)	Spin and Parity	Decay Modes*	Half-Life $t_{1/2}$	Natural Abundance (%)
Eu	63	149	148.917926	$\frac{5}{2}+$	E.C.	93.1 d	
Eu	63	150	149.919702	$0-$	β^+, β^-, E.C.	12.6 h	
Eu	63	151	150.919847	$\frac{5}{2}+$			47.8
Eu	63	152	151.921742	$3-$	β^-, E.C.	13.4 yr	
Eu	63	153	152.921225	$\frac{5}{2}+$			52.2
Eu	63	154	153.922975	$3-$	β^-, E.C.	8.5 yr	
Eu	63	155	154.922889	$\frac{5}{2}+$	β^-	4.73 yr	
Eu	63	156	155.924752	$1+$	β^-	15.2 d	
Eu	63	157	156.925418	$\frac{5}{2}+$	β^-	15.15 h	
Eu	63	158	157.927800	$1-$	β^-	3.5 h	
Eu	63	159	158.929084	$\frac{5}{2}+$	β^-	18 min	
Eu	63	160	159.931880	$0-$	β^-	53 s	
Gd	64	143	142.926490	$\frac{1}{2}+$	β^+, E.C.	39 s	
Gd	64	144	143.922760	$0+$	β^+, E.C.	4.5 min	
Gd	64	145	144.921690	$\frac{1}{2}+$	β^+, E.C.	23 min	
Gd	64	146	145.918304	$0+$	β^+, E.C.	48.3 d	
Gd	64	147	146.918943	$\frac{7}{2}-$	E.C.	38.1 h	
Gd	64	148	147.918113	$0+$	α	75 yr	
Gd	64	149	148.919344	$\frac{7}{2}-$	E.C.	9.3 d	
Gd	64	150	149.918662	$0+$	α	1.8×10^6 yr	
Gd	64	151	150.920346	$\frac{7}{2}-$	E.C.	120 d	
Gd	64	152	151.919786	$0+$			0.20
Gd	64	153	152.921745	$\frac{3}{2}-$	E.C.	241.6 d	
Gd	64	154	153.920861	$0+$			2.18
Gd	64	155	154.922618	$\frac{3}{2}-$			14.80
Gd	64	156	155.922118	$0+$			20.47
Gd	64	157	156.923956	$\frac{3}{2}-$			15.65
Gd	64	158	157.924099	$0+$			24.84
Gd	64	159	158.926384	$\frac{3}{2}-$	β^-	18.6 h	
Gd	64	160	159.927049	$0+$			21.86
Gd	64	161	160.929664	$\frac{5}{2}-$	β^-	3.7 min	
Gd	64	162	161.931010	$0+$	β^-	8.4 min	
Tb	65	145	144.928940		β^+, E.C.	30 s	
Tb	65	146	145.927150	$4-$	β^+, E.C.	23 s	
Tb	65	147	146.923820	$\frac{5}{2}+$	β^+, E.C.	1.6 h	
Tb	65	148	147.924140	$2-$	β^+, E.C.	60 min	
Tb	65	149	148.923248	$\frac{1}{2}+$	α, β^+, E.C.	4.15 h	
Tb	65	150	149.923669	$2-$	β^+, E.C.	3.3 h	
Tb	65	151	150.923100	$\frac{1}{2}+$	β^+, E.C.	17.6 h	
Tb	65	152	151.923919	$2-$	β^+, E.C.	17.6 h	
Tb	65	153	152.923440	$\frac{5}{2}+$	E.C.	2.34 d	
Tb	65	154	153.924690	$0-$	β^+, E.C.	22 h	
Tb	65	155	154.923499	$\frac{3}{2}+$	E.C.	5.3 d	
Tb	65	156	155.924742	$3-$	E.C.	5.3 d	
Tb	65	157	156.924023	$\frac{3}{2}+$	E.C.	150 yr	
Tb	65	158	157.925411	$3-$	β^-, E.C.	150 yr	
Tb	65	159	158.925342	$\frac{3}{2}+$			100
Tb	65	160	159.927163	$3-$	β^-	72.4 d	
Tb	65	161	160.927566	$\frac{3}{2}+$	β^-	6.91 d	
Tb	65	162	161.929510	$1-$	β^-	7.6 min	
Tb	65	163	162.930550	$\frac{3}{2}+$	β^-	19.5 min	

Chemical Symbol	Z	A	Atomic Mass (u)	Spin and Parity	Decay Modes*	Half-Life $t_{1/2}$	Natural Abundance (%)
Tb	65	164	163.933320	5 +	β^-	3.0 min	
Dy	66	147	146.930680	$\frac{1}{2}$ +	β^+, E.C.	80 s	
Dy	66	148	147.927020	0 +	β^+, E.C.	3.1 min	
Dy	66	149	148.927110	$\frac{7}{2}$ −	β^+, E.C.	4.2 min	
Dy	66	150	149.925577	0 +	α, β^+, E.C.	7.17 min	
Dy	66	151	150.926032	$\frac{7}{2}$ −	α, β^+, E.C.	17 min	
Dy	66	152	151.924716	0 +	α, E.C.	2.3 h	
Dy	66	153	152.925769	$\frac{7}{2}$ −	α, β^+, E.C.	6.3 h	
Dy	66	154	153.924429	0 +	α	3×10^6 yr	
Dy	66	155	154.925747	$\frac{3}{2}$ −	β^+, E.C.	10 h	
Dy	66	156	155.925277	0 +			0.06
Dy	66	157	156.925460	$\frac{3}{2}$ −	E.C.	8.1 h	
Dy	66	158	157.924403	0 +			0.10
Dy	66	159	158.925735	$\frac{3}{2}$ −	E.C.	144 d	
Dy	66	160	159.925193	0 +			2.34
Dy	66	161	160.926930	$\frac{5}{2}$ +			18.9
Dy	66	162	161.926795	0 +			25.5
Dy	66	163	162.928728	$\frac{5}{2}$ −			24.9
Dy	66	164	163.929171	0 +			28.2
Dy	66	165	164.931700	$\frac{7}{2}$ +	β^-	2.33 h	
Dy	66	166	165.932803	0 +	β^-	81.6 h	
Dy	66	167	166.935650	$\frac{1}{2}$ −	β^-	6.2 min	
Ho	67	148	147.937340		β^+, E.C.	9 s	
Ho	67	149	148.933600	9 +	β^+, E.C.	21 s	
Ho	67	150	149.933200		β^+, E.C.	88 s	
Ho	67	151	150.931510		α, β^+, E.C.	5.1 min	
Ho	67	152	151.931580	3 +	α, β^+, E.C.	2.4 min	
Ho	67	153	152.930195		β^+, E.C.	9.3 min	
Ho	67	154	153.930610	3 +	β^+, E.C.	12 min	
Ho	67	155	154.929078	$\frac{5}{2}$ +	β^+, E.C.	48 min	
Ho	67	156	155.929640	5 +	β^+, E.C.	2 min	
Ho	67	157	156.928190	$\frac{7}{2}$ −	β^+, E.C.	12.6 min	
Ho	67	158	157.928930	5 +	β^+, E.C.	11.3 min	
Ho	67	159	158.927706		E.C.	33 min	
Ho	67	160	159.928720	5 +	β^+, E.C.	25.6 min	
Ho	67	161	160.927849	$\frac{7}{2}$ −	E.C.	2.5 h	
Ho	67	162	161.929092	1 +	β^+, E.C.	15 min	
Ho	67	163	162.928731	$\frac{7}{2}$ −	E.C.	33 yr	
Ho	67	164	163.930285	1 +	β^-, E.C.	29 min	
Ho	67	165	164.930319	$\frac{7}{2}$ −			100
Ho	67	166	165.932281	0 −	β^-	1.117 d	
Ho	67	167	166.933127	$\frac{7}{2}$ −	β^-	3.1 h	
Ho	67	168	167.935290	3 +	β^-	3.0 min	
Ho	67	169	168.936869	$\frac{7}{2}$ −	β^-	4.7 min	
Ho	67	170	169.939620		β^-	2.8 min	
Er	68	150	149.997710	0 +	β^+, E.C.	20 s	
Er	68	151	150.937200		β^+, E.C.	23 s	
Er	68	152	151.934920	0 +	α, β^+, E.C.	10.3 s	
Er	68	153	152.934870		α, β^+, E.C.	31.7 s	
Er	68	154	153.932772	0 +	α, β^+, E.C.	3.7 min	

(continued)

Chemical Symbol	Z	A	Atomic Mass (u)	Spin and Parity	Decay Modes*	Half-Life $t_{1/2}$	Natural Abundance (%)
Os	76	170	169.963571	0 +	α, β^+, E.C.	7.1 s	
Os	76	171	170.962890		α, β^+, E.C.	7.9 s	
Os	76	172	171.960000	0 +	α, β^+, E.C.	19 s	
Os	76	173	172.957120		α, β^+, E.C.	16 s	
Os	76	174	173.957120	0 +	α, β^+, E.C.	44 s	
Os	76	175	174.956980		β^+, E.C.	1.4 min	
Os	76	176	175.954880	0 +	β^+, E.C.	3.6 min	
Os	76	177	176.954980	$\frac{1}{2}$ −	β^+, E.C.	2.8 min	
Os	76	178	177.953250	0 +	β^+, E.C.	5.0 min	
Os	76	179	178.953830		β^+, E.C.	7 min	
Os	76	180	179.952390	0 +	β^+, E.C.	21.7 min	
Os	76	181	180.953270	$\frac{1}{2}$ −	E.C.	1.75 h	
Os	76	182	181.952120	0 +	E.C.	21.5 h	
Os	76	183	182.953290	$\frac{9}{2}$ +	E.C.	13 h	
Os	76	184	183.952488	0 +			0.02
Os	76	185	184.954041	$\frac{1}{2}$ −	E.C.	93.6 d	
Os	76	186	185.953830	0 +	α	2×10^{15} yr	1.58
Os	76	187	186.955741	$\frac{1}{2}$ −			1.6
Os	76	188	187.955860	0 +			13.3
Os	76	189	188.958137	$\frac{3}{2}$ +			16.1
Os	76	190	189.958436	0 +			26.4
Os	76	191	190.960920	$\frac{9}{2}$ −	β^-	15.4 d	
Os	76	192	191.961467	0 +			41.0
Os	76	193	192.964138	$\frac{3}{2}$ −	β^-	30.6 h	
Os	76	194	193.965173	0 +	β^-	6.0 yr	
Os	76	195	194.968110		β^-	6.5 min	
Os	76	196	195.969620	0 +	β^-	34.9 min	
Ir	77	170	169.974970		α	1.05 s	
Ir	77	171	170.971700		α	1.6 s	
Ir	77	172	171.970550		α	2.1 s	
Ir	77	173	172.967560		α	3.0 s	
Ir	77	174	173.966660		α	4 s	
Ir	77	175	174.964150		α	4.5 s	
Ir	77	176	175.963480		α	8 s	
Ir	77	177	176.961350		α	21 s	
Ir	77	178	177.961050		β^+, E.C.	12 s	
Ir	77	179	178.959190		E.C.	4 min	
Ir	77	180	179.959260		E.C.	1.5 min	
Ir	77	181	180.957640	$\frac{7}{2}$ +	β^+, E.C.	4.9 min	
Ir	77	182	181.957970		β^+, E.C.	15 min	
Ir	77	183	182.956710		β^+, E.C.	56 min	
Ir	77	184	183.957560	5	β^+, E.C.	3.0 h	
Ir	77	185	184.956730	$\frac{5}{2}$ +	β^+, E.C.	14 h	
Ir	77	186	185.957943	5 +	β^+, E.C.	15.7 h	
Ir	77	187	186.957350	$\frac{3}{2}$ +	E.C.	10.5 h	
Ir	77	188	187.958830	2 −	β^+, E.C.	41.4 h	
Ir	77	189	188.958712	$\frac{3}{2}$ +	E.C.	13.2 d	
Ir	77	190	189.960580	4 +	E.C.	11.8 d	
Ir	77	191	190.960584	$\frac{3}{2}$ +			37.3
Ir	77	192	191.962580	4 −	β^-	73.83 d	
Ir	77	193	192.962917	$\frac{3}{2}$ +			62.7
Ir	77	194	193.965069	1 −	β^-	19.2 h	

Chemical Symbol	Z	A	Atomic Mass (u)	Spin and Parity	Decay Modes*	Half-Life $t_{1/2}$	Natural Abundance (%)
Ir	77	195	194.965966	$\frac{3}{2}+$	β^-	2.8 h	
Ir	77	196	195.968370	$0-$	β^-	52 s	
Ir	77	197	196.969629	$\frac{3}{2}+$	β^-	5.85 min	
Ir	77	198	197.972160		β^-	8 s	
Pt	78	172	171.977220	$0+$	α	0.10 s	
Pt	78	173	172.976280		α, β^+, E.C.	0.34 s	
Pt	78	174	173.972811	$0+$	α, β^+, E.C.	0.90 s	
Pt	78	175	174.972130		α, β^+, E.C.	2.52 s	
Pt	78	176	175.968930	$0+$	α, β^+, E.C.	6.3 s	
Pt	78	177	176.968360		α, E.C.	11.2 s	
Pt	78	178	177.965700	$0+$	α, E.C.	21.0 s	
Pt	78	179	178.965270		α, β^+, E.C.	43 s	
Pt	78	180	179.963130	$0+$	α, β^+, E.C.	52 s	
Pt	78	181	180.963100		β^+, E.C.	51 s	
Pt	78	182	181.961160	$0+$	β^+, E.C.	2.7 min	
Pt	78	183	182.961630		β^+, E.C.	7 min	
Pt	78	184	183.959920		β^+, E.C.	17.3 min	
Pt	78	185	184.960700	$\frac{9}{2}+$	β^+, E.C.	71 min	
Pt	78	186	185.959360	$0+$	β^+, E.C.	2.0 h	
Pt	78	187	186.960470	$\frac{3}{2}-$	β^+, E.C.	2.35 h	
Pt	78	188	187.959386	$0+$	E.C.	10.2 d	
Pt	78	189	188.960817		β^+, E.C.	10.9 h	
Pt	78	190	189.959917	$0+$			0.01
Pt	78	191	190.961665	$\frac{3}{2}-$	E.C.	2.96 d	
Pt	78	192	191.961019	$0+$			0.79
Pt	78	193	192.962977	$\frac{1}{2}-$	E.C.	50 yr	
Pt	78	194	193.962655	$0+$			32.9
Pt	78	195	194.964766	$\frac{1}{2}-$			33.8
Pt	78	196	195.964926	$0+$			25.3
Pt	78	197	196.967315	$\frac{1}{2}-$	β^-	18.3 h	
Pt	78	198	197.967869	$0+$			7.2
Pt	78	199	198.970552	$\frac{5}{2}-$	β^-	30.8 min	
Pt	78	200	199.971417	$0+$	β^-	12.5 h	
Pt	78	201	200.974500	$\frac{5}{2}-$	β^-	2.5 min	
Au	79	176	175.980060		β^+, E.C.	1.3 s	
Au	79	177	176.976920		α	1.3 s	
Au	79	178	177.975760		α	2.6 s	
Au	79	179	178.973170		α	7.5 s	
Au	79	180	179.972310		E.C.	8.1 s	
Au	79	181	180.970130		E.C.	11.4 s	
Au	79	182	181.969580		β^+, E.C.	21 s	
Au	79	183	182.967660	$\frac{3}{2}+$	E.C.	42 s	
Au	79	184	183.967540		E.C.	53 s	
Au	79	185	184.965800	$\frac{5}{2}-$	β^+, E.C.	4.3 min	
Au	79	186	185.966100	3	β^+, E.C.	10.7 min	
Au	79	187	186.964460	$\frac{1}{2}+$	β^+, E.C.	8.2 min	
Au	79	188	187.965080	$1-$	β^+, E.C.	8.8 min	
Au	79	189	188.963720	$\frac{1}{2}+$	β^+, E.C.	28.7 min	
Au	79	190	189.964685	$1-$	β^+, E.C.	43 min	
Au	79	191	190.963630	$\frac{3}{2}+$	E.C.	3.2 h	
Au	79	192	191.964793	$1-$	β^+, E.C.	5.0 h	

(continued)

Chemical Symbol	Z	A	Atomic Mass (u)	Spin and Parity	Decay Modes*	Half-Life $t_{1/2}$	Natural Abundance (%)
Au	79	193	192.964050	$\frac{3}{2}+$	E.C.	17.6 h	
Au	79	194	193.965348	$1-$	β^+, E.C.	39.5 h	
Au	79	195	194.965013	$\frac{3}{2}+$	E.C.	186.1 d	
Au	79	196	195.966544	$2-$	β^-, E.C.	6.18 d	
Au	79	197	196.966543	$\frac{3}{2}+$			100
Au	79	198	197.968217	$2-$	β^-	2.693 d	
Au	79	199	198.968740	$\frac{3}{2}+$	β^-	3.14 d	
Au	79	200	199.970670	$1-$	β^-	48.4 min	
Au	79	201	200.971645	$\frac{3}{2}+$	β^-	26 min	
Au	79	202	201.973840	$1-$	β^-	28 s	
Au	79	203	202.975145	$\frac{3}{2}+$	β^-	53 s	
Au	79	204	203.978300		β^-	40 s	
Hg	80	178	177.982476	$0+$	α, E.C.	0.26 s	
Hg	80	179	178.981630		α, E.C.	1.09 s	
Hg	80	180	179.978250	$0+$	α, E.C.	2.9 s	
Hg	80	181	180.977720	$\frac{1}{2}-$	α, β^+, E.C.	3.6 s	
Hg	80	182	181.974750	$0+$	α, β^+, E.C.	11 s	
Hg	80	183	182.974350	$\frac{1}{2}-$	α, β^+, E.C.	8.8 s	
Hg	80	184	183.971810	$0+$	α, β^+, E.C.	30.9 s	
Hg	80	185	184.971900	$\frac{1}{2}-$	β^+, E.C.	50 s	
Hg	80	186	185.969350	$0+$	β^+, E.C.	1.4 min	
Hg	80	187	186.969760	$\frac{3}{2}-$	β^+, E.C.	2.4 min	
Hg	80	188	187.967580	$0+$	β^+, E.C.	3.2 min	
Hg	80	189	188.968230	$\frac{3}{2}-$	E.C.	7.6 min	
Hg	80	190	189.966400	$0+$	E.C.	20 min	
Hg	80	191	190.967220	$\frac{3}{2}-$	β^+, E.C.	49 min	
Hg	80	192	191.965650	$0+$	E.C.	4.9 h	
Hg	80	193	192.966560	$\frac{3}{2}-$	E.C.	3.8 h	
Hg	80	194	193.965391	$0+$	E.C.	520 yr	
Hg	80	195	194.966640	$\frac{1}{2}-$	E.C.	9.5 h	
Hg	80	196	195.965807	$0+$			0.15
Hg	80	197	196.967187	$\frac{1}{2}-$	E.C.	64.1 h	
Hg	80	198	197.966743	$0+$			10.1
Hg	80	199	198.968254	$\frac{1}{2}-$			17.0
Hg	80	200	199.968300	$0+$			23.1
Hg	80	201	200.970277	$\frac{3}{2}+$			13.2
Hg	80	202	201.970617	$0+$			29.65
Hg	80	203	202.972848	$\frac{5}{2}-$	β^-	46.6 d	
Hg	80	204	203.973467	$0+$			6.8
Hg	80	205	204.976047	$\frac{1}{2}-$	β^-	5.2 min	
Hg	80	206	205.977489	$0+$	β^-	8.5 min	
Tl	81	184	183.981670		α, β^+, E.C.	11 s	
Tl	81	186	185.978510		β^+, E.C.	28 s	
Tl	81	187	186.976240	$\frac{1}{2}+$	β^+, E.C.	45 s	
Tl	81	188	187.975880	$2-$	β^+, E.C.	70 s	
Tl	81	189	188.980780	$\frac{1}{2}+$	β^+, E.C.	2.3 min	
Tl	81	190	189.973490	$2-$	β^+, E.C.	2.6 min	
Tl	81	192	191.972120	$2-$	β^+, E.C.	9.4 min	
Tl	81	193	192.970520	$\frac{1}{2}+$	β^+, E.C.	22 min	
Tl	81	194	193.970920	$2-$	β^+, E.C.	33 min	
Tl	81	195	194.969630	$\frac{1}{2}+$	β^+, E.C.	1.13 h	

Chemical Symbol	Z	A	Atomic Mass (u)	Spin and Parity	Decay Modes*	Half-Life $t_{1/2}$	Natural Abundance (%)
Tl	81	196	195.970460	2 −	β^+, E.C.	1.8 h	
Tl	81	197	196.969498	$\frac{1}{2}$ +	β^+, E.C.	2.83 h	
Tl	81	198	197.940460	2 −	β^+, E.C.	5.3 h	
Tl	81	199	198.969870	$\frac{1}{2}$ −	E.C.	7.4 h	
Tl	81	200	199.970934	2 −	E.C.	1.09 d	
Tl	81	201	200.970794	$\frac{1}{2}$ +	E.C.	3.05 d	
Tl	81	202	201.972085	2 −	E.C.	12.23 d	
Tl	81	203	202.972320	$\frac{1}{2}$ +			29.52
Tl	81	204	203.973839	2 −	β^-, E.C.	3.78 yr	
Tl	81	205	204.974401	$\frac{1}{2}$ +			70.48
Tl	81	206	205.976084	0 −	β^-	4.20 min	
Tl	81	207	206.977404	$\frac{1}{2}$ +	β^-	4.77 min	
Tl	81	208	207.981988	5 +	β^-	3.052 min	
Tl	81	209	208.985334	$\frac{1}{2}$ +	β^-	2.20 min	
Tl	81	210	209.990056	5 +	β^-	1.30 min	
Pb	82	184	183.988120	0 +	α	0.6 s	
Pb	82	185	184.987490		α	4.1 s	
Pb	82	186	185.984300	0 +	α, β^+, E.C.	8 s	
Pb	82	187	186.983830	$\frac{1}{2}$ −	β^+, E.C.	15.2 s	
Pb	82	188	187.980970	0 +	α, E.C.	24 s	
Pb	82	189	188.980780		α, E.C.	51 s	
Pb	82	190	189.978070	0 +	α, β^+, E.C.	1.2 min	
Pb	82	191	190.978160		β^+, E.C.	1.3 min	
Pb	82	192	191.975790	0 +	β^+, E.C.	3.5 min	
Pb	82	193	192.976120	$\frac{13}{2}$ +	β^+, E.C.	5.8 min	
Pb	82	194	193.973980	0 +	β^+, E.C.	10 min	
Pb	82	195	194.974480	$\frac{3}{2}$ −	β^+, E.C.	15 min	
Pb	82	196	195.972680	0 +	β^+, E.C.	37 min	
Pb	82	197	196.973360	$\frac{3}{2}$ −	β^+, E.C.	8 min	
Pb	82	198	197.971960	0 +	E.C.	2.4 h	
Pb	82	199	198.972870	$\frac{5}{2}$ −	β^+, E.C.	1.5 h	
Pb	82	200	199.971790	0 +	E.C.	21.5 h	
Pb	82	201	200.972830	$\frac{5}{2}$ −	E.C.	9.33 h	
Pb	82	202	201.972134	0 +	E.C.	5.3×10^4 yr	
Pb	82	203	202.973365	$\frac{5}{2}$ −	E.C.	2.169 d	
Pb	82	204	203.973020	0 +			1.4
Pb	82	205	204.974458	$\frac{5}{2}$ −	E.C.	1.51×10^7 yr	
Pb	82	206	205.974440	0 +			24.1
Pb	82	207	206.975872	$\frac{1}{2}$ −			22.1
Pb	82	208	207.976627	0 +			52.4
Pb	82	209	208.981065	$\frac{9}{2}$ +	β^-	3.25 h	
Pb	82	210	209.984163	0 +	β^-	22.3 yr	
Pb	82	211	210.988735	$\frac{9}{2}$ +	β^-	36.1 min	
Pb	82	212	211.991871	0 +	β^-	10.64 h	
Pb	82	213	212.996510		β^-	10.2 min	
Pb	82	214	213.999798	0 +	β^-	26.8 min	
Bi	83	190	189.988480		α, β^+, E.C.	5.4 s	
Bi	83	191	190.986110		α, β^+, E.C.	13 s	
Bi	83	192	191.985400		α, β^+, E.C.	42 s	
Bi	83	193	192.983180		α, β^+, E.C.	64 s	
Bi	83	194	193.982540	10 −	α, β^+, E.C.	1.8 min	

(continued)

Chemical Symbol	Z	A	Atomic Mass (u)	Spin and Parity	Decay Modes*	Half-Life $t_{1/2}$	Natural Abundance (%)
Bi	83	195	194.980700		α, β^+, E.C.	2.8 min	
Bi	83	196	195.980690		E.C.	4.5 min	
Bi	83	197	196.978880	$\frac{1}{2}+$	β^+, E.C.	10 min	
Bi	83	198	197.979000	$7+$	β^+, E.C.	11.85 min	
Bi	83	199	198.977520	$\frac{9}{2}-$	β^+, E.C.	27 min	
Bi	83	200	199.978090	$7+$	β^+, E.C.	36 min	
Bi	83	201	200.976930	$\frac{9}{2}-$	E.C.	1.80 h	
Bi	83	202	201.977660	$5+$	β^+, E.C.	1.72 h	
Bi	83	203	202.976830	$\frac{9}{2}-$	β^+, E.C.	11.76 h	
Bi	83	204	203.977740	$6+$	E.C.	11.2 h	
Bi	83	205	204.977365	$\frac{9}{2}-$	E.C.	15.31 d	
Bi	83	206	205.978478	$6+$	E.C.	6.243 d	
Bi	83	207	206.978446	$\frac{9}{2}-$	E.C.	32.2 yr	
Bi	83	208	207.979717	$5+$	E.C.	3.68×10^5 yr	
Bi	83	209	208.980374	$\frac{9}{2}-$			100
Bi	83	210	209.984095	$1-$	β^-	5.01 d	
Bi	83	211	210.987255	$\frac{9}{2}-$	α, β^-	2.14 min	
Bi	83	212	211.991255	$1-$	α, β^-	1.009 h	
Bi	83	213	212.994359	$\frac{9}{2}-$	α, β^-	45.6 min	
Bi	83	214	213.998691		β^-	19.9 min	
Bi	83	215	215.001930		β^-	7.4 min	
Po	84	194	193.988180	$0+$	α	0.7 s	
Po	84	195	194.988010		α	4.5 s	
Po	84	196	195.985540	$0+$	α, β^+, E.C.	5.5 s	
Po	84	197	196.985600	$\frac{3}{2}-$	α, β^+, E.C.	56 s	
Po	84	198	197.983360	$0+$	α, β^+, E.C.	1.76 min	
Po	84	199	198.983610	$\frac{3}{2}-$	α, β^+, E.C.	5.2 min	
Po	84	200	199.981700	$0+$	α, β^+, E.C.	11.5 min	
Po	84	201	200.982190	$\frac{3}{2}-$	α, β^+, E.C.	15.3 min	
Po	84	202	201.980680	$0+$	α, β^+, E.C.	44.7 min	
Po	84	203	202.981370	$\frac{5}{2}-$	β^+, E.C.	34.8 min	
Po	84	204	203.980280	$0+$	E.C.	3.53 h	
Po	84	205	204.981150	$\frac{5}{2}-$	β^+, E.C.	1.80 h	
Po	84	206	205.980456	$0+$	α, E.C.	8.8 d	
Po	84	207	206.981570	$\frac{5}{2}-$	α, E.C.	5.83 h	
Po	84	208	207.981222	$0+$	α	2.898 yr	
Po	84	209	208.982404	$\frac{1}{2}-$	α	105 yr	
Po	84	210	209.982848	$0+$	α	138.4 d	
Po	84	211	210.986627	$\frac{9}{2}+$	α	0.52 s	
Po	84	212	211.988842	$0+$	α	0.3 μs	
Po	84	213	212.992833	$\frac{9}{2}+$	α	4.2 μs	
Po	84	214	213.995176	$0+$	α	163 μs	
Po	84	215	214.999419	$\frac{9}{2}+$	α	1.78 ms	
Po	84	216	216.001889	$0+$	α	0.15 s	
Po	84	217	217.006260		α, β^-	<10 s	
At	85	196	195.995730		α	0.3 s	
At	85	197	196.993410	$\frac{9}{2}-$	α, β^+, E.C.	0.4 s	
At	85	198	197.992550		α	4.9 s	
At	85	199	198.990580	$\frac{9}{2}-$	α, β^+, E.C.	7.0 s	
At	85	200	199.990370	$5+$	α, β^+, E.C.	43 s	
At	85	201	200.988440	$\frac{9}{2}-$	α, β^+, E.C.	1.48 s	

Chemical Symbol	Z	A	Atomic Mass (u)	Spin and Parity	Decay Modes*	Half-Life $t_{1/2}$	Natural Abundance (%)
At	85	202	201.988420	5 +	α, β^+, E.C.	3.02 min	
At	85	203	202.986790	$\frac{9}{2}$ −	α, β^+, E.C.	7.4 min	
At	85	204	203.987210	5 +	α, β^+, E.C.	9.2 min	
At	85	205	204.986000	$\frac{9}{2}$ −	α, β^+, E.C.	26.2 min	
At	85	206	205.986580	5 +	α, β^+, E.C.	29.4 min	
At	85	207	206.985730	$\frac{9}{2}$ −	α, β^+, E.C.	1.81 h	
At	85	208	207.986510	6 +	α, β^+, E.C.	1.63 h	
At	85	209	208.986149	$\frac{9}{2}$ −	α, β^+, E.C.	5.41 h	
At	85	210	209.987126	5 +	α, E.C.	8.1 h	
At	85	211	210.987469	$\frac{9}{2}$ −	α, E.C.	7.21 h	
At	85	212	211.990725	1 −	α	0.3 s	
At	85	213	212.992911	$\frac{9}{2}$ −	α	0.11 μs	
At	85	214	213.996347	1 −	α	0.56 μs	
At	85	215	214.998638	$\frac{9}{2}$ −	α	100 μs	
At	85	216	216.002390	1 −	α	300 μs	
At	85	217	217.004694	$\frac{9}{2}$ −	α	32.3 μs	
At	85	218	218.008684		α	1.6 s	
At	85	219	219.011300		α	54 s	
Rn	86	200	199.995700	0 +	α, E.C.	1.0 s	
Rn	86	201	200.995570	$\frac{3}{2}$ −	α, E.C.	7.0 s	
Rn	86	202	201.993230	0 +	α, E.C.	9.9 s	
Rn	86	203	202.993330	$\frac{5}{2}$ −	α, E.C.	45 s	
Rn	86	204	203.991330	0 +	α, E.C.	1.24 min	
Rn	86	205	204.991650	$\frac{5}{2}$ −	α, E.C.	2.83 min	
Rn	86	206	205.990140	0 +	α, E.C.	5.67 min	
Rn	86	207	206.990690	$\frac{5}{2}$ −	α, β^+, E.C.	9.3 min	
Rn	86	208	207.989610	0 +	α, E.C.	24.4 min	
Rn	86	209	208.990370	$\frac{5}{2}$ −	α, β^+	28.5 min	
Rn	86	210	209.989669	0 +	α, E.C.	2.4 h	
Rn	86	211	210.990576	$\frac{1}{2}$ −	α, β^+, E.C.	14.6 h	
Rn	86	212	211.990697	0 +	α	24 min	
Rn	86	213	212.993856	$\frac{9}{2}$ +	α	25.0 ms	
Rn	86	214	213.996347	0 +	α	0.27 μs	
Rn	86	215	214.998720	$\frac{9}{2}$ +	α	2.3 μs	
Rn	86	216	216.000249	0 +	α	45 μs	
Rn	86	217	217.003902	$\frac{9}{2}$ +	α	0.54 ms	
Rn	86	218	218.005580	0 +	α	35 ms	
Rn	86	219	219.009479	$\frac{5}{2}$ +	α	3.96 s	
Rn	86	220	220.011368	0 +	α	55.6 s	
Rn	86	221	221.015470		α, β^-	25 min	
Rn	86	222	222.017570	0 +	α	3.82 d	
Fr	87	201	201.004110		α	48 ms	
Fr	87	202	202.003300		α	0.34 s	
Fr	87	203	203.001000		α	0.55 s	
Fr	87	204	204.000670		α	2.1 s	
Fr	87	205	204.998610		α	3.96 s	
Fr	87	206	205.998460		α	16.0 s	
Fr	87	207	206.996800	$\frac{9}{2}$ −	α	14.8 s	
Fr	87	208	207.997080	7 +	α, E.C.	59 s	
Fr	87	209	208.995878	$\frac{9}{2}$ −	α, E.C.	50.0 s	
Fr	87	210	209.996340	6 +	α	3.2 min	

(continued)

Chemical Symbol	Z	A	Atomic Mass (u)	Spin and Parity	Decay Modes*	Half-Life $t_{1/2}$	Natural Abundance (%)
Fr	87	211	210.995490	$\frac{9}{2}-$	α, E.C.	3.1 min	
Fr	87	212	211.996130	$5+$	α, E.C.	20.0 min	
Fr	87	213	212.996165	$\frac{9}{2}-$	α	34.6 s	
Fr	87	214	213.998948	$1-$	α	5.1 ms	
Fr	87	215	215.000310	$\frac{9}{2}-$	α	0.12 μs	
Fr	87	216	216.003178	$1-$	α	0.7 μs	
Fr	87	217	217.004609	$\frac{9}{2}-$	α	22 μs	
Fr	87	218	218.007553		α	0.7 ms	
Fr	87	219	219.009242	$\frac{9}{2}-$	α	21 s	
Fr	87	220	220.012293	1	α	27.4 s	
Fr	87	221	221.014230	$\frac{5}{2}-$	α	4.9 min	
Fr	87	222	222.017563	2	α, β^-	14.4 min	
Fr	87	223	223.019733	$\frac{3}{2}+$	β^-	21.8 min	
Fr	87	224	224.023220	1	β^-	2.7 min	
Fr	87	225	225.025590		β^-	3.9 min	
Fr	87	226	226.029200		β^-	48 s	
Fr	87	227	227.031770		β^-	2.4 min	
Fr	87	228	228.035570		β^-	39 s	
Ra	88	206	206.003800	$0+$	α	0.4 s	
Ra	88	207	207.003740		α	1.3 s	
Ra	88	208	208.001750	$0+$	α	1.4 s	
Ra	88	209	209.001930		α	4.6 s	
Ra	88	210	210.000430	$0+$	α	3.7 s	
Ra	88	211	211.000860	$\frac{5}{2}-$	α, E.C.	13 s	
Ra	88	212	211.999760	$0+$	α	13.0 s	
Ra	88	213	213.000330	$\frac{1}{2}-$	α, E.C.	2.7 min	
Ra	88	214	214.000079	$0+$	α	2.46 s	
Ra	88	215	215.002695	$\frac{9}{2}+$	α	1.59 ms	
Ra	88	216	216.003509	$0+$	α	0.18 μs	
Ra	88	217	217.006294	$\frac{9}{2}-$	α	1.6 μs	
Ra	88	218	218.007117	$0+$	α	14 μs	
Ra	88	219	219.010053		α	10 ms	
Ra	88	220	220.011004	$0+$	α	23 ms	
Ra	88	221	221.013889		α	28 s	
Ra	88	222	222.015353	$0+$	α	38.0 s	
Ra	88	223	223.018501	$\frac{1}{2}+$	α	11.43 d	
Ra	88	224	224.020186	$0+$	α	3.66 d	
Ra	88	225	225.023604	$\frac{3}{2}+$	β^-	14.8 d	
Ra	88	226	226.025402	$0+$	α	1600 yr	
Ra	88	227	227.029170	$\frac{3}{2}+$	β^-	42.2 min	
Ra	88	228	228.031064	$0+$	β^-	5.75 yr	
Ra	88	229	229.034870	$\frac{3}{2}+$	β^-	4.0 min	
Ra	88	230	230.036990	$0+$	β^-	1.55 h	
Ac	89	210	210.009230		α	0.35 s	
Ac	89	211	211.007590		α	0.25 s	
Ac	89	212	212.007760		α	0.93 s	
Ac	89	213	213.006530	$\frac{9}{2}-$	α	0.8 s	
Ac	89	214	214.006840	$5+$	α, E.C.	8.2 s	
Ac	89	215	215.006410	$\frac{9}{2}-$	α	0.17 s	
Ac	89	216	216.008650	$1-$	α	0.33 ms	
Ac	89	217	217.009322	$\frac{9}{2}-$	α	0.11 μs	

Chemical Symbol	Z	A	Atomic Mass (u)	Spin and Parity	Decay Modes*	Half-Life $t_{1/2}$	Natural Abundance (%)
Ac	89	218	218.001620		α	0.27 μs	
Ac	89	219	219.012390	$\frac{9}{2}$ −	α	7 μs	
Ac	89	220	220.014740		α	26.1 ms	
Ac	89	221	221.015570		α	52 ms	
Ac	89	222	222.017824		α	4.2 s	
Ac	89	223	223.019128	$\frac{5}{2}$ −	α, E.C.	2.2 min	
Ac	89	224	224.021685		α, E.C.	2.9 h	
Ac	89	225	225.023205	$\frac{3}{2}$ −	α	10.0 d	
Ac	89	226	226.026084	1 −	α, β^-, E.C.	1.2 d	
Ac	89	227	227.027750	$\frac{3}{2}$ −	α, β^-	21.77 yr	
Ac	89	228	228.031015	3 +	β^-	6.13 h	
Ac	89	229	229.032980	$\frac{3}{2}$ +	β^-	1.05 h	
Ac	89	230	230.036240	1 +	β^-	2.03 min	
Ac	89	231	231.038550	$\frac{1}{2}$ +	β^-	7.5 min	
Ac	89	232	232.042130	2 −	β^-	35 s	
Th	90	212	212.012890	0 +	α	30 ms	
Th	90	213	213.012940		α	0.14 s	
Th	90	214	214.011430	0 +	α	0.86 s	
Th	90	215	215.011690	$\frac{1}{2}$ −	α	1.2 s	
Th	90	216	216.011030	0 +	α	28 ms	
Th	90	217	217.013050		α	252 μs	
Th	90	218	218.013252	0 +	α	0.11 μs	
Th	90	219	219.015510		α	1.05 μs	
Th	90	220	220.015724	0 +	α	9.7 μs	
Th	90	221	221.018160		α	1.68 ms	
Th	90	222	222.018447	0 +	α	2.8 ms	
Th	90	223	223.020659		α	0.66 s	
Th	90	224	224.021449	0 +	α	1.04 s	
Th	90	225	225.023922	$\frac{3}{2}$ +	α, E.C.	8.0 min	
Th	90	226	226.024885	0 +	α	31 min	
Th	90	227	227.027703	$\frac{3}{2}$ +	α	18.72 d	
Th	90	228	228.028715	0 +	α	1.913 yr	
Th	90	229	229.031755	$\frac{5}{2}$ +	α	7.3×10^3 yr	
Th	90	230	230.033127	0 +	α	7.54×10^4 yr	
Th	90	231	231.036298	$\frac{5}{2}$ +	β^-	25.2 h	
Th	90	232	232.038054	0 +	α	1.4×10^{10} yr	100
Th	90	233	233.041577	$\frac{1}{2}$ +	β^-	22.3 min	
Th	90	234	234.043593	0 +	β^-	24.10 d	
Th	90	235	235.047510		β^-	6.9 min	
Pa	91	216	216.018960		α	0.2 s	
Pa	91	217	217.018250		α	4.9 ms	
Pa	91	218	218.019960		α	0.12 ms	
Pa	91	222	222.023560		α	4.3 ms	
Pa	91	223	223.023950		α	6 ms	
Pa	91	224	224.025530		α	0.95 s	
Pa	91	225	225.026090		α	1.8 s	
Pa	91	226	226.027928		α, E.C.	1.8 s	
Pa	91	227	227.028797	$\frac{5}{2}$ −	α, E.C.	38.3 min	
Pa	91	228	228.030773	3 +	α, E.C.	22 h	
Pa	91	229	229.032073	$\frac{5}{2}$ +	α, E.C.	1.4 d	
Pa	91	230	230.034527	2 −	β^-, E.C.	17.4 d	

(continued)

Chemical Symbol	Z	A	Atomic Mass (u)	Spin and Parity	Decay Modes*	Half-Life $t_{1/2}$	Natural Abundance (%)
Pa	91	231	231.035880	$\frac{3}{2}-$	α	3.27×10^4 yr	
Pa	91	232	232.038565	$2-$	β^-	1.31 d	
Pa	91	233	233.040242	$\frac{3}{2}-$	β^-	27.0 d	
Pa	91	234	234.043303	$4+$	β^-	6.70 h	
Pa	91	235	235.045430	$\frac{3}{2}-$	β^-	24.1 min	
Pa	91	236	236.048890	$1-$	β^-	9.1 min	
Pa	91	237	237.051140	$\frac{1}{2}+$	β^-	8.7 min	
Pa	91	238	238.055040	$3-$	β^-	2.3 min	
U	92	226	226.029170	$0+$	α	0.50 s	
U	92	227	227.030990		α	1.1 min	
U	92	228	228.031356	$0+$	α	9.1 min	
U	92	229	229.033474	$\frac{3}{2}+$	α, E.C.	58 min	
U	92	230	230.033921	$0+$	α	20.8 d	
U	92	231	231.036270	$\frac{5}{2}-$	E.C.	4.2 d	
U	92	232	232.037130	$0+$	α	68.9 yr	
U	92	233	233.039628	$\frac{5}{2}+$	α	1.59×10^5 yr	
U	92	234	234.040946	$0+$	α	2.45×10^5 yr	0.0055
U	92	235	235.043924	$\frac{7}{2}-$	α	7.04×10^8 yr	0.720
U	92	236	236.045562	$0+$	α	2.34×10^7 yr	
U	92	237	237.048724	$\frac{1}{2}+$	β^-	6.75 d	
U	92	238	238.050784	$0+$	α	4.46×10^9 yr	99.2745
U	92	239	239.054289	$\frac{5}{2}+$	β^-	23.54 min	
U	92	240	240.056587	$0+$	β^-	14.1 h	
Np	93	229	229.036230		α	4.2 min	
Np	93	230	230.037810		α, E.C.	4.6 min	
Np	93	231	231.038240	$\frac{5}{2}$	α, E.C.	48.8 min	
Np	93	232	232.040020	$4-$	α, E.C.	14.7 min	
Np	93	233	233.040800	$\frac{5}{2}+$	E.C.	36.2 min	
Np	93	234	234.042888	$0+$	β^+, E.C.	4.4 d	
Np	93	235	235.044056	$\frac{5}{2}+$	α, E.C.	1.08 yr	
Np	93	236	236.046550	$6-$	β^-, E.C.	1.2×10^5 yr	
Np	93	237	237.048167	$\frac{5}{2}+$	α	2.14×10^6 yr	
Np	93	238	238.050941	$2+$	β^-	2.117 d	
Np	93	239	239.052933	$\frac{5}{2}+$	β^-	2.35 d	
Np	93	240	240.056060	$5+$	β^-	1.03 h	
Np	93	241	241.058250	$\frac{5}{2}+$	β^-	13.9 min	
Np	93	242	242.061640	$6+$	β^-	2.2 min	
Pu	94	232	232.041169	$0+$	α, E.C.	34 min	
Pu	94	233	233.042970		α, E.C.	20.9 min	
Pu	94	234	234.043299	$0+$	α, E.C.	8.8 h	
Pu	94	235	235.045260	$\frac{5}{2}+$	α, E.C.	25.6 min	
Pu	94	236	236.046032	$0+$	α	2.85 yr	
Pu	94	237	237.048401	$\frac{7}{2}-$	E.C.	45.1 d	
Pu	94	238	238.049554	$0+$	α	87.74 yr	
Pu	94	239	239.052157	$\frac{1}{2}+$	α	2.411×10^4 yr	
Pu	94	240	240.053808	$0+$	α	6537 yr	
Pu	94	241	241.056845	$\frac{5}{2}+$	α, β^-	14.4 yr	
Pu	94	242	242.058737	$0+$	α	3.76×10^5 yr	
Pu	94	243	243.061998	$\frac{7}{2}+$	β^-	4.95 h	
Pu	94	244	244.064199	$0+$	α, S.F.	8.2×10^7 yr	
Pu	94	245	245.067820	$\frac{9}{2}-$	β^-	10.5 h	

Chemical Symbol	Z	A	Atomic Mass (u)	Spin and Parity	Decay Modes*	Half-Life $t_{1/2}$	Natural Abundance (%)
Pu	94	246	246.070171	0 +	β^-	10.85 d	
Am	95	237	237.050050	$\frac{5}{2}$ −	α, E.C.	1.22 h	
Am	95	238	238.051980	1 +	α, E.C.	1.63 h	
Am	95	239	239.053016	$\frac{5}{2}$ −	α, E.C.	11.901 h	
Am	95	240	240.055278	3 −	α, E.C.	50.9 h	
Am	95	241	241.056823	$\frac{5}{2}$ −	α	432.2 yr	
Am	95	242	242.059541	1 −	β^-, E.C.	16.01 h	
Am	95	243	243.061375	$\frac{5}{2}$ −	α	7.37×10^3 yr	
Am	95	244	244.064279	1 −	β^-	10.1 h	
Am	95	245	245.066444	$\frac{5}{2}$ +	β^-	2.05 h	
Am	95	246	246.069770	7 −	β^-	39 min	
Am	95	247	247.072170		β^-	22 min	
Cm	96	238	238.053020	0 +	α, E.C.	2.4 h	
Cm	96	239	239.054840		E.C.	3 h	
Cm	96	240	240.055503	0 +	α	27 d	
Cm	96	241	241.057465	$\frac{1}{2}$ +	α, E.C.	32.8 d	
Cm	96	242	242.058830	0 +	α	162.9 d	
Cm	96	243	243.061381	$\frac{5}{2}$ +	α	28.5 yr	
Cm	96	244	244.062747	0 +	α	18.11 yr	
Cm	96	245	245.065483	$\frac{7}{2}$ +	α	8.5×10^3 yr	
Cm	96	246	246.067218	0 +	α	4.78×10^3 yr	
Cm	96	247	247.070347	$\frac{9}{2}$ −	α	1.56×10^7 yr	
Cm	96	248	248.072343	0 +	α, S.F.	3.4×10^5 yr	
Cm	96	249	249.075948	$\frac{1}{2}$ +	β^-	64.15 min	
Cm	96	250	250.078352	0 +	α, S.F.	7.4×10^3 yr	
Cm	96	251	251.082290	$\frac{1}{2}$ +	β^-	16.8 min	
Bk	97	242	242.061940		E.C.	7.0 min	
Bk	97	243	243.062997	$\frac{3}{2}$ −	α, E.C.	4.5 h	
Bk	97	244	244.065160	4 −	α, E.C.	4.4 h	
Bk	97	245	245.066357	$\frac{3}{2}$ −	α, E.C.	4.94 d	
Bk	97	246	246.068720	2 −	α, E.C.	1.8 d	
Bk	97	247	247.070300	$\frac{3}{2}$ −	α	1.4×10^3 yr	
Bk	97	248	248.073106	1 −	β^-, E.C.	23.7 h	
Bk	97	249	249.074980	$\frac{7}{2}$ +	α, β^-	320 d	
Bk	97	250	250.078312	2 −	β^-	3.22 h	
Bk	97	251	251.080760	$\frac{3}{2}$ −	β^-	57 min	
Cf	98	240	240.062280	0 +	α	1.06 min	
Cf	98	241	241.063520		α, E.C.	3.8 min	
Cf	98	242	242.063690	0 +	α	3.5 min	
Cf	98	243	243.065390		α, E.C.	10.7 min	
Cf	98	244	244.065979	0 +	α	19.4 min	
Cf	98	245	245.068037		α, E.C.	43.6 min	
Cf	98	246	246.068800	0 +	α	36 h	
Cf	98	247	247.071020	$\frac{7}{2}$ +	α, E.C.	3.11 h	
Cf	98	248	248.072183	0 +	α	334 d	
Cf	98	249	249.074844	$\frac{9}{2}$ −	α	351 d	
Cf	98	250	250.076400	0 +	α	13.1 yr	
Cf	98	251	251.079580	$\frac{1}{2}$ +	α	890 yr	
Cf	98	252	252.081621	0 +	α, S.F.	2.64 yr	
Cf	98	253	253.085127	$\frac{7}{2}$ +	α, β^-	17.8 d	

(continued)

Chemical Symbol	Z	A	Atomic Mass (u)	Spin and Parity	Decay Modes*	Half-Life $t_{1/2}$	Natural Abundance (%)
Cf	98	254	254.087318	0 +	α, S.F.	60.5 d	
Es	99	243	243.069470		α, E.C.	21 s	
Es	99	244	244.070810		α, E.C.	37 s	
Es	99	245	245.071260		α, E.C.	1.33 min	
Es	99	246	246.072920		α, E.C.	7.7 min	
Es	99	247	247.073590		α, E.C.	4.7 min	
Es	99	248	248.075440		α, E.C.	27 min	
Es	99	249	249.076340	$\frac{7}{2}$ +	α, E.C.	1.70 h	
Es	99	250	250.078660		E.C.	8.6 h	
Es	99	251	251.079986	$\frac{3}{2}$ −	α, E.C.	1.38 d	
Es	99	252	252.082944	5 −	α, E.C.	1.29 yr	
Es	99	253	253.084818	$\frac{7}{2}$ +	α	20.47 d	
Es	99	254	254.088019	7 +	α	275 d	
Es	99	255	255.090270	$\frac{7}{2}$ +	α, β^-, S.F.	39.8 d	
Es	99	256	256.093560	1 +	β^-	25 min	
Fm	100	243	243.074460		α	0.18 s	
Fm	100	244	244.074120	0 +	S.F.	3.7 ms	
Fm	100	245	245.076250		α	4 s	
Fm	100	246	246.075290	0 +	α, S.F.	1.1 s	
Fm	100	247	247.076800		α	35 s	
Fm	100	248	248.077171	0 +	α, S.F.	36 s	
Fm	100	249	249.078910	$\frac{7}{2}$ +	α, E.C.	2.6 min	
Fm	100	250	250.079509	0 +	α	30 min	
Fm	100	251	251.081590	$\frac{9}{2}$ −	α, E.C.	5.3 h	
Fm	100	252	252.082466	0 +	α	25.4 h	
Fm	100	253	253.085173	$\frac{1}{2}$ +	α, E.C.	3.0 d	
Fm	100	254	254.086846	0 +	α, S.F.	3.24 h	
Fm	100	255	255.089948	$\frac{7}{2}$ +	α	20.1 h	
Fm	100	256	256.091767	0 +	α, S.F.	2.63 h	
Fm	100	257	257.075099	$\frac{9}{2}$ +	α, S.F.	100.5 d	
Md	101	248	248.082750		α, E.C.	7 s	
Md	101	249	249.082950		α, E.C.	24 s	
Md	101	250	250.084380		α, E.C.	52 s	
Md	101	251	251.084830		α, E.C.	4.0 min	
Md	101	252	252.086470		α, E.C.	2.3 min	
Md	101	254	254.089630		E.C.	10.3 min	
Md	101	255	255.091081	$\frac{7}{2}$ −	α, E.C.	27 min	
Md	101	256	256.093960		α, E.C.	76 min	
Md	101	257	257.095580	$\frac{7}{2}$ −	α, E.C.	5.2 h	
Md	101	258	258.098570	8 −	α	56 d	
No	102	251	251.088870		α	0.8 s	
No	102	252	252.088949	0 +	α, S.F.	2.3 s	
No	102	253	253.090530	$\frac{9}{2}$ −	α	1.7 min	
No	102	254	254.090953	0 +	α	55 s	
No	102	255	255.093260	$\frac{1}{2}$ +	α, E.C.	3.1 min	
No	102	256	256.094252	0 +	α	3.2 s	
No	102	257	257.096850	$\frac{7}{2}$ +	α	25 s	
No	102	258	258.098150	0 +	S.F.	1.2 ms	
No	102	259	259.100931	$\frac{9}{2}$ +	α, E.C.	58 min	

Chemical Symbol	Z	A	Atomic Mass (u)	Spin and Parity	Decay Modes*	Half-Life $t_{1/2}$	Natural Abundance (%)
Lw	103	253	253.095150		α	1.4 s	
Lw	103	254	254.096320		α	20 s	
Lw	103	255	255.096670		α	22 s	
Lw	103	256	256.098490		α, S.F.	28 s	
Lw	103	257	257.099480	$\frac{7}{2}$ +	α	0.65 s	
Lw	103	258	258.101710		α	4.3 s	
Lw	103	259	259.102900		α	5.4 s	
Lw	103	260	260.105320		α	3 min	
Rf	104	257	257.102950		α	4.8 s	
Rf	104	258	258.103430	0 +	S.F.	11 ms	
Rf	104	259	259.105530		α, S.F.	65 s	
Rf	104	260	260.106300	0 +	S.F.	20 ms	
Rf	104	261	261.108690		α, S.F.	65 s	
Ha	105	257	257.107770		α, S.F.	1 s	
Ha	105	258	258.109020		α	4 s	
Ha	105	259	259.109580		S.F.	1.2 s	
Ha	105	260	260.111040		α, S.F.	1.5 s	
Ha	105	261	261.111820		α, S.F.	1.8 s	
Ha	105	262	262.113760		α, S.F.	34 s	

Appendix C Nobel Laureates in Physics

Year	Nobel Laureate		Recognized for
1901	Wilhelm Konrad Röntgen	1845–1923	discovery of X rays
1902	Hendrik Antoon Lorentz	1853–1928	their researches into the influence of
	Pieter Zeeman	1865–1943	magnetism upon radiation phenomena
1903	Antoine Henri Becquerel	1852–1908	his discovery of spontaneous radioactivity
	Pierre Curie	1859–1906	their joint researches on the radiation phenomena
	Marie Sklodowska Curie	1867–1934	discovered by Prof. Henri Becquerel
1904	John William Strutt, Lord Rayleigh	1842–1919	investigations of the densities of the most important gases and his discovery of argon
1905	Phillipp Eduard Anton Lenard	1862–1947	his work on cathode rays
1906	Joseph John Thomson	1856–1940	his theoretical and experimental investigations on the conduction of electricity by gases
1907	Albert Abraham Michelson	1852–1931	his optical precision instruments and metrological investigations carried out with their aid
1908	Gabriel Lippmann	1845–1921	his method of producing colors photographically based on the phenomena of interference
1909	Gugliemo Marconi	1874–1937	their contributions to the development of
	Karl Ferdinand Braun	1850–1918	wireless telegraphy
1910	Johannes Diderik van der Waals	1837–1923	his work on the equation of state of gases and liquids
1911	Wilhelm Wien	1864–1928	his discoveries regarding the laws governing the radiation of heat
1912	Nils Gustaf Dalén	1869–1937	his invention of automatic regulators for use in conjunction with gas accumulators for illuminating lighthouses and buoys
1913	Heike Kamerlingh Onnes	1853–1926	his investigations of the properties of matter at low temperatures which led, *inter alia,* to the production of liquid helium
1914	Max Theodor Felix von Laue	1879–1960	his discovery of the diffraction of Röntgen rays by crystals
1915	William Henry Bragg	1862–1942	their services in the analysis of crystal
	William Lawrence Bragg	1890–1971	structure by means of X rays
1917	Charles Glover Barkla	1877–1944	his discovery of the characteristic X rays of the elements
1918	Max Planck	1858–1947	his discovery of energy quanta
1919	Johannes Stark	1874–1957	his discovery of the Doppler effect in canal rays and the splitting of spectral lines in electric fields
1920	Charles Édouard Guillaume	1861–1938	the service he has rendered to precision measurement in Physics by his discovery of anomalies in nickel steel alloys
1921	Albert Einstein	1879–1955	his services to Theoretical Physics, and especially for his discovery of the laws of the photoelectric effect
1922	Niels Henrik David Bohr	1885–1962	the investigation of the structure of atoms, and of the radiation emanating from them
1923	Robert Andrews Millikan	1868–1953	his work on the elementary charge of electricity and on the photoelectric effect

Year	Nobel Laureate		Recognized for
1924	Karl Manne Georg Siegbahn	1886–1979	his discovery and research in the field of X-ray spectroscopy
1925	James Franck	1882–1964	their discovery of the laws governing the impact of an electron upon an atom
	Gustav Hertz	1887–1975	
1926	Jean Baptiste Perrin	1870–1942	his work on the discontinuous structure of matter, and especially for his discovery of sedimentation equilibrium
1927	Arthur Holly Compton	1892–1962	his discovery of the effect named after him
	Charles Thomson Rees Wilson	1869–1959	his method of making the paths of electrically charged particles visible by condensation of vapor
1928	Owen Willans Richardson	1879–1959	his work on the thermionic phenomenon, and especially for the discovery of the law named after him
1929	Prince Louis Victor de Broglie	1892–1987	his discovery of the wave nature of electrons
1930	Sir Chandrasekhara Venkata Raman	1888–1970	his work on the scattering of light and for the discovery of the effect named after him
1931	(no award)		
1932	Werner Heisenberg	1901–1976	the creation of quantum mechanics, the application of which has, among other things, led to the discovery of the allotropic forms of hydrogen
1933	Erwin Schrödinger	1887–1961	the discovery of new, productive forms of atomic theory
	Paul Adrien Maurice Dirac	1902–1984	
1934	(no award)		
1935	James Chadwick	1891–1974	his discovery of the neutron
1936	Victor Franz Hess	1883–1964	the discovery of cosmic radiation
	Carl David Anderson	1905–1984	his discovery of the positron
1937	Clinton Joseph Davisson	1881–1958	their experimental discovery of the diffraction of electrons by crystals
	George Paget Thomson	1892–1975	
1938	Enrico Fermi	1901–1954	his demonstrations of the existence of new radioactive elements produced by neutron irradiation, and for his related discovery of nuclear reactions brought about by slow neutrons
1939	Ernest Orlando Lawrence	1901–1958	the invention and development of the cyclotron and for results obtained with it, especially for artificial radioactive elements
1943	Otto Stern	1888–1969	his contribution to the development of the molecular ray method and his discovery of the magnetic moment of the proton
1944	Isidor Isaac Rabi	1898–1988	his resonance method for recording the magnetic properties of atomic nuclei
1945	Wolfgang Pauli	1900–1958	his discovery of the Exclusion Principle (Pauli Principle)
1946	Percy Williams Bridgman	1881–1961	the invention of an apparatus to produce extremely high pressures, and for the discoveries he made therewith in the field of high-pressure physics
1947	Sir Edward Victor Appleton	1892–1965	his investigations of the physics of the upper atmosphere, especially for the discovery of the so-called Appleton layer
1948	Patrick Maynard Stuart Blackett	1897–1974	his development of the Wilson cloud chamber method, and his discoveries therewith in nuclear physics and cosmic radiation

(continued)

Year	Nobel Laureate		Recognized for
1949	Hideki Yukawa	1907–1981	his prediction of the existence of mesons on the basis of theoretical work on nuclear forces
1950	Cecil Frank Powell	1903–1969	his development of the photographic method of studying nuclear processes and his discoveries regarding mesons made with this method
1951	Sir John Douglas Cockcroft	1897–1967	their pioneer work on the transmutation of atomic nuclei by artificially accelerated atomic particles
	Ernest Thomas Sinton Walton	b. 1903	
1952	Felix Bloch	1905–1983	their development of new methods for nuclear magnetic precision measurements and discoveries in connection therewith
	Edward Mills Purcell	b. 1912	
1953	Frits Zernike	1888–1966	his demonstration of the phase-contrast method, especially for his invention of the phase-contrast microscope
1954	Max Born	1882–1970	his fundamental research in quantum mechanics especially for his statistical interpretation of the wave function
	Walther Bothe	1891–1957	the coincidence method and his discoveries made therewith
1955	Willis Eugene Lamb	b. 1913	his discoveries concerning the fine structure of the hydrogen spectrum
	Polykarp Kusch	b. 1911	his precision determination of the magnetic moment of the electron
1956	William Bradford Shockley	1910–1989	their researches on semiconductors and their discovery of the transistor effect
	John Bardeen	1908–1991	
	Walter Houser Brattain	1902–1987	
1957	Chen Ning Yang	b. 1922	their penetrating investigation of the parity laws which has led to important discoveries regarding elementary particles
	Tsung Dao Lee	b. 1926	
1958	Pavel Alekseyevich Cherenkov	b. 1904	the discovery and interpretation of the Cherenkov effect
	Ilya Mikhaylovich Frank	b. 1908	
	Igor Yevgenyevich Tamm	1895–1971	
1959	Emilio Gino Segrè	b. 1905	their discovery of the antiproton
	Owen Chamberlain	b. 1920	
1960	Donald Arthur Glaser	b. 1926	the invention of the bubble chamber
1961	Robert Hofstadter	b. 1915	his pioneering studies of electron scattering in atomic nuclei and for his thereby-achieved discoveries concerning the structure of the nucleons
	Rudolf Ludwig Mössbauer	b. 1929	his researches concerning the resonance absorption of γ rays and his discovery in this connection of the effect which bears his name
1962	Lev Davidovich Landau	1908–1968	his pioneering theories of condensed matter, especially liquid helium
1963	Eugene Paul Wigner	b. 1902	his contributions to the theory of the atomic nucleus and the elementary particles, particularly through the discovery and application of fundamental symmetry principles
	Maria Goeppert Mayer	1906–1972	their discoveries concerning nuclear shell structure
	J. Hans D. Jensen	1907–1973	
1964	Charles Hard Townes	b. 1915	fundamental work in the field of quantum electronics which has led to the construction of oscillators and amplifiers based on the maser-laser principle
	Nikolay Gennadiyevich Basov	b. 1922	
	Alekanndr Mikhazlovich Prokhorov	b. 1916	

Year	Nobel Laureate		Recognized for
1965	Shinichiro Tomonaga	1906–1979	their fundamental work in quantum electrodynamics, with profound consequences for the physics of elementary particles
	Julian Schwinger	b. 1918	
	Richard Phillips Feynman	1918–1988	
1966	Alfred Kastler	1902–1984	the discovery and development of optical methods for studying Hertzian resonance in atoms
1967	Hans Albrecht Bethe	b. 1906	his contributions to the theory of nuclear reactions, especially his discoveries concerning the energy production in stars
1968	Luis Walter Alvarez	1911–1988	his decisive contribution to elementary particle physics, in particular the discovery of a large number of resonance states, made possible through his development of the technique of using the hydrogen bubble chamber and data analysis
1969	Murray Gell-Mann	b. 1929	his contributions and discoveries concerning the classification of elementary particles and their interactions
1970	Hannes Olof Gösta Alfvén	b. 1908	fundamental work and discoveries in magnetohydrodynamics with fruitful applications in different parts of plasma physics
	Louis Eugène Félix Néel	b. 1904	fundamental work and discoveries concerning antiferromagnetism and ferrimagnetism which have led to important applications in solid-state physics
1971	Dennis Gabor	1900–1979	his discovery of the principles of holography
1972	John Bardeen	1908–1991	their development of a theory of superconductivity
	Leon Neil Cooper	b. 1930	
	John Robert Schrieffer	b. 1931	
1973	Leo Esaki	b. 1925	his discovery of tunneling in semiconductors
	Ivar Giaever	b. 1929	his discovery of tunneling in superconductors
	Brian David Josephson	b. 1940	his theoretical prediction of the properties of a supercurrent through a tunnel barrier
1974	Anthony Hewish	b. 1924	the discovery of pulsars
	Sir Martin Ryle	1918–1984	his pioneering work in radio astronomy
1975	Aage Bohr	b. 1922	the discovery of the connection between collective motion and particle motion and the development of the theory of the structure of the atomic nucleus based on this connection
	Ben Mottelson	b. 1926	
	Leo James Rainwater	b. 1917	
1976	Burton Richter	b. 1931	their (independent) discovery of an important fundamental particle
	Samuel Chao Chung Ting	b. 1936	
1977	Philip Warren Anderson	b. 1923	their fundamental theoretical investigations of the electronic structure of magnetic and disordered systems
	Sir Nevill Francis Mott	b. 1905	
	John Hasbrouck Van Vleck	1899–1980	
1978	Pyotr Leonidovich Kapitsa	1894–1984	his basic inventions and discoveries in low-temperature physics
	Arno Allan Penzias	b. 1933	their discovery of cosmic microwave background radiation
	Robert Woodrow Wilson	b. 1936	
1979	Sheldon Lee Glashow	b. 1932	their unified model of the action of the weak and electromagnetic forces and for their prediction of the existence of neutral currents
	Abdus Salam	b. 1926	
	Steven Weinberg	b. 1933	
1980	James Watson Cronin	b. 1931	the discovery of violations of fundamental symmetry principles in the decay of neutral *K* mesons
	Val L. Fitch	b. 1923	

(continued)

Year	Nobel Laureate		Recognized for
1981	Nicolaas Bloembergen	b. 1920	their contribution to the development of laser spectroscopy
	Arthur Leonard Schawlow	b. 1921	
	Kai M. Siegbahn	b. 1918	his contribution to high-resolution electron microscopy
1982	Kenneth Geddes Wilson	b. 1936	his method of analyzing the critical phenomena inherent in the changes of matter under the influence of pressure and temperature
1983	Subrahmanyan Chandrasekhar	b. 1910	his theoretical studies of the structure and evolution of stars
	William A. Fowler	b. 1911	his studies of the formation of chemical elements in the universe
1984	Carlo Rubbia	b. 1934	their decisive contributions to the large project, which led to the discovery of the field particles *W* and *Z*, communicators of the weak interaction
	Simon van der Meer	b. 1925	
1985	Klaus von Klitzing	b. 1943	his discovery of the quantized Hall resistance
1986	Ernst August Friedrich Ruska	1906–1988	his invention of the electron microscope
	Gerd Binnig	b. 1947	their invention of the scanning-tunneling electron microscope
	Heinrich Rohrer	b. 1933	
1987	Karl Alex Müller	b. 1927	their discovery of a new class of superconductors
	Johannes Georg Bednorz	b. 1950	
1988	Leon Max Lederman	b. 1922	the neutrino-beam method and the demonstration of the doublet structure of the leptons through the discovery of the muon neutrino
	Melvin Schwartz	b. 1932	
	Jack Steinberger	b. 1921	
1989	Norman Foster Ramsey, Jr.	b. 1915	his invention of the separated-oscillatory-field method and its use in the hydrogen maser and other atomic clocks
	Hans Georg Dehmelt	b. 1922	their development of the ion trap technique
	Wilhelm Paul	b. 1913	
1990	Jerome I. Friedman	b. 1930	their experiment which led to a breakthrough in our understanding of matter
	Henry W. Kendall	b. 1926	
	Richard E. Taylor	b. 1929	
1991	Pierre-Gilles de Gennes	b. 1932	his work on polymers and liquid crystals

Answers to Odd-Numbered Problems*

CHAPTER 1

1.1 3.33 s
1.5 6×10^7 m
1.7 28 m; 107 m
1.9 (*a*) 1.42 ns (*b*) 4×10^8 $(\text{cm}^2\cdot\text{s})^{-1}$
(*c*) 2.26×10^8 and 1.28×10^8 $(\text{cm}^2\cdot\text{s})^{-1}$
1.11 (*a*) $0.946c$; $0.385c$ (*b*) $0.877c$ at 47°
1.13 $f' = f_0(1 - \beta^2)^{1/2}$
1.19 (*a*) 0.9804 μs (*b*) 2.86 μs (*c*) 1.647 μs
(*d*) 731 m
1.21 (*a*) 0.936 μs (*b*) 189.5 m
1.23 $f_c = f_0[(1 \pm \beta)/(1 \mp \beta)]^{1/2}$ from A; (*a*) resp.
(*b*) $f_c = f_0[(1 \pm \beta)/(1 \mp \beta)]$
1.25 3639 nm

CHAPTER 2

2.1 $0.941c$
2.3 $(\sqrt{3}/2)c$
2.5 2.35×10^8 V
2.7 (*a*) 25.55 eV (*b*) 46.55 keV (*c*) 35.62 MeV
2.11 $\sqrt{15}m_0c$; $\sqrt{5}m_0c^2$
2.13 $\beta_\pm = \{(2\sqrt{2}/3) \pm [1 - (2m_0/M_0)^2]^{1/2}\}/\{1 \pm (2\sqrt{2}/3)[1 - (2m_0/M_0)^2]^{1/2}\}$
2.15 (*a*) $\beta(t) = (eEt/m_0c)/[1 + (eEt/m_0c)^2]^{1/2}$
(*b*) $x(t) = (m_0c^2/eE)\{[1 + (eEt/m_0c)^2]^{1/2} - 1\}$
2.17 (*a*) $\beta_{\text{CM}} = 2\sqrt{2}/3$ (*b*) $16m_0c^2$ (*c*) 36.9°
2.19 $E = 5m_0c^2$; $p = \sqrt{15}m_0c$; $M_0 = \sqrt{10}m_0$
2.23 109.3 MeV
2.25 139.7 MeV/c^2 (pion)
2.27 KE = 0.784 MeV; $v = 0.919c$
2.29 28.27 MeV

CHAPTER 3

3.1 (*a*) 0.5×10^{-6} nm (*b*) 1.39×10^{-3} nm
(*c*) Yes (but only with great care)
3.3 $\Delta f = -1.77$ Hz; $f = 2{,}999{,}999{,}998.23$ Hz
3.5 4.26×10^{-6} radians = 0.88 seconds of arc
3.7 $\Delta\nu/\nu = \frac{1}{2}(R\omega/c)^2$

CHAPTER 4

4.1 (*a*) 1.0×10^{26} (*b*) 5.0×10^{18} (*c*) 1.0×10^{14}
4.5 1497 K
4.9 0.00243 nm
4.11 2.9 K
4.15 6.76×10^{15} photons/m^3
4.17 $\phi = 1.82$ eV; $h = 6.76 \times 10^{-34}$ J·s
4.19 1033 nm
4.27 (*a*) $C_{4.2} = 5.87 \times 10^{-4}R$
(*b*) $C_{1.0} = 7.92 \times 10^{-6}R$
4.29 (*a*) 72.7 nm (*b*) 3.6×10^{-483}

CHAPTER 5

5.1 (*a*) 302 V (*b*) 7.55 cm
5.3 $R_p = 36.5$ cm, $R_\alpha = 73$ cm, $R_{\text{He}-} = 146$ cm, $R_{\text{H}-} = 73$ cm
5.5 $d = (2Ze^2/4\pi\epsilon_0E)[M + m)/M]$
5.7 1.1×10^{-14} m
5.9 23 MeV
5.11 (*a*) 1.76×10^{-14} m; 2.0 MeV
(*b*) 1.27×10^{-14} m; 0.43 MeV
5.17 (*a*) 1.05 mA (*b*) 12.5 T (*c*) 0.132 mA, 0.388 T
5.19 (*a*) $\text{H}(3\rightarrow2) \simeq \text{He}^+(6\rightarrow4)$;
$\text{H}(4\rightarrow2) \simeq \text{He}^+(8\rightarrow4)$ (*b*) $\Delta\lambda = 0.268$ nm
5.21 −6.8 eV; −1.7 eV; −0.756 eV
5.23 434.2 nm
5.25 (*a*) 22.9 keV (*b*) 72.5 keV
5.27 titanium
5.29 $4 \rightarrow 2$; $3 \rightarrow 2$; $9 \rightarrow 3$; $5 \rightarrow 3$; $7 \rightarrow 4$; $6 \rightarrow 4$

CHAPTER 6

6.1 (*a*) 0.027 nm (*b*) 640 fm (*c*) 320 fm
6.3 (*a*) 15 keV (*b*) 8.2 eV (*c*) 2.1 eV
6.7 $R \geq 0.13 \times 10^{-10}$ m
6.13 Approximately 2×10^{-14} m
6.17 0.72 MeV; 0.33 MeV

* Solutions to problems requiring derivations or proofs of relations or graphs are not included.

CHAPTER 7

7.1 2.04 MeV, 8.18 MeV
7.3 $\sqrt{2/a}$
7.5 $0.568\hbar$, $1.670\hbar$
7.7 $X \cot X = -\sqrt{Y^2 - X^2}$
7.9 340 eV
7.11 $R = [(1 - k'/k)/(1 + k'/k)]^2$; $T = 4kk'/(1 + k'/k)^2$
7.13 $R = 0.0294$, $T = 0.9706$
7.19 $\Delta x\, \Delta p = 3\hbar/2$
7.21 (*a*) Particle moves more rapidly inside potential; $P_{in}/P_{out} = [E/(E + V_0)]^{1/2}$ (*c*) $R = 0$ and $T = 1$ if $E = (nh^2/8ma^2) - V_0$

CHAPTER 8

8.11 $P_{2,0} = P_{2,1} = 0.053$
8.15 1.6×10^{-14}

CHAPTER 9

9.1 He, $1s2p \rightarrow 1s^2$; Be, $2s2p \rightarrow 2s^2$; C, $2s^22p3s \rightarrow 2s^22p^2$.
9.3 For $3^2S_{1/2}$ $\Delta E = 0.174$ meV; for $3P_{1/2}$ $\Delta E = 0.058$ meV; for $3^2P_{3/2}$ $\Delta E = 0.116$ meV. $L{\cdot}S$ splitting is 2.14 meV, about 20 times larger
9.5 Assuming proportional to $1/r_n{}^3$, obtain $\Delta E_{so} = 0.38$ meV; measured $\Delta E_{so} = 0.68$ meV; reason, $\langle 1/r^3 \rangle \neq 1/\langle r^3 \rangle$
9.9 3^2D, split into $3^2D_{3/2}$ and $3^2D_{5/2}$
9.11 $4^2F_{5/2}$, 3^1P_1
9.13 $\Delta E = 0.232$ meV; $T = 33.6$ K; $\nu = 5.6 \times 10^{10}$ Hz At $T = 300$ K, $E = 0.052$ J/mol; at $T = 4$ K, $E = 3.61$ J/mol

CHAPTER 10

10.1 $F_c = 2.9 \times 10^{-9}$ N; $V_c = 0.226$ eV; $A \simeq 2 \times 10^{-13}$ eV·nm^{23}; $n \simeq 22$
10.3 $V_c(\text{KBr})/V_c(\text{RbCl}) = 1.32$
10.5 $\nu \simeq 1.65 \times 10^{13}$ Hz; $\lambda = 1.82 \times 10^4$ nm
10.9 $k = 379$ N/m; $r_0 = 0.14$ nm; $T \simeq 290$ K
10.11 For H_2: 0.273 eV, 0.819 eV, 1.365 eV
For HD: 0.237 eV, 0.709 eV, 1.181 eV
For D_2: 0.193 eV, 0.579 eV, 0.965 eV
10.13 If by "significant" we mean 1 percent or more, then $T \geq 252$ K
10.15 $L_m = 46$

CHAPTER 11

11.3

	(a)	(b)	(c)
n_0	2.30	2.45	2.0
n_1	1.53	1.55	1.67
n_2	0.979	0.90	1.0
n_3	0.587	0.45	1.0
n_4	0.327	0.35	0.33
n_5	0.163	0.15	0
n_6	0.070	0.10	0
n_7	0.023	0.05	0
n_8	0.005	0.05	0

(*d*) $A \simeq 4.0$, $\beta \simeq 0.75/\epsilon$

11.5

1	2	3	4	5	6	7
7	2	2	1	1	0	0
6	3	2	1	1	0	0
5	4	2	1	1	0	0
5	3	3	1	1	0	0
5	3	2	2	1	0	0
4	4	3	1	1	0	0
4	4	2	2	1	0	0
4	3	3	2	1	0	0
4	3	2	2	1	1	0

$n_0 = 1.89$, $n_1 = 1.67$, $n_2 = 1.22$, $n_3 = 0.889$, $n_4 = 0.778$, $n_5 = 0.333$, $n_6 = 0.111$, $n_7 = 0.111$. E_F is that energy for which $n_i = 1.0$; hence $E_F \simeq 2.7\epsilon$, somewhat lower than for the case when the total energy is a minimum, corresponding to $T = 0$ K
11.7 $v_{max} = (2kT/m)^{1/2}$, $(v^2)_{max} = kT/m$; both values are less than their averages
11.11 $32\pi^5k^4T^3/15h^3c^3$
11.13 $(3/5)E_F$
11.15 1.37 K
11.17 1.36×10^8 eV

CHAPTER 12

12.1 11.1 eV (using $n = 9$)
12.3 6.34 eV
12.5 $\alpha \simeq 1.1$
12.7 $d = 0.141n$ nm, where n is integer
12.9 At $T = 293$ K, $\Delta c/c = 2.76 \times 10^{-23}$; at $T = 1350$, $\Delta c/c = 1.4 \times 10^{-4}$
12.15 For isotropy, $m = 2$

CHAPTER 13

13.1 $m^* = 1.4m_e$; $\theta_D = 346$ K
13.3 $\theta_D(\text{K}) = 91$ K; $\theta_D(\text{Rb}) = 56$ K; $\theta_D(\text{Cs}) = 39$ K
13.5 1.07×10^6
13.7 (*a*) 32.3 mg (*b*) 1.32 $\mu\Omega$·cm
13.9 (*a*) 70 (*b*) 12
13.11 $\tau = 4.44 \times 10^{-13}$ s; $v_d = 3.9 \times 10^3$ m/s; energy $< 10^{-3}kT$; $E = 2100$ V/cm
13.13 0.674 eV; σ not proportional to n; n not exactly proportional to $e^{-E_G/2kT}$
13.17 About one flux quantum
13.19 2.418 kHz

CHAPTER 14

14.1 ^{28}Si: 14,14; ^{131}Xe: 54,77; ^{145}Nd: 60,85; ^{183}W: 74,109

14.3 ^{107}Ag, 2.34×10^{-14} m; ^{14}N, 4.32×10^{-15} m; cannot neglect nuclear recoil; α particle within range of nuclear force for ^{14}N

14.5 ^{3}H, 8.48 MeV; ^{3}He, 7.72 MeV; ^{9}Be, 58.17 MeV; ^{204}Hg, 1609 MeV

14.7 (*a*) I = 0; (*b*) 2.406 MeV/α-bond (*c*) ^{12}C, 92.11 MeV (comp. 92.16); ^{20}Ne, 165.5 (comp. 160.65)

14.9 7.2 MeV

14.11 ^{40}Ca, 8.551 MeV; ^{40}K, 8.538 MeV ^{48}Ca, 8.667 MeV; ^{48}K, 8.431 MeV; ^{48}Sc, 8.656 MeV ^{208}Pb, 7.868 MeV; ^{208}Tl, 7.847 MeV; ^{208}Bi, 7.850 MeV

CHAPTER 15

15.1 8 kBq

15.3 8 h; 4 h

15.7 33 min

15.9 2.87 mg ^{226}Ra; 2.75 mg ^{231}Pa

15.11 6.7 MeV

15.13 0.434 keV

15.15 0.156 MeV

15.17 p; γ; $\beta^- + \bar{\nu}$

15.19 ^{15}N*; 13.7 MeV

15.21 3.42×10^{8} s^{-1}; 3.65×10^{-6}

15.23 1.94 MeV; 1.20 MeV; 7.55 MeV; 7.30 MeV; 1.73 MeV; 4.97 MeV

15.25 About 3.7%

15.27 5.24 MeV; 1.9 mW

15.29 0.89

15.31 7.75×10^{7} s^{-1}

15.33 1.5×10^{4} yr

15.35 2.45×10^{9} yr

15.37 286 Bq

15.39 9.0×10^{37} s^{-1}; 1.35×10^{17} kg/yr; 6.75×10^{-14}

CHAPTER 16

16.1 (*a*) $\ell_1 = 4.46$ cm; $\ell_5 = 6.20$ cm; $\ell_{10} = 7.33$ cm; $\ell_{50} = 9.44$ cm
(*b*) $N \simeq (10^3 \text{ m})/(10^{-1} \text{ m}) = 10^4$;
$E = (0.02 \text{ MeV})(10^4) = 200$ MeV

16.3 $B = 0.375$ T; $f = 5.62$ MHz

16.5 $E_\alpha = 60$ MeV

16.7 $D = 86.6$ m; $E = 14.7$ GeV

16.9 $B_{per} = 2.46$ T; $B_{cen} = 2.22$ T; f = 8.55 MHz

16.11 64 V

CHAPTER 17

17.1 (*a*) Weak (*b*) strong (*c*) weak (*d*) weak (*e*) electromagnetic (*f*) strong

17.3 (*a*) Forbidden (*b*) forbidden (*c*) Not strong interaction (*d*) not strong interaction (*e*) allowed (*f*) forbidden

17.5 (*a*) γ (*b*) ν_e (*c*) n (*d*) $\bar{\nu}_\mu$ (*e*) Λ

17.7 (*a*) $K^+ + p$ (*b*) $K^+ + \pi^+$ (*c*) K^0 (*d*) $K^0 + \bar{n}$

17.9 (*a*) 2.5×10^{-6} eV (*b*) 1.1 keV (*c*) 8.2×10^{-6} eV (*d*) 8.0×10^{-6} eV

17.11 (*a*) $\bar{u}d \rightarrow \mu^- + \bar{\nu}_\mu$ (*b*) $udd \rightarrow uud + \beta^- + \bar{\nu}_e$ (*c*) $uus \rightarrow uds + u\bar{d}$

17.13 $\pi^+(u\bar{d})$ and $\pi^-(\bar{u}d)$ are each other's antiparticles, so mass same
$\Sigma^+(uus)$ and $\Sigma^-(dds)$ are not mutual antiparticles, so mass not the same

Index